VOLUME ONE HUNDRED AND TWENTY SIX

INTERNATIONAL REVIEW OF NEUROBIOLOGY

Animal Models for Medications Screening to Treat Addiction

INTERNATIONAL REVIEW OF NEUROBIOLOGY

VOLUME 126

VOLUME ONE HUNDRED AND TWENTY SIX

INTERNATIONAL REVIEW OF NEUROBIOLOGY

Animal Models for Medications Screening to Treat Addiction

Edited by

RICHARD L. BELL

Institute of Psychiatric Research
Indiana University School of Medicine
Indianapolis, IN, United States

SHAFIQUR RAHMAN

Department of Pharmaceutical Sciences
South Dakota State University
Brookings, SD, United States

ELSEVIER

AMSTERDAM • BOSTON • HEIDELBERG • LONDON
NEW YORK • OXFORD • PARIS • SAN DIEGO
SAN FRANCISCO • SINGAPORE • SYDNEY • TOKYO
Academic Press is an imprint of Elsevier

Academic Press is an imprint of Elsevier
50 Hampshire Street, 5th Floor, Cambridge, MA 02139, USA
525 B Street, Suite 1800, San Diego, CA 92101-4495, USA
The Boulevard, Langford Lane, Kidlington, Oxford OX5 1GB, UK
125 London Wall, London, EC2Y 5AS, UK

First edition 2016

ISBN: 978-0-12-804013-3
ISSN: 0074-7742

For information on all Academic Press publications
visit our website at http://store.elsevier.com/

CONTENTS

CONTRIBUTORS

J.B. Acri
National Institute on Drug Abuse, National Institutes of Health, Bethesda, MD, United States

B. Andraka-Christou
564 E Graham PL, Bloomington, IN, United States

M.A. Arends
Committee on the Neurobiology of Addictive Disorders, The Scripps Research Institute, La Jolla, CA, United States

S. Barron
University of Kentucky, Lexington, KY, United States

H.C. Becker
Charleston Alcohol Research Center; Medical University of South Carolina; RHJ Department of Veterans Affairs Medical Center, Charleston, SC, United States

R.L. Bell
Institute of Psychiatric Research, Indiana University School of Medicine, Indianapolis, IN, United States

Y.A. Blednov
Waggoner Center for Alcohol and Addiction Research, The University of Texas at Austin, Austin, TX, United States

Y. Buhidma
Institute of Neuroscience, Newcastle University, Newcastle upon Tyne, United Kingdom

T.R. Buske
University of Texas at Austin, Austin, TX, United States

J.C. Crabbe
Oregon Health & Science University; Portland Alcohol Research Center; VA Portland Health Care System, Portland, OR, United States

M. Egli
National Institute on Alcohol Abuse and Alcoholism, National Institutes of Health, Bethesda, MD, United States

E.A. Engleman
Institute of Psychiatric Research, Indiana University School of Medicine, Indianapolis, IN, United States

A. Facciol
University of Toronto Mississauga, Mississauga, ON, Canada

L. Fields
University of Kentucky, Lexington, KY, United States

R. Gerlai
University of Toronto Mississauga, Mississauga, ON, Canada

G.D. Greenberg
Oregon Health & Science University; Portland Alcohol Research Center; VA Portland Health Care System, Portland, OR, United States

C.N. Haile
Texas Institute for Measurement, Evaluation and Statistics (TIMES), University of Houston, Houston, TX, United States

R.A. Harris
Waggoner Center for Alcohol and Addiction Research, The University of Texas at Austin, Austin, TX, United States

S. Hauser
Institute of Psychiatric Research, Indiana University School of Medicine, Indianapolis, IN, United States

A. Hawkey
University of Kentucky, Lexington, KY, United States

C.M. Hostetler
Oregon Health & Science University, Portland, OR, United States

Z.M. Jeanes
University of Texas at Austin, Austin, TX, United States

Z. Justinova
Preclinical Pharmacology Section, Behavioral Neuroscience Research Branch, NIDA, NIH, DHHS, Baltimore, MD, United States

D.M. Kircher
University of Texas at Austin, Austin, TX, United States

T.A. Kosten
Texas Institute for Measurement, Evaluation and Statistics (TIMES), University of Houston, Houston, TX, United States

B. Le Foll
Translational Addiction Research Laboratory, Campbell Family Mental Health Research Institute, Centre for Addiction and Mental Health (CAMH); Addiction Medicine Service, Ambulatory Care and Structured Treatments; Institute of Medical Sciences, University of Toronto, Toronto, ON, Canada

T. Liang
Indiana University School of Medicine, Indianapolis, IN, United States

J.M. Littleton
University of Kentucky; Naprogenix, Inc., Lexington, KY, United States

M.F. Lopez
Charleston Alcohol Research Center, Charleston, SC, United States

E.Y. Maier
University of Texas at Austin, Austin, TX, United States

R.A. Mangieri
University of Texas at Austin, Austin, TX, United States

J. Mayfield
Waggoner Center for Alcohol and Addiction Research, The University of Texas at Austin, Austin, TX, United States

J. McClintick
Center for Medical Genomics, Indiana University School of Medicine, Indianapolis, IN, United States

J.R.K. Mootz
Methamphetamine Abuse Research Center, Oregon Health & Science University, Portland, OR, United States

R.A. Morrisett
University of Texas at Austin, Austin, TX, United States

O. Ohia-Nwoko
Texas Institute for Measurement, Evaluation and Statistics (TIMES), University of Houston, Houston, TX, United States

L.V. Panlilio
Preclinical Pharmacology Section, Behavioral Neuroscience Research Branch, NIDA, NIH, DHHS, Baltimore, MD, United States

J.A. Peck
State University of New York College at Cortland, Cortland, NY, United States

T.J. Phillips
Methamphetamine Abuse Research Center, Oregon Health & Science University; Veterans Affairs Portland Health Care System, Portland, OR, United States

S. Rahman
Department of Pharmaceutical Sciences, South Dakota State University, Brookings, SD, United States

C. Reed
Methamphetamine Abuse Research Center, Oregon Health & Science University, Portland, OR, United States

R. Renteria
University of Texas at Austin, Austin, TX, United States

Z.A. Rodd
Institute of Psychiatric Research, Indiana University School of Medicine, Indianapolis, IN, United States

A.E. Ryabinin
Oregon Health & Science University, Portland, OR, United States

Y. Sari
University of Toledo, Toledo, OH, United States

M. Shoaib
Institute of Neuroscience, Newcastle University, Newcastle upon Tyne, United Kingdom

S. Tran
University of Toronto Mississauga, Mississauga, ON, Canada

J.M. Trigo
Translational Addiction Research Laboratory, Campbell Family Mental Health Research Institute, Centre for Addiction and Mental Health (CAMH), Toronto, ON, Canada

D.A. White
National Institute on Drug Abuse, National Institutes of Health, Bethesda, MD, United States

PREFACE

Drug abuse and addiction to a variety of substances, including alcohol, nicotine, opiates, and psychostimulants, are a significant public health concern for all nations. Substantial evidence indicates that the medical, social, and economic cost poses a monumental challenge to the health and economic stability of developed and developing nations across the globe. In the United States alone, it is estimated that the annual cost for addiction to drugs of abuse, nicotine, and alcohol approximates a half trillion dollars. Therefore, prevention and treatment of alcohol and drug addiction, or addictive behaviors, is paramount for all global citizens. Animal research has greatly enhanced our understanding of the genetic, behavioral, and neurobiological mechanisms mediating these addictions. This special issue sought to bring together researchers and the animal models they use to investigate the central and peripheral effects of alcohol and drugs of abuse. Because the focus was on establishing effective treatments for alcohol and drug dependence first, and on the procedures for testing these pharmacological treatments second, some of the topics are novel and not discussed elsewhere. We believe this makes this work distinctive in nature. In addition, this volume addresses models for medications screening, rather than the often used term "medications development." The latter subject has been covered persuasively elsewhere, but the former is discussed cogently herein for the first time, again making this edition unique in focus.

The chapters included in this volume provide multiple animal models of alcohol and drug addiction, which, respectively, highlight preclinical and clinical findings on the neurobiology of alcohol- and/or drug dependence. This neurobiological background is used to present the behavioral procedures, and an in vitro assay, with which one can screen medications targeting the neurochemicals and/or neuromodulators mediating addiction and dependence. The authors of each chapter are experts in their field, including respective drug-of-abuse, species, and/or model system. The format of the book is that of a compendium, with each chapter, for the most part, being able to stand alone. This was done so that, if used as a reference, an individual can proceed to the chapter which would be of most benefit at the moment. The goal of the editors and authors, however, was for researchers to also read chapters that are not directly related to their area of expertise. That is why this book did not focus on only one drug-of-abuse, or one species, or, for that matter, on behavioral procedures only. We believe the reader will gain much from looking at other perspectives as we have while developing this volume.

The decision to focus on medications screening resulted from the observation that pharmacotherapies to treat addiction are not a high priority for the pharmaceutical industry. This fact and possible ways to address it are discussed in the first two chapters we solicited from nonacademic and nonindustry representatives. The first chapter *Considerations in the evaluation of potential efficacy of medications for alcohol and drug use disorders*, by Drs. Egli, White, and Acri, discusses the fact that though the pharmaceutical industry does not consider medications development to treat addiction a priority, industry does have an arsenal of FDA-approved, or in advanced clinical trials, compounds that might be "repurposed" for treating addiction. And, it is for this reason the focus of this volume is on medications screening, rather than medications development. The second chapter *A pressing need for pharmacotherapy development to treat addiction: an editorial from a legal perspective*, by Dr. Andraka-Christou, discusses the lack of medications development to treat addiction and ways the pharmaceutical industry can be "incentivized" to pursue medications development or at least agree to produce significant quantities of compounds found to be effective in treating addiction. The third chapter *Identification of treatment targets in a genetic mouse model of voluntary methamphetamine drinking*, by Drs. Phillips, Mootz, and Reed, discusses the selectively bred mouse lines they have developed to determine the genetics, as well as behavioral and neurobiological factors mediating methamphetamine abuse and dependence, which by extension allows for monitoring changes induced by compounds to treat methamphetamine dependence. The fourth chapter *Medication screening for compounds to treat cannabis use disorder*, by Drs. Panlilio, Justinova, Trigo, and Le Foll, discusses some of the deleterious effects of cannabis use disorder (CUD), animal model findings on treatment targets for CUD, and their translatability to the human laboratory and finally into the treatment setting. The fifth chapter *How can we improve on modeling nicotine addiction to develop better smoking cessation treatments?*, by Drs. Shoaib and Buhidma, discusses some of the factors, including genetics and comorbidity, that have made it difficult to find and/or develop more efficacious treatments for smoking cessation. In addition, they point out the benefits and shortcomings of the most commonly used intravenous self-administration animal models, as well as possible modifications to operant procedures that will increase the scientific validity and reliability of future findings. The sixth chapter *An animal model of alcohol dependence to screen medications for treating alcoholism,* by Drs. Becker and Lopez, discusses their well-established technique to induce ethanol dependence, associated dependent measures focusing on this end-point, findings from this model, as well as the need to recognize much of the ethanol research literature in the past can be construed as addressing individuals who have not progressed to advanced stages of dependence and, therefore, the dependent-state needs to receive more attention. The seventh chapter *A genetic animal model of alcoholism for screening medications to treat addiction,* by Drs. Bell, Hauser, Rodd, Liang, Sari, McClintick, Rahman, and Engleman, discusses the selectively bred alcohol-preferring P rat and its usefulness to address the genetic, behavioral, and neurobiological factors mediating alcoholism with a focus on the major neurotransmitter and neuropeptide systems examined thus far. These authors also sought to place the findings with P rats into the context of similar clinical observations, with a discussion on polymorphisms and pharmacogenetics. The eighth chapter *Animal models and the development of vaccines to treat substance use disorders,* by Drs. Ohia-Nwoko, Kosten, and Haile, discusses the concepts and methods for developing vaccines against multiple drugs of abuse, individually and in concert, and their use, as well as observed efficacy, at different stages of the addiction cycle. The ninth chapter *Genes and alcohol consumption: studies with mutant mice,* by Drs. Mayfield, Arends, Harris, and Blednov, discusses 155 genes that have been manipulated in the mouse and the effects of these gene manipulations on ethanol drinking and/or self-administration with a discussion on some gene target findings that have been translated into positive findings in the human laboratory setting. The tenth chapter *Gene targeting studies of hyperexcitability and affective states of alcohol withdrawal in rodents,* by Drs. Greenberg and Crabbe, discusses gene manipulations that have altered ethanol-induced as well as ethanol withdrawal-induced effects, highlighting the fact that addiction is more than only ingesting a drug or alcohol; and pharmacotherapies, as well as behavioral therapies, need to be developed or modified to address the constellation of symptoms present in the alcohol- and/or drug-dependent individual. The eleventh chapter *Abstinence-conflict model: toward an optimal animal model for screening medications promoting drug abstinence,* by Dr. Peck, discusses a model that introduces conflict into the paradigm of modeling drug-taking and drug-seeking behavior. He highlights the strengths and weaknesses of many existing animal models and how the introduction of choice to receive or avoid punishment in these models more closely mimics the clinical condition. The twelfth chapter *Prairie voles as a model to screen medications for the treatment of alcoholism and addictions,* by Drs. Ryabinin and Hostetler, discusses the socially monogamous prairie vole as a model of human affiliation and how this social construct is missing from practically all animal models in use today. The chapter also highlights findings, using this model, on how alcohol disrupts social bonding in a sex-dependent and time-dependent manner; and how future medications for, at least some, alcoholics and addicts will need to facilitate positive social bonding and disrupt negative social bonding in order to support long-term abstinence. The thirteenth chapter *Animal models for medication development and application to treat fetal alcohol effects,* by Drs. Barron, Hawkey, Fields, and Littleton, discusses findings from their laboratory on developing and screening medications to ameliorate the deleterious effects of in utero alcohol exposure as well as noting efficacious effects for certain compounds delivered at different stages of development. The fourteenth chapter *Using* in vitro *electrophysiology to screen medications: accumbal plasticity as an engram of alcohol dependence,* by Drs. Renteria, Jeanes, Mangieri, Maier, Kircher, Buske, and Morrisett, introduces this novel electrophysiology screening procedure, which takes advantage of known long-term plasticity changes in glutamatergic activity that differentiate alcohol-dependent from nondependent rodents. These authors provide examples of in vitro findings from their laboratory and how they parallel in vivo findings in rodent and other model systems. The fifteenth chapter *The zebrafish, a novel model organism for screening compounds affecting acute and chronic ethanol-induced effects,* by Drs. Tran, Facciol, and Gerlai, discusses the growing field of alcohol research in zebrafish, some of the behavioral and neurochemical findings from this research, and how experimental techniques and neurobiological measures are becoming more sophisticated with a focus on high-throughput screening of medications targeting alcohol abuse/dependence and, putatively, drug abuse/dependence.

Collectively, this Special Issue provides a detailed overview of "where we're at" and insight into "where we're going" as a field examining the neuroscience of addiction. It is clear from the material presented here that the animal models and techniques used by addiction neuroscientists have become more sophisticated, more valid, and more reliable over time. While it has been believed that animal models can never mimic the complex social aspects and cognitive processes underlying addiction, the refinement of animal models has proven this belief to be false. Herein, we have discussed the need to treat more than the act of consuming a drug in order to foster long-term abstinence. And, the animal models described by the authors in this volume confirm the field has advanced significantly in addressing cognitive, social, and physiological aspects of addiction that were not recognized and/or considered in the past. Moreover, one common theme found in the enclosed chapters is the role of genetics in a predisposition to abuse alcohol and/or drugs and develop dependence upon these substances. Thus, pharmacogenetics is a facet of addiction that will have to be carefully considered as we move forward in solving this monumental, yet age-old public health problem. Finally, despite the limitations and future challenges outlined in this volume, it is our sincere hope that readers will be motivated to continue exploring the fascinating neurobiological details associated with various animal models of addiction, and new directions and possibilities for prevention and treatment of alcohol and drug addiction in the years to come.

We would like to thank all of the contributing authors for their excellent contributions to this volume. We also thank Dr. R. Adron Harris, the Editor-in-Chief of the *International Review of Neurobiology*, for his invitation and guidance. We also thank Ms. Poppy Garraway and Ms. Hannah Colford and all of the other fine people at Elsevier for their assistance and support in bringing this volume to fruition.

February 2016

Richard L. Bell

Shafiqur Rahman

CHAPTER ONE

Considerations in the Evaluation of Potential Efficacy of Medications for Alcohol and Drug Use Disorders: An Editorial☆

M. Egli*, D.A. White†,1, J.B. Acri†
*National Institute on Alcohol Abuse and Alcoholism, National Institutes of Health, Bethesda, MD, United States
†National Institute on Drug Abuse, National Institutes of Health, Bethesda, MD, United States
[1]Corresponding author: e-mail address: whitedav@nida.nih.gov

Contents

Abstract

The societal burden created by alcohol and drug use disorders is estimated to be on the order of hundreds of billions of dollars, creating a need for effective medications to reduce use and prevent relapse. While there are FDA-approved medications to facilitate abstinence and prevent relapse for some indications including, alcohol, tobacco, and opiate use disorders, there are no approved treatments for other abused substances, including cocaine, methamphetamine, and cannabis, leaving these critical medical needs unmet. The development of such medications has fallen largely to the government with efforts spearheaded by the National Institute on Drug Abuse and the National Institute on Alcoholism and Alcohol Abuse. Both agencies have medication development programs with preclinical components that include the standardized evaluation of compounds using animal models. This chapter describes the rationale and considerations involved in the use of such models, including reinstatement of drug self-administration.

☆*Disclaimer*: The views expressed herein are the personal views of the authors and may not represent the views of the National Institute on Alcohol Abuse and Alcoholism, the National Institute on Drug Abuse, or the National Institutes of Health.

International Review of Neurobiology, Volume 126 2016 Published by Elsevier Inc.
ISSN 0074-7742
http://dx.doi.org/10.1016/bs.irn.2016.02.007

The development of screening assays to discover medications for the treatment of substance and alcohol use disorders is an important ongoing challenge, in part because of the societal burden imposed by these disorders, estimated at $700 billion annually resulting from crime, lost work productivity, and health care (http://www.drugabuse.gov/related-topics/trends-statistics). For alcohol, tobacco, and opiate use disorders, there are FDA-approved medications to facilitate abstinence and prevent relapse. However, there are no approved medications for cocaine, methamphetamine, and cannabis use disorders. Because the pharmaceutical industry has not embraced the treatment of substance use disorders, it has fallen to the government to fund much of the research and development through grants and contracts. Internal programs within both the National Institute on Drug Abuse (NIDA) and the National Institute on Alcoholism and Alcohol Abuse (NIAAA) have been created for the purpose of working with industry partners to evaluate and advance potential medications for substance/alcohol use disorders. To that end, both institutes have established programs that include the standardized evaluation of compounds using animal models at contract sites. Under these extramural contracts, compounds can be blinded to the researchers to insure objectivity and testing using standardized protocols tailored to the solubility and pharmacokinetics of each compound. Many of these protocols use some form of drug or alcohol self-administration. The following sections describe the rationale and considerations involved in the use of these paradigms for evaluation of potential medications.

When it was shown that animals will self-administer most drugs abused by humans (Balster & Lukas, 1985; Schuster & Thompson, 1969), it became evident that abused drugs reinforce behavior much like food and other naturally occurring reinforcers. Moreover, these findings indicate that drug taking is under the same pharmacological controlling conditions in animals and humans. Thus, drug taking in laboratory animals may be considered a homologous behavior, a parallel phenomenon rather than a model per se, in the same way that naturally occurring cancers in animals such as dogs are natural manifestations of the same conditions and processes that occur in other mammals, including man. Indeed, drug taking can occur across a number of species, and it has been described as a phylogenetically preserved behavior (Deroche-Gamonet, Belin, & Piazza, 2004).

Drug self-administration studies typically use the intravenous route; however, the oral route is preferred for alcohol self-administration studies because alcohol is typically consumed in the form of beverages. In alcohol self-administration studies, when animals drink to pharmacologically effective levels, alcohol functions as a positive reinforcer; however, consumption and resulting blood alcohol levels seldom reach levels permitting conclusions about alcohol dependence as distinct from socially acceptable alcohol drinking levels (ie, one or two drinks). Therefore, investigators have developed voluntary drinking paradigms to engender high intake (>5 g/kg) and intoxicating blood alcohol concentrations (80–100 mg/dL or higher). The models include genetic (see Barkley-Levenson & Crabbe, 2014; Bell et al., 2012; Blednov et al., 2010; Ciccocioppo, 2013), scheduled-access (Carnicella, Ron, & Barak, 2014; Rhodes, Best, Belknap, Finn, & Crabbe, 2005; Simms et al., 2008; Spanagel & Holter, 1999; Thiele & Navarro, 2014), and chronic exposure/extended access paradigms (Knapp & Breese, 2012; Vendruscolo & Roberts, 2014). In addition to the excessive drinking criterion that serves as the basis of these diverse models, each displays additional behavioral and neurophysiological traits associated with alcohol addiction. Important for evaluating medications, ethanol drinking in these models is characterized by a sensitivity to various pharmacological agents, suggesting that drinking is influenced by distinct neurobiological mechanisms of alcohol dependence (Meinhardt & Sommer, 2015; Sprow & Thiele, 2012). Similar specialized models have also been developed with other addictive drugs. For example, binge cocaine paradigms may more clearly elucidate drug-induced changes to receptors and transporters (Spangler, Unterwald, & Kreek, 1993; Unterwald, Ho, Rubenfeld, & Kreek, 1994), and short-access vs long-access escalation may enable study of drug-induced neural plasticity (Kallupi et al., 2013) or of the dysregulation of stress circuitry (Koob et al., 2014).

Models focusing on neural dysregulation associated with dependence and addiction have led to the discovery of gene networks and brain neural adaptations associated with dependence and addiction—including the identification of drugable targets that can be further validated pharmacologically. For the purposes of evaluation of the potential efficacy of drugs, it is sufficient that the models have biological and behavioral relevance to addiction beyond the initial reinforcing effects. Metaphorically, each model captures a significant portion of the neurobiological landscape of addiction; changes in the behavioral "read-out"—alcohol or drug intake—suggest that neurobiological targets relevant to addiction are engaged by the test drug in a way that is sufficient to affect drug taking or drinking as well as related phenotypes in humans. Does this response, by extension, differentiate clinically effective from clinically ineffective medications?

1. PREDICTIVE VALIDITY

Predictive validity, or more specifically, the ability to predict medication effects—both positive and negative—is the most salient of the types of model validity (face, construct, and predictive) for the evaluation of potential medications. For example, for depression, tests like the forced swim, whether or not they meet criteria for other types of validity, are useful because they are predictive of medication effects (Castagne, Moser, Roux, & Porsolt, 2011; Porsolt, Le, & Jalfre, 1977).

Assessing the predictive value of animal models may be achieved by reviewing how clinically tested medications perform in preclinical models (see Egli, 2005; Koob, Lloyd, & Mason, 2009; Mello, 1992). The approved medications for alcohol dependence, naltrexone, acamprosate, nalmefene, and medications that have generated positive effects clinically such as varenicline, gabapentin, topiramate, baclofen, and ondansetron have generally yielded positive effects in animal models of excessive alcohol drinking. Preclinical tests of notable failures in clinical trials [such as bromocriptine, ritanserin, levetiracetam, and quetiapine (Fertig et al., 2012; Johnson et al., 1996; Litten et al., 2012; Naranjo, Dongier, & Bremner, 1997; Powell et al., 1995)] have also been performed in excessive drinking models and shown to be ineffective (H. Becker & R.L. Bell, unpublished data; see also Egli, 2005). Indeed, levetiracetam has been shown to increase drinking in rats (R.L. Bell, unpublished data), mice (Fish et al., 2014), and in nontreatment-seeking alcohol abusers (Mitchell, Grossman, Coker, & Messing, 2012). Despite the emerging predictive validity of animal models of excessive alcohol drinking, exceptions have occurred requiring a nuanced consideration of preclinical test results. For example, SSRI medications have largely been ineffective in clinical trials for alcohol dependence yet reduce alcohol drinking in animal models. Thus, positive findings for SSRIs and similar drugs should be interpreted with caution.

For cocaine and methamphetamine, there are no clinically effective medications with which to validate preclinical testing procedures, but for nicotine and opiates, there are effective medications for reducing use and promoting abstinence. Do these medications show efficacy in self-administrations models? For the most part, the answer is yes. Varenicline, an approved treatment for smoking cessation, reduced both nicotine self-administration and nicotine reinstatement in rats (O'Connor, Parker, Rollema, & Mead, 2010) and squirrel monkeys (Z. Justinova &

S. Goldberg, unpublished data). Similarly, lorcaserin reduced nicotine self-administration in rats (Higgins et al., 2012; Levin et al., 2011) and was effective in a smoking cessation trial reported by Esai and Arena Pharmaceuticals (http://invest.arenapharm.com/releasedetail.cfm?ReleaseID=879743). Bupropion, another approved treatment, reduced nicotine self-administration in rats following repeated dosing (Rauhut, Dwoskin, & Bardo, 2005), but was ineffective following acute doses (Shoaib, Sidhpura, & Shafait, 2003) or more variable in its effects (Bruijnzeel & Markou, 2003). For opiates, buprenorphine has been shown to reduce heroin and morphine self-administration in monkeys (Mello, Bree, & Mendelson, 1983) as has methadone in rats (Peng et al., 2010) and dogs (Jones & Prada, 1977). Despite these positive findings, there are probably as many negative findings in the published literature, in part, because of variability in methodology, leaving us to conclude that they may be predictive.

2. METHODOLOGY

With the exception of tests performed by pharmaceutical companies and targeted medication development programs, most preclinical studies of medication targets for drug and alcohol addiction are supported by research grants. Grant-supported preclinical research is often focused on identifying reliable and robust effects and revealing their underlying mechanisms rather than predicting clinical outcomes and informing medication development decisions. In the course of performing such work, protocol refinements may be implemented to maximize the size and reproducibility of the effect under investigation. Failures along the way are quickly abandoned and are seldom reported; conditions critical to successful demonstration of the drug's effect—if they are reported—may not be obvious. The result may be an incomplete view of both the target's validity and a medication's efficacy. Although the published findings may obtain for the experimental conditions reported, it is unknown whether the reported positive results represent an accurate clinical picture with "translational fidelity," especially when conditions failing to confirm an effect are underreported (see Rosenthal, 1979). To make conclusive decisions about medication efficacy in preclinical tests as in clinical testing, it is important that testing be performed under blinded conditions, such that—the identity of the compound should be unknown to the investigator, and the investigator should have no stake in the outcome of the tests. In addition, it is important to confirm that the medication's intended target has been engaged (especially important when results are negative), include routine comparisons with positive and negative control medications, and measure alternate behaviors to determine the specificity of the drug effect.

Determining the effects of a potential antiaddictive therapy using the dose-response function for drug self-administration provides crucial information about its potential clinical efficacy. Most self-administered drugs have a characteristic inverted U-shaped dose-response curve. Evaluating test compound effects on self-administration of a single peak drug dose could be misinterpreted since decreased self-administration can result from a downward, leftward, or rightward shift in the overall dose-effect function (Mello & Negus, 1996). Thus, unless additional doses are evaluated, decreased self-administration of a single peak dose could be mistaken for a positive treatment effect, when in fact the decrease could result from the undesirable enhancement of the drug's potency (see Mello & Negus, 1996). Following the reasoning that vertical (downward) shifts in drug self-administration dose-response curves correspond to the motivation to take drug—as opposed to horizontal shifts, which indicate changes in drug potency—potential therapies should produce a vertical downward shift of the overall dose-response curve (Ahmed & Koob, 1998; Piazza, Deroche-Gamonent, Rouge-Pont, & Le, 2000). Interestingly, concentration-response functions are seldom studied in oral alcohol self-administration studies because of concerns that taste aversions at higher concentrations could bias the dose-response curve. It has been demonstrated, however, that full, reliable inverted U-shaped dose-response curves can be obtained in rodents with a history of excessive alcohol intake at concentrations ranging from 2.5% to 60% (Carnicella, Yowell, & Ron, 2011). Providing concurrent access to multiple ethanol concentrations in drinking tubes may yield comparable information (see Rodd-Henricks et al., 2001), although this approach requires further validation.

3. EVALUATING SELECTIVE EFFECTS

When evaluating the efficacy of a potential medication, it may be important to distinguish treatment effects from possible motor-impairing and sedative effects as well as general effects on natural reward-related behaviors such as feeding. Many CNS-active substances will reduce self-administration through sedation or behavioral impairment. Unless appropriate controls for other types of reinforced behavior are used, it is easy to mistake sedation for a treatment effect as exemplified by butorphanol (Mello & Negus, 1996), bromocriptine (Kleven & Woolverton, 1990), or pimozide (Woolverton & Virus, 1989). In addition, it is useful to determine whether reductions in drug-reinforced behavior are selective or represent more general effects on motivated behaviors. Direct comparisons of drug and nondrug reinforcers are challenging in that equalizing the baseline reinforcing value of drug and the control substance may be difficult. A potentially informative approach is found in models comparing effects on drug or alcohol taking under conditions modeling addiction and dependence (eg, escalation, chronic intermittent ethanol exposure-induced drinking, drinking in the dark) with intake under similar conditions that do not model addiction and dependence. As discussed earlier, not only do selective effects in the dependence model suggest that the test medication is engaging targets relevant to addiction but also rules out other nonselective effects. Absence of specificity (ie, reduction of drug-reinforced and nondrug-reinforced behavior) does not necessarily rule out a therapeutic effect; the FDA-approved medication naltrexone not only reduces alcohol and heroin intake but also reduces consumption of nonalcoholic beverages and food in animals and humans (Shelton & Grant, 2001; Yeomans & Gray, 1997). Nevertheless, having some indicator of potential lack of selectivity due to general effects on behavior is an important starting point for fully evaluating novel medications.

4. NONHUMAN PRIMATES

While most evaluations of medication efficacy are usually initiated in mice and rats, when sufficient test drug supplies are available, there is value and multiple reasons for using nonhuman primates in the evaluation of potential medications. Advantages include longer life span, with the possibility of extended studies, and drug pharmacokinetic and metabolism profiles more similar to humans (Weerts, Fantegrossi, & Goodwin, 2007). Monkeys can be maintained for years using limited access cocaine self-administration, recapitulating the neurobiological changes that occur in humans. Limited access to cocaine is necessary because without it, monkeys will overdose within days (Deneau, Yanagita, & Seevers, 1969; Johanson, Balster, & Bonese, 1976). It is sometimes argued that a longer period of drug use is required to produce "compulsive" drug taking (Vanderschuren & Everitt, 2004) or that dose escalation of drug taking occurring over time requires longer sessions or a longer period of exposure to self-administration. Under limited access conditions in the laboratory, rhesus monkeys will develop stable self-administration behavior for years. Unless one is specifically studying the transition from casual to compulsive use, monkeys with years of cocaine self-administration exposure would seemingly manifest the biological changes associated with addiction. Nonhuman primates have also been used to examine medications for alcohol, although the majority focused on validating paradigms using medications with known clinical effects. To that end, naltrexone, baclofen, and varenicline have been shown to reduce drinking or diminish responses to cues associated with alcohol reinforcement in monkeys (Barr et al., 2010; Boyle et al., 1998; Duke, Kaminski, & Weerts, 2014; Kaminski, Duke, & Weerts, 2012; Kaminski & Weerts, 2014; Shelton & Grant, 2001; Vallender, Ruedi-Bettschen, Miller, & Platt, 2010).

5. CUE EFFECTS AND REINSTATEMENT

So far, we have reviewed studies of medication effects on drug and alcohol intake. We recognize that medications may be critically important to ameliorate relapse-precipitating conditions after a period of abstinence. In animal models, stimuli proximal to drug reinforcement acquire conditioned effects that strengthen and perpetuate drug-seeking behavior. Cue-induced reinstatement of drug-reinforced behavior develops quite rapidly and is extremely persistent in animals (Grimm, Hope, Wise, & Shaham, 2001), and is likely to activate the same neurobiological mechanisms in both animals (DiChiara & Imperato, 1988; Wise, 2002) and humans (Childress et al., 1999; Volkow et al., 2006). Thus, the maintaining stimuli and the endogenous reward systems that drugs and alcohol activate (see Everitt & Wolf, 2002 for review) are also likely preserved across species (Deroche-Gamonet et al., 2004). Similarly, stress-reinstatement models have enabled study of the role of the hypothalamic-pituitary-adrenal axis to drug seeking (Shaham & Stewart, 1995), and cue-reinstatement models have illustrated the power (Gipson et al., 2013) and persistence (Weiss et al., 2001) of conditioned cues to stimulate drug seeking.

Reinstatement models in rats and nonhuman primates appear to hold promise for allowing the evaluation of effects of cues, stress, and drugs on behavior in models using extinction or forced abstinence. Naltrexone and acamprosate are effective in alcohol reinstatement studies and conditioned place preference studies, as are a number of clinically untested drugs (Egli, 2005). Similarly, the approved smoking treatments varenicline and bupropion are effective in reducing nicotine cue- and prime-induced reinstatement (Dwoskin, Rauhut, King-Pospisil, & Bardo, 2006; Le et al., 2012; O'Connor et al., 2010), while other clinically available drugs (eg, naltrexone and lorcaserin) have demonstrated efficacy against reinstatement of stimulant seeking (Gerrits, Kuzmin, & van Ree, 2005; Haggkvist, Lindholm, & Franck, 2009; Harvey-Lewis, Li, Higgins, & Fletcher, 2015). Whether the reinstatement paradigm predicts amelioration of specific clinical indications such as craving or relapse by medications is currently unknown, although other types of validity (construct and face) have been described in detail (Epstein, Preston, Stewart, & Shaham, 2006; Katz & Higgins, 2003). However, as with self-administration in general, it is important to have appropriate controls for general depressant effects on behavior, and if animals are to be tested multiple times, controls to preclude effects on learning.

6. CONCLUSION

Medication effects observed in animal studies can also be demonstrated in human laboratory studies (Haney & Spealman, 2008; Zimmermann, O'Connor, & Ramchandani, 2013). In addition, concordance between human laboratory outcomes and Phase 2 clinical trials strengthens our confidence that novel medications having positive effects in human laboratory studies will be effective in the clinic (Mason & Higley, 2013). Nevertheless, positive findings from human laboratory studies often evaporate in the face of large, multisite clinical trials. There are many possible reasons behind the lack of clinical success of potential drug candidates (Arrowsmith & Miller, 2013; McCann et al., 2015). Factors such as social environment, economic opportunities, familial support, availability of health and mental health services, education, and academic achievement contribute to recovery in addition to the biological effects of pharmacological treatment. A key question remains whether the neurophysiological and behavioral actions of a medication will have a measurable influence on recovery in the larger therapeutic milieu where patients have greater heterogeneity in factors influencing their recovery. This concern is particularly evident in the widely observed placebo effect, in which notable therapeutic gains are observed in patients receiving no active drug and the active medication provides modest, albeit statistically significant, benefit beyond the placebo. From this perspective, current evidence suggests that medication effects in animal models are indeed translatable to humans. It must be recognized, however, that any translational research program is an iterative process. As additional medications are evaluated in humans and as recovery mechanisms become better understood, new and existing animal models may be used more effectively to guide future medication development decisions.

REFERENCES

Ahmed, S. H., & Koob, G. F. (1998). Transition from moderate to excessive drug intake: Change in hedonic set point. *Science*, *282*, 298–300.

Arrowsmith, J., & Miller, P. (2013). Trial watch: Phase II and phase III attrition rates 2011–2012. *Nature Reviews. Drug Discovery*, *12*, 569.

Balster, R. L., & Lukas, S. E. (1985). Review of self-administration. *Drug and Alcohol Dependence*, *14*, 249–261.

Barkley-Levenson, A. M., & Crabbe, J. C. (2014). High drinking in the dark mice: A genetic model of drinking to intoxication. *Alcohol*, *48*, 217–223.

Barr, C. S., Chen, S. A., Schwandt, M. L., Lindell, S. G., Sun, H., Suomi, S. J., et al. (2010). Suppression of alcohol preference by naltrexone in the rhesus macaque: A critical role of genetic variation at the micro-opioid receptor gene locus. *Biological Psychiatry*, *67*, 78–80.

Bell, R. L., Sable, H. J., Colombo, G., Hyytia, P., Rodd, Z. A., & Lumeng, L. (2012). Animal models for medications development targeting alcohol abuse using selectively bred rat lines: Neurobiological and pharmacological validity. *Pharmacology, Biochemistry, and Behavior*, *103*, 119–155.

Blednov, Y. A., Ozburn, A. R., Walker, D., Ahmed, S., Belknap, J. K., & Harris, R. A. (2010). Hybrid mice as genetic models of high alcohol consumption. *Behavior Genetics*, *40*, 93–110.

Boyle, A. E., Stewart, R. B., Macenski, M. J., Spiga, R., Johnson, B. A., & Meisch, R. A. (1998). Effects of acute and chronic doses of naltrexone on ethanol self-administration in rhesus monkeys. *Alcoholism, Clinical and Experimental Research*, *22*, 359–366.

Bruijnzeel, A. W., & Markou, A. (2003). Characterization of the effects of bupropion on the reinforcing properties of nicotine and food in rats. *Synapse*, *50*, 20–28.

Carnicella, S., Ron, D., & Barak, S. (2014). Intermittent ethanol access schedule in rats as a preclinical model of alcohol abuse. *Alcohol*, *48*, 243–252.

Carnicella, S., Yowell, Q. V., & Ron, D. (2011). Regulation of operant oral ethanol self-administration: A dose–response curve study in rats. *Alcoholism, Clinical and Experimental Research*, *35*, 116–125.

Castagne, V., Moser, P., Roux, S., & Porsolt, R. D. (2011). Rodent models of depression: Forced swim and tail suspension behavioral despair tests in rats and mice. *Current Protocols in Neuroscience*, *55*, 8.10A.1–8.10A.14, chapter 8, Unit.

Childress, A. R., Mozley, P. D., McElgin, W., Fitzgerald, J., Reivich, M., & O'Brien, C. P. (1999). Limbic activation during cue-induced cocaine craving. *The American Journal of Psychiatry*, *156*, 11–18.

Ciccocioppo, R. (2013). Genetically selected alcohol preferring rats to model human alcoholism. *Current Topics in Behavioral Neurosciences*, *13*, 251–269.

Deneau, G., Yanagita, T., & Seevers, M. H. (1969). Self-administration of psychoactive substances by the monkey. *Psychopharmacologia*, *16*, 30–48.

Deroche-Gamonet, V., Belin, D., & Piazza, P. V. (2004). Evidence for addiction-like behavior in the rat. *Science*, *305*, 1014–1017.

DiChiara, G., & Imperato, A. (1988). Drugs abused by humans preferentially increase synaptic dopamine concentrations in the mesolimbic system of freely moving rats. *Proceedings of the National academy of Sciences of the United States of America*, *85*, 5274–5278.

Duke, A. N., Kaminski, B. J., & Weerts, E. M. (2014). Baclofen effects on alcohol seeking, self-administration and extinction of seeking responses in a within-session design in baboons. *Addiction Biology*, *19*, 16–26.

Dwoskin, L. P., Rauhut, A. S., King-Pospisil, K. A., & Bardo, M. T. (2006). Review of the pharmacology and clinical profile of bupropion, an antidepressant and tobacco use cessation agent. *CNS Drug Reviews*, *12*, 178–207.

Egli, M. (2005). Can experimental paradigms and animal models be used to discover clinically effective medications for alcoholism? *Addiction Biology*, *10*, 309–319.

Epstein, D. H., Preston, K. L., Stewart, J., & Shaham, Y. (2006). Toward a model of drug relapse: An assessment of the validity of the reinstatement procedure. *Psychopharmacology (Berlin)*, *189*, 1–16.

Everitt, B. J., & Wolf, M. E. (2002). Psychomotor stimulant addiction: A neural systems perspective. *The Journal of Neuroscience*, *22*, 3312–3320.

Fertig, J. B., Ryan, M. L., Falk, D. E., Litten, R. Z., Mattson, M. E., Ransom, J., et al. (2012). A double-blind, placebo-controlled trial assessing the efficacy of levetiracetam extended-release in very heavy drinking alcohol-dependent patients. *Alcoholism, Clinical and Experimental Research*, *36*, 1421–1430.

Fish, E. W., Agoglia, A. E., Krouse, M. C., Muller, R. G., Robinson, J. E., & Malanga, C. J. (2014). Levetiracetam results in increased and decreased alcohol drinking with different access procedures in C57BL/6J mice. *Behavioural Pharmacology*, *25*, 61–70.

Gerrits, M. A., Kuzmin, A. V., & van Ree, J. M. (2005). Reinstatement of cocaine-seeking behavior in rats is attenuated following repeated treatment with the opioid receptor antagonist naltrexone. *European Neuropsychopharmacology*, *15*, 297–303.

Gipson, C. D., Kupchik, Y. M., Shen, H., Reissner, K. J., Thomas, C. A., & Kalivas, P. W. (2013). Relapse induced by cues predicting cocaine depends on rapid, transient synaptic potentiation. *Neuron*, *77*, 867–872.

Grimm, J. W., Hope, B. T., Wise, R. A., & Shaham, Y. (2001). Neuroadaptation. Incubation of cocaine craving after withdrawal. *Nature*, *412*, 141–142.

Haggkvist, J., Lindholm, S., & Franck, J. (2009). The opioid receptor antagonist naltrexone attenuates reinstatement of amphetamine drug-seeking in the rat. *Behavioural Brain Research*, *197*, 219–224.

Haney, M., & Spealman, R. (2008). Controversies in translational research: Drug self-administration. *Psychopharmacology (Berlin)*, *199*, 403–419.

Harvey-Lewis, C., Li, Z., Higgins, G. A., & Fletcher, P. J. (2015). The 5-HT receptor agonist lorcaserin reduces cocaine self-administration, reinstatement of cocaine-seeking and cocaine induced locomotor activity. *Neuropharmacology*, *101*, 237–245.

Higgins, G. A., Silenieks, L. B., Rossmann, A., Rizos, Z., Noble, K., Soko, A. D., et al. (2012). The 5-HT2C receptor agonist lorcaserin reduces nicotine self-administration, discrimination, and reinstatement: Relationship to feeding behavior and impulse control. *Neuropsychopharmacology*, *37*, 1177–1191.

Johanson, C. E., Balster, R. L., & Bonese, K. (1976). Self-administration of psychomotor stimulant drugs: The effects of unlimited access. *Pharmacology, Biochemistry, and Behavior*, *4*, 45–51.

Johnson, B. A., Jasinski, D. R., Galloway, G. P., Kranzler, H., Weinreib, R., Anton, R. F., et al. (1996). Ritanserin in the treatment of alcohol dependence—A multi-center clinical trial. Ritanserin Study Group. *Psychopharmacology (Berlin)*, *128*, 206–215.

Jones, B. E., & Prada, J. A. (1977). Effects of methadone and morphine maintenance on drug-seeking behavior in the dog. *Psychopharmacology (Berlin)*, *54*, 109–112.

Kallupi, M., Wee, S., Edwards, S., Whitfield, T. W., Jr., Oleata, C. S., Luu, G., et al. (2013). Kappa opioid receptor-mediated dysregulation of gamma-aminobutyric acidergic transmission in the central amygdala in cocaine addiction. *Biological Psychiatry*, *74*, 520–528.

Kaminski, B. J., Duke, A. N., & Weerts, E. M. (2012). Effects of naltrexone on alcohol drinking patterns and extinction of alcohol seeking in baboons. *Psychopharmacology (Berlin)*, *223*, 55–66.

Kaminski, B. J., & Weerts, E. M. (2014). The effects of varenicline on alcohol seeking and self-administration in baboons. *Alcoholism, Clinical and Experimental Research*, *38*, 376–383.

Katz, J. L., & Higgins, S. T. (2003). The validity of the reinstatement model of craving and relapse to drug use. *Psychopharmacology (Berlin)*, *168*, 21–30.
Kleven, M. S., & Woolverton, W. L. (1990). Effects of bromocriptine and desipramine on behavior maintained by cocaine or food presentation in rhesus monkeys. *Psychopharmacology (Berlin)*, *101*, 208–213.
Knapp, D. J., & Breese, G. R. (2012). Models of chronic alcohol exposure and dependence. *Methods in Molecular Biology*, *829*, 205–230.
Koob, G. F., Buck, C. L., Cohen, A., Edwards, S., Park, P. E., Schlosburg, J. E., et al. (2014). Addiction as a stress surfeit disorder. *Neuropharmacology*, *76*(Pt. B), 370–382.
Koob, G. F., Lloyd, G. K., & Mason, B. J. (2009). Development of pharmacotherapies for drug addiction: A Rosetta stone approach. *Nature Reviews Drug Discovery*, *8*, 500–515.
Le, F. B., Chakraborty-Chatterjee, M., Lev-Ran, S., Barnes, C., Pushparaj, A., Gamaleddin, I., et al. (2012). Varenicline decreases nicotine self-administration and cue-induced reinstatement of nicotine-seeking behaviour in rats when a long pretreatment time is used. *The International Journal of Neuropsychopharmacology*, *15*, 1265–1274.
Levin, E. D., Johnson, J. E., Slade, S., Wells, C., Cauley, M., Petro, A., et al. (2011). Lorcaserin, a 5-HT2C agonist, decreases nicotine self-administration in female rats. *The Journal of Pharmacology and Experimental Therapeutics*, *338*, 890–896.
Litten, R. Z., Fertig, J. B., Falk, D. E., Ryan, M. L., Mattson, M. E., Collins, J. F., et al. (2012). A double-blind, placebo-controlled trial to assess the efficacy of quetiapine fumarate XR in very heavy-drinking alcohol-dependent patients. *Alcoholism, Clinical and Experimental Research*, *36*, 406–416.
Mason, B. J., & Higley, A. E. (2013). A translational approach to novel medication development for protracted abstinence. *Current Topics in Behavioral Neurosciences*, *13*, 647–670.
McCann, D. J., Petry, N. M., Bresell, A., Isacsson, E., Wilson, E., & Alexander, R. C. (2015). Medication nonadherence, "professional subjects," and apparent placebo responders: Overlapping challenges for medications development. *Journal of Clinical Psychopharmacology*, *35*, 566–573.
Meinhardt, M. W., & Sommer, W. H. (2015). Postdependent state in rats as a model for medication development in alcoholism. *Addiction Biology*, *20*, 1–21.
Mello, N. K. (1992). Behavioral strategies for the evaluation of new pharmacotherapies for drug abuse treatment. *NIDA Research Monograph*, *119*, 150–154.
Mello, N. K., Bree, M. P., & Mendelson, J. H. (1983). Comparison of buprenorphine and methadone effects on opiate self-administration in primates. *The Journal of Pharmacology and Experimental Therapeutics*, *225*, 378–386.
Mello, N. K., & Negus, S. S. (1996). Preclinical evaluation of pharmacotherapies for treatment of cocaine and opioid abuse using drug self-administration procedures. *Neuropsychopharmacology*, *14*, 375–424.
Mitchell, J. M., Grossman, L. E., Coker, A. R., & Messing, R. O. (2012). The anticonvulsant levetiracetam potentiates alcohol consumption in non-treatment seeking alcohol abusers. *Journal of Clinical Psychopharmacology*, *32*, 269–272.
Naranjo, C. A., Dongier, M., & Bremner, K. E. (1997). Long-acting injectable bromocriptine does not reduce relapse in alcoholics. *Addiction*, *92*, 969–978.
O'Connor, E. C., Parker, D., Rollema, H., & Mead, A. N. (2010). The alpha4beta2 nicotinic acetylcholine-receptor partial agonist varenicline inhibits both nicotine self-administration following repeated dosing and reinstatement of nicotine seeking in rats. *Psychopharmacology (Berlin)*, *208*, 365–376.
Peng, X. Q., Xi, Z. X., Li, X., Spiller, K., Li, J., Chun, L., et al. (2010). Is slow-onset long-acting monoamine transport blockade to cocaine as methadone is to heroin? Implication for anti-addiction medications. *Neuropsychopharmacology*, *35*, 2564–2578.

Piazza, P. V., Deroche-Gamonent, V., Rouge-Pont, F., & Le, M. M. (2000). Vertical shifts in self-administration dose–response functions predict a drug-vulnerable phenotype predisposed to addiction. *The Journal of Neuroscience*, *20*, 4226–4232.

Porsolt, R. D., Le, P. M., & Jalfre, M. (1977). Depression: A new animal model sensitive to antidepressant treatments. *Nature*, *266*, 730–732.

Powell, B. J., Campbell, J. L., Landon, J. F., Liskow, B. I., Thomas, H. M., Nickel, E. J., et al. (1995). A double-blind, placebo-controlled study of nortriptyline and bromocriptine in male alcoholics subtyped by comorbid psychiatric disorders. *Alcoholism, Clinical and Experimental Research*, *19*, 462–468.

Rauhut, A. S., Dwoskin, L. P., & Bardo, M. T. (2005). Tolerance does not develop to the decrease in nicotine self-administration produced by repeated bupropion administration. *Nicotine & Tobacco Research*, 7, 901–907.

Rhodes, J. S., Best, K., Belknap, J. K., Finn, D. A., & Crabbe, J. C. (2005). Evaluation of a simple model of ethanol drinking to intoxication in C57BL/6J mice. *Physiology & Behavior*, *84*, 53–63.

Rodd-Henricks, Z. A., Bell, R. L., Kuc, K. A., Murphy, J. M., McBride, W. J., Lumeng, L., et al. (2001). Effects of concurrent access to multiple ethanol concentrations and repeated deprivations on alcohol intake of alcohol-preferring rats. *Alcoholism, Clinical and Experimental Research*, *25*, 1140–1150.

Rosenthal, R. (1979). The file drawer problem and tolerance for null results. *Psychological Bulletin*, *86*, 638–641.

Schuster, C. R., & Thompson, T. (1969). Self administration of and behavioral dependence on drugs. *Annual Review of Pharmacology*, *9*, 483–502.

Shaham, Y., & Stewart, J. (1995). Stress reinstates heroin-seeking in drug-free animals: An effect mimicking heroin, not withdrawal. *Psychopharmacology (Berlin)*, *119*, 334–341.

Shelton, K. L., & Grant, K. A. (2001). Effects of naltrexone and Ro 15–4513 on a multiple schedule of ethanol and Tang self-administration. *Alcoholism, Clinical and Experimental Research*, *25*, 1576–1585.

Shoaib, M., Sidhpura, N., & Shafait, S. (2003). Investigating the actions of bupropion on dependence-related effects of nicotine in rats. *Psychopharmacology (Berlin)*, *165*, 405–412.

Simms, J. A., Steensland, P., Medina, B., Abernathy, K. E., Chandler, L. J., Wise, R., et al. (2008). Intermittent access to 20% ethanol induces high ethanol consumption in Long-Evans and Wistar rats. *Alcoholism, Clinical and Experimental Research*, *32*, 1816–1823.

Spanagel, R., & Holter, S. M. (1999). Long-term alcohol self-administration with repeated alcohol deprivation phases: An animal model of alcoholism? *Alcohol and Alcoholism*, *34*, 231–243.

Spangler, R., Unterwald, E. M., & Kreek, M. J. (1993). 'Binge' cocaine administration induces a sustained increase of prodynorphin mRNA in rat caudate-putamen. *Brain Research Molecular Brain Research*, *19*, 323–327.

Sprow, G. M., & Thiele, T. E. (2012). The neurobiology of binge-like ethanol drinking: Evidence from rodent models. *Physiology & Behavior*, *106*, 325–331.

Thiele, T. E., & Navarro, M. (2014). "Drinking in the dark" (DID) procedures: A model of binge-like ethanol drinking in non-dependent mice. *Alcohol*, *48*, 235–241.

Unterwald, E. M., Ho, A., Rubenfeld, J. M., & Kreek, M. J. (1994). Time course of the development of behavioral sensitization and dopamine receptor up-regulation during binge cocaine administration. *The Journal of Pharmacology and Experimental Therapeutics*, *270*, 1387–1396.

Vallender, E. J., Ruedi-Bettschen, D., Miller, G. M., & Platt, D. M. (2010). A pharmacogenetic model of naltrexone-induced attenuation of alcohol consumption in rhesus monkeys. *Drug and Alcohol Dependence*, *109*, 252–256.

Vanderschuren, L. J., & Everitt, B. J. (2004). Drug seeking becomes compulsive after prolonged cocaine self-administration. *Science*, *305*, 1017–1019.

Vendruscolo, L. F., & Roberts, A. J. (2014). Operant alcohol self-administration in dependent rats: Focus on the vapor model. *Alcohol, 48*, 277–286.

Volkow, N. D., Wang, G. J., Telang, F., Fowler, J. S., Logan, J., Childress, A. R., et al. (2006). Cocaine cues and dopamine in dorsal striatum: Mechanism of craving in cocaine addiction. *The Journal of Neuroscience, 26*, 6583–6588.

Weerts, E. M., Fantegrossi, W. E., & Goodwin, A. K. (2007). The value of nonhuman primates in drug abuse research. *Experimental and Clinical Psychopharmacology, 15*, 309–327.

Weiss, F., Martin-Fardon, R., Ciccocioppo, R., Kerr, T. M., Smith, D. L., & Ben-Shahar, O. (2001). Enduring resistance to extinction of cocaine-seeking behavior induced by drug-related cues. *Neuropsychopharmacology, 25*, 361–372.

Wise, R. A. (2002). Brain reward circuitry: Insights from unsensed incentives. *Neuron, 36*, 229–240.

Woolverton, W. L., & Virus, R. M. (1989). The effects of a D1 and a D2 dopamine antagonist on behavior maintained by cocaine or food. *Pharmacology, Biochemistry, and Behavior, 32*, 691–697.

Yeomans, M. R., & Gray, R. W. (1997). Effects of naltrexone on food intake and changes in subjective appetite during eating: Evidence for opioid involvement in the appetizer effect. *Physiology & Behavior, 62*, 15–21.

Zimmermann, U. S., O'Connor, S., & Ramchandani, V. A. (2013). Modeling alcohol self-administration in the human laboratory. *Current Topics in Behavioral Neurosciences, 13*, 315–353.

CHAPTER TWO

A Pressing Need for Pharmacotherapy Development to Treat Drug Addiction: An Editorial from a Legal Perspective

B. Andraka-Christou[1]
564 E Graham PL, Bloomington, IN, United States
[1]Corresponding author: e-mail address: bandraka@indiana.edu

Contents

Abstract

Only three FDA-approved pharmacological treatments exist for treating opiate dependence, all of which are underprescribed and underused. No FDA-approved pharmacological treatments exist for cocaine or methamphetamine dependence. More evidence-based, FDA-approved treatments are needed for treating drug dependence, but pharmaceutical companies are unlikely to pursue such research without government incentives. Today pharmaceutical companies primarily conduct research and development (R&D) related to "blockbuster" and rare diseases; drug dependence does not fall into either category. Further compounding the problem, pharmaceutical

International Review of Neurobiology, Volume 126
ISSN 0074-7742
http://dx.doi.org/10.1016/bs.irn.2016.02.003

companies have been recently slashing risky areas of research, rather than adopting new areas. Fortunately, the government has a number of options to incentivize pharmaceutical R&D relating to drug dependence treatment, including the following: market exclusivity for new medications, tax breaks, priority review vouchers, liability reduction, and an advanced market commitment.

1. INTRODUCTION

The social costs of drug addiction (dependence) are profound, including prosecution of drug-related crimes, incarceration, death, emergency room (ER) treatments, hospitalization, and productivity loss. According to the National Institute on Drug Abuse (NIDA) of the National Institutes of Health (NIH), the abuse of illicit drugs costs the United States $224 billion and the abuse of alcohol costs the United States $224 billion annually. In particular, opiate dependence has become a public health crisis in the United States, where more deaths resulted from opiate overdoses than from car accidents in 2014 (Chen, Hedegaard, & Warner, 2014). Likewise, methamphetamine and cocaine dependence ruin lives and cost significant government resources.

The Substance Abuse and Mental Health Services Administration (SAMHSA) expects the government to pay for 70% of substance abuse treatment expenditures by 2020 (USDHHS, 2014d, p. 31). Thus, it is in the government's best interest to support the development of better treatments for drug dependence. For instance, between 11% and 15% of all Medicaid beneficiaries have substance abuse treatment needs (Boozang, Bachrach, & Detty, 2014). State and Federal governments also fund and provide drug addiction treatment through the criminal justice system, including drug courts and in prisons (National Association of Drug Court Professionals, 2015).

The Affordable Care Act (ACA) expanded Medicaid eligibility thresholds in many states, allowing more individuals than ever before to obtain substance abuse treatment through Medicaid. Additionally, all health insurance providers on the newly established ACA Marketplace must cover substance abuse treatment as an essential benefit for enrollees. These benefits must conform to the Mental Health Parity and Addiction Equity Act (MHPAEA), which requires that insurers apply the same standards to covering and managing mental health and substance abuse treatment as they apply to other medical and surgical benefits. As a result, more Americans are gaining access to substance abuse treatment, often with premiums and copays subsidized by the Federal government (USDHHS, 2014d).

The standard of care for treating opiate dependence is a combination of mental health counseling and medication (Amato et al., 2005; Boldt, 2010a; Hesse & Pederson, 2008; Jones, Campopiano, Baldwin, & McCance-Katz, 2015; Kastelic, Pont, & Stöver, 2008; Kjome & Moeller, 2011; Magura, Rosenblum, Lewis, & Joseph, 1993; Nicholls, Bragaw, & Ruetsch, 2010; Stotts, Carrie Dodrill, & Kosten, 2009; USDHHS, 2015c; WHO, 2009). Unfortunately, only three FDA-approved pharmacological treatments exist for treating opiate dependence, all of which are underprescribed and underused (Jones et al., 2015). On the other hand, there are no FDA-approved treatments for cocaine dependence (USDHHS, 2010d) or for methamphetamine dependence (USDHHS, 2013a). Mental health counseling from licensed therapists, particularly cognitive behavioral therapy, may help some individuals suffering from cocaine or methamphetamine dependence, but the cost of therapy and the severe undersupply of therapists are a barrier to widespread treatment (Vestal, 2015).

More evidence-based, FDA-approved treatments are needed for treating drug dependence, but pharmaceutical companies are unlikely to pursue such research without government incentives. Today pharmaceutical companies primarily conduct research and development (R&D) related to "blockbuster" and rare diseases. Blockbuster diseases are those with a large profit potential, usually due to large disease population size. Rare diseases, also known as orphan diseases, are usually defined as those affecting less than 200,000 individuals in the United States. However, drug dependence is neither a "blockbuster disease" nor a rare disease. Fortunately, the government has a number of options to incentivize pharmaceutical R&D relating to under-researched diseases. Congress could pass legislation including one or more of the following incentives: market exclusivity for new medicines, tax breaks, priority review vouchers, liability reduction, and an advanced market commitment. The present editorial begins with a discussion of the current drug dependence crisis and existing pharmacological treatments. Next, the article provides an overview of R&D trends in the pharmaceutical industry. Finally, the article suggests legislation for incentivizing pharmaceutical R&D.

2. THE NEED FOR LEGISLATIVE INCENTIVES FOR R&D OF DEPENDENCE MEDICATION

2.1 Opiate Dependence and Treatment

Opiate overdoses have quadrupled over the last decade in the United States (Chen et al., 2014). The Centers for Disease Control and Prevention (CDC) reported that more people died from opiate overdoses in the United States in 2014 than from car accidents (Chen et al., 2014). Some states have been hit especially hard, such as Kentucky, where deaths from heroin overdoses increased by 500% from 2011 to 2012 (Drug Free Kentucky, 2013). Most individuals who overdose on opiates in the United States are dependent on heroin or prescription painkillers (Chen et al., 2014). As the cost of prescription painkillers has increased, the percentage of drug users using heroin (which is relatively cheaper) has increased (Maxwell, 2015). Approximately 2.1 million individuals in the United States are estimated to be dependent on opiate prescription pain killers, and 467,000 are dependent on heroin (Volkow, 2014). Opiate dependence has many negative (often tragic) consequences for the dependent individual, his or her family, and society at large (Birnbaum et al., 2011; USDHHS, 2015c). Costs to the dependent individual include productivity loss, increased medical costs (especially from ER visits and hospitalization), mental distress, and sometimes death. Costs to society include decreased economic productivity, increased medical costs, law enforcement resources, incarceration, and drug-related crimes.

According to the World Health Organization, the most effective treatment for opiate dependence is medication combined with counseling (United Nations, Office on Drugs and Crime, World Health Organization, 2009). This treatment combination is referred to as medication-assisted treatment, or MAT. However, only three effective medications for treating opiate dependence have been approved by the FDA: buprenorphine (commonly known as Suboxone®), Vivitrol® (extended-release naltrexone), and methadone (USDHHS, 2015a, 2015b, 2015c, 2015d, 2015e). Each of these three medications has been proven significantly more effective at preventing drug use relapse than a placebo in rigorous, double-blind experimental studies (Amato et al., 2005; Boldt, 2010b; Jones et al., 2015; Kjome & Moeller, 2011; Stotts et al., 2009; USDHHS, 2015a, 2015c, 2015d). Importantly, experimental studies have found that the combination of medication and counseling is more effective than counseling alone at preventing relapse (Hesse & Pederson, 2008; USDHHS, 2015c). Also, the retention rate for MAT is greater than the retention rate for either counseling or 12-step groups (Caplehorn, 1994; Caplehorn, MacNeil, & Kleinbaum, 1993; Trautmann & Wittchen, 2012; Willenbring, Hagedorn, & Postier, 2004). Unfortunately, all medications for treating opiate dependence are underutilized by opiate-dependent individuals in the United States, underprescribed by physicians, rarely available within in-patient rehabilitation centers, rarely used within prisons (Nunn et al., 2009), and underused within drug courts (Matusow et al., 2013; NADCP, 2011; SAMHSA, 2014).

2.1.1 Methadone

Methadone is a full agonist that activates the opiate receptor, diminishing cravings for opiates and preventing euphoria if the patient abuses opiates (SAMHSA, 2014). Unfortunately, even though most individuals who obtain methadone use it for treatment and do not abuse it, methadone can be dangerous if diverted and improperly used (Herman, Stancliff, & Langrod, 2000). As a result, methadone is only available at certified methadone treatment centers to which the patient must usually return daily in order to continue treatment (Federal Opioid Treatment Standards, 2003). If the patient has continuously undergone methadone maintenance treatment for a period of time and has met the "take-home" eligibility criteria in the Drug Enforcement Agency (DEA) regulation, then the methadone clinic may permit him or her to take some methadone home. The take-home amount ranges from 1 day's worth to 1 week's worth (if the patient has been in treatment for at least two years.). Methadone treatment centers tend to be heavily visible and stigmatized by city governments and residents (Herman et al., 2000). Therefore, it is difficult for opiate-dependent individuals to maintain anonymity during methadone treatment, as coworkers or friends may see them entering the methadone clinic.

2.1.2 Vivitrol®

Vivitrol® is a once-per-month injection (SAMHSA, 2012). It does not contain any opiate ingredient, it is not a controlled substance, and it may be prescribed by any licensed physician. Vivitrol® contains extended-release naltrexone, which is a complete mu-receptor antagonist, meaning it completely blocks the mu-receptor. As a result, Vivitrol® prevents an individual from experiencing euphoria if he or she abuses any opiate, helping to prevent relapse (Krupitsky et al., 2001). Unfortunately, if a patient receives Vivitrol® prior to complete detoxification, then the patient will experience immediate and distressing withdrawals. Therefore, patients must first detox completely before taking Vivitrol®. However, for many individuals dependent on opiates, the need to completely physically detoxify prior to treatment may act as a disincentive to beginning treatment, because opiate withdrawals can be very distressing. Withdrawal symptoms often include up to 2 weeks of diarrhea, muscle aches, excessive sweating, jerking or twitching muscles, fatigue, and nausea (The National Alliance of Advocates for Buprenorphine Treatment, 2008). In-patient rehabilitation centers allow the patient to undergo complete detoxification under supervision, while preventing relapse and providing treatment for these painful withdrawal symptoms. However, in-patient detoxification costs approximately \$31,500 for 30 days (Cherkis, 2015). Unfortunately, insurance companies, Medicare and Medicaid do not cover the cost of hospitalization for opioid withdrawal. Finally, Vivitrol® is very expensive, costing around \$1000 per month for an individual lacking health insurance coverage (American Association for the Treatment of Opioid Dependence, Inc., 2015), a common scenario for people suffering from substance abuse (Chicago Business, 2013).

2.1.3 Buprenorphine (Suboxone®)

Buprenorphine is a partial mu-agonist (USDHHS, 2015b). It prevents the patient from going into withdrawals or experiencing cravings, while preventing euphoria from any opiate used (including too much buprenorphine). The patient typically receives a monthly prescription, which he or she fills at a local pharmacy, and then takes one or two pills or sublingual films at home daily. Buprenorphine can be prescribed at any licensed physician's office, as long as the physician obeys DEA rules for Schedule III substances, and as long as the physician has a "DATA waiver" from SAMHSA, as required under the Drug Dependence and Treatment Act of 2000. However, in 2012 only 46.6% of United States counties had at least one physician with a DATA waiver (Rosenblatt, Andrilla, Catlin, & Larson, 2015, p. 25). Only 2% of all United States physicians had a DATA waiver, and only 3% of primary care physicians (the largest group of physicians in the nation) have a DATA waiver (Rosenblatt et al., 2015, p. 25). Surveys of physicians suggest that the lack of DATA waivers is due to a combination of the following factors: lack of addiction treatment education in medical school, lack of knowledge about the existence of pharmacological treatments for addiction and DATA waiver requirements, lack of institutional support, reputation of drug addiction patients as being "difficult" patients, and regulatory concerns (Walley et al., 2008). Unfortunately, because methadone and buprenorphine contain an opiate ingredient, many treatment providers oppose their use, including many Narcotics Anonymous group members (White et al., 2013) and a large minority of mental health therapists (Knudsen, Ducharme, Roman, & Link, 2005). Even though it is possible to abuse buprenorphine, buprenorphine overdoses are extremely rare relative to heroin, oxycodone, or methadone overdoses (USDHHS, 2015b).

2.2 Cocaine Dependence and Treatment

Cocaine use, including crack cocaine use, has been declining somewhat over the past decade but continues to remain a pressing public health concern (USDHHS, 2010c). As of 2008, approximately 1.2 million individuals in the United States are estimated to abuse or be dependent on cocaine. Annually, almost 500,000 ER visits are related to cocaine abuse or dependence, equaling about one in four drug-related ER visits (USDHHS, 2010c). Cocaine use frequently leads to other diseases, such as hepatitis C, which is prevalent in 40–50% of injection drug users (USDHHS, 2010b). Cocaine use also significantly increases the risk for HIV/AIDS and viral hepatitis (USDHHS, 2010a). Cocaine dependence relapse rates are very high, frequently occurring after years of abstinence. Long-term cocaine use diminishes the functioning of certain receptors in the brain, resulting in "diminishing sensitivity to natural rewards" (USDHHS, 2010b). No FDA-approved medications currently exist for the treatment of cocaine dependence. However, like opiate dependence, cocaine dependence involves known receptors in the brain, suggesting potential targets for treatment discovery. For example, according to NIDA, the dopamine D2 and dopamine D3 receptors, both of which function in natural reward pathways, may be effective targets for pharmacological treatments (USDHHS, 2010b). Additionally, an FDA-approved medication for treating alcoholism, topiramate, shows promise in helping cocaine-dependent individuals (Science Daily, 2013). NIDA also expects that a vaccine could be developed to prevent cocaine from taking effect in the brain (USDHHS, 2010b).

2.3 Methamphetamine Dependence

Approximately 1 million Americans use methamphetamine annually (USDHHS, 2013c). Even though methamphetamine use is declining nationally, significant regional variation exists. In particular, the West and Midwest continue to feel strong economic effects from methamphetamine use. In 2011, methamphetamine use was related to approximately 100,000 ER visits (USDHHS, 2013c). The RAND Corporation estimates that the annual economic loss in the United States resulting from methamphetamine use equals \$16.2–\$48.3 billion (Nicosia, Pacula, Kilmer, Lundberg, & Chiesa, 2005). The percentage of methamphetamine users who are dependent is unknown, but long-term use of methamphetamine frequently leads to dependence and other diseases (USDHHS, 2013b). Regular methamphetamine use significantly reduces dopamine binding to dopamine receptors and increases neurotoxicity. Former methamphetamine users may experience neurotoxic effects after years of abstinence, due to overactivation of cells in the brain that regulate inflammation[1] (Wang et al., 2004). Additionally, former methamphetamine users may exhibit psychosis in the form of paranoia, hallucinations, and repetitive motor activity. Finally, long-term methamphetamine usage may lead to increased violent activity, memory loss, distractibility, severe tooth decay, skin sores, and malnutrition.

No FDA-approved treatment exists for the treatment of methamphetamine dependence (USDHHS, 2013a). However, NIDA has pointed to two potentially promising areas of research: suppression of neuro-inflammatory actions of brain cells that regulate inflammation, and utilization of the immune system to neutralize the effects of methamphetamine before it reaches the brain. One promising medication, Ibudilast, is undergoing Phase II clinical trials to treat methamphetamine addiction (United States National Institutes of Health, ClinicalTrials.gov, 2015). Preliminary results suggest Ibudilast may dampen cravings and improve cognition (Johnson, Matsuda, & Iwaki, 2014). Additionally, the FDA has fast-tracked[2] Ibudilast for treatment of methamphetamine dependence (Miles, 2013).

3. THE DECLINE OF R&D IN PHARMACEUTICAL COMPANIES

3.1 General Trends

The number of medications developed each year in the United States is declining (measured by the number of annual New Drug Applications (NDAs) submitted to the FDA) (Scannell, Blanckley, Boldon, & Warrington, 2012). A number of reasons for this decline have been suggested, including loss of "the low hanging fruit" (Mundasad, 2012). In other words, the easiest medications to develop have already been created. As a result, pharmaceutical companies must either develop "me too" medications or develop innovative medications for complex diseases (which naturally have high research failure rates). Ninety percent of FDA-approved drugs have been submitted to the FDA by pharmaceutical companies (Bartfai & Lees, 2013). Business administrators and marketers, rather than doctors or scientists, decide which diseases will receive the R&D resources

[1] Microglial cells regulate inflammatory responses in the brain. Overactivation of microglial cells, as occurs with long-term use of methamphetamine, can lead to an uncontrolled inflammatory response in the brain. Uncontrolled inflammation in the brain can cause neurodegenerative diseases.

[2] The FDA fast-track provides pharmaceutical sponsors with additional FDA communication and meetings, a rolling review of newly completed NDA sections, and the potential for Accelerated Approval and Priority Review of the medication (FDA, 2015).

of pharmaceutical companies. Market demand and access is a prime consideration. In particular, major industry players are worried about the upcoming "patent cliff" when many of the best-selling pharmaceuticals' patents will expire (Management Centre Europe, 2012). Under pressure from stockholders, pharmaceutical companies have aggressively cut R&D expenditures over the last decade with R&D spending growth significantly decreasing between 1995 and 2015 (LaMattina, 2014; Pharmaceutical Research and Manufacturers of America, 2015, p. 35).

Additionally, major pharmaceutical companies have cut risky R&D areas from their profiles, such as antibacterial development, treatment for depression, schizophrenia, and vaccine development (CBS News, 2010; Childress, 2013; LaMattina, 2014; Offit, 2005; Plumridge, 2014). Vaccine R&D slowed over the last 50 years, as pharmaceutical companies found the market for vaccines to be too small and the cost of vaccine development to be too high. Since the late 1990, many projects in neurology, cardiovascular medicine, and psychiatry have fallen out of favor due to failures in clinical trials, making the medications too financially risky to develop.[3] At the same time, pharmaceutical companies are diversifying into other types of products, such as health supplements, animal health, generics, and diagnostics (Management Centre Europe, 2012). Diversification has been occurring, in part, through a wave of mergers and acquisitions throughout the pharmaceutical industry (Mittra, 2007).

Even if decisions to scrap potentially unprofitable research areas are rational from a business perspective (and perhaps even required under corporate law), these decisions are detrimental to society as a whole. First, even if a new drug is ultimately not approved by the FDA, the scientific knowledge gleaned from R&D can be reused in the development of future medications. For example, the process of identifying chemical pathways and targets for one disease often assists in developing treatment for other diseases, because many diseases share similar chemical pathways (this is especially common in oncology and inflammation; Bartfai & Lees, 2013). Second, one drug may be used to treat multiple conditions, thus expanding its social benefits. For example, the medication Gleevec® was initially projected to be only used

[3] One prominent feature in the new R&D landscape is that many pharmas have withdrawn from therapeutic categories where clinical development is lengthy, expensive, and fraught with regulatory hurdles. So for example, some companies have reduced or entirely quit cardiovascular research where trials often need to enroll 20,000 or more patients in order to show a statistically meaningful improvement over existing medications. Instead research budgets have flowed into oncology where regulators often accepted registration trials with only a few hundred test subjects (Hoffman, 2014).

to treat leukemia but now is routinely used to treat a range of cancers (Querido, 2012).

R&D of major pharmaceuticals predominantly falls into two broad categories: "blockbuster" disease treatment (usually developed in-house by major pharmaceutical companies) and rare disease treatment (often acquired or licensed by major pharmaceutical companies from smaller biotechnology companies) (Mittra, 2007). Blockbuster disease treatments target a large population size with relatively predictable profit. Profitability is also expected if the basic science behind the disease is relatively well understood and if clinical trials for treatment of the disease tend to fare well.

Rare diseases are considered pharma's "darlings," attracting significant R&D resources (Bartfai & Lees, 2013). In 2014 one-third of approved NDAs were for treatment of rare diseases (orphan diseases) (Pharmaceutical Research and Manufacturers of America, 2015). There are three primary reasons for this trend. First, development of treatments for orphan diseases results in significant benefits from the government under the Orphan Drug Act (1983). Second, most rare diseases are genetic diseases, which have clear diagnoses; diseases with clear diagnoses have relatively higher treatment response rates than diseases without clear diagnoses. As a result, genetic diseases tend to have shorter clinical trial times with clearer results relative to nongenetic diseases (Segalat, 2007). As a whole, targeting a genetic disease is extremely efficient for pharmaceutical companies and may include the development of a diagnostic product in addition to the treatment. Third, biologics are easier to develop for genetic diseases, a category into which most rare diseases fall, and biologics are extremely profitable (Gill, 2015). For example, enzyme replacement therapy may cost up to $150,000 per year (Sharma, Jacob, Tandon, & Kumar, 2010). There remains a significant gap of R&D between "blockbuster diseases"/orphan diseases and drug addiction. Without government incentives or significant interest group pressure, pharmaceutical companies are unlikely to devote R&D resources to diseases that fall into this gap, including drug addiction.

3.2 Reasons for Pharmaceutical Lack of R&D Into Drug Dependence Treatment

Considering the significant individual and societal costs of drug dependence, more research into opiate, cocaine, and methamphetamine dependence treatment is critically needed, including basic scientific and translational research. Currently, NIDA is researching and testing possible pharmacological treatments for cocaine dependence (USDHHS, 2010d) and methamphetamine dependence (USDHHS, 2013a). However, the resources of academia and government-funded translational research are dwarfed by the resources of pharmaceutical companies. For example, in December 2015, President Obama signed into law the Consolidated Appropriations Act 2016, providing approximately $685 million for its translational research center for the year 2016 (Consolidated Appropriations Act 2016, 2015; USDHHS, 2015b). By comparison, in 2014 the United States pharmaceutical industry spent approximately $51 billion in R&D, almost all of which is translational rather than basic research (Pharmaceutical Research and Manufacturers of America, 2015). The pharmaceutical industry's resource advantage includes translational research knowledge, experience, revenue, employees, and infrastructure. Therefore, the United States must continue to look to the pharmaceutical industry for R&D of most new medications.

A number of possible explanations exist for the current lack of R&D into drug addiction treatment by pharmaceutical companies. First, pharmaceutical companies may be concerned that the targeted disease population would be unable to afford medical treatment, precluding a return on financial investment. The concern is understandable given the strong correlation between drug dependence and low socioeconomic status (Chicago Business, 2013). On the other hand, Medicaid programs in every state cover buprenorphine, and some Medicaid programs cover methadone and/or Vivitrol® (American Society of Addiction Medicine, 2013). Therefore, even if patients themselves cannot afford treatment, the government may pick up the bill, so long as the medication is effective, safe, and cost saving. However, relative to commercial insurance providers, Medicaid reimburses pharmaceutical companies at a lower rate.

Second, because pharmaceutical companies are narrowing existing research areas (Crow, 2013; Weisman, 2015), they are unlikely to open up new research areas for which they have little preexisting knowledge (excepting acquisition of the research area through a merger). A steep learning curve exists for each new research area that a pharmaceutical company undertakes (Hu, Schultz, Sheu, & Tschopp, 2007). If limited basic research exists regarding the targets and mechanisms of the disease, then the learning curve is especially steep. In such a case, the pharmaceutical company will usually prefer to expand its existing, commercially successful research areas rather than invest capital into the new research area.

Finally, bias against medication-assisted treatment for drug addiction may exist within the corporate structure, especially given the existing bias against medication-assisted treatment in society as a whole. Investors may misconstrue bias as lack of demand for medication-assisted treatment (Volkow & Li, 2004). Of course, pharmaceutical companies could begin to overcome such bias by educating the public about the existence of medication-assisted treatments through educational campaigns or through direct-to-consumer advertising. Likewise, the government could begin a mass media educational campaign to destigmatize medication-assisted treatment specifically and addiction treatment more generally.

4. LEGISLATION SPECIFICALLY INCENTIVIZING R&D INTO DEPENDENCE TREATMENT

The United States government could provide market-based incentives for R&D into drug addiction. Congress could pass legislation specifically targeting drug dependence treatment R&D, including some combination of push and pull incentives. Historically, Congress has provided market exclusivity, tax credits, priority review vouchers, clinical trial funding, advanced market commitment, and liability reduction to incentivize research into specific disease categories. Clinical trial funding is a "pull mechanism," as it is awarded prior to completed development; market exclusivity, tax credits, priority review vouchers, and advanced market commitment are considered "push mechanisms," because they reward development after the fact (Hamming, 2013).

4.1 Market Exclusivity

The government could provide market exclusivity for new medications targeting drug dependence. Such an incentive would be modeled on the exclusivity provided for rare disease treatments under the Orphan Drug Act (ODA). The ODA is widely considered the most successful incentive-based pharmaceutical legislation in the United States, and the Department of Health and Human Services has called the exclusivity provision the most influential incentive in the ODA (United States Department of Health and Human Services, Office of the Inspector General [USDHHS], 2001). Due to its success, the ODA has served as a model for pediatric treatment incentives (Food & Drug Modernization Act, 1997).

Under the ODA, the first corporation (sponsor) to market an orphan drug (meaning a drug intended to treat a disease affecting less than 200,000 individuals in the United States) with a new active ingredient is granted a 7-year marketing exclusivity. The marketing exclusivity period is provided whether or not the drug is patented, providing investment security. In practice, the exclusivity may elongate an existing patent or protect the sponsor if the patent is invalidated. During the exclusivity period, no other drug for the same disease with the same active ingredient may be approved by the FDA, unless one of the following occurs: the sponsor does not manufacture the product, the sponsor cannot produce sufficient quantities for market demand, or the sponsor gives its consent to allow another corporation to market their product (Orphan Drug Act, 1983). A similar marketing exclusivity could be provided for drug dependence treatment.

The current marketing exclusivity provided for rare diseases under the ODA has been criticized in situations where the sponsor of the new treatment makes a significant profit from the treatment (Rin-Laures & Janofsky, 1991). However, legislation incentivizing drug dependence treatment could include an exclusivity provision similar to that in the European Union's version of the ODA, which shortens the European Union's 10-year marketing exclusivity to 6 years if at the end of the 5th year it is shown "that the product is sufficiently profitable not to justify maintenance of market exclusivity" (EC, 1999, at Article 8, § 2).

4.2 Tax Credit

The government could provide a tax credit to corporations that develop treatments for drug dependence. For example, the ODA provides a 50% tax credit for "qualified clinical testing expenses" (ODA, 1983, at § 4(b)). A tax credit for clinical testing may be particularly important in the case of treatments for drug dependence, because recruiting patients with stigmatized diseases is especially difficult and expensive (Corrigan, 2004; Woodall, Morgan, Sloan, & Howard, 2010). More generally, the government could provide a tax credit for research and experimentation, including clinical trial expenditures. Such a tax credit would be especially useful for companies without any basic research experience in drug addiction treatment, because discovery, extraction, and synthesis of alternative molecules takes approximately 3–6 years to complete and is very expensive (Crimm, 1994, p. 1019). Furthermore, such basic research often proves futile. Research and experimentation credits would not apply to expenditures for marketing, sales, or manufacturing. Providing tax credits to pharmaceutical companies for developing treatment for drug dependence may be politically contentious. However, the tax credit could be politically palatable if made progressive, with the total credit being inversely related to sales revenue.

4.3 Priority Review Vouchers

Currently, a new treatment may obtain priority review designation if the medication would result in "significant improvements in the safety or effectiveness of the treatment, diagnosis, or prevention of serious conditions when compared to standard applications" (USDHHS, 2014c). Priority review designation shortens the time period during which the FDA reviews the NDA, allowing the sponsor to make a profit more quickly. However, standard priority review designation only applies to FDA approval of that particular medication. The incentive to develop drug addiction treatment would be stronger if a priority review *voucher* was provided that could be used not at that moment, but during the FDA approval process of a future blockbuster disease treatment. For example, Congress has already established a priority review voucher system for other classes of diseases, e.g., neglected tropical diseases (Food & Drug Administration Amendment Act, 2007, at § 1102). Under that scheme, the sponsor who obtains a priority review voucher for a neglected disease treatment may use the voucher to obtain priority review of any subsequent NDA, guaranteeing review within 6 months of using the voucher (USDHHS, 2014b). As a result, a sponsor may use the voucher during approval of a later blockbuster drug, potentially saving 7 months of review time (Thaul, 2008). Even though pharmaceutical companies will not expect substantial profit from drug dependence treatments, if the voucher is used later for review of a blockbuster treatment, then the voucher could save the pharmaceutical company hundreds of millions of dollars (Hamming, 2013). Alternatively, the sponsor may sell the voucher to another pharmaceutical company at a significant profit. For example, in 2015 a sponsor of a rare pediatric disease treatment sold its priority review voucher for $350 million to another pharmaceutical company (United Therapeutics, 2015).

4.4 Advanced Market Commitment

Perhaps most controversially, the government could promise to purchase a certain quantity of medication for treating drug dependence, under the condition that the FDA approves the medication after reviewing it for efficacy and safety. For example, the United States Department of Homeland Security (DHS) is authorized by Congress in the Project Bioshield Act of 2004 to purchase and stockpile vaccines against anthrax and smallpox in order to protect public health in the event of terrorism (Project Bioshield Act, 2004). Specifically, the Project Bioshield Act authorizes the Secretary of the DHS to negotiate the terms of contracts with vaccine developers and manufactures, including negotiating "quantity, production schedule, and price," as well as carrying out "other activities that may be reasonably required, including advanced R&D" (Project Bioshield Act, 2004, § 319F-2(7)(c)(i)(I)). Pharmaceutical companies may legitimately fear that individuals suffering from drug dependence will be unable to afford medication, limiting profit potential (O'Reilly, 2007). A guaranteed government purchase would help alleviate such concerns. However, in order for pharmaceutical companies to alter their behavior, they must feel that they can rely on the government's promise, meaning that they should see a long-term government commitment in the legislation.[4]

4.5 Liability Reduction

Some scholars have argued that product liability claims against pharmaceutical companies have become so frequent and expensive that they deter potential R&D (Crimm, 1994). Liability risk may especially deter R&D into disease areas with which companies have little research experience. The government could incentivize research (or at least not disincentivize research) by reducing liability for certain types of products or treatments. For example, Congress reduced liability for smallpox vaccines and anthrax vaccines under the Project Bioshield Act in order to incentivize vaccine development (Project Bioshield Act, 2004).

First, states and the Federal government could ban strict liability claims for product defects in pharmaceutical treatments for drug dependence. The FDA already reviews preapproval testing, manufacturing, labeling, advertising, and postmarket surveillance (Jackson, 1992). However, consumers may still sue pharmaceutical manufacturers for a product defect under strict liability theory, despite FDA approval. Under strict liability theory, plaintiffs need only prove that an unreasonably dangerous defect exists in the product in order to claim damages; the plaintiffs do not need to prove a negligent act on the part of the defendant. As a result, it is easier for plaintiffs to bring suits under strict liability theory than negligence. Even though strict liability theory may provide important consumer protection for many other types of products, it is less necessary for pharmaceuticals, because mandatory FDA review already provides an extra layer of consumer protection.

[4] The Project Bioshield Act has been criticized for not sufficiently persuading pharmaceutical companies that the government is a long-term partner. The Act has also been criticized for failing to contain an indemnification provision (O'Reilly, 2007).

Second, states and the Federal government could ban claims for punitive damages in the case of pharmaceutical treatments for drug dependence. As a result, plaintiffs would only be compensated for actual harms resulting from any unreasonably defective product. Relative to compensatory damages, punitive damage awards are more likely to result in corporate insolvency, making corporations more risk averse. The risk of punitive damages usually serves an important purpose: it deters corporations from engaging in egregiously unethical behavior that may harm consumers. However, because mandatory FDA review already exists for pharmaceuticals, the risk of punitive damages may serve more as an R&D disincentive than as a protection for consumers.

4.6 Political Considerations

By targeting drug dependence treatment as worthy of government incentives, the government would send a powerful message to the United States population: dependence is a disease that should be taken seriously and treated effectively. Currently, widespread stigma against persons suffering from drug dependence exists, along with the misconception that drug dependence is an immoral choice rather than a chronic disease. Using social constructionism theory, this section discusses the political difficulties of helping persons who are dependent on drugs. Social constructionism theory is commonly used by political scientists to explain why legislation passes or fails to pass, depending on whom the legislation benefits or harms (Sabatier, 2007). The theory posits the existence of "target groups": groups chosen to be benefitted or burdened by a policy design. The allocation of benefits and burdens to target groups depends on the power of the groups and whether they are construed as deserving or undeserving by society. There are four categories of target groups: the advantaged, the contenders, the dependents, and the deviants.

Advantaged groups are high-powered groups who are positively construed as "deserving," such as the military. Providing benefits to advantaged groups creates political gain for policymakers. Contender groups are high-powered groups who are negatively construed as "undeserving," such as oil companies. They often receive benefits "sub rosa": through details buried within legislation. Dependent groups are low-powered and positively construed as "deserving," such as low-income children. They are viewed with sympathy, but their lack of economic and political power curtails their receipt of benefits. Finally, deviant groups are low-powered and negatively construed as "undeserving," such as felons. They have few advocacy groups, limited resources, difficulty organizing politically, and make up a kind of "permanent underclass." Policymakers may gain political capital by punishing them. Conversely, policymakers may lose political capital by benefiting them (Sabatier, 2007).

Even though ill individuals usually fall into the "dependent" target group, society construes drug-dependent individuals as "deviants." As a result, drug-dependent individuals are often blamed for their disease. Additionally, because criminals are considered "deviants" and drug dependence may lead to incarceration (e.g., through conviction of drug possession), drug-dependent individuals have been socially construed as criminals. Because drug-dependent individuals are often construed as "deviants," Congress has little pressure to pass legislation benefiting them and plenty of pressure to pass legislation indirectly punishing them (such as by increasing mandatory sentences for possession of drugs). Therefore, any future legislation that allocates resources to help drug-dependent individuals (such as by incentivizing R&D for treatment) must address the negative social construction of the target group. The government should try to reconstrue drug-dependent individuals as "dependents" (individuals suffering from a chronic disease) instead of "deviants."

Unfortunately, reconstruction of a target group may take a long time. For example, homosexuals for many decades in the United States were considered and treated as "deviants" under legislation; only recently has this social construction begun to change. If policy makers wish to reconstrue the definition of drug-dependent individuals from "deviants" to "dependents," then they could take the following steps. Policy makers could encourage the formation of an advocacy coalition between researchers, medical doctors, the media, and public figures in order to create a cultural shift in the understanding of the nature of drug dependence. In particular, drug dependence would be repeatedly, publically described as a disease, not a moral failing. A similar education process has been proposed by the UN for reducing stigma associated with HIV/AIDS (UNAIDS, Joint United Nations Programme on HIV/AIDS, 2012). Additionally, policy makers could decriminalize drug use and divert resources from incarceration to treatment. Significant stigma of drug dependence has resulted from increased criminalization of drug use, especially from the War on Drugs, leading the public to view drug dependence from a moral rather than a medical standpoint (Amundson, Zajicek, & Hunt, 2014).

5. CONCLUSION

New medications for treating opiate dependence, cocaine dependence, and methamphetamine dependence are needed. However, pharmaceutical companies are narrowing their research focus, focusing capital on areas of high profit, in particular "blockbuster" and rare diseases that can be treated with expensive biologics. Drug dependence does not fall into either of these two categories. As a result, the government should provide incentives to pharmaceutical companies for researching and developing medication for treating drug dependence. Fortunately, the government has experience in incentivizing R&D into other historically under-researched areas: rare diseases, neglected tropical diseases, and vaccines related to bioterrorism. Congress should pass legislation providing a variety of incentives for drug dependence treatment R&D, including market exclusivity, tax credits for clinical research, priority review vouchers, advanced market commitment, and liability reduction.

However, drug dependence is a stigmatized disease, and drug-dependent individuals are often socially construed as deviants. As a result, politicians may find it difficult to garner the political support to pass a bill providing research incentives ultimately benefiting drug-dependent individuals. Therefore, the government should raise awareness about drug dependence as a disease, rather than a moral failing, needing effective treatment. For example, the Department of Health and Human Services could develop and execute a mass media educational campaign fighting existing stigma. Additionally, the government could begin to decriminalize drug possession, diverting resources from incarceration to treatment. Under the ACA, more Americans than ever before will be eligible for substance abuse treatment through Medicaid or private health insurance, providing another reason for society to demand development of more effective treatments for drug dependence.

REFERENCES

Amato, L., Davoli, M., Perucci, C., Ferri, M., Faggiano, F., et al. (2005). An overview of systematic reviews of the effectiveness of opiate maintenance therapies: Available evidence to inform clinical practice and research. *Journal of Substance Abuse Treatment*, *28*, 321–329.

American Association for the Treatment of Opioid Dependence, Inc. (2015). *AAOTD guidelines for using naltrexone (vivitrol) in OTPs*. Retrieved from, http://www.aatod.org/policies/policy-statements/aatod-guidelines-for-using-naltrexone-vivitrol-in-otps/.

American Society of Addiction Medicine (2013). *Advancing access to addiction medications*. Retrieved from, http://www.asam.org/docs/default-source/advocacy/aaam_implications-for-opioid-addiction-treatment_final.

Amundson, K., Zajicek, A., & Hunt, V. (2014). Pathologies of the poor: What do the War on drugs and welfare reform have in common? *Journal of Sociology and Social Welfare*, *XLI*(1), 5–28.

Bartfai, T., & Lees, G. (2013). *The future of drug discovery: Who decides which diseases to treat?* Oxford, UK: Elsevier.

Birnbaum, H. G., White, A. G., Schiller, M., Waldman, T., Cleveland, J. M., & Roland, C. L. (2011). Societal costs of prescription opioid abuse, dependence, and misuse in the United States. *Pain Medicine, 12*(4), 657–667.

Boldt, R. (2010a). The "tomahawk" and the "healing balm": Drug treatment courts in theory and practice. *University of Maryland Law Journal of Race, Religion, Gender, and Class, 9*(1), 45–71.

Boldt, R. (2010b). The "tomahawk" and the "healing balm": Drug treatment courts in theory and practice. *University of Maryland Law Journal of Race, Religion, Gender, and Class, 10*(45), 59–60.

Boozang, P., Bachrach, D., & Detty, A. (2014). *Coverage and delivery of adult substance abuse services in medicaid managed care.* Medicaid Managed Care Information Resource Center, Centers for Medicare and Medicaid Services. Technical Assistance Brief 2 (2).

Caplehorn, J. R. (1994). A comparison of abstinence-oriented and indefinite methadone maintenance treatment. *The International Journal of Addictions, 11*, 1361–1375.

Caplehorn, J. R., MacNeil, D. R., & Kleinbaum, D. G. (1993). Clinic policy and retention in methadone maintenance. *International Journal of Addictions, 28*(1), 73–89.

CBS News (2010). *Merck closing 8 plants, 8 research sites.* Retrieved from, http://www.cbsnews.com/news/merck-closing-8-plants-8-research-sites/.

Chen, L. H., Hedegaard, H., & Warner, M. (2014). *Drug-poisoning deaths involving opioid analgesics: United States, 1999–2011.* Centers for Disease Control and Prevention. NCHS Data Brief (No. 166). Retrieved from, http://www.cdc.gov/nchs/data/databriefs/db166.htm.

Cherkis, J. (2015). Dying to be free. *The Huffington Post.* Retrieved from, http://projects.huffingtonpost.com/dying-to-be-free-heroin-treatment.

Chicago Business (2013). *Health care overhaul will add coverage for millions of addicts.* Retrieved from, http://www.chicagobusiness.com/article/20130911/NEWS03/130919947/health-care-overhaul-will-add-coverage-for-millions-of-addicts.

Childress, S. (2013). Dr. Charles Knirsch: "These are not ruthless decisions" *Frontline, PBS News.* Retrieved from, http://www.pbs.org/wgbh/pages/frontline/health-science-technology/hunting-the-nightmare-bacteria/dr-charles-knirsch-these-are-not-ruthless-decisions/.

Consolidated Appropriations Act, 2016 (2015). Public Law No. 114–113.

Corrigan, P. (2004). How stigma interferes with mental health care. *American Psychologist, 59*(7), 614–625.

Crimm, N. (1994). A tax proposal to promote pharmacologic research, to encourage conventional prescription drug innovation and improvement, and to reduce product liability claims. *Wake Forest Law Review, 29*, 1007–1099.

Crow, J. (2013). Non-profit pharma. *Chemistry World.* Retrieved from, http://www.rsc.org/chemistryworld/2013/09/pharma-medicinal-chemistry-non-profit-rare-disease.

Drug Free Kentucky (2013). *Northern Kentucky's collective response to the heroin epidemic.* Retrieved from, http://drugfreenky.org/wp-content/uploads/2013/11/Northern-Kentuckys-Collective-Response-Final.pdf.

European Commission (1999). *Regulation [EC] No 141/2000 of the European parliament and of the council.* Article 8, § 2.

Federal Opioid Treatment Standards (2003). 42 C.F.R. § 8.12 (h)(3)(i)(II).

Food and Drug Administration Amendment Act (2007). Public Law 110–85, 121 Stat. 823 § 1102, codified as amended at 21 U.S.C. 301.

Food and Drug Modernization Act (1997). Pub. L. No. 105–115, § 111, 111 Stat. 2296, codified as amended at 21 U.S.C. § 351.

Gill, V. (2015). The rise of the biologic. *Royal Society of Chemistry*. Retrieved from, http://www.rsc.org/chemistryworld/Issues/2008/January/TheRiseOfBiologic.asp.

Hamming, L. (2013). The promise of priority review vouchers as a legislative tool to encourage drugs for neglected diseases. *Duke Law & Technology Review*, *11*, 390–412.

Herman, J., Stancliff, S., & Langrod, J. (2000). Methadone maintenance treatment: A review of historical and clinical issues. *The Mount Sinai Journal of Medicine*, *67*(5), 347. 361.

Hesse, M., & Pederson, M. (2008). Easy-access services in low-threshold opiate agonist maintenance. *International Journal of Mental Health and Addiction*, *6*, 316.

Hoffman, D. (2014). What's driving the pharmaceutical industry's drug development process? *Philly News*. Retrieved from, http://www.philly.com/philly/blogs/healthcare/Whats-driving-the-pharmaceutical-industrys-drug-development-process.html.

Hu, M., Schultz, K., Sheu, J., & Tschopp, D. (2007). *The innovation gap in pharmaceutical drug discovery and new models for R&D success*. Kellog School of Management. Retrieved from, *http://www.kellogg.northwestern.edu/biotech/faculty/articles/newrdmodel.pdf*.

Jackson, C. (1992). Comment: Pharmaceutical product liability may be hazardous to your health: A no-fault alternative to concurrent regulation. *American University Law Review*, *42*, 199.

Johnson, K. W., Matsuda, K., & Iwaki, Y. (2014). Ibudilast for the treatment of drug addiction and other neurological conditions. *Clinical Investigation*, *4*(3), 269–279.

Jones, C. M., Campopiano, M., Baldwin, G., & McCance-Katz, E. (2015). National and state treatment need and capacity for opioid agonist medication-assisted treatment. *American Journal of Public Health*, *105*(8), e55–e63.

Kastelic, A., Pont, J., & Stöver, H. (2008). *Opioid substitution treatment in custodial settings: A practical guide*. United Nations Office on Drugs and Crime/World Health Organization. Retrieved from, https://www.unodc.org/documents/hiv-aids/OST_in_Custodial_Settings.pdf (Last visited Jan. 4, 2016).

Kjome, K., & Moeller, F. (2011). Long-acting injectable naltrexone for the management of patients with opioid dependence. *Substance Abuse: Research and Treatment*, *5*, 1–9.

Knudsen, H. K., Ducharme, L. J., Roman, P. M., & Link, T. (2005). Buprenorphine diffusion: The attitudes of substance abuse treatment counselors. *Journal of Substance Abuse Treatment*, *29*, 95–106.

Krupitsky, E., Nunes, E. V., Ling, W., Illeperuma, A., Gastfriend, D. R., et al. (2001). Injectable extended-release naltrexone for opioid dependence: A double-blind, placebo-controlled, multicentre randomised trial. *The Lancet*, *377*(9776), 1506–1513.

LaMattina, J. (2014). Pharma R&D cuts hurting U.S. competitive standing. *Forbes*. Retrieved from, http://www.forbes.com/sites/johnlamattina/2014/01/03/pharma-rd-cuts-hurting-u-s-competitive-standing/.

Magura, S., Rosenblum, A., Lewis, C., & Joseph, H. (1993). The effectiveness of in-jail methadone maintenance. *Journal of Drug Issues*, *23*, 75–99.

Management Centre Europe (2012). New opportunities and strategies in the pharmaceutical industry. *The Executive Issue*, *38*(6). Retrieved from, http://www.mce-ama.com/executive-issue-38-pharma-industry-2012/.

Matusow, H., et al. (2013). Medication assisted treatment in US drug courts: Results from a nationwide survey of availability, barriers and attitudes. *Journal of Substance Abuse Treatment*, *44*(5), 473–480.

Maxwell, J. (2015). The pain reliever and heroin epidemic in the United States: Shifting winds in the perfect storm. *Journal of Addictive Diseases*, *34*(2–3), 127–140.

Miles, K. (2013). Meth addiction cure: UCLA tests ibudilast on human addicts. *The Huffington Post*. Retrieved from, http://www.huffingtonpost.com/2013/04/03/meth-addiction-cure-ucla-ibudilast_n_2863126.html.

Mittra, J. (2007). Life science innovation and the restructuring of the pharmaceutical industry: Merger, acquisition and strategic alliance behavior of large firms. *Technology Analysis and Strategic Management, 19*(3), 279282–283.

Mundasad, S. (2012). The end of drug discovery. *Health, BBC News*. Retrieved from, http://www.bbc.com/news/health-18095669.

National Association of Drug Court Professionals (2011). *Resolution of the board of directors on the availability of medically assisted treatment (M.A.T.) for addiction in drug courts*. Retrieved from, http://www.nadcp.org/sites/default/files/nadcp/NADCP%20Board%20Statement%20on%20MAT.pdf.

National Association of Drug Court Professionals (2015). *What are drug courts?* Retrieved from, http://www.nadcp.org/learn/what-are-drug-courts.

Nicholls, L., Bragaw, L., & Ruetsch, C. (2010). Supplement opioid dependence treatment and guidelines. *Supplement to Journal of Managed Care Pharmacy, 16*(1-b), S14–S21.

Nicosia, N., Pacula, R., Kilmer, B., Lundberg, R., & Chiesa, J. (2005). *The economic cost of methamphetamine abuse in the U.S.* The Rand Corporation. Retrieved from, http://www.rand.org/pubs/monographs/MG829.html.

Nunn, A., Zaller, N., Dickman, S., Trimbur, C., Nijhawan, A., et al. (2009). Methadone and buprenorphine prescribing and referral practices in US prison systems: Results from a nationwide survey. *Drug and Alcohol Dependence, 105*(1–2), 83–88.

Offit, P. A. (2005). Why are pharmaceutical companies gradually abandoning vaccines? *Health Affairs, 24*(3), 622–630.

O'Reilly, M. (2007). Case brief: The failures of project bioshield & congressional attempts to remedy it. *DePaul Journal of Health Care, 10*, 503–511.

Orphan Drug Act (1983). Pub. L. No. 97–414, 96 Stat. 2049, codified as amended at 21 U.S.C., 26 U.S.C., 35 U.S.C., 42 U.S.C.

Pharmaceutical Research and Manufacturers of America (2015). *2015 Profile biopharmaceutical research industry*. Retrieved from, http://www.phrma.org/sites/default/files/pdf/2014_PhRMA_PROFILE.pdf.

Plumridge, H. (2014). Drug makers tiptoe back into antibiotic R&D. *The Wall Street Journal* Retrieved from, http://www.wsj.com/articles/SB10001424052702303465004579322601579895822.

Project Bioshield Act (2004). Pub. L. No. 108–276, 118 Stat. 835 § 2(a), codified as amended at 42 U.S.C. 243; Pub. L. No. 110–85, 121 Stat. 823, codified as amended at 21 U.S.C. & 42 U.S.C.

Querido, J. (2012). *Imatinib—The dawn of targeted treatments*. Cancer Research UK. Retrieved from, http://scienceblog.cancerresearchuk.org/2012/10/25/imatinib-the-dawn-of-targeted-treatments/.

Rin-Laures, L.-H., & Janofsky, D. (1991). Recent developments concerning the Orphan Drug Act. *Harvard Journal of Law & Technology, 4*, 269–297.

Rosenblatt, R., Andrilla, C. H., Catlin, M., & Larson, E. H. (2015). Geographic and specialty distribution of U.S. physicians trained to treat opioid use disorder. *Annals of Family Medicine, 13*(1), 23–26.

Sabatier, P. (2007). *Theories of the policy process* (2nd ed.). New York, NY: Westview Press.

Scannell, J., Blanckley, A., Boldon, H., & Warrington, B. (2012). Diagnosing the decline in pharmaceutical R&D efficiency. *Nature Reviews. Drug Discovery, 11*, 191–200.

Science Daily (2013). *Study reveals first effective medicine to treat cocaine dependence*. Retrieved from, http://www.sciencedaily.com/releases/2013/10/131025113741.htm.

Segalat, L. (2007). Loss-of-function genetic diseases and the concept of pharmaceutical targets. *Orphanet Journal of Rare Disease, 2*, 30.

Sharma, A., Jacob, A., Tandon, M., & Kumar, D. (2010). Orphan drug: Development trends and strategies. *Journal of Pharmacy and Bioallied Sciences, 290*(4), 290–299.

Stotts, A., Carrie Dodrill, C., & Kosten, T. (2009). Opioid dependence treatment: Options in pharmacotherapy. *Expert Opinion on Pharmacotherapy, 10*(11), 1727.

Substance Abuse and Mental Health Services Administration (2012). An introduction to extended-release injectable naltrexone for the treatment of people with opioid dependence. *Advisory, 11*(1). Retrieved from, https://store.samhsa.gov/shin/content/SMA12-4682/SMA12-4682.pdf.

Substance Abuse and Mental Health Services Administration (2014). Adult drug courts and medication-assisted treatment for opioid dependence. *In Brief, 8*(1). Retrieved from, http://dsamh.utah.gov/pdf/SAMHSA%20MAT.pdf.

Thaul, S. (2008). *FDA fast track and priority review programs*. Congressional Research Service. Retrieved from, https://www.fas.org/sgp/crs/misc/R41983.pdf.

The National Alliance of Advocates for Buprenorphine Treatment (2008). *What is withdrawal?* Retrieved from, http://www.naabt.org/faq_answers.cfm?ID=16.

Trautmann, S., & Wittchen, H.-U. (2012). An analysis of German settings providing opioid maintenance therapy. *Substance Use and Misuse, 47*, 2223.

UNAIDS, Joint United Nations Programme on HIV/AIDS (2012). *Key programmes to reduce stigma and discrimination and increase access to justice in national HIV responses*. Retrieved from, http://www.unaids.org/sites/default/files/media_asset/Key_Human_Rights_Programmes_en_May2012_0.pdf.

United Nations, Office on Drugs and Crime, World Health Organization (2009). Discussion paper: Principles of drug dependence treatment. https://www.unodc.org/documents/drug-treatment/UNODC-WHO-Principles-of-Drug-Dependence-Treatment-March08.pdf.

United States Department of Health and Human Services [USDHHS] (2014a). U.S. Food and Drug Administration. *Priority Review*. Retrieved from, http://www.fda.gov/ForPatients/Approvals/Fast/ucm405405.htm.

United States Department of Health and Human Services [USDHHS] (2014b). In *U.S. Food and Drug Administration. Projections of National Expenditures for Treatment of Mental & Substance Use Disorders, 2010–2020*. Retrieved from, http://store.samhsa.gov/shin/content//SMA14-4883/SMA14-4883.pdf.

United States Department of Health and Human Services [USDHHS] (2015a). *About buprenorphine treatment*. Retrieved from, http://buprenorphine.samhsa.gov/about.html.

United States Department of Health and Human Services [USDHHS] (2015b). *Executive summary: Opioid abuse in the U.S. and HHS actions to address opioid-drug related overdoses and deaths*. Retrieved from, http://aspe.hhs.gov/sp/reports/2015/OpioidInitiative/es_OpioidInitiative.pdf.

United States Department of Health and Human Services, Food and Drug Administration [USDHHS] (2014). *Guidance for industry neglected tropical diseases of the developing world: Developing drugs for treatment or prevention*. Retrieved from, http://www.fda.gov/downloads/drugs/guidancecomplianceregulatoryinformation/guidances/ucm269221.pdf.

United States Department of Health and Human Services, National Institute on Drug Abuse [USDHHS] (2010a). *Are cocaine abusers at risk for contracting HIV/AIDS and hepatitis B and C?* Retrieved from, http://www.drugabuse.gov/publications/research-reports/cocaine/are-cocaine-abusers-risk-contracting-hivaids-hepatitis-b-c.

United States Department of Health and Human Services, National Institute on Drug Abuse [USDHHS] (2010b). *What treatments are effective for cocaine abusers?* Retrieved from, http://www.drugabuse.gov/publications/research-reports/cocaine/what-treatments-are-effective-cocaine-abusers.

United States Department of Health and Human Services, National Institute on Drug Abuse [USDHHS] (2013a). *What treatments are effective for methamphetamine abusers?* Retrieved from, http://www.drugabuse.gov/publications/research-reports/methamphetamine/what-treatments-are-effective-methamphetamine-abusers.

United States Department of Health and Human Services, National Institute on Drug Abuse [USDHHS] (2013b). *What is the scope of methamphetamine abuse in the U.S.?* Retrieved from, http://www.drugabuse.gov/publications/research-reports/methamphetamine/what-scope-methamphetamine-abuse-in-united-states.

United States Department of Health and Human Services, National Institute on Drug Abuse [USDHHS] (2010c). *What is the scope of cocaine use in the U.S.?* Retrieved from, http://www.drugabuse.gov/publications/research-reports/cocaine/what-scope-cocaine-use-in-united-states.

United States Department of Health and Human Services, National Institute on Drug Abuse [USDHHS] (2013c). *What are the long-term effects of methamphetamine abuse?* Retrieved from, http://www.drugabuse.gov/publications/research-reports/methamphetamine/what-are-long-term-effects-methamphetamine-abuse.

United States Department of Health and Human Services, National Institute on Drug Abuse [USDHHS] (2015). *Principles of drug addiction: A research-based guide. What is drug addiction treatment?* Retrieved from, http://www.drugabuse.gov/publications/principles-drug-addiction-treatment-research-based-guide-third-edition/frequently-asked-questions/what-drug-addiction-treatment.

United States Department of Health and Human Services, National Institutes of Health [USDHHS] (2015a). *About NCATS, budget.* Retrieved from, http://www.ncats.nih.gov/about/budget/budget.html.

United States Department of Health and Human Services, National Institutes of Health [USDHHS] (2015b). *Trends & Statistics.* Retrieved from, http://www.drugabuse.gov/related-topics/trends-statistics.

United States Department of Health and Human Services, Office of the Inspector General [USDHHS] (2001). *The Orphan Drug Act implementation & impact.* Retrieved from, https://oig.hhs.gov/oei/reports/oei-09-00-00380.pdf.

United States Department of Health of Health and Human Services, National Institute on Drug Abuse [USDHHS] (2010). *What are the long-term effects of cocaine use?* National Institute on Drug Abuse. Retrieved from, http://www.drugabuse.gov/publications/research-reports/cocaine/what-are-long-term-effects-cocaine-use.

United States Food and Drug Administration (2015). *Fast track, breakthrough therapy, accelerated approval, priority review.* Retrieved from, http://www.fda.gov/ForPatients/Approvals/Fast/default.htm.

United States National Institutes of Health, ClinicalTrials.gov (2015). *Trial of ibudilast for methamphetamine dependence (IBUD ph II).* Retrieved from, https://clinicaltrials.gov/ct2/show/NCT01860807.

United Therapeutics (2015). *United therapeutics corporation agrees to sell priority review voucher to AbbVie for $350 million.* Retrieved from, http://ir.unither.com/releasedetail.cfm?ReleaseID=928100.

Vestal, C. (April 1, 2015). *How severe is the shortage of substance abuse specialists?* The Pew Charitable Trusts. Retrieved from, *http://www.pewtrusts.org/en/research-and-analysis/blogs/stateline/2015/4/01/how-severe-is-the-shortage-of-substance-abuse-specialists.*

Volkow, N. D. (2014). *America's dependence to opioids: Heroin and prescription drug abuse.* U.S. Department of Health and Human Services, National Institute on Drug Abuse. Retrieved from, http://www.drugabuse.gov/about-nida/legislative-activities/testimony-to-congress/2015/americas-addiction-to-opioids-heroin-prescription-drug-abuse.

Volkow, N. D., & Li, T.-K. (2004). Drug addiction: The neurobiology of behavior gone awry. *Nature Reviews. Neuroscience*, *5*(963), 969.

Walley, A., Alperen, J., Cheng, D., Botticelli, M., Castro-Donlan, C., Samet, J., et al. (2008). Office-based management of opioid dependence with buprenorphine: Clinical practices and barriers. *Journal of General Internal Medicine*, *23*(9), 1393–1398.

Wang, J., Qian, W., Liu, J., Zhao, J., Yu, P., et al. (2004). Effect of methamphetamine on the microglial damage: Role of potassium channel Kv1. *PLoS One*, *9*(2), 7–9.

Weisman, R. (2015). Research cuts ripple across biopharma sector. *Boston Globe*. Retrieved from, http://www.bostonglobe.com/business/2015/03/18/some-fear-wave-biopharma-research-cuts-could-slow-progress-fighting-infectious-diseases/10lDVFIuUAJs9oLrqA3AIO/story.html.

White, W., et al. (2013). Coparticipation in 12-step mutual aid groups and methadone maintenance treatment: A survey of 322 patients. *Journal of Groups in Addiction & Recovery*, *8*(4), 294–308.

Willenbring, M., Hagedorn, H., & Postier, A. (2004). Variations in evidence-based clinical practices in nine United States veterans administration opioid agonist therapy clinics. *Drug and Alcohol Dependence*, *75*(1), 97–106.

Woodall, A., Morgan, C., Sloan, C., & Howard, L. (2010). Barriers to participation in mental health research: Are there specific gender, ethnicity, and age related barriers? *BMC Psychiatry*, *10*, 103.

World Health Organization (2009). *Guidelines for the psychosocially assisted pharmacological treatment of opioid dependence*. Retrieved from, http://www.who.int/substance_abuse/publications/opioid_dependence_guidelines.pdf.

CHAPTER THREE

Identification of Treatment Targets in a Genetic Mouse Model of Voluntary Methamphetamine Drinking

T.J. Phillips*,†,1, J.R.K. Mootz*, C. Reed*

*Methamphetamine Abuse Research Center, Oregon Health & Science University, Portland, OR, United States
†Veterans Affairs Portland Health Care System, Portland, OR, United States
[1]Corresponding author: e-mail address: phillipt@ohsu.edu

Contents

International Review of Neurobiology, Volume 126
ISSN 0074-7742
http://dx.doi.org/10.1016/bs.irn.2016.02.001

Abstract

Methamphetamine has powerful stimulant and euphoric effects that are experienced as rewarding and encourage use. Methamphetamine addiction is associated with debilitating illnesses, destroyed relationships, child neglect, violence, and crime; but after many years of research, broadly effective medications have not been identified. Individual differences that may impact not only risk for developing a methamphetamine use disorder but also affect treatment response have not been fully considered. Human studies have identified candidate genes that may be relevant, but lack of control over drug history, the common use or coabuse of multiple addictive drugs, and restrictions on the types of data that can be collected in humans are barriers to progress. To overcome some of these issues, a genetic animal model comprised of lines of mice selectively bred for high and low voluntary methamphetamine intake was developed to identify risk and protective alleles for methamphetamine consumption, and identify therapeutic targets. The mu opioid receptor gene was supported as a target for genes within a top-ranked transcription factor network associated with level of methamphetamine intake. In addition, mice that consume high levels of methamphetamine were found to possess a nonfunctional form of the trace amine-associated receptor 1 (TAAR1). The *Taar1* gene is within a mouse chromosome 10 quantitative trait locus for methamphetamine consumption, and TAAR1 function determines sensitivity to aversive effects of methamphetamine that may curb intake. The genes, gene interaction partners, and protein products identified in this genetic mouse model represent treatment target candidates for methamphetamine addiction.

1. METHAMPHETAMINE ACTIONS AND ABUSE

Amphetamine-like stimulants, including methamphetamine (MA), are used and abused around the world. It has been estimated that within the United States (US), 0.5–0.6% of the population are broadly defined as users (United Nations Office on Drug and Crime, 2015). Illicit drug use, again broadly defined, is estimated to cost the United States approximately $193 billion dollars annually, which amounts to about $1 billion lost to MA abuse per year (U.S. Department of Justice National Drug Intelligence Center, 2011). In addition to substantial societal costs, there are numerous social and health ramifications. For example, excessive MA use is associated with higher rates of homelessness and incarceration (Milloy, Kerr, Buxton, Montaner, & Wood, 2009; Rommel, Rohleder, Wagenpfeil, Haertel-Petri, & Kesting, 2015), child neglect and violence (Watanabe-Galloway et al., 2009), and unprotected sex resulting in higher rates of sexually transmitted infections (Cheng et al., 2010; Colfax & Shoptaw, 2005; Gonzales, Marinelli-Casey, Shoptaw, Ang, & Rawson, 2006; Milloy et al., 2009).

Chronic MA exposure also affects cardiovascular function (Varga, Ferdinandy, Liaudet, & Pacher, 2015) and has neurotoxic effects (Davidson, Gow, Lee, & Ellinwood, 2001; Johnson, Venters, Guarraci, & Zewail-Foote, 2015; Krasnova & Cadet, 2009; Panenka et al., 2013). Furthermore, purported immunosuppressant effects of MA may exacerbate existing health problems or diseases contracted in the course of MA abuse (Gavrilin, Mathes, & Podell, 2002; Talloczy et al., 2008; Yu, Montes, Larson, & Watson, 2002).

To better understand why MA is addictive, considerable research has focused on elucidating its mechanism(s) of action. MA acts via the norepinephrine (NE), dopamine (DA), and serotonin (5-HT) transporters (NET, DAT, and SERT, respectively) and increases the levels of these monoamines in the synapse (Panenka et al., 2013; Rothman et al., 2001). Comprehensive reviews have covered these mechanisms (Fleckenstein, Volz, Riddle, Gibb, & Hanson, 2007; Haile, 2007; Sora et al., 2009) and Fig. 1 summarizes presynaptic, synaptic, and postsynaptic consequences of acute MA exposure, using the DA synapse as an example. Briefly, MA serves as a monoamine transporter substrate, competing with the neurotransmitter for transporter binding, and preventing monoamine reuptake. This increases the probability of monoamine binding to postsynaptic and presynaptic receptors. MA is transported into the cell by the transporter or by diffusion through the plasma membrane (Fleckenstein et al., 2007), and once inside the cell, it displaces vesicular and intracellular monoamines, facilitating their release via reverse transport (Fleckenstein et al., 2007; Kahlig et al., 2005; Sulzer et al., 1995). Release of vesicular monoamine into the cytosol occurs via interactions of MA with the vesicular monoamine transporter 2 (VMAT2) (Sulzer, Sonders, Poulsen, & Galli, 2005). Due to the crucial role of DA in reward circuitry, this modulation of DA transport is a leading contributor to MA addiction (Panenka et al., 2013; Volkow & Morales, 2015). Finally, a lesser studied target of MA is the trace amine-associated receptor 1 (TAAR1), a G protein-coupled receptor that binds endogenous trace amines (Lam, Espinoza, Gerasimov, Gainetdinov, & Salahpour, 2015), and also MA (Bunzow et al., 2001). TAAR1 is a negative regulator of monoamines, so that animals with a nonfunctional receptor exhibit a heightened monoaminergic response to amphetamine (Wolinsky et al., 2007).

Knowing how MA acts at the cellular level is only part of the larger picture relevant to MA dependence. Environmental factors and genetic differences contribute to an individual's pattern of experimentation with illicit substances and to whether or not their substance use becomes habitual

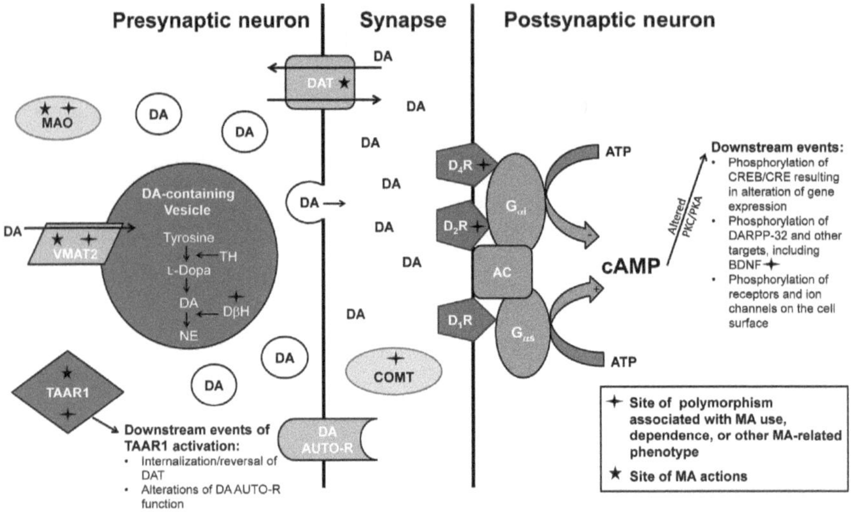

Fig. 1 Sites of action of MA and genetic polymorphisms within the dopamine (DA) synapse associated with human MA-related phenotypes. After entering the cell via diffusion or the DA transporter (DAT), MA has multiple sites of action, indicated by the *five-pointed stars*. MA can bind to monoamine oxidase (MAO) to increase its ability to break down monoamines, to the vesicular monoamine transporter 2 (VMAT2), which results in neurotransmitter accumulation in the cytosol and potential transport via the DAT into the synapse. MA can also bind to the trace amine-associated receptor 1 (TAAR1), which may result in alterations of DA tone via modification of DAT and DA autoreceptor (DA AUTO-R) functioning. Genetic differences may lead to functional variation for multiple entities within the DA synapse and alter an individual's susceptibility to MA use or dependence or MA-related psychosis. These are noted by the *four-pointed stars* and include MAO, VMAT2, dopamine beta-hydroxylase (DβH), TAAR1, catechol-*O*-methyltransferase (COMT), DA receptor type 2 (D_2R), DA receptor type 4 (D_4R), and brain-derived neurotrophic factor (BDNF). Details about specific polymorphisms and their association with MA-related phenotypes are discussed in the text or can be found in the following references: Chen, Lipska, et al. (2004), Li et al. (2004), Sery, Vojtova, and Zvolsky (2001), and Ujike et al. (2009). *TH*, tyrosine hydroxylase; *L-Dopa*, L-3,4-dihydroxyphenylalanine; *NE*, norepinephrine; D_1R, DA receptor type 1; *AC*, adenylate cyclase; $G_{\alpha s}$, G protein alpha subunit, stimulatory; $G_{\alpha i}$, G protein alpha subunit, inhibitory; *cAMP*, cyclic AMP.

(Le Moal, 2009). Studies in animal models of MA use have served to elucidate some of these influences. For example, in rats, environmental enrichment reduced MA preference in a two-bottle choice procedure (Hajheidari, Miladi-Gorji, & Bigdeli, 2015) and reduced cue-induced reinstatement in an operant procedure (Hofford, Darna, Wilmouth, Dwoskin, & Bardo, 2014). However, genetic factors have a noteworthy impact on vulnerability to addiction (Crabbe, 2002; Kreek, Nielsen, Butelman, & La Forge, 2005;

Le Moal, 2009; Phillips, Kamens, & Wheeler, 2008), and a number of genes associated with differences in MA consumption have been identified in animals and humans, as reviewed later. These genes are involved in a number of neural pathways and MA-specific responses. Understanding the contributions of these genes to the effects of MA and to MA abuse is an important goal in the search for more effective treatments, and animal models of addiction-related traits are important contributors to both genetic and therapeutic discovery. In the following sections, we first review the medications literature for MA dependence and then examine data that support the importance of genetic factors in risk for MA dependence. Finally, we describe the characteristics of a mouse model that was created to identify genetic factors associated with risk for MA consumption and therapeutic discovery, summarize existing findings, and speculate about treatment targets.

2. OVERVIEW OF EXISTING MA MEDICATIONS STUDIES

A paper within the last 5-year period reviewed the literature on pharmacological MA treatment options and concluded that "no single medication has demonstrated consistent efficacy" (Brackins, Brahm, & Kissack, 2011). We update this information later and largely draw the same conclusion. However, an important factor that deserves more attention is the role of individual differences in treatment efficacy (Li et al., 2014).

2.1 Dopamine

Since MA increases synaptic DA levels, it follows that blocking DA receptors might attenuate the effects of MA. The DA D_2 receptor (D_2R) subtype antagonist, eticlopride, significantly reduced amphetamine-seeking behavior (Izzo, Orsini, Koob, & Pulvirenti, 2001), and the antipsychotic drug, risperidone, an antagonist at both D_2R and 5-HT_2 receptors, decreased MA use in humans (Meredith et al., 2009; Nejtek et al., 2008). However, there are conflicting data regarding the viability of DA receptor antagonists as potential therapeutics for MA addiction. For example, one study found that risperidone did not decrease MA preference relative to a food choice in monkeys (Banks & Blough, 2015). Also, although the mechanism by which cocaine increases synaptic monoamine levels is somewhat different from MA, it is noteworthy that in two placebo-controlled double-blind trials, risperidone did not reduce cocaine use and had undesirable side effects (Grabowski et al., 2000; Loebl et al., 2008). Further analysis suggests that antipsychotics in general may be undesirable treatments for MA addiction, partly due to their intolerable side effects (Kishi, Matsuda, Iwata, & Correll, 2013).

But is there a potential role for DA receptor partial agonists, which have low intrinsic activity and antagonist effects when DA levels are high and may be less prone to produce adverse side effects? The D_2R partial agonist, terguride, had effects similar to eticlopride on amphetamine seeking in rats (Izzo et al., 2001) and increased rates of responding for cocaine, following cocaine self-administration training (Pulvirenti, Balducci, Piercy, & Koob, 1998). One interpretation is that terguride attenuated the reinforcing effects of the psychostimulants. Another D_2R partial agonist, aripiprazole, reduced subjective drug effects and participants were less able to discriminate between dextroamphetamine (D-amphetamine) and placebo. However, impaired digit-symbol substitution test performance also occurred, particularly when aripiprazole (20 mg) was given without D-amphetamine (Lile et al., 2005). A lower dose of aripiprazole (10 mg) did not reduce the discriminative stimulus effects of D-amphetamine, but still reduced subjective drug effects (Stoops, Lile, Glaser, & Rush, 2006). However, when tested against methylphenidate and a placebo, participants taking aripiprazole (15 mg) had the highest number of urine analyses positive for MA (Tiihonen et al., 2007). On the other hand, when subjects were given the opportunity to self-administer MA on a progressive-ratio schedule, compared to placebo, aripiprazole (15 mg) reduced MA self-administration and attenuated some positive subjective effects (Stoops, Bennett, Lile, Sevak, & Rush, 2013). These data suggest that aripiprazole and perhaps other D_2R partial agonists may decrease rewarding and reinforcing effects of MA under controlled, but not nonlaboratory, conditions. Finally, it should be noted that both aripiprazole and terguride have agonist or antagonist activity at different 5-HT receptor subtypes (Newman-Tancredi et al., 2002), which could have played a role in previous findings.

2.2 Vesicular Monoamine Transporter 2

MA achieves vesicular release of DA into the cytosol through interaction with VMAT2 (Fleckenstein et al., 2007; Liu & Edwards, 1997; Nickell, Siripurapu, Vartak, Crooks, & Dwoskin, 2014; Sulzer et al., 2005; Vocci & Appel, 2007), which has been investigated as a pharmacotherapeutic target. Tetrabenazine (TBZ), an FDA-approved drug used to treat symptoms of Huntington's disease, is a reversible VMAT2 inhibitor (Chen, Ondo, Dashtipour, & Swope, 2012; Meyer et al., 2011;

Yero & Rey, 2008). In rats, higher doses of TBZ attenuated MA self-administration and suppressed locomotor activity, regardless of whether the rats received saline or MA; tolerance did not develop with repeated TBZ treatment (Meyer et al., 2011). However, a lower dose of TBZ increased MA self-administration. TBZ also decreased food-maintained lever responding, but did not attenuate MA discrimination. Although TBZ has a higher affinity for VMAT2, it also inhibits DA uptake by binding to DAT, increasing its risk for abuse (Meyer et al., 2011). To our knowledge, the effects of TBZ on MA-related behaviors have not been tested in humans, but its FDA-approved status removes some barriers.

Lobeline is a VMAT2 inhibitor (Teng, Crooks, & Dwoskin, 1998) originally used as a smoking cessation aid (Schwartz, 1969). A recent review summarized some of the neurochemical and MA-related behavioral actions of lobeline, and of its analog, lobelane (Nickell et al., 2014). Lobeline decreased amphetamine-induced DA release in rat striatal slices and decreased MA-induced hyperactivity, behavioral sensitization (a measure of neuroadaptation), and self-administration in rats. Lobelane inhibited MA-induced DA efflux and attenuated MA self-administration, although rats developed some tolerance to lobelane's effects. Lobeline has gone through phase 1 testing (ClinicalTrial.gov ID NCT00439504) and was deemed safe in MA users. Pharmacotherapeutic research is ongoing with lobeline (and its analogs), which remains one of the more promising MA abuse therapeutic candidates to date.

2.3 Gamma-Aminobutyric Acid

Manipulations of gamma-aminobutyric acid (GABA) systems have been considered as viable treatments for MA addiction, in part, due to their connections with DAergic systems (Karila et al., 2010). Gama-vinyl-GABA (GVG; aka, vigabatrin) irreversibly inhibits GABA-transaminase (Jung, Lippert, Metcalf, Bohlen, & Schechter, 1977), thereby increasing GABA levels, which have the potential to inhibit DAergic neurons in the nucleus accumbens (NAcc) and striatum (Ashby et al., 1999; Dewey et al., 1992; Gardner et al., 2002; Morgan & Dewey, 1998). Pretreatment with GVG inhibited MA-induced increases in DA levels in the NAcc of rats (Gerasimov et al., 1999). Additionally, GVG can block cocaine-induced conditioned place preference (CPP), a measure of drug reward (Dewey et al., 1998). GVG may also impact neuroadaptations associated with repeated psychostimulant treatment, as it attenuated both the acquisition and expression of cocaine-induced sensitization (Gardner et al., 2002). Finally, regarding clinical investigation, in individuals addicted to MA, 16/18 individuals treated with GVG tested negative for MA during the last 6 weeks of the 9-week trial; however, there was no non-GVG-treated control for comparison (Brodie, Figueroa, Laska, & Dewey, 2005). A more recent study found no decrease in the subjective effects of MA in nontreatment-seeking MA-dependent individuals treated with GVG, compared to placebo (De La Garza et al., 2009). Thus, GVG may not reduce positive subjective effects of MA that play an important role in addiction.

Topiramate also facilitates GABAergic activity and thereby reduces mesocorticolimbic DA activity (White, Smith, & Wilcox, 2007), along with other effects (Karila et al., 2010). In a placebo-controlled trial in MA-dependent individuals, acute topiramate enhanced the positive "stimulate" effect of MA, but not "euphoria," and had no effect on "crave" (Johnson et al., 2007). Transcriptome profiling in human white blood cells was recently used to examine differentially expressed genes in MA-dependent individuals with or without a positive response to topiramate treatment and represents another approach to identifying potentially efficacious treatment mechanisms (Li et al., 2014). Some identified genes were involved in GABA and glutamate (Glu) receptor signaling, which may suggest differences in responders vs nonresponders in sensitivity to topiramate. Pathways of differentially expressed genes between the placebo and topiramate treatment groups included those involved in synaptic plasticity, signal transduction, and inflammation/immune function, among others. Additional data are needed to determine whether this approach will be beneficial for MA dependence treatment development.

2.4 Cyclic AMP

Due to persistent changes in the brain induced by repeated drug exposure, transcriptional mechanisms that regulate gene expression have been implicated in addiction (Nestler, 2012). Such MA-induced mechanisms have the potential to change the function of several neurotransmitter systems and impact cellular morphology. One transcriptional element, cyclic AMP (cAMP), has received considerable attention for its possible role in drug addiction (Haile, Kosten, & Kosten, 2009; Mamiya et al., 2001; Nestler, 2001; Shichinohe et al., 2001; Terwilliger, Beitner-Johnson, Sevarino, Crain, & Nestler, 1991; Yan et al., 2006), and a few relevant studies have focused on MA. For example, rats treated with rolipram or nefiracetam, modulators of cAMP levels, were less able to discriminate between MA and a control injection (Yan et al., 2006). Pretreatment with rolipram also inhibited behavioral sensitization to MA (Iyo et al., 1996). Ibudilast, which inhibits phosphodiesterase, the enzyme that degrades cAMP, attenuated stress- and priming-induced reinstatement of MA-reinforced lever-pressing in rats (Beardsley, Shelton, Hendrick, & Johnson, 2010) and also attenuated MA-induced locomotor activation and sensitization (Snider et al., 2012). Ibudilast is also known to modulate glial cell activity, perhaps through its action on cAMP (Kawanokuchi, Mizuno, Kato, Mitsuma, & Suzumura, 2004; Suzumura, Ito, & Mizuno, 2003), suggesting that glia may play a role in some MA effects. A recent review discusses the potential of glial modulators as psychostimulant abuse therapies (Beardsley & Hauser, 2014) and relates to neuroimmune therapy development discussed later.

2.5 Replacement Therapy and Treating Withdrawal

The pharmacological interventions discussed up to this point are ones intended to block the rewarding effects of MA and thereby attenuate drug seeking. Another proposed therapy focuses on replacing MA with a less addictive substance and ameliorating withdrawal symptoms without producing sufficient euphoria to form an addiction. We review some of the more well-investigated options, and the reader is referred to several thorough reviews of stimulant-replacement therapies (Grabowski, Shearer, Merrill, & Negus, 2004; Herin, Rush, & Grabowski, 2010). Modafinil has been studied as a potential replacement therapy (Brensilver, Heinzerling, & Shoptaw, 2013; Grabowski et al., 2004; Haile et al., 2009; Herin et al., 2010) and is an FDA-approved wakefulness-promoting drug used to treat narcolepsy and other sleep-related disorders (Ballon & Feifel, 2006; Karila et al., 2010; Vocci & Appel, 2007) that has been considered as a treatment for ADHD (Greenhill et al., 2006; Kahbazi et al., 2009). Modafinil has actions in the brain that, like MA, results in increases in monoamine levels (Korotkova et al., 2007; Loland et al., 2012; Minzenberg & Carter, 2008; Nguyen, Tian, You, Lee, & Jang, 2011; Quisenberry & Baker, 2015). In one study, modafinil alleviated the symptoms of MA withdrawal (McGregor, Srisurapanont, Mitchell, Wickes, & White, 2008) and was comparable in efficacy to mirtazapine, an antidepressant also studied for its potential to treat MA withdrawal (Graves & Napier, 2011; Kongsakon, Papadopoulos, & Saguansiritham, 2005). Yet, in larger outpatient studies, modafinil did not increase abstinence rates or significantly ameliorate withdrawal symptoms (Anderson et al., 2012; Shearer et al., 2009). Mirtazapine studies have yielded similarly inconsistent results (for review, see Graves, Rafeyan, Watts, & Napier, 2012). Further, in mice, modafinil induced a CPP (Nguyen et al., 2011), indicating that it may have abuse potential.

Because of its actions as a monoamine uptake inhibitor, bupropion has also been studied as a possible therapeutic, again with mixed results. Bupropion reduced the subjective effects of MA (Newton et al., 2006) and reduced MA consumption (Elkashef et al., 2008), but had little to no effect on several measures of MA craving or abstinence (Anderson et al., 2015; Elkashef et al., 2008; Heinzerling et al., 2014; Shoptaw et al., 2008). However, there was evidence that bupropion was effective in specific subsets of participants, including lower baseline MA users and men (Elkashef et al., 2008; Heinzerling et al., 2014; Shoptaw et al., 2008). Thus, the efficacy of bupropion may be dependent on the level of MA use, which may also impact the severity of MA withdrawal symptoms.

Methylphenidate and D-amphetamine have actions similar to MA, but their use in replacement therapy is complicated by high abuse potential. Methylphenidate is currently used to treat ADHD (Karila et al., 2010) and has shown promise in treating MA addiction (Tiihonen et al., 2007). Also, in a double-blind clinical trial, D-amphetamine treatment resulted in a reduction of cocaine use in participants (Grabowski et al., 2001), and when D-amphetamine elixir or tablets were substituted for intravenous or "street" amphetamine, amphetamine use was reduced in active users (Charnaud & Griffiths, 1998; Klee, Wright, Carnwath, & Merrill, 2001). However, sustained release D-amphetamine did not reduce intranasal MA use in a human laboratory study of eight nontreatment-seeking adults who met the criteria for stimulant dependence (Pike, Stoops, Hays, Glaser, & Rush, 2014). These laboratory results are consistent with two clinical trials that found no effect of D-amphetamine on MA use (Galloway et al., 2011; Longo et al., 2010).

2.6 Immunotherapy and Neuroimmune Therapy

Rather than targeting the sites of action of MA, immunotherapies target the MA molecule. Administration of existing antibodies (passive immunotherapy) and immunization against a particular substance (active immunotherapy) are both being explored as alternative treatments for substance abuse. We only briefly cover this approach, but an excellent overview of the history and studies of immunotherapies for drug abuse can be found in Kosten and Owens (2005).

Anti-MA monoclonal antibodies (mAbs) altered the discriminative stimulus properties of MA in rats and pigeons (McMillan, Hardwick, Li, & Owens, 2002) and attenuated the locomotor activating effects of lower doses (0.3–1.0 mg/kg), but not a higher dose (3 mg/kg), of MA (Byrnes-Blake et al., 2003; Byrnes-Blake, Laurenzana, Landes, Gentry, & Owens, 2005). In fact, in rats treated with one mAb, an increased locomotor activation response occurred to 3 mg/kg MA (Byrnes-Blake et al., 2003). A more recent study found reduced MA-induced hyperactivity in rats that had been injected with MA 30 min prior to mAb treatment, indicating that mAbs could be effective post-MA use and possibly be used to treat MA overdose (Laurenzana et al., 2014). In addition, an anti-MA mAb was found to be safe in humans and had similar pharmacokinetics as other non-MA mAbs tested in humans (Stevens, Henry, Owens, Schutz, & Gentry, 2014). Thus, anti-MA mAbs appear to provide some protection against MA effects, but the potentially transient nature of those effects will require further consideration (Byrnes-Blake et al., 2005; Hambuchen et al., 2015; Laurenzana et al., 2009).

Active immunotherapy offers a longer lasting treatment solution, compared to mAbs (Kosten & Owens, 2005). Development of an MA vaccine remains in an early stage, but some results have been promising. Rats were able to develop MA antibodies even while being administered MA; however, the vaccine did not impact MA-induced locomotor activity (Byrnes-Blake, Carroll, Abraham, & Owens, 2001), and MA self-administration in rats actually increased following an MA vaccination (Duryee et al., 2009). However, rats treated with a recently developed vaccine had attenuated MA-induced locomotor activation and thermoregulation disruptions (Miller et al., 2013). Inoculated rats also had delayed acquisition of MA self-administration and reduced rates of acquisition, relative to controls (66% vs 100%) (Miller et al., 2015).

Immunotherapy development for MA dependence is encouraging, but is accompanied by a few drawbacks that deserve mention. Active immunotherapies require several weeks before sufficient antibodies are produced (Duryee et al., 2009; Miller et al., 2013), and in addition to duration of action, therapeutic efficacy is dependent on mAb affinity in the passive approach (Byrnes-Blake et al., 2005). One suggested approach is to coadminister an anti-MA mAb and an MA vaccine, which could provide initial resistance to MA while the active immunization is taking effect. When this was tried in rats, no significant alterations to the anti-MA immunological effects of either treatment were found and the effects were persistent (Hambuchen et al., 2015). The next step will be to examine the effect of this combined therapy on one or more MA traits.

Finally, there has been significant recent interest in the development of neuroimmune therapies for addiction, based on considerable evidence of the involvement of neuroinflammatory processes in the development of drug dependence and toxicity (Loftis & Janowsky, 2014). This field of research is rapidly growing and the reader is referred to a recent review of neuroimmune therapies for addiction, including MA dependence (Ray, Roche, Heinzerling, & Shoptaw, 2014). Importantly, in human MA users, proinflammatory cytokines and chemokines were associated with neurocognitive deficits (Loftis, Choi, Hoffman, & Huckans, 2011) and a marker for activated microglia was increased in MA users in remission, compared to nonusing controls, with binding levels inversely correlated with duration of MA abstinence (Sekine et al., 2008). Medications that counteract MA-induced neuroinflammatory effects, for example, ibudilast, could attenuate neurodegeneration, as discussed in greater detail by Ray et al. (2014), and one such peptide has been studied for its ability to treat MA-induced neuropsychiatric impairments (Loftis, Wilhelm, Vandenbark, & Huckans, 2013). Ibudilast has been used for over 2 decades in Japan as an asthma treatment and for poststroke complications (Rolan, Hutchinson, & Johnson, 2009). In 2013 the FDA fast-tracked human tests of ibudilast as a treatment for MA addiction and a study is underway at the University of California Los Angeles, with an estimated completion date of June 2017 (U.S. National Institutes of Health, 2015; identifier NCT01860807).

3. SUPPORT FOR THE IMPORTANCE OF GENETIC FACTORS IN HUMAN RISK FOR MA ABUSE

MA acts via multiple mechanisms to initiate changes that underlie its positive and negative motivational effects and impact risk for further MA use (see Fig. 1). Differences in genetic makeup have the potential to dictate subjective response and sensitivity to psychomotor stimulation or physiological response to MA (eg, increased heart rate or blood pressure), which may impact the probability of dependence development. Genetically determined individual difference has been a "hot" topic in pharmacotherapeutics for many diseases, from cancer (Maring, Tawadros, Steer, & Lee, 2015) to addiction (Ferraguti, Pascale, & Lucarelli, 2015), and individual differences could be harnessed in the development of medications that are more effective. Thus, knowledge of genetic differences that influence MA-related responses is important for understanding both the basic pharmacological effects of MA and risk for dependence.

Data for MA-related genetic influences in human populations are limited, compared to that for other drugs of abuse (ie, alcohol and nicotine). However, existing data support heritability for drug dependence. Heritability for nonsubstance-specific addiction was reported at approximately 40% (Koob, Arends, & Le Moal, 2014), whereas heritability specifically for stimulants (cocaine and MA) has been reported in the range of 33–79% (Ehlers, Gizer, Gilder, & Wilhelmsen, 2011). In general, genetic variation appears to play a moderate to large role in addiction. Broad, genome-wide association and linkage approaches, as well as candidate gene approaches, have been used to investigate genetic influence. Genes for which there is evidence of association with MA dependence or dependence-related phenotypes have been recently summarized (Bousman, Glatt, Everall, & Tsuang, 2009; Grant et al., 2012). Some promising candidate genes have been nominated, based on preclinical evidence, but have yet to be followed up in human studies (Harkness, Shi, Janowsky, & Phillips, 2015; Lohr et al., 2015). The protein products of these genes and modulatory partners are potential targets for pharmacotherapies. Some intersect with medication development studies reviewed earlier, most of which were initiated on the basis of MA mechanisms of action, rather than genetic information. Comprehensive discussion of all relevant genes is beyond the scope of this review, but a few are highlighted here and potential polymorphisms that could have an impact on MA actions are represented in Fig. 1.

3.1 Brain-Derived Neurotrophic Factor

Preclinical studies have implicated brain-derived neurotrophic factor (BDNF) in several neurological processes, including learning-associated changes and conditioned drug effects, which require learning (Shen et al., 2014; Tyler, Alonso, Bramham, & Pozzo-Miller, 2002; Zhong, Liu, Hu, Wang, & Zhao, 2015). A recent paper summarized the preclinical data for signaling pathways that could represent cognition-enhancing drug targets, including BDNF (Ménard, Gaudreau, & Quirion, 2015). The BDNF rs6265 (val66met) polymorphism may impact risk for MA dependence in south Asian and Thai populations (Haerian, 2013; Iamjan, Thanoi,

Watiktinkorn, Nudmamud-Thanoi, & Reynolds, 2015), and may also play a role in sex differences in MA use in Caucasian populations (Heinzerling & Shoptaw, 2012). Plasma BDNF levels were significantly higher in 30-day abstinent individuals diagnosed as MA dependent, when compared to controls (Kim et al., 2005), and were speculated to play a role in MA neurotoxicity. Although the impact of the val66met polymorphism was not examined in that study, there is evidence that depolarization-induced plasma BDNF level is partly regulated by that polymorphism (Egan et al., 2003), and may, via this mechanism, impact risk for MA dependence. A preliminary study found a greater treatment effect for MA dependence in association with this polymorphism in Hispanics, but not Caucasians; however, it did not matter whether the treatment was placebo or modafinil (Heinzerling, McCracken, Swanson, Ray, & Shoptaw, 2012).

3.2 Dopamine Beta-Hydroxylase

Dopamine beta-hydroxylase (DβH) is a synaptic vesicular enzyme that converts DA into NE and is coreleased with catecholamines (Kaufman & Friedman, 1965). Individuals that do not produce DβH usually perish at birth; however, there are several viable polymorphisms that result in low levels of this enzyme and are associated with physiological abnormalities, due to high plasma levels of DA and low levels of NE and epinephrine (Senard & Rouet, 2006). The C-1021T (rs1611115) DβH gene polymorphism results in decreased DβH levels and altered cerebral spinal fluid monoamine concentrations (Jonsson et al., 2004). Recent studies identified an association of this polymorphism with MA-induced paranoia (Kalayasiri, Verachai, Gelernter, Mutirangura, & Malison, 2014), and successful replacement therapy treatment with the DA-increasing drugs, levodopa/carbidopa for cocaine dependence (Liu et al., 2014).

3.3 Catechol-*O*-Methyltransferase

Catechol-*O*-methyltransferase (COMT) participates in the enzymatic degradation of catecholamines. There are several studied variants of this gene in humans; however, the Val158Met polymorphism is thought to be associated with MA abuse, dependence, and psychosis (Li et al., 2004; Suzuki et al., 2006). This specific polymorphism produces a large decrease in the functionality of the enzyme that results in higher synaptic DA activity (Chen, Hu, et al., 2004; Zhu et al., 2004), providing a potential mechanism for its influence. A preliminary investigation examined modafinil treatment outcome for MA dependence in relation to *COMT* genotype and obtained modest evidence for a relationship in Hispanics; however, sample sizes were small (Heinzerling et al., 2012).

3.4 Fatty Acid Amide Hydrolase

Fatty acid amide hydrolase is a membrane-associated serine hydrolase located in brain and liver that inactivates endocannabinoids (Cravatt et al., 2001). The Pro129Thr variant results in normal catalytic activity, but enhanced sensitivity to proteolytic degradation and was associated with drug dependence in earlier research (Sipe, Chiang, Gerber, Beutler, & Cravatt, 2002). Further investigation in MA-dependent populations noted no association in one case (Morita et al., 2005), and a significant association in another (Sim, Hatim, Reynolds, & Mohamed, 2013). Differences in outcome could be population based, as the former study was in a Japanese population and the latter in a Malaysian population.

3.5 Mu Opioid Receptor

The mu opioid receptor gene (*OPRM1*) produces a receptor (the MOP-r) that is a site of action for commonly used opioids, many of which have high abuse potential. MOP-r agonists influence DA neurotransmission (Vecchiola et al., 1999) and attenuate MA-induced increases in DA (El Daly, Chefer, Sandill, & Shippenberg, 2000; Pereira et al., 2011). Multiple polymorphisms in *OPRM1* have been investigated for their association with MA dependence and psychosis (Ide et al., 2004, 2006). The Asn40Asp (aka, A118G or rs1799971) polymorphism produces a MOP-r that binds opioids more tightly and exhibits greater opioid binding-induced activation of G protein-coupled potassium channels, compared to other variants (Bond et al., 1998). This variant was associated with MA dependence and psychosis latency in a Japanese population (Ide et al., 2004), but no MA-related associations were found in a Caucasian sample (Bousman et al., 2010). Asn40Asp was not associated with treatment outcome for MA dependence in a preliminary placebo-controlled investigation of modafinil (Heinzerling et al., 2012). Finally, a recent meta-analysis of 25 datasets with over 28,000 individuals of European ancestry examined the association of Asn40Asp with liability to substance dependence and included subjects dependent on alcohol, opioid, cannabis, cocaine, and nicotine, but not MA (Schwantes-An et al., 2015). A modest protective effect on substance dependence was found, rather than the increased liability found for MA dependence.

3.6 Monoamine Oxidase A

Monoamine oxidase A (MAOA) is a monoamine-metabolizing enzyme. The gene that encodes MAOA is located on the X chromosome, and a polymorphism located within the eighth exon (941T>G) is associated with low (941T) or high (941G) MAOA activity (Hotamisligil & Breakefield, 1991). There is also a 30-bp variable number tandem repeat upstream of this point mutation, which can result in one of five different alleles that appear to impact level of transcriptional activity (Sabol, Hu, & Hamer, 1998; Saify & Saadat, 2015). Lower MAOA activity has been related to behavioral risk factors associated with addiction (Caspi et al., 2002; Gerra et al., 2004; Moeller et al., 2014) and with increased alcohol consumption (Tikkanen et al., 2009, 2010). The MAOA 941T>G polymorphism was associated with severity of the consequences (gray matter loss or personality disorders) of cocaine and heroin dependence (Alia-Klein et al., 2011; Yang, Kavi, Wang, Wu, & Hao, 2012). Finally, in MA-dependent subjects, the allele for high MAOA activity was associated with increased risk of developing MA psychosis (Nakamura et al., 2009). This result for MA suggests that low MAOA activity may be a protective factor, whereas existing data suggest that it is a risk factor for other drugs of abuse.

4. DEVELOPMENT OF SELECTIVELY BRED LINES FOR HIGH AND LOW MA INTAKE

Human genetic investigation has identified potential risk alleles. However, lack of control over drug history, the use or coabuse of multiple addictive drugs, and restrictions on the types of data that can be collected in humans are barriers to progress. Animal models, though they cannot completely replicate the human experience, can help in the search for risk factors and in therapeutic development. Although outstanding animal models using operant self-administration procedures exist to study the reinforcing effects of MA and other pertinent factors, such as extinction and reinstatement, missing has been an animal model that provides the opportunity for virtually unlimited intake and takes genetic influences on intake into consideration. Our goal was to create such a model, which would then be used to identify genetic risk factors and mechanisms involved in determining differential MA intake, and ultimately lead to the identification of treatment targets.

To create this genetic model, we used bidirectional selective breeding for high and low levels of voluntary MA intake in a two-bottle choice procedure in which MA was offered vs water for 18 h per day. MA was dissolved in tap water at a concentration of 20 or 40 mg/L and these concentrations were consecutively offered for 4-day periods. Collectively, we call these lines the MA drinking or MADR lines, and lines bred for high MA intake are referred to as MA high drinking or MAHDR, whereas those bred for low MA intake are referred to as MA low drinking or MALDR. Breeding of the first set of MADR lines was initiated in 2007, and divergence in MA consumption was bidirectional across generations of selection, so that the mean consumption of the originating parent population was intermediate to that of the two selected lines (Wheeler et al., 2009). Selection results were similar for a second set of lines created about 2 years later (Shabani, McKinnon, Reed, Cunningham, & Phillips, 2011). Short-term, mass selection procedures (Belknap, Richards, O'Toole, Helms, & Phillips, 1997) have been used to create the MADR lines. This method results in a rapid response, but also more inbreeding per generation than does within-family selection (Falconer & Mackay, 1996). For this reason, short-term selected lines are bred for few generations (four to five) and then replaced with newly created lines for correlated response testing to determine traits that influence or share genetic regulation with voluntary MA consumption. At this point in time, we have created four sets of MADR lines, at approximately 2-year intervals. When replicate lines are consecutively created, it is possible not only to confirm results from former lines but also to develop new hypotheses based on existing results that can be followed up in later replicates. Methods, selective breeding and correlated response results and justification for the drinking model, including relevance to human MA use, have been detailed in previous publications (Phillips, Eastwood, & Harkness, 2014; Phillips & Shabani, 2015; Shabani et al., 2011; Wheeler et al., 2009).

4.1 Existing Behavioral and Physiological Data for the MA Drinking Lines

4.1.1 Nondrug-Related Behavioral Traits

Several nondrug-related behavioral traits have been examined in the MADR lines for which no significant differences have been detected. These include consumption of inert tastants, including saccharin, quinine, potassium chloride, and sodium chloride; baseline nociceptive response; baseline body temperature and body temperature after saline injection; adult body weight; baseline anxiety level; exploratory behavior; swim speed; free-running circadian period; and several learning-related traits (Eastwood & Phillips, 2012; Harkness et al., 2015; Moschak, Stang, Phillips, &

Mitchell, 2012; Olsen et al., 2013; Shabani et al., 2011; Shabani, Dobbs, et al., 2012; Shabani, McKinnon, Cunningham, & Phillips, 2012; Wheeler et al., 2009). The absence of a difference between the MADR lines in the consumption and preference for quinine is particularly relevant, because MA has a bitter taste (Volkow, 2013) that could impact intake in a genotype-specific manner. The absence of a line difference for quinine intake and comparable initial levels of MA intake for the two lines (Eastwood, Barkley-Levenson, & Phillips, 2014; Shabani, Dobbs, et al., 2012) suggest that avoidance or insensitivity to bitter taste did not play a significant role in the response to selective breeding for MA intake.

Significant differences have been found for total fluid intake during some MA drinking studies (Eastwood & Phillips, 2014; Wheeler et al., 2009); MAHDR mice consumed ~0.5 mL more fluid than MALDR mice. This could be due to their higher MA intake, which would be expected to stimulate behavior and perhaps generate thirst. In fact, a significant positive correlation was found in one study between amount of MA consumed during a 1-h period and level of locomotor behavior (Shabani, Dobbs, et al., 2012). Total volume consumed during a morphine drinking study also differed, with MAHDR mice consuming more than MALDR mice, but only during a later phase of the study, and a second morphine drinking study did not detect a difference (Eastwood & Phillips, 2014). Baseline locomotor activity level differences were detected in some studies (Eastwood & Phillips, 2012), but not in the majority (Gubner, Reed, McKinnon, & Phillips, 2013; Olsen et al., 2013; Shabani, McKinnon, et al., 2012; Shabani et al., 2011; Wheeler et al., 2009). The rarity of this finding across studies suggests that the differences were idiosyncratic to particular sets of mice, rather than genetically correlated with the selection trait. Finally, although no learning differences have been found, in a Morris water maze learning retention test, a memory deficit was detected in MAHDR mice (Olsen et al., 2013); this finding requires replication.

4.1.2 Drug Reward-Related Traits

In addition to voluntary two-bottle choice MA intake, MA reward-related phenotypes examined in the MADR lines include MA-induced CPP and operant oral and intracerebroventricular (ICV) MA self-administration. Mice were conditioned with 0.5 mg/kg MA and then tested in a nondrugged condition for CPP. MAHDR mice exhibited preference for MA-paired tactile cues, whereas MALDR mice spent an equal amount of time on both floor types (Wheeler et al., 2009). This study was repeated using conditioning doses of 0.5, 2, and 4 mg/kg MA in a second set of MADR lines. MAHDR mice exhibited preference, whereas MALDR mice remained neutral, regardless of dose (Shabani et al., 2011). When an additional preference test was administered under the influence of the same doses of MA to which the mice had been conditioned, the 0.5 mg/kg MA MAHDR group of mice exhibited preference, whereas the 2 and 4 mg/kg MA groups of MAHDR mice did not. MALDR mice exhibited significant aversion for the MA-paired floor, regardless of dose (Shabani et al., 2011).

Operant self-administration procedures were used to examine sensitivity to the reinforcing effects of MA in MADR mice. In one study, mice were trained to bar-press to gain access to a saccharin drinking tube, then saccharin was faded out as MA was faded in. The instrumental behavior and MA intake were measured across trials. In another study, mice were trained to press a bar to receive a food reward, ICV cannulae were implanted, and the instrumental behavior was then measured when it delivered ICV MA infusions of various concentrations. MA available for consumption or ICV delivery maintained higher levels of instrumental behavior in MAHDR than MALDR mice. Furthermore, MAHDR mice consumed more MA than MALDR mice, except during the first trial when MA was present in the drinking solution; MA intake was comparable during the initial MA trial and higher in MAHDR mice thereafter (Shabani, Dobbs, et al., 2012).

The temporal pattern and various appetitive and consummatory characteristics of MA drinking have been examined to gain insight into the microstructure of MA intake in this model of differential genetic risk (Eastwood et al., 2014). A lickometer system recorded cumulative licks from simultaneously available water and MA-containing tubes; total volume consumed was also recorded. MAHDR and MALDR mice consumed similar amounts of MA on the first day of access, and MAHDR mice consumed significantly more MA thereafter. MALDR mice consumed more water than MAHDR mice, but total fluid consumption was comparable for the two types of mice. MAHDR mice had a shorter latency to the first bout of MA drinking, a larger number of MA bouts, longer bout duration, and shorter interbout interval. No statistically significant differences between the lines of mice were found in the microstructural analysis of water intake. These results, along with the CPP and operant findings, indicate a strong genetic correlation between genetic risk for MA intake and sensitivity to the rewarding and reinforcing effects of MA, which indicates that some of the same genes impact MA intake and sensitivity to MA reward and reinforcement. Thus, treatments that block the experience of MA-induced reward would be expected to impact MA intake.

Finally, two published studies in the MADR lines have queried the genetic relationship between risk for MA intake and reward-related phenotypes for other addictive drugs. Sensitivity to cocaine-induced CPP was comparable for MAHDR and MALDR mice, indicating different underlying mechanisms for these two psychostimulant drugs (Gubner et al., 2013). When two-bottle choice morphine consumption was measured, MALDR mice consumed more morphine than MAHDR mice (Eastwood & Phillips, 2014). The difference in morphine intake between the MADR lines suggests that opioid systems may have been altered by selective breeding for MA intake. This could be due to common genetic regulation or to genes that impact morphine intake, which are linked to those that influence MA intake. The possibility that opioid drugs could impact MA intake is discussed later.

4.1.3 Drug Aversion-Related Traits

The MADR lines are differentially sensitive to negative motivational effects of MA. Two phenotypes have been studied. The first was conditioned taste aversion (CTA). To measure CTA, the perceived effect of the drug is paired with a novel tasting solution by administering the drug immediately after consumption of the novel solution (the conditioned cue). If the perceived effect is aversive, then consumption of the solution when it is next offered will decline. MADR mice were tested by treating them with MA after consumption of a novel sodium chloride-containing solution. MALDR mice exhibited CTA induced by multiple doses of MA, including the lowest dose tested (1 mg/kg). MAHDR mice were insensitive to MA-induced CTA up to the highest dose tested (4 mg/kg) (Shabani, McKinnon, et al., 2012; Wheeler et al., 2009). The second phenotype was conditioned place aversion (CPA). In the CPP procedure, MA (or saline) is administered immediately before exposure to the conditioned cues. However, in the CTA and CPA procedures, MA (or saline) is given immediately after conditioned cue exposure. It has been suggested that the postcue exposure treatment procedure produces robust place aversion, because the cues are paired with the transition from a sober to drugged state, which is experienced as aversive (Cunningham, Smith, & McMullin, 2003). MALDR mice exhibited CPA to both 2 and 4 mg/kg MA, whereas MAHDR mice exhibited CPA only to the 4 mg/kg MA dose. Also, as mentioned earlier, when mice were tested in the CPP procedure in a drug-present condition (Shabani et al., 2011), MALDR but not MAHDR mice exhibited aversion. In this situation, the conditioning drug has perceived effects like those experienced during conditioning, and thus, state-dependent learning may impact the response (Cunningham & Noble, 1992). Overall, MALDR mice are more sensitive to aversive effects of MA, which may protect them from consuming higher amounts of MA, and treatments that increase sensitivity to aversive effects could be beneficial in humans.

4.1.4 Locomotor Activation and Sensitization

Psychostimulant-induced locomotor activation and sensitization are traits of interest in the addiction field, because data support modulation by components of the brain reward system (Aguilar-Rivera, Casanova, Gatica, Quirk, & Fuentealba, 2015; Jerlhag, Egecioglu, Dickson, & Engel, 2010; Kerstetter et al., 2015; Lominac et al., 2014; Meyer, Meshul, & Phillips, 2009). Furthermore, behavioral sensitization serves as a surrogate for underlying neuroadaptations that have relevance to relapse (Blum et al., 2009; Davidson et al., 2007; Kerstetter et al., 2015), and neuroplasticity underlying drug-induced sensitization contributes to the transition from controlled to compulsive drug seeking and use (Kalivas, Volkow, & Seamans, 2005; Kourrich, Calu, & Bonci, 2015; Robinson & Berridge, 2003). The MADR lines exhibited similar levels of MA dose-dependent locomotor activation after acute MA treatment (Shabani et al., 2011; Wheeler et al., 2009). Magnitude of locomotor sensitization did not differ between the two selected lines after treatment with 0.5 or 2 mg/kg MA; however, only MAHDR mice exhibited sensitization to a 4 mg/kg dose (Shabani et al., 2011). Thus, risk for MA intake appears to be genetically related to magnitude of sensitization, but not acute activation. This suggests that treatments which prevent the development or expression of sensitization may have an impact on MA intake.

The effects on locomotor activity of acutely administered cocaine and opioid receptor agonist drugs have also been examined in the MADR lines. Cocaine (5–30 mg/kg) induced comparable locomotor stimulation in the two mouse lines (Gubner et al., 2013). However, MALDR mice exhibited greater activation than MAHDR mice to the locomotor stimulant effects of the mixed opioid receptor agonist, morphine, and the more specific MOP-r agonist, fentanyl (Eastwood & Phillips, 2012). The MADR lines did not differ in sensitivity to the analgesic effects of opioid receptor agonists (Eastwood & Phillips, 2012).

4.1.5 Impulsivity

Considerable evidence indicates that heightened levels of impulsivity are associated with the use of addictive drugs (Tang, Posner, Rothbart, & Volkow, 2015), but only a few studies have examined individual associations or the genetic relationship between psychostimulant use and impulsive traits (Anker, Gliddon, & Carroll, 2008; Dalley et al., 2007; Marusich, Darna, Charnigo, Dwoskin, & Bardo, 2011; Yates, Marusich, Gipson, Beckmann, & Bardo, 2012). When Sprague Dawley rats were separated into groups that exhibited high and low impulsivity in a delay discounting task, the high-impulsivity group exhibited MA-induced CPP, whereas the low group did not, suggesting that high impulsivity is associated with greater sensitivity to the rewarding effects of MA (Yates et al., 2012). However, a multivariate approach that included individual differences in delay discounting behavior did not find a relationship with amphetamine self-administration (Marusich et al., 2011). When MADR mice were examined for level of behavioral inhibition in a go/no-go task, no significant differences were found. MA administered prior to the task decreased responding overall and did not increase impulsive-like behaviors in either mouse line (Moschak et al., 2012). Thus, no evidence for a genetic relationship between risk for MA intake and impulsive-like behavior or sensitivity to the effect of MA on impulsive-like behavior was obtained, using this task.

4.1.6 Circadian Period

MA use can disturb sleep (Hasler, Smith, Cousins, & Bootzin, 2012), perhaps via alterations of circadian period (Eastwood, Allen, & Raber, 2012). Circadian period was examined in MADR mice before and after voluntary drinking of 25 and then 50 mg/L MA for 24 h per day for 6 days each. Prior to MA consumption, the MADR lines did not differ in free-running period. MAHDR mice consumed about three and four times more MA from the two concentration solutions than MALDR mice. Although the MALDR line consumed significantly less MA, their free-running period increased after consumption of both MA concentrations, whereas a shift in MAHDR mice occurred only after they achieved a higher dose by consuming the 50 mg/L solution. These data support a genetic relationship between MA intake and sensitivity to the circadian disrupting effect of MA, which could be one effect of MA that results in lower MA intake in the MALDR line.

4.1.7 Body Temperature

Physiological effects of MA could contribute to the overall subjective experience that impacts MA use. MA can induce both hypothermia and hyperthermia, depending upon dose, external temperature, and time after administration (eg, Bowyer et al., 1994; Sabol, Yancey, Speaker, & Mitchell, 2013). High ambient temperature increased intravenous self-administration of MA in rats (Cornish et al., 2008). Body temperature responses to MA were examined in MADR mice under normal ambient temperature for MA doses from 1 to 16 mg/kg (Harkness et al., 2015). Hypothermia was the predominant response in MALDR mice, whereas hyperthermia was the predominant response in MAHDR mice. These different temperature responses in the MADR lines, which may be experienced during initial sessions of voluntary MA consumption, could play a role in their difference in continued intake. MA-induced hypothermia may be subjectively experienced as aversive and limit MA intake. It has been suggested that hypothermia may increase the duration during which an association can occur between a conditioned stimulus and unconditioned stimulus (Christianson, Anderson, Misanin, & Hinderliter, 2005). Thus, in MALDR mice, hypothermia could enhance the association between some other unconditioned unpleasant effect of MA and the taste of MA, which would serve as a conditioned stimulus. That the taste of MA is not initially aversive in these mice is evidenced by their intake during the first drinking session, which is comparable to that of MAHDR mice (Eastwood et al., 2014; Shabani, Dobbs, et al., 2012). When MADR mice were tested for their body temperature response to ethanol, both lines exhibited a hypothermic response that was comparable (Harkness et al., 2015). Thus, the mechanisms underlying ethanol- vs MA-induced hypothermia appear to be different. Additional research is needed to examine body temperature during voluntary MA drinking and to determine whether MA drinking is altered by changes in ambient temperature. Gene expression analysis suggests differential regulation of neuroimmune pathways by MA in MAHDR and MALDR mice (Wheeler et al., 2009), but it is currently unknown whether neuroimmune factors are involved in the differential thermal response to MA.

4.2 Blood Drug Levels

Published data for blood MA levels after intraperitoneal MA injection indicated higher peak MA levels in MAHDR mice (about 25% higher), but a comparable rate of clearance for the two lines (Shabani, McKinnon, et al., 2012). Higher levels of MA in the blood seem unlikely to explain the reduced aversion to MA of MAHDR mice, but could play a role in increased reward. Brain levels have not been examined, and data have been published for only one replicate set of the MADR lines. Blood MA data collected during an MA drinking study corresponded with MA intake. Both intake and blood level were comparable in the MADR lines on the first day of MA access, and intake and blood levels were higher in MAHDR mice at subsequent time periods and when MA was offered at higher concentrations (Eastwood et al., 2014). Morphine and fentanyl blood levels and clearance after intraperitoneal injection were comparable in the MADR lines (Eastwood & Phillips, 2012).

4.3 Neurochemical Information

DA and 5-HT levels in the NAcc and medial prefrontal cortex (mPFC) were measured in MADR mice before and after intraperitoneal injection of saline or 2 mg/kg MA (Lominac et al., 2014). Basal measures of these neurotransmitters by no net-flux in vivo microdialysis detected higher DA levels in the NAcc and mPFC of MALDR compared to MAHDR mice, and higher levels of 5-HT in MAHDR compared to MALDR mice in the NAcc, but not mPFC. Extracellular levels of DA and 5-HT after MA treatment, measured by conventional in vivo microdialysis, did not differ between the lines of mice in the NAcc. However, the DA response was larger in the mPFC in MAHDR compared to MALDR mice, and the 5-HT response was larger in the mPFC in MALDR compared to MAHDR mice. Glu was measured in the mPFC using similar procedures, and MAHDR mice had twice the basal extracellular Glu content of MALDR mice (Lominac et al., 2016). After 2 mg/kg MA treatment, Glu levels changed in opposite directions, increasing over time in MALDR mice, while decreasing in MAHDR mice.

Protein expression analyses were performed using immunoblotting for DAT, SERT, D_2R, and the 5-HT 1B receptor (5-$HT_{1B}R$) in NAcc core, NAcc shell and mPFC tissue from MADR mice treated with saline or 2 mg/kg MA (Lominac et al., 2014); levels of several Glu-related proteins in the mPFC have also been examined (Lominac et al., 2016). Regardless of treatment, protein expression was significantly higher in MAHDR mice for DAT and SERT in the NAcc shell and higher for DAT in the NAcc core, with a strong statistical trend in the same direction for SERT. Protein expression was higher in MALDR mice for the D_2R in the NAcc shell and for the 5-$HT_{1B}R$ in the NAcc core. The DAT was not detectable in the mPFC, but MALDR mice had higher SERT levels in this brain region. In saline-injected mice, lower Homer2a/b levels were found in mPFC tissue from MAHDR compared to MALDR mice, but there were no line differences for Homer1b/c, mGlu1/5, GluN1/2a, or EAAT1/2/3. The p-ERK/ERK ratio was measured as a marker of cellular activity. This ratio did not significantly differ between the MADR lines after saline treatment; however, after MA treatment, MAHDR mice exhibited a lower p-ERK/ERK ratio. This suggests blunted MA-stimulated mPFC neural activity, consistent with the blunted Glu response to MA in MAHDR mice. Both the microdialysis and protein expression arms of these studies should be repeated in a second set of replicate lines, but if reliable, the results have implications for neurochemical processes associated with risk for voluntary MA consumption.

4.4 Genetic Findings and Treatment

In part, the MADR lines were created to identify genetic factors relevant to voluntary MA intake. Gene expression results for NAcc tissue taken from MADR mice 4 h after injection with saline or 2 mg/kg MA, suggested the importance of neuroinflammatory/neuroimmune signaling processes in differential MA sensitivity (Wheeler et al., 2009). Certain genes that were differentially expressed under saline conditions between the selected lines were oppositely affected by MA treatment, moving expression in both selected lines toward intermediate levels, consistent with an initial allostatic state (Koob et al., 2014) associated with differential risk for MA intake that was in some way normalized by MA treatment.

In a subsequent study, quantitative trait locus (QTL) mapping data for the MA drinking trait were generated from DNA samples, along with microarray gene expression data from RNA samples from three different brain regions in MA-naïve MADR mice (Belknap et al., 2013). This study in MA-naïve mice addressed genetic risk for MA consumption. A QTL was identified on mouse chromosome 10 that accounted for more than 50% of the genetic variance in MA intake. Although many genes reside within the confidence interval of this QTL, *Oprm1* was considered as a candidate for the relevant gene because MALDR mice exhibited higher *Oprm1* expression in the mPFC (but not ventral midbrain or NAcc), compared to MAHDR mice. Greater density of MOP-r only in the mPFC in MALDR mice was confirmed by receptor radioligand binding (E.C. Eastwood et al., unpublished data). However, gene network analysis based on differentially expressed genes did not identify *Oprm1* as a key player in the most significant network, which was highly enriched in transcription factor genes, some of which are involved in neuroinflammatory/neuroimmune signaling. When *Oprm1* expression information was added to this transcription factor network, it attained hub status, meaning that it had a high level of interaction with other differentially expressed genes in the network, such that other genes in the network impacted *Oprm1* expression. These results suggest that *Oprm1* polymorphisms may not determine risk for MA intake, but that genes that alter *Oprm1* expression may play a role in genetic risk for MA consumption.

Based on both the genetic findings and other data supporting a role for opioid systems in MA effects, the MADR lines were tested for the impact of MOP-r agonists and an antagonist on MA drinking. Because MALDR mice are more sensitive to the locomotor activating effects of MOP-r agonists, and consume less MA than MAHDR mice (Eastwood & Phillips, 2012, 2014), higher opioid system activity could be protective against risk for MA intake. We thus hypothesized that a MOP-r agonist would reduce MA intake in MAHDR mice. The effect of buprenorphine, a partial agonist with less addiction potential than a full MOP-r agonist, was tested. Buprenorphine dose dependently reduced MA intake, whereas the MOP-r antagonist, naltrexone had no significant effect (Eastwood & Phillips, 2014).

Increasing opioid system activity reduced MA intake; however, some other gene(s) within the chromosome 10 QTL must be responsible for risk. Another gene residing in that region is *Taar1*. However, *Taar1* was not differentially expressed between the MADR lines and did not show up in the network analysis (Belknap et al., 2013), and thus, was initially ignored. When a database search led to the discovery of a nonsynonymous single-nucleotide polymorphism in that gene that differentiated the DBA/2J and C57BL/6J progenitor strains of the MADR lines, interest was increased. Subsequent studies in MADR mice and also *Taar1* knockout and wild-type mice, confirmed a role for *Taar1* in MA intake, as well as sensitivity to MA-induced thermal and aversive effects (Harkness et al., 2015). The absence of *Taar1* in the network analysis can be explained by the fact that both alleles are expressed, although only one allele codes for a functional receptor (Harkness et al., 2015).

5. SPECULATION ABOUT MECHANISMS AND TREATMENT TARGETS

Fig. 1 summarizes the targets of MA and genetic risk factors from human investigation. Interventions could be considered that impact these systems via receptors, receptor signaling, or metabolic processing and be extended to other monoamines. Some of these targets have already been considered in existing pharmacotherapeutic research. In this section, we focus on results from our genetic animal model and highlight the neurochemical and genetic findings.

Fig. 2 represents naïve and MA state neurocircuitry, receptor, and transporter information in the MADR lines for data collected so far. In addition,

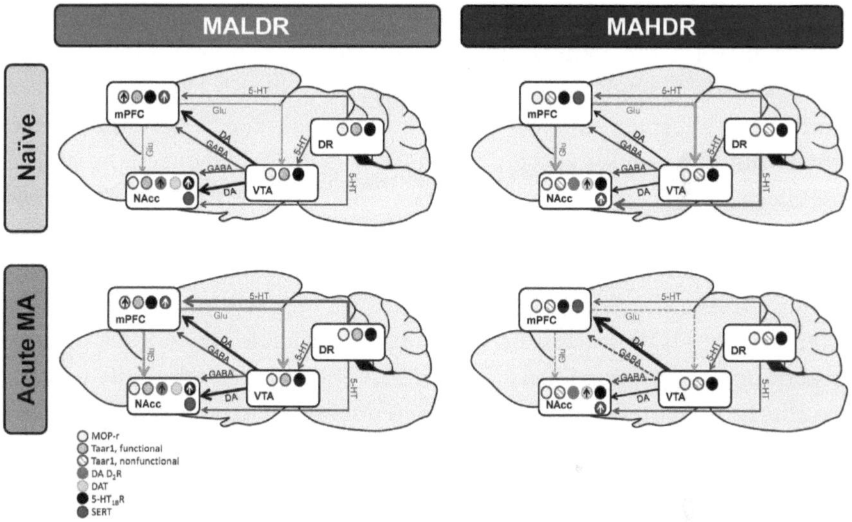

Fig. 2 Neurobiological alterations associated with selection for high vs low MA consumption. Shown are known and predicted neurotransmitter pathway resting state conditions for MALDR and MAHDR mice (Naïve), and states of these pathways after MA administration (Acute MA). Several brain regions important in responses to MA are represented and include the prefrontal cortex (mPFC), nucleus accumbens (NAcc), ventral tegmental area (VTA), and dorsal raphe (DR). The information shown is based on published data (citations below) with some predicted outcomes based on known changes in other systems. Labeled lines and arrows indicate a neurotransmitter projection between brain regions, including dopamine (DA), gamma-aminobutyric acid (GABA), glutamate (Glu), and serotonin (5-HT). Characteristics of the line used to represent a projection indicate baseline levels (*regular line*), lower than baseline (*dotted line*), a slight enhancement (*slightly bolded line*), or greater degree of enhancement (*heavily bolded line*). Circles within each brain region denote MA-relevant receptors and transporters that are found in that region, with shading and patterns defined in the figure. An upward arrow within the symbol indicates a higher number of that receptor. For additional information, see Di Cara et al. (2011), Espinoza et al. (2015), Lam et al. (2015), Latagliata et al. (2014), Lindemann et al. (2008), Lominac et al. (2014, 2016), Mantsch, Baker, Funk, Le and Shaham (2016), Mylecharane (1996), Pasternak and Pan (2013), Revel et al. (2013), and Taylor, Lewis, and Olive (2013). Additional description of brain neurochemistry associated with the information illustrated in this figure can be found in the text.

some predicted differences between the lines are included, based on the existing information. For example, microdialysis data indicate a DA response to MA in MAHDR mice that is absent in MALDR mice (Lominac et al., 2014). This increase in DA activity is associated with a reduction in Glu from baseline levels in MAHDR mice after MA treatment (Lominac et al., 2016). A reduced excitatory effect of Glu in the VTA in MAHDR mice does not appear to reduce the DA response to MA, but it could reduce GABAergic activity (as suggested in Fig. 2). Recent data indicate higher mGluR1 mRNA in VTA DA neurons than VTA GABA neurons in GAD67-GFP mice (Merrill, Friend, Newton, Hopkins, & Edwards, 2015); however, another Glu receptor subtype could be more relevant in MAHDR mice. Alprazolam, the positive allosteric modulator of GABA, reduced the subjective effects of oral amphetamine (Marks, Lile, Stoops, & Rush, 2014), and stimulation of GABA neurons induced place aversion (Tan et al., 2012) and disrupted reward consumption (sucrose) by inhibiting DA activity (van Zessen, Phillips, Budygin, & Stuber, 2012). Thus, reduced activity of this system in MAHDR mice is predicted based on their greater sensitivity to conditioned and motivational effects of MA and reduced sensitivity to aversive effects. Repeated cocaine exposure reduced the activity of VTA GABA neurons and disinhibited DA neurons (Bocklisch et al., 2013). Selective breeding for high MA intake could have produced a comparable response to MA in MAHDR mice that occurs even after acute MA exposure. Interventions that coordinately increase the GABA response and reduce the DA response to MA would be predicted to impact MA use by reducing sensitivity to MA reward and increasing sensitivity to MA aversion.

Taar1 is expressed in VTA DA neurons (Di Cara et al., 2011; Lindemann et al., 2008) and in the mPFC (Espinoza et al., 2015). Furthermore, there is evidence to suggest that TAAR1 influences Glu-related neuroactivity (Revel et al., 2011). This regulation may occur via alteration of NMDA receptor activity, resulting in increased Glu release (Espinoza et al., 2015). Reduced Glu response to MA in MAHDR mice is consistent with the absence of TAAR1 function in these mice. Because MAHDR mice express only the nonfunctional receptor, which is insensitive to agonists, pharmacological intervention designed to directly increase TAAR1 activity and examine the impact on MA intake is not an option. There may be TAAR1 interaction partners that also have the potential to impact MA intake. Of course, the most obvious are monoamines, since TAAR1 is a negative regulator of monoamine systems. Other impacts of TAAR1 activation are illustrated in Fig. 3 and could be considered, including manipulations of Glu

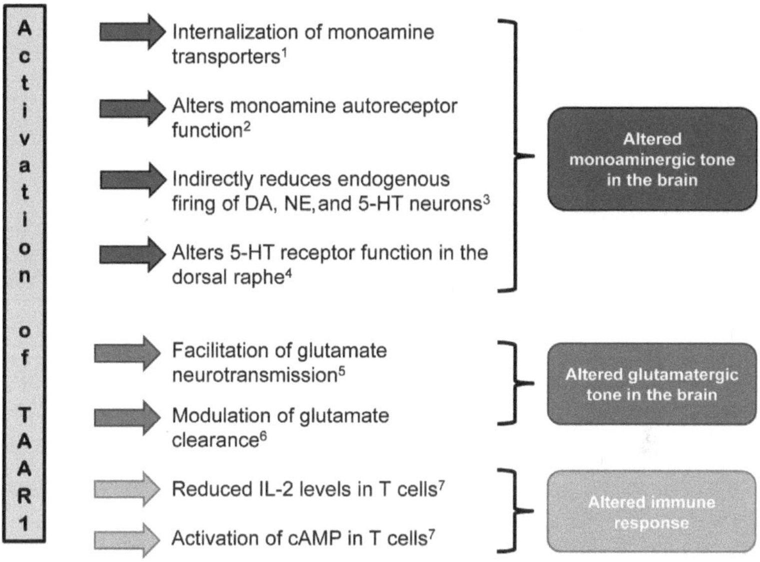

Fig. 3 Influence of TAAR1 activation on neurological functioning. Represented is the variety of events that can occur with TAAR1 activation. Details are presented in the text. Alterations of monoaminergic and glutamatergic tone in the brain, as well as altered immune response, are processes relevant to MA-related differences between the MADR selected lines of mice, based on existing data. [1]Revel et al. (2011), Wolinsky et al. (2007), Xie, Westmoreland, and Miller (2008), and Xie and Miller (2007, 2008, 2009a, 2009b); [2]Miller (2012) and Xie et al. (2007, 2008); [3]Lindemann et al. (2008), Revel et al. (2011), and Wolinsky et al. (2007); [4]Revel et al. (2012, 2011) and Xie et al. (2007, 2008); [5]Espinoza et al. (2015); [6]Cisneros and Ghorpade (2014); and [7]Sriram et al. (2015).

systems and neuroimmune responses. In addition, genetic risk factors should be considered. We are currently analyzing human DNA samples to determine whether there are polymorphisms that result in different levels of TAAR1 function and whether TAAR1 function is associated with risk for MA dependence.

In conclusion, discovery of broadly effective pharmacotherapies or robust genetic risk factors for MA dependence has been elusive. Existing data support the MADR lines as a powerful animal model for the study of differential genetically determined sensitivity to MA reward, reinforcement, and aversion. Outcomes for multiple measures have been reproducible and have eliminated alternative explanations for the large differences in voluntary MA intake in these lines of mice. Future studies will complete the development of a genetic binge model of MA intake and will test mechanistic hypotheses derived from the existing data that could lead to new treatment targets for human translational studies.

ACKNOWLEDGMENTS

The authors were supported by NIH grants P50 DA018165, P60 AA010769, R24 AA020245, U01 AA013519, and the Department of Veterans Affairs during the writing of this paper.

REFERENCES

Aguilar-Rivera, M. I., Casanova, J. P., Gatica, R. I., Quirk, G. J., & Fuentealba, J. A. (2015). Amphetamine sensitization is accompanied by an increase in prelimbic cortex activity. *Neuroscience*, *288*, 1–9. http://dx.doi.org/10.1016/j.neuroscience.2014.12.027.

Alia-Klein, N., Parvaz, M. A., Woicik, P. A., Konova, A. B., Maloney, T., Shumay, E., … Goldstein, R. Z. (2011). Gene x disease interaction on orbitofrontal gray matter in cocaine addiction. *Archives of General Psychiatry*, *68*(3), 283–294. http://dx.doi.org/10.1001/archgenpsychiatry.2011.10.

Anderson, A. L., Li, S. H., Biswas, K., McSherry, F., Holmes, T., Iturriaga, E., … Elkashef, A. M. (2012). Modafinil for the treatment of methamphetamine dependence. *Drug and Alcohol Dependence*, *120*(1–3), 135–141. http://dx.doi.org/10.1016/j.drugalcdep.2011.07.007.

Anderson, A. L., Li, S. H., Markova, D., Holmes, T. H., Chiang, N., Kahn, R., … Elkashef, A. M. (2015). Bupropion for the treatment of methamphetamine dependence in non-daily users: A randomized, double-blind, placebo-controlled trial. *Drug and Alcohol Dependence*, *150*, 170–174. http://dx.doi.org/10.1016/j.drugalcdep.2015.01.036.

Anker, J. J., Gliddon, L. A., & Carroll, M. E. (2008). Impulsivity on a Go/No-go task for intravenous cocaine or food in male and female rats selectively bred for high and low saccharin intake. *Behavioural Pharmacology*, *19*(5–6), 615–629. http://dx.doi.org/10.1097/FBP.0b013e32830dc0ae.

Ashby, C. R., Jr., Rohatgi, R., Ngosuwan, J., Borda, T., Gerasimov, M. R., Morgan, A. E., … Dewey, S. L. (1999). Implication of the GABA(B) receptor in gamma vinyl-GABA's inhibition of cocaine-induced increases in nucleus accumbens dopamine. *Synapse*, *31*(2), 151–153. http://dx.doi.org/10.1002/(sici)1098-2396(199902)31:2<151::aid-syn8>3.0.co;2-w.

Ballon, J. S., & Feifel, D. (2006). A systematic review of modafinil: Potential clinical uses and mechanisms of action. *The Journal of Clinical Psychiatry*, *67*(4), 554–566.

Banks, M. L., & Blough, B. E. (2015). Effects of environmental manipulations and treatment with bupropion and risperidone on choice between methamphetamine and food in rhesus monkeys. *Neuropsychopharmacology*, *40*(9), 2198–2206. http://dx.doi.org/10.1038/npp.2015.63.

Beardsley, P. M., & Hauser, K. F. (2014). Glial modulators as potential treatments of psychostimulant abuse. *Advances in Pharmacology*, *69*, 1–69. http://dx.doi.org/10.1016/b978-0-12-420118-7.00001-9.

Beardsley, P. M., Shelton, K. L., Hendrick, E., & Johnson, K. W. (2010). The glial cell modulator and phosphodiesterase inhibitor, AV411 (ibudilast), attenuates prime- and stress-induced methamphetamine relapse. *European Journal of Pharmacology*, *637*(1–3), 102–108. http://dx.doi.org/10.1016/j.ejphar.2010.04.010.

Belknap, J. K., McWeeney, S., Reed, C., Burkhart-Kasch, S., McKinnon, C. S., Li, N., … Phillips, T. J. (2013). Genetic factors involved in risk for methamphetamine intake and sensitization. *Mammalian Genome*, *24*(11–12), 446–458. http://dx.doi.org/10.1007/s00335-013-9484-9.

Belknap, J. K., Richards, S. P., O'Toole, L. A., Helms, M. L., & Phillips, T. J. (1997). Short-term selective breeding as a tool for QTL mapping: Ethanol preference drinking in mice. *Behavior Genetics*, *27*(1), 55–66.

Blum, K., Chen, T. J., Downs, B. W., Bowirrat, A., Waite, R. L., Braverman, E. R., ... Gold, M. (2009). Neurogenetics of dopaminergic receptor supersensitivity in activation of brain reward circuitry and relapse: Proposing "deprivation-amplification relapse therapy" (DART). *Postgraduate Medicine*, *121*(6), 176–196. http://dx.doi.org/10.3810/pgm.2009.11.2087.

Bocklisch, C., Pascoli, V., Wong, J. C. Y., House, D. R. C., Yvon, C., de Roo, M., ... Lüscher, C. (2013). Cocaine disinhibits dopamine neurons by potentiation of GABA transmission in the ventral tegmental area. *Science*, *341*(6153), 1521–1525. http://dx.doi.org/10.1126/science.1237059.

Bond, C., LaForge, K. S., Tian, M., Melia, D., Zhang, S., Borg, L., ... Yu, L. (1998). Single-nucleotide polymorphism in the human mu opioid receptor gene alters beta-endorphin binding and activity: Possible implications for opiate addiction. *Proceedings of the National Academy of Sciences of the United States of America*, *95*(16), 9608–9613.

Bousman, C. A., Glatt, S. J., Cherner, M., Atkinson, J. H., Grant, I., Tsuang, M. T., & Everall, I. P. (2010). Preliminary evidence of ethnic divergence in associations of putative genetic variants for methamphetamine dependence. *Psychiatry Research*, *178*(2), 295–298. http://dx.doi.org/10.1016/j.psychres.2009.07.019.

Bousman, C. A., Glatt, S. J., Everall, I. P., & Tsuang, M. T. (2009). Genetic association studies of methamphetamine use disorders: A systematic review and synthesis. *American Journal of Medical Genetics. Part B, Neuropsychiatric Genetics*, *150b*(8), 1025–1049. http://dx.doi.org/10.1002/ajmg.b.30936.

Bowyer, J. F., Davies, D. L., Schmued, L., Broening, H. W., Newport, G. D., Slikker, W., Jr., & Holson, R. R. (1994). Further studies of the role of hyperthermia in methamphetamine neurotoxicity. *The Journal of Pharmacology and Experimental Therapeutics*, *268*(3), 1571–1580.

Brackins, T., Brahm, N. C., & Kissack, J. C. (2011). Treatments for methamphetamine abuse: A literature review for the clinician. *Journal of Pharmacy Practice*, *24*(6), 541–550. http://dx.doi.org/10.1177/0897190011426557.

Brensilver, M., Heinzerling, K. G., & Shoptaw, S. (2013). Pharmacotherapy of amphetamine-type stimulant dependence: An update. *Drug and Alcohol Review*, *32*(5), 449–460. http://dx.doi.org/10.1111/dar.12048.

Brodie, J. D., Figueroa, E., Laska, E. M., & Dewey, S. L. (2005). Safety and efficacy of gamma-vinyl GABA (GVG) for the treatment of methamphetamine and/or cocaine addiction. *Synapse*, *55*(2), 122–125. http://dx.doi.org/10.1002/syn.20097.

Bunzow, J. R., Sonders, M. S., Arttamangkul, S., Harrison, L. M., Zhang, G., Quigley, D. I., ... Grandy, D. K. (2001). Amphetamine, 3,4-methylenedioxymethamphetamine, lysergic acid diethylamide, and metabolites of the catecholamine neurotransmitters are agonists of a rat trace amine receptor. *Molecular Pharmacology*, *60*(6), 1181–1188.

Byrnes-Blake, K. A., Carroll, F. I., Abraham, P., & Owens, S. M. (2001). Generation of anti-(+)methamphetamine antibodies is not impeded by (+)methamphetamine administration during active immunization of rats. *International Immunopharmacology*, *1*(2), 329–338.

Byrnes-Blake, K. A., Laurenzana, E. M., Carroll, F. I., Abraham, P., Gentry, W. B., Landes, R. D., & Owens, S. M. (2003). Pharmacodynamic mechanisms of monoclonal antibody-based antagonism of (+)-methamphetamine in rats. *European Journal of Pharmacology*, *461*(2–3), 119–128. http://dx.doi.org/10.1016/S0014-2999(03)01313-X.

Byrnes-Blake, K. A., Laurenzana, E. M., Landes, R. D., Gentry, W. B., & Owens, S. M. (2005). Monoclonal IgG affinity and treatment time alters antagonism of (+)-methamphetamine effects in rats. *European Journal of Pharmacology*, *521*(1–3), 86–94. http://dx.doi.org/10.1016/j.ejphar.2005.08.016.

Caspi, A., McClay, J., Moffitt, T. E., Mill, J., Martin, J., Craig, I. W., ... Poulton, R. (2002). Role of genotype in the cycle of violence in maltreated children. *Science*, *297*(5582), 851–854. http://dx.doi.org/10.1126/science.1072290.

Charnaud, B., & Griffiths, V. (1998). Levels of intravenous drug misuse among clients prescribed oral dexamphetamine or oral methadone: A comparison. *Drug and Alcohol Dependence*, *52*(1), 79–84.

Chen, C. K., Hu, X., Lin, S. K., Sham, P. C., Loh el, W., Li, T., ... Ball, D. M. (2004). Association analysis of dopamine D2-like receptor genes and methamphetamine abuse. *Psychiatric Genetics*, *14*(4), 223–226. http://dx.doi.org/10.1097/00041444-200412000-00011.

Chen, J., Lipska, B. K., Halim, N., Ma, Q. D., Matsumoto, M., Melhem, S., ... Weinberger, D. R. (2004). Functional analysis of genetic variation in catechol-O-methyltransferase (COMT): Effects on mRNA, protein, and enzyme activity in postmortem human brain. *American Journal of Human Genetics*, *75*(5), 807–821. http://dx.doi.org/10.1086/425589.

Chen, J. J., Ondo, W. G., Dashtipour, K., & Swope, D. M. (2012). Tetrabenazine for the treatment of hyperkinetic movement disorders: A review of the literature. *Clinical Therapeutics*, *34*(7), 1487–1504. http://dx.doi.org/10.1016/j.clinthera.2012.06.010.

Cheng, W. S., Garfein, R. S., Semple, S. J., Strathdee, S. A., Zians, J. K., & Patterson, T. L. (2010). Increased drug use and STI risk with injection drug use among HIV-seronegative heterosexual methamphetamine users. *Journal of Psychoactive Drugs*, *42*(1), 11–18. http://dx.doi.org/10.1080/02791072.2010.10399781.

Christianson, J. P., Anderson, M. J., Misanin, J. R., & Hinderliter, C. F. (2005). Effect of low body temperature on associative interference in conditioned taste aversion. *Perceptual and Motor Skills*, *100*(3 Pt. 2), 913–919. http://dx.doi.org/10.2466/pms.100.3c.913-919.

Cisneros, I. E., & Ghorpade, A. (2014). Methamphetamine and HIV-1-induced neurotoxicity: Role of trace amine associated receptor 1 cAMP signaling in astrocytes. *Neuropharmacology*, *85*, 499–507. http://dx.doi.org/10.1016/j.neuropharm.2014.06.011.

Colfax, G., & Shoptaw, S. (2005). The methamphetamine epidemic: Implications for HIV prevention and treatment. *Current HIV/AIDS Reports*, *2*(4), 194–199. http://dx.doi.org/10.1007/s11904-005-0016-4.

Cornish, J. L., Clemens, K. J., Thompson, M. R., Callaghan, P. D., Dawson, B., & McGregor, I. S. (2008). High ambient temperature increases intravenous methamphetamine self-administration on fixed and progressive ratio schedules in rats. *Journal of Psychopharmacology*, *22*(1), 100–110. http://dx.doi.org/10.1177/0269881107082286.

Crabbe, J. C. (2002). Genetic contributions to addiction. *Annual Review of Psychology*, *53*, 435–462. http://dx.doi.org/10.1146/annurev.psych.53.100901.135142.

Cravatt, B. F., Demarest, K., Patricelli, M. P., Bracey, M. H., Giang, D. K., Martin, B. R., & Lichtman, A. H. (2001). Supersensitivity to anandamide and enhanced endogenous cannabinoid signaling in mice lacking fatty acid amide hydrolase. *Proceedings of the National Academy of Sciences of the United States of America*, *98*(16), 9371–9376. http://dx.doi.org/10.1073/pnas.161191698.

Cunningham, C. L., & Noble, D. (1992). Methamphetamine-induced conditioned place preference or aversion depending on dose and presence of drug. *Annals of the New York Academy of Sciences*, *654*, 431–433.

Cunningham, C. L., Smith, R., & McMullin, C. (2003). Competition between ethanol-induced reward and aversion in place conditioning. *Learning & Behavior*, *31*(3), 273–280.

Dalley, J. W., Fryer, T. D., Brichard, L., Robinson, E. S., Theobald, D. E., Lääne, K., ... Robbins, T. W. (2007). Nucleus accumbens D2/3 receptors predict trait impulsivity and cocaine reinforcement. *Science*, *315*(5816), 1267–1270. http://dx.doi.org/10.1126/science.1137073.

Davidson, C., Gopalan, R., Ahn, C., Chen, Q., Mannelli, P., Patkar, A. A., ... Ellinwood, E. H. (2007). Reduction in methamphetamine induced sensitization and reinstatement after combined pergolide plus ondansetron treatment during withdrawal. *European Journal of Pharmacology, 565*(1–3), 113–118. http://dx.doi.org/10.1016/j.ejphar.2007.02.056.

Davidson, C., Gow, A. J., Lee, T. H., & Ellinwood, E. H. (2001). Methamphetamine neurotoxicity: Necrotic and apoptotic mechanisms and relevance to human abuse and treatment. *Brain Research. Brain Research Reviews, 36*(1), 1–22.

De La Garza, R., 2nd., Zorick, T., Heinzerling, K. G., Nusinowitz, S., London, E. D., Shoptaw, S., ... Newton, T. F. (2009). The cardiovascular and subjective effects of methamphetamine combined with gamma-vinyl-gamma-aminobutyric acid (GVG) in non-treatment seeking methamphetamine-dependent volunteers. *Pharmacology, Biochemistry, and Behavior, 94*(1), 186–193. http://dx.doi.org/10.1016/j.pbb.2009.08.007.

Dewey, S. L., Morgan, A. E., Ashby, C. R., Jr., Horan, B., Kushner, S. A., Logan, J., ... Brodie, J. D. (1998). A novel strategy for the treatment of cocaine addiction. *Synapse, 30*(2), 119–129. http://dx.doi.org/10.1002/(sici)1098-2396(199810)30:2<119::aid-syn1->3.0.co;2-f.

Dewey, S. L., Smith, G. S., Logan, J., Brodie, J. D., Yu, D. W., Ferrieri, R. A., ... Meller, E. (1992). GABAergic inhibition of endogenous dopamine release measured in vivo with 11C-raclopride and positron emission tomography. *The Journal of Neuroscience, 12*(10), 3773–3780.

Di Cara, B., Maggio, R., Aloisi, G., Rivet, J. M., Lundius, E. G., Yoshitake, T., ... Millan, M. J. (2011). Genetic deletion of trace amine 1 receptors reveals their role in auto-inhibiting the actions of ecstasy (MDMA). *The Journal of Neuroscience, 31*(47), 16928–16940. http://dx.doi.org/10.1523/jneurosci.2502-11.2011.

Duryee, M. J., Bevins, R. A., Reichel, C. M., Murray, J. E., Dong, Y., Thiele, G. M., & Sanderson, S. D. (2009). Immune responses to methamphetamine by active immunization with peptide-based, molecular adjuvant-containing vaccines. *Vaccine, 27*(22), 2981–2988. http://dx.doi.org/10.1016/j.vaccine.2009.02.105.

Eastwood, E., Allen, C. N., & Raber, J. (2012). Effects of neonatal methamphetamine and thioperamide exposure on spatial memory retention and circadian activity later in life. *Behavioural Brain Research, 230*(1), 229–236. http://dx.doi.org/10.1016/j.bbr.2012.02.003.

Eastwood, E. C., Barkley-Levenson, A. M., & Phillips, T. J. (2014). Methamphetamine drinking microstructure in mice bred to drink high or low amounts of methamphetamine. *Behavioural Brain Research, 272*, 111–120. http://dx.doi.org/10.1016/j.bbr.2014.06.035.

Eastwood, E. C., & Phillips, T. J. (2012). Opioid sensitivity in mice selectively bred to consume or not consume methamphetamine. *Addiction Biology, 19*(3), 370–379. http://dx.doi.org/10.1111/adb.12003.

Eastwood, E. C., & Phillips, T. J. (2014). Morphine intake and the effects of naltrexone and buprenorphine on the acquisition of methamphetamine intake. *Genes, Brain, and Behavior, 13*(2), 226–235. http://dx.doi.org/10.1111/gbb.12100.

Egan, M. F., Kojima, M., Callicott, J. H., Goldberg, T. E., Kolachana, B. S., Bertolino, A., ... Weinberger, D. R. (2003). The BDNF val66met polymorphism affects activity-dependent secretion of BDNF and human memory and hippocampal function. *Cell, 112*(2), 257–269.

Ehlers, C. L., Gizer, I. R., Gilder, D. A., & Wilhelmsen, K. C. (2011). Linkage analyses of stimulant dependence, craving, and heavy use in American Indians. *American Journal of Medical Genetics. Part B, Neuropsychiatric Genetics, 156b*(7), 772–780. http://dx.doi.org/10.1002/ajmg.b.31218.

El Daly, E., Chefer, V., Sandill, S., & Shippenberg, T. S. (2000). Modulation of the neurotoxic effects of methamphetamine by the selective kappa-opioid receptor agonist U69593. *Journal of Neurochemistry*, *74*(4), 1553–1562.

Elkashef, A. M., Rawson, R. A., Anderson, A. L., Li, S. H., Holmes, T., Smith, E. V., ... Weis, D. (2008). Bupropion for the treatment of methamphetamine dependence. *Neuropsychopharmacology*, *33*(5), 1162–1170. http://dx.doi.org/10.1038/sj.npp.1301481.

Espinoza, S., Lignani, G., Caffino, L., Maggi, S., Sukhanov, I., Leo, D., ... Gainetdinov, R. R. (2015). TAAR1 modulates cortical glutamate NMDA receptor function. *Neuropsychopharmacology*, *40*(9), 2217–2227. http://dx.doi.org/10.1038/npp.2015.65.

Falconer, D. S., & Mackay, T. F. C. (1996). *Introduction to quantitative genetics* (4th ed.). Essex, UK: Longman Group Ltd.

Ferraguti, G., Pascale, E., & Lucarelli, M. (2015). Alcohol addiction: A molecular biology perspective. *Current Medicinal Chemistry*, *22*(6), 670–684. http://dx.doi.org/10.2174/0929867321666141229103158.

Fleckenstein, A. E., Volz, T. J., Riddle, E. L., Gibb, J. W., & Hanson, G. R. (2007). New insights into the mechanism of action of amphetamines. *Annual Review of Pharmacology and Toxicology*, *47*, 681–698. http://dx.doi.org/10.1146/annurev.pharmtox.47.120505.105140.

Galloway, G. P., Buscemi, R., Coyle, J. R., Flower, K., Siegrist, J. D., Fiske, L. A., ... Mendelson, J. (2011). A randomized, placebo-controlled trial of sustained-release dextroamphetamine for treatment of methamphetamine addiction. *Clinical Pharmacology and Therapeutics*, *89*(2), 276–282. http://dx.doi.org/10.1038/clpt.2010.307.

Gardner, E. L., Schiffer, W. K., Horan, B. A., Highfield, D., Dewey, S. L., Brodie, J. D., & Ashby, C. R., Jr. (2002). Gamma-vinyl GABA, an irreversible inhibitor of GABA transaminase, alters the acquisition and expression of cocaine-induced sensitization in male rats. *Synapse*, *46*(4), 240–250. http://dx.doi.org/10.1002/syn.10138.

Gavrilin, M. A., Mathes, L. E., & Podell, M. (2002). Methamphetamine enhances cell-associated feline immunodeficiency virus replication in astrocytes. *Journal of Neurovirology*, *8*(3), 240–249. http://dx.doi.org/10.1080/13550280290049660.

Gerasimov, M. R., Ashby, C. R., Jr., Gardner, E. L., Mills, M. J., Brodie, J. D., & Dewey, S. L. (1999). Gamma-vinyl GABA inhibits methamphetamine, heroin, or ethanol-induced increases in nucleus accumbens dopamine. *Synapse*, *34*(1), 11–19. http://dx.doi.org/10.1002/(sici)1098-2396(199910)34:1<11::aid-syn2>3.0.co;2-5.

Gerra, G., Garofano, L., Bosari, S., Pellegrini, C., Zaimovic, A., Moi, G., ... Donnini, C. (2004). Analysis of monoamine oxidase A (MAO-A) promoter polymorphism in male heroin-dependent subjects: Behavioural and personality correlates. *Journal of Neural Transmission*, *111*(5), 611–621. http://dx.doi.org/10.1007/s00702-004-0129-8.

Gonzales, R., Marinelli-Casey, P., Shoptaw, S., Ang, A., & Rawson, R. A. (2006). Hepatitis C virus infection among methamphetamine-dependent individuals in outpatient treatment. *Journal of Substance Abuse Treatment*, *31*(2), 195–202. http://dx.doi.org/10.1016/j.jsat.2006.04.006.

Grabowski, J., Rhoades, H., Schmitz, J., Stotts, A., Daruzska, L. A., Creson, D., & Moeller, F. G. (2001). Dextroamphetamine for cocaine-dependence treatment: A double-blind randomized clinical trial. *Journal of Clinical Psychopharmacology*, *21*(5), 522–526.

Grabowski, J., Rhoades, H., Silverman, P., Schmitz, J. M., Stotts, A., Creson, D., & Bailey, R. (2000). Risperidone for the treatment of cocaine dependence: Randomized, double-blind trial. *Journal of Clinical Psychopharmacology*, *20*(3), 305–310.

Grabowski, J., Shearer, J., Merrill, J., & Negus, S. S. (2004). Agonist-like, replacement pharmacotherapy for stimulant abuse and dependence. *Addictive Behaviors*, *29*(7), 1439–1464. http://dx.doi.org/10.1016/j.addbeh.2004.06.018.

Grant, K. M., LeVan, T. D., Wells, S. M., Li, M., Stoltenberg, S. F., Gendelman, H. E., ... Bevins, R. A. (2012). Methamphetamine-associated psychosis. *Journal of Neuroimmune Pharmacology*, 7(1), 113–139. http://dx.doi.org/10.1007/s11481-011-9288-1.

Graves, S. M., & Napier, T. C. (2011). Mirtazapine alters cue-associated methamphetamine seeking in rats. *Biological Psychiatry*, *69*(3), 275–281. http://dx.doi.org/10.1016/j.biopsych.2010.09.032.

Graves, S. M., Rafeyan, R., Watts, J., & Napier, T. C. (2012). Mirtazapine, and mirtazapine-like compounds as possible pharmacotherapy for substance abuse disorders: Evidence from the bench and the bedside. *Pharmacology & Therapeutics*, *136*(3), 343–353. http://dx.doi.org/10.1016/j.pharmthera.2012.08.013.

Greenhill, L. L., Biederman, J., Boellner, S. W., Rugino, T. A., Sangal, R. B., Earl, C. Q., ... Swanson, J. M. (2006). A randomized, double-blind, placebo-controlled study of modafinil film-coated tablets in children and adolescents with attention-deficit/hyperactivity disorder. *Journal of the American Academy of Child and Adolescent Psychiatry*, *45*(5), 503–511. http://dx.doi.org/10.1097/01.chi.0000205709.63571.c9.

Gubner, N. R., Reed, C., McKinnon, C. S., & Phillips, T. J. (2013). Unique genetic factors influence sensitivity to the rewarding and aversive effects of methamphetamine versus cocaine. *Behavioural Brain Research*, *256*, 420–427. http://dx.doi.org/10.1016/j.bbr.2013.08.035.

Haerian, B. S. (2013). BDNF rs6265 polymorphism and drug addiction: A systematic review and meta-analysis. *Pharmacogenomics*, *14*(16), 2055–2065. http://dx.doi.org/10.2217/pgs.13.217.

Haile, C. N. (2007). Neurochemical and neurobehavioral consequences of methamphetamine abuse. In S. B. Karch (Ed.), *Neurochemistry of abused drugs* (pp. 53–79). Boca Raton: CRC Press.

Haile, C. N., Kosten, T. R., & Kosten, T. A. (2009). Pharmacogenetic treatments for drug addiction: Cocaine, amphetamine and methamphetamine. *The American Journal of Drug and Alcohol Abuse*, *35*(3), 161–177. http://dx.doi.org/10.1080/00952990902825447.

Hajheidari, S., Miladi-Gorji, H., & Bigdeli, I. (2015). Effect of the environmental enrichment on the severity of psychological dependence and voluntary methamphetamine consumption in methamphetamine withdrawn rats. *Neuroscience Letters*, *584*, 151–155. http://dx.doi.org/10.1016/j.neulet.2014.10.017.

Hambuchen, M. D., Carroll, F. I., Ruedi-Bettschen, D., Hendrickson, H. P., Hennings, L. J., Blough, B. E., ... Owens, S. M. (2015). Combining active immunization with monoclonal antibody therapy to facilitate early initiation of a long-acting anti-methamphetamine antibody response. *Journal of Medicinal Chemistry*, *58*(11), 4665–4677. http://dx.doi.org/10.1021/acs.jmedchem.5b00220.

Harkness, J. H., Shi, X., Janowsky, A., & Phillips, T. J. (2015). Trace amine-associated receptor 1 regulation of methamphetamine intake and related traits. *Neuropsychopharmacology*, *40*(9), 2175–2184. http://dx.doi.org/10.1038/npp.2015.61.

Hasler, B. P., Smith, L. J., Cousins, J. C., & Bootzin, R. R. (2012). Circadian rhythms, sleep, and substance abuse. *Sleep Medicine Reviews*, *16*(1), 67–81. http://dx.doi.org/10.1016/j.smrv.2011.03.004.

Heinzerling, K. G., McCracken, J. T., Swanson, A. N., Ray, L. A., & Shoptaw, S. J. (2012). COMT Val158Met, BDNF Val66Met, and OPRM1 Asn40Asp and methamphetamine dependence treatment response: Preliminary investigation. *Journal of Clinical Psychopharmacology*, *32*(1), 135–137. http://dx.doi.org/10.1097/JCP.0b013e318240a48e.

Heinzerling, K. G., & Shoptaw, S. (2012). Gender, brain-derived neurotrophic factor Val66Met, and frequency of methamphetamine use. *Gender Medicine*, *9*(2), 112–120. http://dx.doi.org/10.1016/j.genm.2012.02.005.

Heinzerling, K. G., Swanson, A. N., Hall, T. M., Yi, Y., Wu, Y., & Shoptaw, S. J. (2014). Randomized, placebo-controlled trial of bupropion in methamphetamine-dependent participants with less than daily methamphetamine use. *Addiction*, *109*(11), 1878–1886. http://dx.doi.org/10.1111/add.12636.

Herin, D. V., Rush, C. R., & Grabowski, J. (2010). Agonist-like pharmacotherapy for stimulant dependence: Preclinical, human laboratory, and clinical studies. *Annals of the New York Academy of Sciences*, *1187*, 76–100. http://dx.doi.org/10.1111/j.1749-6632.2009.05145.x.

Hofford, R. S., Darna, M., Wilmouth, C. E., Dwoskin, L. P., & Bardo, M. T. (2014). Environmental enrichment reduces methamphetamine cue-induced reinstatement but does not alter methamphetamine reward or VMAT2 function. *Behavioural Brain Research*, *270*, 151–158. http://dx.doi.org/10.1016/j.bbr.2014.05.007.

Hotamisligil, G. S., & Breakefield, X. O. (1991). Human monoamine oxidase A gene determines levels of enzyme activity. *American Journal of Human Genetics*, *49*(2), 383–392.

Iamjan, S. A., Thanoi, S., Watiktinkorn, P., Nudmamud-Thanoi, S., & Reynolds, G. P. (2015). BDNF (Val66Met) genetic polymorphism is associated with vulnerability for methamphetamine dependence. *Pharmacogenomics*, *16*(14), 1541–1545. http://dx.doi.org/10.2217/pgs.15.96.

Ide, S., Kobayashi, H., Tanaka, K., Ujike, H., Sekine, Y., Ozaki, N., … Sora, I. (2004). Gene polymorphisms of the mu opioid receptor in methamphetamine abusers. *Annals of the New York Academy of Sciences*, *1025*, 316–324. http://dx.doi.org/10.1196/annals.1316.039.

Ide, S., Kobayashi, H., Ujike, H., Ozaki, N., Sekine, Y., Inada, T., … Sora, I. (2006). Linkage disequilibrium and association with methamphetamine dependence/psychosis of mu-opioid receptor gene polymorphisms. *The Pharmacogenomics Journal*, *6*(3), 179–188. http://dx.doi.org/10.1038/sj.tpj.6500355.

Iyo, M., Bi, Y., Hashimoto, K., Tomitaka, S. I., Inada, T., & Fukui, S. (1996). Does an increase of cyclic AMP prevent methamphetamine-induced behavioral sensitization in rats? *Annals of the New York Academy of Science*, *801*, 377–383.

Izzo, E., Orsini, C., Koob, G. F., & Pulvirenti, L. (2001). A dopamine partial agonist and antagonist block amphetamine self-administration in a progressive ratio schedule. *Pharmacology, Biochemistry, and Behavior*, *68*(4), 701–708.

Jerlhag, E., Egecioglu, E., Dickson, S. L., & Engel, J. A. (2010). Ghrelin receptor antagonism attenuates cocaine- and amphetamine-induced locomotor stimulation, accumbal dopamine release, and conditioned place preference. *Psychopharmacology*, *211*(4), 415–422. http://dx.doi.org/10.1007/s00213-010-1907-7.

Johnson, B. A., Roache, J. D., Ait-Daoud, N., Wells, L. T., Wallace, C. L., Dawes, M. A., … Wang, X. Q. (2007). Effects of acute topiramate dosing on methamphetamine-induced subjective mood. *The International Journal of Neuropsychopharmacology*, *10*(1), 85–98. http://dx.doi.org/10.1017/s1461145705006401.

Johnson, Z., Venters, J., Guarraci, F. A., & Zewail-Foote, M. (2015). Methamphetamine induces DNA damage in specific regions of the female rat brain. *Clinical and Experimental Pharmacology & Physiology*, *42*(6), 570–575. http://dx.doi.org/10.1111/1440-1681.12404.

Jonsson, E. G., Bah, J., Melke, J., Abou Jamra, R., Schumacher, J., Westberg, L., … Sedvall, G. C. (2004). Monoamine related functional gene variants and relationships to monoamine metabolite concentrations in CSF of healthy volunteers. *BMC Psychiatry*, *4*, 4. http://dx.doi.org/10.1186/1471-244x-4-4.

Jung, M. J., Lippert, B., Metcalf, B. W., Bohlen, P., & Schechter, P. J. (1977). Gamma-vinyl GABA (4-amino-hex-5-enoic acid), a new selective irreversible inhibitor of GABA-T: Effects on brain GABA metabolism in mice. *Journal of Neurochemistry*, *29*(5), 797–802.

Kahbazi, M., Ghoreishi, A., Rahiminejad, F., Mohammadi, M. R., Kamalipour, A., & Akhondzadeh, S. (2009). A randomized, double-blind and placebo-controlled trial of modafinil in children and adolescents with attention deficit and hyperactivity disorder.

Psychiatry Research, *168*(3), 234–237. http://dx.doi.org/10.1016/j.psychres.2008.06.024.

Kahlig, K. M., Binda, F., Khoshbouei, H., Blakely, R. D., McMahon, D. G., Javitch, J. A., & Galli, A. (2005). Amphetamine induces dopamine efflux through a dopamine transporter channel. *Proceedings of the National Academy of Sciences of the United States of America*, *102*(9), 3495–3500. http://dx.doi.org/10.1073/pnas.0407737102.

Kalayasiri, R., Verachai, V., Gelernter, J., Mutirangura, A., & Malison, R. T. (2014). Clinical features of methamphetamine-induced paranoia and preliminary genetic association with DBH-1021C→T in a Thai treatment cohort. *Addiction*, *109*(6), 965–976. http://dx.doi.org/10.1111/add.12512.

Kalivas, P. W., Volkow, N., & Seamans, J. (2005). Unmanageable motivation in addiction: A pathology in prefrontal-accumbens glutamate transmission. *Neuron*, *45*(5), 647–650. http://dx.doi.org/10.1016/j.neuron.2005.02.005.

Karila, L., Weinstein, A., Aubin, H. J., Benyamina, A., Reynaud, M., & Batki, S. L. (2010). Pharmacological approaches to methamphetamine dependence: A focused review. *British Journal of Clinical Pharmacology*, *69*(6), 578–592. http://dx.doi.org/10.1111/j.1365-2125.2010.03639.x.

Kaufman, S., & Friedman, S. (1965). Dopamine-beta-hydroxylase. *Pharmacological Reviews*, *17*, 71–100.

Kawanokuchi, J., Mizuno, T., Kato, H., Mitsuma, N., & Suzumura, A. (2004). Effects of interferon-beta on microglial functions as inflammatory and antigen presenting cells in the central nervous system. *Neuropharmacology*, *46*(5), 734–742. http://dx.doi.org/10.1016/j.neuropharm.2003.11.007.

Kerstetter, K. A., Wunsch, A. M., Nakata, K. G., Donckels, E., Neumaier, J. F., & Ferguson, S. M. (2015). Corticostriatal afferents modulate responsiveness to psychostimulant drugs and drug-associated stimuli. *Neuropsychopharmacology*. http://dx.doi.org/10.1038/npp.2015.253. Advanced online publication.

Kim, D. J., Roh, S., Kim, Y., Yoon, S. J., Lee, H. K., Han, C. S., & Kim, Y. K. (2005). High concentrations of plasma brain-derived neurotrophic factor in methamphetamine users. *Neuroscience Letters*, *388*(2), 112–115. http://dx.doi.org/10.1016/j.neulet.2005.06.042.

Kishi, T., Matsuda, Y., Iwata, N., & Correll, C. U. (2013). Antipsychotics for cocaine or psychostimulant dependence: Systematic review and meta-analysis of randomized, placebo-controlled trials. *The Journal of Clinical Psychiatry*, *74*(12), e1169–e1180. http://dx.doi.org/10.4088/JCP.13r08525.

Klee, H., Wright, S., Carnwath, T. O. M., & Merrill, J. (2001). The role of substitute therapy in the treatment of problem amphetamine use. *Drug and Alcohol Review*, *20*(4), 417–429. http://dx.doi.org/10.1080/09595230120092715.

Kongsakon, R., Papadopoulos, K. I., & Saguansiritham, R. (2005). Mirtazapine in amphetamine detoxification: A placebo-controlled pilot study. *International Clinical Psychopharmacology*, *20*(5), 253–256.

Koob, G. F., Arends, M. A., & Le Moal, M. (2014). What is addiction? In G. F. Koob, M. A. Arends, & M. Le Moal (Eds.), *Drugs, addiction, and the brain* (pp. 1–27). San Diego: Academic Press. Chapter 1.

Korotkova, T. M., Klyuch, B. P., Ponomarenko, A. A., Lin, J. S., Haas, H. L., & Sergeeva, O. A. (2007). Modafinil inhibits rat midbrain dopaminergic neurons through D2-like receptors. *Neuropharmacology*, *52*(2), 626–633. http://dx.doi.org/10.1016/j.neuropharm.2006.09.005.

Kosten, T., & Owens, S. M. (2005). Immunotherapy for the treatment of drug abuse. *Pharmacology & Therapeutics*, *108*(1), 76–85. http://dx.doi.org/10.1016/j.pharmthera.2005.06.009.

Kourrich, S., Calu, D. J., & Bonci, A. (2015). Intrinsic plasticity: An emerging player in addiction. *Nature Reviews. Neuroscience*, *16*(3), 173–184. http://dx.doi.org/10.1038/nrn3877.

Krasnova, I. N., & Cadet, J. L. (2009). Methamphetamine toxicity and messengers of death. *Brain Research Reviews*, *60*(2), 379–407. http://dx.doi.org/10.1016/j.brainresrev.2009.03.002.

Kreek, M. J., Nielsen, D. A., Butelman, E. R., & La Forge, K. S. (2005). Genetic influences on impulsivity, risk taking, stress responsivity and vulnerability to drug abuse and addiction. *Nature Neuroscience*, *8*(11), 1450–1457. http://dx.doi.org/10.1038/nn1583.

Lam, V. M., Espinoza, S., Gerasimov, A. S., Gainetdinov, R. R., & Salahpour, A. (2015). In-vivo pharmacology of trace-amine associated receptor 1. *European Journal of Pharmacology*, *763*(Pt. B), 136–142. http://dx.doi.org/10.1016/j.ejphar.2015.06.026.

Latagliata, E. C., Valzania, A., Pascucci, T., Campus, P., Cabib, S., & Puglisi-Allegra, S. (2014). Stress-induced activation of ventral tegmental mu-opioid receptors reduces accumbens dopamine tone by enhancing dopamine transmission in the medial prefrontal cortex. *Psychopharmacology*, *231*(21), 4099–4108. http://dx.doi.org/10.1007/s00213-014-3549-7.

Laurenzana, E. M., Hendrickson, H. P., Carpenter, D., Peterson, E. C., Gentry, W. B., West, M., … Owens, S. M. (2009). Functional and biological determinants affecting the duration of action and efficacy of anti-(+)-methamphetamine monoclonal antibodies in rats. *Vaccine*, *27*(50), 7011–7020. http://dx.doi.org/10.1016/j.vaccine.2009.09.072.

Laurenzana, E. M., Stevens, M. W., Frank, J. C., Hambuchen, M. D., Hendrickson, H. P., White, S. J., … Gentry, W. B. (2014). Pharmacological effects of two anti-methamphetamine monoclonal antibodies. Supporting data for lead candidate selection for clinical development. *Human Vaccines & Immunotherapeutics*, *10*(9), 2638–2647. http://dx.doi.org/10.4161/hv.29707.

Le Moal, M. (2009). Drug abuse: Vulnerability and transition to addiction. *Pharmacopsychiatry*, *42*(Suppl. 1), S42–S55. http://dx.doi.org/10.1055/s-0029-1216355.

Li, T., Chen, C. K., Hu, X., Ball, D., Lin, S. K., Chen, W., … Collier, D. A. (2004). Association analysis of the DRD4 and COMT genes in methamphetamine abuse. *American Journal of Medical Genetics. Part B, Neuropsychiatric Genetics*, *129b*(1), 120–124. http://dx.doi.org/10.1002/ajmg.b.30024.

Li, M. D., Wang, J., Niu, T., Ma, J. Z., Seneviratne, C., Ait-Daoud, N., … Johnson, B. A. (2014). Transcriptome profiling and pathway analysis of genes expressed differentially in participants with or without a positive response to topiramate treatment for methamphetamine addiction. *BMC Medical Genomics*, 7, 65. http://dx.doi.org/10.1186/s12920-014-0065-x.

Lile, J. A., Stoops, W. W., Vansickel, A. R., Glaser, P. E., Hays, L. R., & Rush, C. R. (2005). Aripiprazole attenuates the discriminative-stimulus and subject-rated effects of D-amphetamine in humans. *Neuropsychopharmacology*, *30*(11), 2103–2114. http://dx.doi.org/10.1038/sj.npp.1300803.

Lindemann, L., Meyer, C. A., Jeanneau, K., Bradaia, A., Ozmen, L., Bluethmann, H., … Hoener, M. C. (2008). Trace amine-associated receptor 1 modulates dopaminergic activity. *The Journal of Pharmacology and Experimental Therapeutics*, *324*(3), 948–956. http://dx.doi.org/10.1124/jpet.107.132647.

Liu, Y., & Edwards, R. H. (1997). The role of vesicular transport proteins in synaptic transmission and neural degeneration. *Annual Review of Neuroscience*, *20*, 125–156. http://dx.doi.org/10.1146/annurev.neuro.20.1.125.

Liu, S., Green, C. E., Lane, S. D., Kosten, T. R., Moeller, F. G., Nielsen, D. A., & Schmitz, J. M. (2014). The influence of dopamine beta-hydroxylase gene polymorphism rs1611115 on levodopa/carbidopa treatment for cocaine dependence: A preliminary study. *Pharmacogenetics and Genomics*, *24*(7), 370–373. http://dx.doi.org/10.1097/fpc.0000000000000055.

Loebl, T., Angarita, G. A., Pachas, G. N., Huang, K. L., Lee, S. H., Nino, J., … Evins, A. E. (2008). A randomized, double-blind, placebo-controlled trial of long-acting risperidone in cocaine-dependent men. *The Journal of Clinical Psychiatry*, *69*(3), 480–486.

Loftis, J. M., Choi, D., Hoffman, W., & Huckans, M. S. (2011). Methamphetamine causes persistent immune dysregulation: A cross-species, translational report. *Neurotoxicity Research*, *20*(1), 59–68. http://dx.doi.org/10.1007/s12640-010-9223-x.

Loftis, J. M., & Janowsky, A. (2014). Neuroimmune basis of methamphetamine toxicity. *International Review of Neurobiology*, *118*, 165–197. http://dx.doi.org/10.1016/b978-0-12-801284-0.00007-5.

Loftis, J. M., Wilhelm, C. J., Vandenbark, A. A., & Huckans, M. (2013). Partial MHC/neuroantigen peptide constructs: A potential neuroimmune-based treatment for methamphetamine addiction. *PLoS One*, *8*(2), e56306. http://dx.doi.org/10.1371/journal.pone.0056306.

Lohr, K. M., Stout, K. A., Dunn, A. R., Wang, M., Salahpour, A., Guillot, T. S., & Miller, G. W. (2015). Increased vesicular monoamine transporter 2 (VMAT2; Slc18a2) protects against methamphetamine toxicity. *ACS Chemical Neuroscience*, *6*(5), 790–799. http://dx.doi.org/10.1021/acschemneuro.5b00010.

Loland, C. J., Mereu, M., Okunola, O. M., Cao, J., Prisinzano, T. E., Mazier, S., … Newman, A. H. (2012). R-modafinil (armodafinil): A unique dopamine uptake inhibitor and potential medication for psychostimulant abuse. *Biological Psychiatry*, *72*(5), 405–413. http://dx.doi.org/10.1016/j.biopsych.2012.03.022.

Lominac, K. D., McKenna, C. L., Schwartz, L. M., Ruiz, P. N., Wroten, M. G., Miller, B. W., … Szumlinski, K. K. (2014). Mesocorticolimbic monoamine correlates of methamphetamine sensitization and motivation. *Frontiers in Systems Neuroscience*, *8*, 70. http://dx.doi.org/10.3389/fnsys.2014.00070.

Lominac, K. D., Quadir, S. G., Barrett, H. M., McKenna, C. L., Schwartz, L. M., Ruiz, P. N., … Szumlinski, K. K. (2016). Prefrontal glutamate correlates of methamphetamine sensitization and preference. *The European Journal of Neuroscience*, (in press).

Longo, M., Wickes, W., Smout, M., Harrison, S., Cahill, S., & White, J. M. (2010). Randomized controlled trial of dexamphetamine maintenance for the treatment of methamphetamine dependence. *Addiction*, *105*(1), 146–154. http://dx.doi.org/10.1111/j.1360-0443.2009.02717.x.

Mamiya, T., Noda, Y., Ren, X., Hamdy, M., Furukawa, S., Kameyama, T., … Nabeshima, T. (2001). Involvement of cyclic AMP systems in morphine physical dependence in mice: Prevention of development of morphine dependence by rolipram, a phosphodiesterase 4 inhibitor. *British Journal of Pharmacology*, *132*(5), 1111–1117. http://dx.doi.org/10.1038/sj.bjp.0703912.

Mantsch, J. R., Baker, D. A., Funk, D., Le, A. D., & Shaham, Y. (2016). Stress-induced reinstatement of drug seeking: 20 years of progress. *Neuropsychopharmacology*, *41*(1), 335–356. http://dx.doi.org/10.1038/npp.2015.142.

Maring, E. D., Tawadros, P. S., Steer, C. J., & Lee, J. T. (2015). Systematic review of candidate single-nucleotide polymorphisms as biomarkers for responsiveness to neoadjuvant chemoradiation for rectal cancer. *Anticancer Research*, *35*(7), 3761–3766.

Marks, K. R., Lile, J. A., Stoops, W. W., & Rush, C. R. (2014). Separate and combined impact of acute naltrexone and alprazolam on subjective and physiological effects of oral d-amphetamine in stimulant users. *Psychopharmacology*, *231*(14), 2741–2750. http://dx.doi.org/10.1007/s00213-014-3449-x.

Marusich, J. A., Darna, M., Charnigo, R. J., Dwoskin, L. P., & Bardo, M. T. (2011). A multivariate assessment of individual differences in sensation seeking and impulsivity as predictors of amphetamine self-administration and prefrontal dopamine function in rats. *Experimental and Clinical Psychopharmacology, 19*(4), 275–284. http://dx.doi.org/10.1037/a0023897.

McGregor, C., Srisurapanont, M., Mitchell, A., Wickes, W., & White, J. M. (2008). Symptoms and sleep patterns during inpatient treatment of methamphetamine withdrawal: A comparison of mirtazapine and modafinil with treatment as usual. *Journal of Substance Abuse Treatment, 35*(3), 334–342. http://dx.doi.org/10.1016/j.jsat.2007.12.003.

McMillan, D. E., Hardwick, W. C., Li, M., & Owens, S. M. (2002). Pharmacokinetic antagonism of (+)-methamphetamine discrimination by a low-affinity monoclonal anti-methamphetamine antibody. *Behavioural Pharmacology, 13*(5–6), 465–473.

Ménard, C., Gaudreau, P., & Quirion, R. (2015). Signaling pathways relevant to cognition-enhancing drug targets. *Handbook of Experimental Pharmacology, 228*, 59–98. http://dx.doi.org/10.1007/978-3-319-16522-6_3.

Meredith, C. W., Jaffe, C., Cherrier, M., Robinson, J. P., Malte, C. A., Yanasak, E. V., … Saxon, A. J. (2009). Open trial of injectable risperidone for methamphetamine dependence. *Journal of Addiction Medicine, 3*(2), 55–65. http://dx.doi.org/10.1097/ADM.0b013e31818e2185.

Merrill, C. B., Friend, L. N., Newton, S. T., Hopkins, Z. H., & Edwards, J. G. (2015). Ventral tegmental area dopamine and GABA neurons: Physiological properties and expression of mRNA for endocannabinoid biosynthetic elements. *Scientific Reports, 5*, 16176. http://dx.doi.org/10.1038/srep16176.

Meyer, A. C., Horton, D. B., Neugebauer, N. M., Wooters, T. E., Nickell, J. R., Dwoskin, L. P., & Bardo, M. T. (2011). Tetrabenazine inhibition of monoamine uptake and methamphetamine behavioral effects: Locomotor activity, drug discrimination and self-administration. *Neuropharmacology, 61*(4), 849–856. http://dx.doi.org/10.1016/j.neuropharm.2011.05.033.

Meyer, P. J., Meshul, C. K., & Phillips, T. J. (2009). Ethanol- and cocaine-induced locomotion are genetically related to increases in accumbal dopamine. *Genes, Brain, and Behavior, 8*(3), 346–355. http://dx.doi.org/10.1111/j.1601-183X.2009.00481.x.

Miller, G. M. (2012). Avenues for the development of therapeutics that target trace amine associated receptor 1 (TAAR1). *Journal of Medicinal Chemistry, 55*(5), 1809–1814. http://dx.doi.org/10.1021/jm201437t.

Miller, M. L., Aarde, S. M., Moreno, A. Y., Creehan, K. M., Janda, K. D., & Taffe, M. A. (2015). Effects of active anti-methamphetamine vaccination on intravenous self-administration in rats. *Drug and Alcohol Dependence, 153*, 29–36. http://dx.doi.org/10.1016/j.drugalcdep.2015.06.014.

Miller, M. L., Moreno, A. Y., Aarde, S. M., Creehan, K. M., Vandewater, S. A., Vaillancourt, B. D., … Taffe, M. A. (2013). A methamphetamine vaccine attenuates methamphetamine-induced disruptions in thermoregulation and activity in rats. *Biological Psychiatry, 73*(8), 721–728. http://dx.doi.org/10.1016/j.biopsych.2012.09.010.

Milloy, M. J., Kerr, T., Buxton, J., Montaner, J., & Wood, E. (2009). Methamphetamine use and rates of incarceration among street-involved youth in a Canadian setting: A cross-sectional analysis. *Substance Abuse Treatment, Prevention, and Policy, 4*, 17. http://dx.doi.org/10.1186/1747-597x-4-17.

Minzenberg, M. J., & Carter, C. S. (2008). Modafinil: A review of neurochemical actions and effects on cognition. *Neuropsychopharmacology, 33*(7), 1477–1502. http://dx.doi.org/10.1038/sj.npp.1301534.

Moeller, S. J., Parvaz, M. A., Shumay, E., Wu, S., Beebe-Wang, N., Konova, A. B., … Goldstein, R. Z. (2014). Monoamine polygenic liability in health and cocaine dependence: Imaging genetics study of aversive processing and associations with depression symptomatology. *Drug and Alcohol Dependence, 140*, 17–24. http://dx.doi.org/10.1016/j.drugalcdep.2014.04.019.

Morgan, A. E., & Dewey, S. L. (1998). Effects of pharmacologic increases in brain GABA levels on cocaine-induced changes in extracellular dopamine. *Synapse*, *28*(1), 60–65. http://dx.doi.org/10.1002/(sici)1098-2396(199801)28:1<60::aid-syn7>3.0.co;2-a.

Morita, Y., Ujike, H., Tanaka, Y., Uchida, N., Nomura, A., Ohtani, K., … Kuroda, S. (2005). A nonsynonymous polymorphism in the human fatty acid amide hydrolase gene did not associate with either methamphetamine dependence or schizophrenia. *Neuroscience Letters*, *376*(3), 182–187. http://dx.doi.org/10.1016/j.neulet.2004.11.050.

Moschak, T. M., Stang, K. A., Phillips, T. J., & Mitchell, S. H. (2012). Behavioral inhibition in mice bred for high vs. low levels of methamphetamine consumption or sensitization. *Psychopharmacology*, *222*(2), 353–365. http://dx.doi.org/10.1007/s00213-012-2650-z.

Mylecharane, E. J. (1996). Ventral tegmental area 5-HT receptors: Mesolimbic dopamine release and behavioural studies. *Behavioural Brain Research*, *73*(1–2), 1–5.

Nakamura, K., Sekine, Y., Takei, N., Iwata, Y., Suzuki, K., Anitha, A., … Mori, N. (2009). An association study of monoamine oxidase A (MAOA) gene polymorphism in methamphetamine psychosis. *Neuroscience Letters*, *455*(2), 120–123. http://dx.doi.org/10.1016/j.neulet.2009.02.048.

Nejtek, V. A., Avila, M., Chen, L. A., Zielinski, T., Djokovic, M., Podawiltz, A., … Rush, A. J. (2008). Do atypical antipsychotics effectively treat co-occurring bipolar disorder and stimulant dependence? A randomized, double-blind trial. *The Journal of Clinical Psychiatry*, *69*(8), 1257–1266.

Nestler, E. J. (2001). Molecular basis of long-term plasticity underlying addiction. *Nature Reviews. Neuroscience*, *2*(2), 119–128. http://dx.doi.org/10.1038/35053570.

Nestler, E. J. (2012). Transcriptional mechanisms of drug addiction. *Clinical Psychopharmacology and Neuroscience*, *10*(3), 136–143. http://dx.doi.org/10.9758/cpn.2012.10.3.136.

Newman-Tancredi, A., Cussac, D., Quentric, Y., Touzard, M., Verriele, L., Carpentier, N., & Millan, M. J. (2002). Differential actions of antiparkinson agents at multiple classes of monoaminergic receptor. III. Agonist and antagonist properties at serotonin, 5-HT(1) and 5-HT(2), receptor subtypes. *The Journal of Pharmacology and Experimental Therapeutics*, *303*(2), 815–822. http://dx.doi.org/10.1124/jpet.102.039883.

Newton, T. F., Roache, J. D., De La Garza, R., 2nd., Fong, T., Wallace, C. L., Li, S. H., … Kahn, R. (2006). Bupropion reduces methamphetamine-induced subjective effects and cue-induced craving. *Neuropsychopharmacology*, *31*(7), 1537–1544. http://dx.doi.org/10.1038/sj.npp.1300979.

Nguyen, T. L., Tian, Y. H., You, I. J., Lee, S. Y., & Jang, C. G. (2011). Modafinil-induced conditioned place preference via dopaminergic system in mice. *Synapse*, *65*(8), 733–741. http://dx.doi.org/10.1002/syn.20892.

Nickell, J. R., Siripurapu, K. B., Vartak, A., Crooks, P. A., & Dwoskin, L. P. (2014). The vesicular monoamine transporter-2: An important pharmacological target for the discovery of novel therapeutics to treat methamphetamine abuse. *Advances in Pharmacology*, *69*, 71–106. http://dx.doi.org/10.1016/b978-0-12-420118-7.00002-0.

Olsen, R. H., Allen, C. N., Derkach, V. A., Phillips, T. J., Belknap, J. K., & Raber, J. (2013). Impaired memory and reduced sensitivity to the circadian period lengthening effects of methamphetamine in mice selected for high methamphetamine consumption. *Behavioural Brain Research*, *256*, 197–204. http://dx.doi.org/10.1016/j.bbr.2013.08.015.

Panenka, W. J., Procyshyn, R. M., Lecomte, T., MacEwan, G. W., Flynn, S. W., Honer, W. G., & Barr, A. M. (2013). Methamphetamine use: A comprehensive review of molecular, preclinical and clinical findings. *Drug and Alcohol Dependence*, *129*(3), 167–179. http://dx.doi.org/10.1016/j.drugalcdep.2012.11.016.

Pasternak, G. W., & Pan, Y.-X. (2013). Mu opioids and their receptors: Evolution of a concept. *Pharmacological Reviews*, *65*(4), 1257–1317. http://dx.doi.org/10.1124/pr.112.007138.

Pereira, F. C., Gough, B., Macedo, T. R., Ribeiro, C. F., Ali, S. F., & Binienda, Z. K. (2011). Buprenorphine modulates methamphetamine-induced dopamine dynamics in the rat caudate nucleus. *Neurotoxicity Research, 19*(1), 94–101. http://dx.doi.org/10.1007/s12640-009-9143-9.

Phillips, T. J., Eastwood, E. C., & Harkness, J. H. (2014). Drug abuse: Amphetamines. In S. Pietropaolo, F. Sluyter, & W. E. Crusio (Eds.), *Handbook of behavioral genetics of the mouse II. Genetic mouse models of neurobehavioral disorders. Section 6: Substance dependence and abuse* (pp. 330–349). Cambridge: Cambridge University Press.

Phillips, T. J., Kamens, H. M., & Wheeler, J. M. (2008). Behavioral genetic contributions to the study of addiction-related amphetamine effects. *Neuroscience and Biobehavioral Reviews, 32*(4), 707–759. http://dx.doi.org/10.1016/j.neubiorev.2007.10.008.

Phillips, T. J., & Shabani, S. (2015). An animal model of differential genetic risk for methamphetamine intake. *Frontiers in Neuroscience, 9*, 327. http://dx.doi.org/10.3389/fnins.2015.00327.

Pike, E., Stoops, W. W., Hays, L. R., Glaser, P. E., & Rush, C. R. (2014). Methamphetamine self-administration in humans during D-amphetamine maintenance. *Journal of Clinical Psychopharmacology, 34*(6), 675–681. http://dx.doi.org/10.1097/jcp.0000000000000207.

Pulvirenti, L., Balducci, C., Piercy, M., & Koob, G. F. (1998). Characterization of the effects of the partial dopamine agonist terguride on cocaine self-administration in the rat. *The Journal of Pharmacology and Experimental Therapeutics, 286*(3), 1231–1238.

Quisenberry, A. J., & Baker, L. E. (2015). Dopaminergic mediation of the discriminative stimulus functions of modafinil in rats. *Psychopharmacology, 232*(24), 4411–4419. http://dx.doi.org/10.1007/s00213-015-4065-0.

Ray, L. A., Roche, D. J., Heinzerling, K., & Shoptaw, S. (2014). Opportunities for the development of neuroimmune therapies in addiction. *International Review of Neurobiology, 118*, 381–401. http://dx.doi.org/10.1016/b978-0-12-801284-0.00012-9.

Revel, F. G., Meyer, C. A., Bradaia, A., Jeanneau, K., Calcagno, E., Andre, C. B., … Hoener, M. C. (2012). Brain-specific overexpression of trace amine-associated receptor 1 alters monoaminergic neurotransmission and decreases sensitivity to amphetamine. *Neuropsychopharmacology, 37*(12), 2580–2592. http://dx.doi.org/10.1038/npp.2012.109.

Revel, F. G., Moreau, J. L., Gainetdinov, R. R., Bradaia, A., Sotnikova, T. D., Mory, R., … Hoener, M. C. (2011). TAAR1 activation modulates monoaminergic neurotransmission, preventing hyperdopaminergic and hypoglutamatergic activity. *Proceedings of the National Academy of Sciences of the United States of America, 108*(20), 8485–8490. http://dx.doi.org/10.1073/pnas.1103029108.

Revel, F. G., Moreau, J. L., Pouzet, B., Mory, R., Bradaia, A., Buchy, D., … Hoener, M. C. (2013). A new perspective for schizophrenia: TAAR1 agonists reveal antipsychotic- and antidepressant-like activity, improve cognition and control body weight. *Molecular Psychiatry, 18*(5), 543–556. http://dx.doi.org/10.1038/mp.2012.57.

Robinson, T. E., & Berridge, K. C. (2003). Addiction. *Annual Review of Psychology, 54*, 25–53. http://dx.doi.org/10.1146/annurev.psych.54.101601.145237.

Rolan, P., Hutchinson, J., & Johnson, K. (2009). Ibudilast: A review of its pharmacology efficacy and safety in respiratory and neurological disease. *Expert Opinion on Pharmacotherapy, 10*(17), 2897–2904. http://dx.doi.org/10.1517/14656560903426189.

Rommel, N., Rohleder, N. H., Wagenpfeil, S., Haertel-Petri, R., & Kesting, M. R. (2015). Evaluation of methamphetamine-associated socioeconomic status and addictive behaviors, and their impact on oral health. *Addictive Behaviors, 50*, 182–187. http://dx.doi.org/10.1016/j.addbeh.2015.06.040.

Rothman, R. B., Baumann, M. H., Dersch, C. M., Romero, D. V., Rice, K. C., Carroll, F. I., & Partilla, J. S. (2001). Amphetamine-type central nervous system stimulants release norepinephrine more potently than they release dopamine and serotonin. *Synapse*, *39*(1), 32–41. http://dx.doi.org/10.1002/1098-2396(20010101)39:1<32::aid-syn5>3.0.co;2-3.

Sabol, S. Z., Hu, S., & Hamer, D. (1998). A functional polymorphism in the monoamine oxidase A gene promoter. *Human Genetics*, *103*(3), 273–279.

Sabol, K. E., Yancey, D. M., Speaker, H. A., & Mitchell, S. L. (2013). Methamphetamine and core temperature in the rat: Ambient temperature, dose, and the effect of a D2 receptor blocker. *Psychopharmacology*, *228*(4), 551–561. http://dx.doi.org/10.1007/s00213-013-3059-z.

Saify, K., & Saadat, M. (2015). Association between variable number of tandem repeats (VNTR) polymorphism in the promoter region of monoamine oxidase A (MAOA) gene and susceptibility to heroin dependence. *Psychiatry Research*, *229*(3), 1055–1056. http://dx.doi.org/10.1016/j.psychres.2015.08.017.

Schwantes-An, T. H., Zhang, J., Chen, L. S., Hartz, S. M., Culverhouse, R. C., Chen, X., … Saccone, N. L. (2015). Association of the OPRM1 variant rs1799971 (A118G) with non-specific liability to substance dependence in a collaborative de novo meta-analysis of European-ancestry cohorts. *Behavior Genetics*. http://dx.doi.org/10.1007/s10519-015-9737-3. Advanced online publication.

Schwartz, J. L. (1969). A critical review and evaluation of smoking control methods. *Public Health Reports*, *84*(6), 483–506.

Sekine, Y., Ouchi, Y., Sugihara, G., Takei, N., Yoshikawa, E., Nakamura, K., … Cadet, J. L. (2008). Methamphetamine causes microglial activation in the brains of human abusers. *The Journal of Neuroscience*, *28*(22), 5756–5761. http://dx.doi.org/10.1523/jneurosci.1179-08.2008.

Senard, J. M., & Rouet, P. (2006). Dopamine beta-hydroxylase deficiency. *Orphanet Journal of Rare Diseases*, *1*, 7. http://dx.doi.org/10.1186/1750-1172-1-7.

Sery, O., Vojtova, V., & Zvolsky, P. (2001). The association study of DRD2, ACE and AGT gene polymorphisms and methamphetamine dependence. *Physiological Research*, *50*(1), 43–50.

Shabani, S., Dobbs, L. K., Ford, M. M., Mark, G. P., Finn, D. A., & Phillips, T. J. (2012). A genetic animal model of differential sensitivity to methamphetamine reinforcement. *Neuropharmacology*, *62*(7), 2169–2177. http://dx.doi.org/10.1016/j.neuropharm.2012.01.002.

Shabani, S., McKinnon, C. S., Cunningham, C. L., & Phillips, T. J. (2012). Profound reduction in sensitivity to the aversive effects of methamphetamine in mice bred for high methamphetamine intake. *Neuropharmacology*, *62*(2), 1134–1141. http://dx.doi.org/10.1016/j.neuropharm.2011.11.005.

Shabani, S., McKinnon, C. S., Reed, C., Cunningham, C. L., & Phillips, T. J. (2011). Sensitivity to rewarding or aversive effects of methamphetamine determines methamphetamine intake. *Genes, Brain, and Behavior*, *10*(6), 625–636. http://dx.doi.org/10.1111/j.1601-183X.2011.00700.x.

Shearer, J., Darke, S., Rodgers, C., Slade, T., van Beek, I., Lewis, J., … Wodak, A. (2009). A double-blind, placebo-controlled trial of modafinil (200 mg/day) for methamphetamine dependence. *Addiction*, *104*(2), 224–233. http://dx.doi.org/10.1111/j.1360-0443.2008.02437.x.

Shen, Y. L., Chang, T. Y., Chang, Y. C., Tien, H. H., Yang, F. C., Wang, P. Y., & Liao, R. M. (2014). Elevated BDNF mRNA expression in the medial prefrontal cortex after d-amphetamine reinstated conditioned place preference in rats. *Neuroscience*, *263*, 88–95. http://dx.doi.org/10.1016/j.neuroscience.2014.01.015.

Shichinohe, S., Ozawa, H., Hashimoto, E., Tatschner, T., Riederer, P., & Saito, T. (2001). Changes in the cAMP-related signal transduction mechanism in postmortem human brains of heroin addicts. *Journal of Neural Transmission*, *108*(3), 335–347.

Shoptaw, S., Heinzerling, K. G., Rotheram-Fuller, E., Steward, T., Wang, J., Swanson, A. N., … Ling, W. (2008). Randomized, placebo-controlled trial of bupropion for the treatment of methamphetamine dependence. *Drug and Alcohol Dependence*, *96*(3), 222–232. http://dx.doi.org/10.1016/j.drugalcdep.2008.03.010.

Sim, M. S., Hatim, A., Reynolds, G. P., & Mohamed, Z. (2013). Association of a functional FAAH polymorphism with methamphetamine-induced symptoms and dependence in a Malaysian population. *Pharmacogenomics*, *14*(5), 505–514. http://dx.doi.org/10.2217/pgs.13.25.

Sipe, J. C., Chiang, K., Gerber, A. L., Beutler, E., & Cravatt, B. F. (2002). A missense mutation in human fatty acid amide hydrolase associated with problem drug use. *Proceedings of the National Academy of Sciences of the United States of America*, *99*(12), 8394–8399. http://dx.doi.org/10.1073/pnas.082235799.

Snider, S. E., Vunck, S. A., van den Oord, E. J., Adkins, D. E., McClay, J. L., & Beardsley, P. M. (2012). The glial cell modulators, ibudilast and its amino analog, AV1013, attenuate methamphetamine locomotor activity and its sensitization in mice. *European Journal of Pharmacology*, *679*(1–3), 75–80. http://dx.doi.org/10.1016/j.ejphar.2012.01.013.

Sora, I., Li, B., Fumushima, S., Fukui, A., Arime, Y., Kasahara, Y., … Ikeda, K. (2009). Monoamine transporter as a target molecule for psychostimulants. *International Review of Neurobiology*, *85*, 29–33. http://dx.doi.org/10.1016/s0074-7742(09)85003-4.

Sriram, U., Cenna, J. M., Haldar, B., Fernandes, N. C., Razmpour, R., Fan, S., … Potula, R. (2015). Methamphetamine induces trace amine-associated receptor 1 (TAAR1) expression in human T lymphocytes: Role in immunomodulation. *Journal of Leukocyte Biology*. http://dx.doi.org/10.1189/jlb.4A0814-395RR.

Stevens, M. W., Henry, R. L., Owens, S. M., Schutz, R., & Gentry, W. B. (2014). First human study of a chimeric anti-methamphetamine monoclonal antibody in healthy volunteers. *MAbs*, *6*(6), 1649–1656. http://dx.doi.org/10.4161/19420862.2014.976431.

Stoops, W. W., Bennett, J. A., Lile, J. A., Sevak, R. J., & Rush, C. R. (2013). Influence of aripiprazole pretreatment on the reinforcing effects of methamphetamine in humans. *Progress in Neuropsychopharmacology and Biological Psychiatry*, *47*, 111–117. http://dx.doi.org/10.1016/j.pnpbp.2013.08.007.

Stoops, W. W., Lile, J. A., Glaser, P. E., & Rush, C. R. (2006). A low dose of aripiprazole attenuates the subject-rated effects of d-amphetamine. *Drug and Alcohol Dependence*, *84*(2), 206–209. http://dx.doi.org/10.1016/j.drugalcdep.2006.02.004.

Sulzer, D., Chen, T. K., Lau, Y. Y., Kristensen, H., Rayport, S., & Ewing, A. (1995). Amphetamine redistributes dopamine from synaptic vesicles to the cytosol and promotes reverse transport. *The Journal of Neuroscience*, *15*(5 Pt. 2), 4102–4108.

Sulzer, D., Sonders, M. S., Poulsen, N. W., & Galli, A. (2005). Mechanisms of neurotransmitter release by amphetamines: A review. *Progress in Neurobiology*, *75*(6), 406–433. http://dx.doi.org/10.1016/j.pneurobio.2005.04.003.

Suzuki, A., Nakamura, K., Sekine, Y., Minabe, Y., Takei, N., Suzuki, K., … Mori, N. (2006). An association study between catechol-O-methyl transferase gene polymorphism and methamphetamine psychotic disorder. *Psychiatric Genetics*, *16*(4), 133–138. http://dx.doi.org/10.1097/01.ypg.0000218613.35139.cd.

Suzumura, A., Ito, A., & Mizuno, T. (2003). Phosphodiesterase inhibitors suppress IL-12 production with microglia and T helper 1 development. *Multiple Sclerosis*, *9*(6), 574–578.

Talloczy, Z., Martinez, J., Joset, D., Ray, Y., Gacser, A., Toussi, S., … Santambrogio, L. (2008). Methamphetamine inhibits antigen processing, presentation, and phagocytosis. *PLoS Pathogens*, *4*(2), e28. http://dx.doi.org/10.1371/journal.ppat.0040028.

Tan, K. R., Yvon, C., Turiault, M., Mirzabekov, J. J., Doehner, J., Labouèbe, G., … Lüscher, C. (2012). GABA neurons of the VTA drive conditioned place aversion. *Neuron*, *73*(6), 1173–1183. http://dx.doi.org/10.1016/j.neuron.2012.02.015.

Tang, Y. Y., Posner, M. I., Rothbart, M. K., & Volkow, N. D. (2015). Circuitry of self-control and its role in reducing addiction. *Trends in Cognitive Sciences*, *19*(8), 439–444. http://dx.doi.org/10.1016/j.tics.2015.06.007.

Taylor, S. B., Lewis, C. R., & Olive, M. F. (2013). The neurocircuitry of illicit psychostimulant addiction: Acute and chronic effects in humans. *Substance Abuse and Rehabilitation*, *4*, 29–43. http://dx.doi.org/10.2147/sar.s39684.

Teng, L., Crooks, P. A., & Dwoskin, L. P. (1998). Lobeline displaces [3H] dihydrotetrabenazine binding and releases [3H]dopamine from rat striatal synaptic vesicles: Comparison with d-amphetamine. *Journal of Neurochemistry*, *71*(1), 258–265.

Terwilliger, R. Z., Beitner-Johnson, D., Sevarino, K. A., Crain, S. M., & Nestler, E. J. (1991). A general role for adaptations in G-proteins and the cyclic AMP system in mediating the chronic actions of morphine and cocaine on neuronal function. *Brain Research*, *548*(1–2), 100–110.

Tiihonen, J., Kuoppasalmi, K., Fohr, J., Tuomola, P., Kuikanmaki, O., Vorma, H., … Meririnne, E. (2007). A comparison of aripiprazole, methylphenidate, and placebo for amphetamine dependence. *The American Journal of Psychiatry*, *164*(1), 160–162. http://dx.doi.org/10.1176/ajp.2007.164.1.160.

Tikkanen, R., Ducci, F., Goldman, D., Holi, M., Lindberg, N., Tiihonen, J., & Virkkunen, M. (2010). MAOA alters the effects of heavy drinking and childhood physical abuse on risk for severe impulsive acts of violence among alcoholic violent offenders. *Alcoholism, Clinical and Experimental Research*, *34*(5), 853–860. http://dx.doi.org/10.1111/j.1530-0277.2010.01157.x.

Tikkanen, R., Sjoberg, R. L., Ducci, F., Goldman, D., Holi, M., Tiihonen, J., & Virkkunen, M. (2009). Effects of MAOA-genotype, alcohol consumption, and aging on violent behavior. *Alcoholism, Clinical and Experimental Research*, *33*(3), 428–434. http://dx.doi.org/10.1111/j.1530-0277.2008.00853.x.

Tyler, W. J., Alonso, M., Bramham, C. R., & Pozzo-Miller, L. D. (2002). From acquisition to consolidation: On the role of brain-derived neurotrophic factor signaling in hippocampal-dependent learning. *Learning & Memory*, *9*(5), 224–237. http://dx.doi.org/10.1101/lm.51202.

U.S. Department of Justice National Drug Intelligence Center. (2011). National Drug Threat Assessment. NDIC Publication, Product No. 2011–Q0317-001.

U.S. National Institutes of Health. (2015). *Trial of ibudilast for methamphetamine dependence (IBUD ph II)*. Retrieved from, https://clinicaltrials.gov/ct2/show/NCT01860807.

Ujike, H., Katsu, T., Okahisa, Y., Takaki, M., Kodama, M., Inada, T., … Kuroda, S. (2009). Genetic variants of D2 but not D3 or D4 dopamine receptor gene are associated with rapid onset and poor prognosis of methamphetamine psychosis. *Progress in Neuro-Psychopharmacology & Biological Psychiatry*, *33*(4), 625–629. http://dx.doi.org/10.1016/j.pnpbp.2009.02.019.

United Nations Office on Drug and Crime. (2015). World Drug Report. United Nations Publication, Sales No. E.15.XI.6.

van Zessen, R., Phillips, J. L., Budygin, E. A., & Stuber, G. D. (2012). Activation of VTA GABA neurons disrupts reward consumption. *Neuron*, *73*(6), 1184–1194. http://dx.doi.org/10.1016/j.neuron.2012.02.016.

Varga, Z. V., Ferdinandy, P., Liaudet, L., & Pacher, P. (2015). Drug-induced mitochondrial dysfunction and cardiotoxicity. *American Journal of Physiology. Heart and Circulatory Physiology*, *309*(9), H1453–H1467. http://dx.doi.org/10.1152/ajpheart.00554.2015.

Vecchiola, A., Collyer, P., Figueroa, R., Labarca, R., Bustos, G., & Magendzo, K. (1999). Differential regulation of mu-opioid receptor mRNA in the nucleus accumbens shell and core accompanying amphetamine behavioral sensitization. *Brain Research. Molecular Brain Research*, *69*(1), 1–9.

Vocci, F. J., & Appel, N. M. (2007). Approaches to the development of medications for the treatment of methamphetamine dependence. *Addiction*, *102*(Suppl. 1), 96–106. http://dx.doi.org/10.1111/j.1360-0443.2007.01772.x.

Volkow, N. D. (2013). Methamphetamine. National Institute on Drug Abuse, *Research report series*: *Vol. 9*. U.S. Department of Health and Human Services. National Institutes of Health Publications 13-4210; 10 pages, http//www.drugabuse.gov/sites/default/files/methrrs.pdf.

Volkow, N. D., & Morales, M. (2015). The brain on drugs: From reward to addiction. *Cell*, *162*(4), 712–725. http://dx.doi.org/10.1016/j.cell.2015.07.046.

Watanabe-Galloway, S., Ryan, S., Hansen, K., Hullsiek, B., Muli, V., & Malone, A. C. (2009). Effects of methamphetamine abuse beyond individual users. *Journal of Psychoactive Drugs*, *41*(3), 241–248. http://dx.doi.org/10.1080/02791072.2009.10400534.

Wheeler, J. M., Reed, C., Burkhart-Kasch, S., Li, N., Cunningham, C. L., Janowsky, A., … Phillips, T. J. (2009). Genetically correlated effects of selective breeding for high and low methamphetamine consumption. *Genes, Brain, and Behavior*, *8*(8), 758–771. http://dx.doi.org/10.1111/j.1601-183X.2009.00522.x.

White, H. S., Smith, M. D., & Wilcox, K. S. (2007). Mechanisms of action of antiepileptic drugs. *International Review of Neurobiology*, *81*, 85–110. http://dx.doi.org/10.1016/s0074-7742(06)81006-8.

Wolinsky, T. D., Swanson, C. J., Smith, K. E., Zhong, H., Borowsky, B., Seeman, P., … Gerald, C. P. (2007). The trace amine 1 receptor knockout mouse: An animal model with relevance to schizophrenia. *Genes, Brain, and Behavior*, *6*(7), 628–639. http://dx.doi.org/10.1111/j.1601-183X.2006.00292.x.

Xie, Z., & Miller, G. M. (2007). Trace amine-associated receptor 1 is a modulator of the dopamine transporter. *The Journal of Pharmacology and Experimental Therapeutics*, *321*(1), 128–136. http://dx.doi.org/10.1124/jpet.106.117382.

Xie, Z., & Miller, G. M. (2008). Beta-phenylethylamine alters monoamine transporter function via trace amine-associated receptor 1: Implication for modulatory roles of trace amines in brain. *The Journal of Pharmacology and Experimental Therapeutics*, *325*(2), 617–628. http://dx.doi.org/10.1124/jpet.107.134247.

Xie, Z., & Miller, G. M. (2009a). A receptor mechanism for methamphetamine action in dopamine transporter regulation in brain. *The Journal of Pharmacology and Experimental Therapeutics*, *330*(1), 316–325. http://dx.doi.org/10.1124/jpet.109.153775.

Xie, Z., & Miller, G. M. (2009b). Trace amine-associated receptor 1 as a monoaminergic modulator in brain. *Biochemical Pharmacology*, *78*(9), 1095–1104. http://dx.doi.org/10.1016/j.bcp.2009.05.031.

Xie, Z., Westmoreland, S. V., Bahn, M. E., Chen, G. L., Yang, H., Vallender, E. J., … Miller, G. M. (2007). Rhesus monkey trace amine-associated receptor 1 signaling: Enhancement by monoamine transporters and attenuation by the D2 autoreceptor in vitro. *The Journal of Pharmacology and Experimental Therapeutics*, *321*(1), 116–127. http://dx.doi.org/10.1124/jpet.106.116863.

Xie, Z., Westmoreland, S. V., & Miller, G. M. (2008). Modulation of monoamine transporters by common biogenic amines via trace amine-associated receptor 1 and monoamine autoreceptors in human embryonic kidney 293 cells and brain synaptosomes. *The Journal of Pharmacology and Experimental Therapeutics*, *325*(2), 629–640. http://dx.doi.org/10.1124/jpet.107.135079.

Yan, Y., Nitta, A., Mizuno, T., Nakajima, A., Yamada, K., & Nabeshima, T. (2006). Discriminative-stimulus effects of methamphetamine and morphine in rats are attenuated by cAMP-related compounds. *Behavioural Brain Research*, *173*(1), 39–46. http://dx.doi.org/10.1016/j.bbr.2006.05.029.

Yang, M., Kavi, V., Wang, W., Wu, Z., & Hao, W. (2012). The association of 5-HTR2A-1438A/G, COMTVal158Met, MAOA-LPR, DATVNTR and 5-HTTVNTR gene polymorphisms and antisocial personality disorder in male heroin-dependent Chinese subjects. *Progress in Neuro-Psychopharmacology & Biological Psychiatry, 36*(2), 282–289. http://dx.doi.org/10.1016/j.pnpbp.2011.11.009.

Yates, J. R., Marusich, J. A., Gipson, C. D., Beckmann, J. S., & Bardo, M. T. (2012). High impulsivity in rats predicts amphetamine conditioned place preference. *Pharmacology, Biochemistry, and Behavior, 100*(3), 370–376. http://dx.doi.org/10.1016/j.pbb.2011.07.012.

Yero, T., & Rey, J. A. (2008). Tetrabenazine (xenazine), an FDA-approved treatment option for Huntington's disease-related chorea. *P & T, 33*(12), 690–694.

Yu, Q., Montes, S., Larson, D. F., & Watson, R. R. (2002). Effects of chronic methamphetamine exposure on heart function in uninfected and retrovirus-infected mice. *Life Sciences, 71*(8), 953–965.

Zhong, P., Liu, Y., Hu, Y., Wang, T., & Zhao, Y. P. (2015). BDNF interacts with endocannabinoids to regulate cocaine-induced synaptic plasticity in mouse midbrain dopamine neurons. *The Journal of Neuroscience, 35*(10), 4469–4481. http://dx.doi.org/10.1523/jneurosci.2924-14.2015.

Zhu, G., Lipsky, R. H., Xu, K., Ali, S., Hyde, T., Kleinman, J., … Goldman, D. (2004). Differential expression of human COMT alleles in brain and lymphoblasts detected by RT-coupled 5′ nuclease assay. *Psychopharmacology, 177*(1–2), 178–184. http://dx.doi.org/10.1007/s00213-004-1938-z.

CHAPTER FOUR

Screening Medications for the Treatment of Cannabis Use Disorder

L.V. Panlilio*, Z. Justinova*, J.M. Trigo†, B. Le Foll†,‡,§,¶,||,1

*Preclinical Pharmacology Section, Behavioral Neuroscience Research Branch, NIDA, NIH, DHHS, Baltimore, MD, United States
†Translational Addiction Research Laboratory, Campbell Family Mental Health Research Institute, Centre for Addiction and Mental Health (CAMH), Toronto, ON, Canada
‡Addiction Medicine Service, Ambulatory Care and Structured Treatments, Centre for Addiction and Mental Health, Toronto, ON, Canada
§Campbell Family Mental Health Research Institute, Centre for Addiction and Mental Health, Toronto, ON, Canada
¶University of Toronto, Toronto, ON, Canada
||Institute of Medical Sciences, University of Toronto, Toronto, ON, Canada
[1]Corresponding author: e-mail address: bernard.lefoll@camh.ca

Contents

Abstract

Cannabis use has been increasingly accepted legally and in public opinion. However, cannabis has the potential to produce adverse physical and mental health effects, and cannabis use disorder (CUD) occurs in a substantial percentage of both occasional and daily cannabis users. Many people have difficulty discontinuing use despite receiving treatment. Therefore, it would be beneficial to develop safe and effective medications for treating CUD. To achieve this, methods have been developed for screening and evaluating potential medications using animal models and controlled experimental protocols in human volunteers. In this chapter, we describe: (1) animal models available for assessing the effect of potential medications on specific aspects of CUD, (2) the main findings obtained so far with these animal models, (3) the approaches used to assess potential medications in humans in laboratory experiments and clinical trials, and (4) the effectiveness of several potential pharmacotherapies on particular aspects of CUD modeled in these human studies.

International Review of Neurobiology, Volume 126
ISSN 0074-7742
http://dx.doi.org/10.1016/bs.irn.2016.02.005

ABBREVIATIONS

5-HT$_1$ and 5-HT$_{1B}$ serotonin receptor type 1A and 1B
A$_{2A}$ adenosine receptor type A2A
CB$_1$ cannabinoid receptor type 1
CUD Cannabis use disorder
DSM-5 *Diagnostic and Statistical Manual of Mental Disorders*, Fifth Edition (2013)
FAAH fatty acid amide hydrolase
GABA gamma-aminobutyric acid
NAC *N*-acetylcysteine
PF-04457845 a FAAH inhibitor
THC Δ^9-tetrahydrocannabinol
URB597 a FAAH inhibitor
WIN55-212 a synthetic CB$_1$ agonist
α7 nicotinic acetylcholine type alpha 1

1. INTRODUCTION

Cannabis has long been used for medicinal and recreational purposes, and both of these uses are becoming increasingly accepted legally and in public opinion. However, these societal trends should not obscure the fact that—like other medicinal/recreational drugs—cannabis has the potential to produce adverse effects on physical and mental health. It should also be recognized that currently available strains of high-potency cannabis and synthetic cannabinoid "spice" drugs can have adverse effects that are substantially worse than those associated with traditional strains of marijuana.

Cannabis-derived materials such as marijuana and hashish contain a large number of chemical constituents, but the abuse-related euphorigenic effects result from the actions of Δ^9-tetrahydrocannabinol (THC) on CB$_1$ receptors in the brain (Le Foll & Goldberg, 2005; Williams & Kirkham, 2002). Like other drugs of abuse including opioids and psychostimulants, THC and synthetic cannabinoids increase dopamine signaling in neurons projecting from the ventral tegmental area to the nucleus accumbens and prefrontal cortex (Oleson & Cheer, 2012). These effects on dopamine are believed to underlie the rewarding/reinforcing effects that induce users to repeatedly seek out and take the drug.

Cannabis, THC, and synthetic cannabinoids can also produce adverse emotional effects (anxiety, panic), cognitive impairment (amnesia, difficulty concentrating), schizophrenia-like effects (paranoia, delusions), and cardiovascular effects (hypertension, tachycardia) (Hall, 2015; Karila et al., 2014; Panlilio, Goldberg, & Justinova, 2015). Cannabis smoking also has adverse effects on the respiratory system (Gates, Jaffe, & Copeland, 2014). Synthetic cannabinoids are typically more potent than THC (Gates et al., 2014) and have stronger effects at the CB_1 receptor because they are full agonists, unlike THC, which is only a partial agonist (Fantegrossi, Moran, Radominska-Pandya, & Prather, 2014). Furthermore, the ingredients and doses in spice drug preparations are highly variable, and these preparations have been associated with large numbers of emergency room visits.

In addition to these direct adverse effects, cannabis can have indirect adverse effects associated with its illegality and the fact that many employers do not tolerate its use. These many direct and indirect adverse effects exert a limiting effect on cannabis use, and in many cases may be sufficient to lead to abstinence. As with other drugs of abuse (Heyman, 2009), it is likely that most users either control or discontinue their cannabis use without the aid of psychosocial or pharmacologic intervention. However, many individuals have difficulty controlling their use of cannabis. Cannabis use disorder (CUD) is recognized in the *Diagnostic and Statistical Manual of Mental Disorders*, Fifth Edition (2013) (DSM-5) (American Psychiatric Association, 2013) and diagnosed in terms that mirror those of other substance use disorders, including inability to curtail use despite a desire to do so, continued use despite the consequent loss of important social, occupational, and recreational activities, and showing symptoms of drug craving, tolerance, and withdrawal. As of 2010, over 760,000 people per year were being treated specifically for cannabis use in the United States (Batts et al., 2014). On average, individuals who seek treatment do so after more than 10 years of daily use and more than six serious attempts to quit (Budney, Roffman, Stephens, & Walker, 2007).

There are currently no medications approved specifically for treating CUD. The purpose of this chapter is to review basic methods that are used to evaluate compounds that might be developed into such medications. Animal models are useful for understanding the basic neurobiology that underlies the effect of cannabinoids and for screening novel drugs for potential therapeutic effects. Controlled studies in human volunteers are critical for determining whether a medication can be effective. So far, most of this research in humans has involved testing medications that are already approved for other purposes, including pharmaceutical formulations of THC, which might function as a replacement therapy analogous to the use of methadone as a treatment for heroin addiction. The goals of this human and animal research are to provide treatments that can help decrease or stop cannabis use, reduce withdrawal symptoms, and prevent craving and relapse.

2. ANIMAL MODELS

Most animal research relevant to medications development for the treatment of CUD has been conducted in rats, mice, and squirrel monkeys. In general, the basic mechanisms and functions of the central nervous system in rodents and nonhuman primates are similar enough to humans to provide useful information concerning the effects of drugs on human behavior. Consequently, animal models have been indispensable to the fields of psychopharmacology and behavioral neuroscience. However, there are obviously differences in behavior across species, and the differences between rodents and primates happen to be more striking with respect to the rewarding effects of cannabinoids than they are with other drugs of abuse. That is, rats will self-administer most of the drugs that are abused by humans, but they do not show robust self-administration of THC. The reasons for this difference are not clear, but rodents still play an important role in cannabinoid research because they do self-administer the synthetic cannabinoid WIN 55,212-2 (Fattore, Cossu, Martellotta, & Fratta, 2001; Lefever, Marusich, Antonazzo, & Wiley, 2014; Mendizabal, Zimmer, & Maldonado, 2006), and they also display other abuse-relevant effects of THC. In contrast with rats, squirrel monkeys readily self-administer THC and other cannabinoids (Justinova, Solinas, Tanda, Redhi, & Goldberg, 2005; Justinova, Tanda, Redhi, & Goldberg, 2003; Justinova, Yasar, Redhi, & Goldberg, 2011; Tanda, Munzar, & Goldberg, 2000), and we believe this behavior represents the best available animal model of human cannabis use for testing potential medications (Panlilio, Justinova, & Goldberg, 2010).

Drug discrimination procedures are used to model the subjective, perceived effects of drugs (Solinas, Panlilio, Justinova, Yasar, & Goldberg, 2006). In a typical procedure, animals (usually rats) are trained to detect the effects of a drug. They are given an injection before each daily training session, during which they can obtain food pellets by pressing one of two levers. Some days the injection contains THC, and some days it contains only vehicle (placebo), and the type of injection signals which of the two levers produces food on that day. For example, a rat might always receive food for pressing the left lever on THC days and the right lever on vehicle days. Once this discrimination has been established, tests can be performed with novel compounds to determine whether they block, enhance, or mimic the effects of THC. Although it cannot be determined exactly what aspects of the drug effect the rat is responsive to, this type of procedure can provide reliable information about the pharmacological specificity of a drug for receptor classes and subtypes, and it can indicate whether a test compound can block or mimic the effects of THC. Training for this procedure can take weeks or months, but well-trained rats can be tested repeatedly, which is ideal for screening novel compounds.

Place-conditioning procedures are typically used as a relatively simple means of studying the rewarding effects of drugs in rodents. Training takes place in an apparatus with two distinctive compartments. During training, one of the compartments is paired with the effects of THC or another cannabinoid agonist by injecting the rat before confining it to the compartment. The other compartment is paired with vehicle in a similar manner. During a test, the rat is given access to both compartments, and the relative amount of time spent in each compartment is measured. If the drug has a rewarding effect, this typically imparts a conditioned rewarding effect to the drug-paired compartment through Pavlovian conditioning, and the rat spends more time there (ie, exhibits a conditioned place preference). In contrast, if the drug has aversive effects, the rat avoids the drug-paired compartment and spends more time in the vehicle-paired compartment (ie, exhibits a conditioned place aversion). In most studies with cannabinoid agonists, rats have shown either conditioned place avoidance or no clear preference for either compartment (Cheer, Kendall, & Marsden, 2000; Klein et al., 2011; Polissidis et al., 2009; Robinson, Hinder, Pertwee, & Riedel, 2003; Vlachou, Nomikos, Stephens, & Panagis, 2007). This suggests that cannabinoid reward is weak in rats, or that it is masked by aversive effects of THC (eg, anxiety).

Drug self-administration procedures provide the most direct evidence of rewarding effects of a drug (Panlilio & Goldberg, 2007). Typically, a rat or monkey is given an intravenous catheter and allowed to press a lever that activates a syringe pump that delivers a drug through the catheter. Squirrel monkeys self-administer THC at rates similar to the rates at which they self-administer cocaine, methamphetamine, and nicotine (Justinova, Ferre, et al., 2011; Justinova et al., 2015; Panlilio et al., 2015; Schindler et al., 2010). In many ways, the basic drug self-administration procedure is a close analog of human drug use, modeling the contingencies that critically influence drug-seeking behavior. For example, in both the drug abuse environment and the animal model, drug seeking can be induced by environmental cues that signal the availability of a drug, and it can be maintained over time by presenting cues that have been associated with the effects of the drug. Thus, the basic drug self-administration procedure can be modified to study treatments that might have therapeutic treatment on specific phases of addiction. In the study of potential treatments for CUD, four such procedures have been used with THC self-administration in squirrel monkeys: maintenance tests, drug-induced reinstatement, cue-induced reinstatement, and second-order schedules.

Maintenance tests involve establishing a stable baseline level of THC self-administration, then testing the effects of a treatment drug on this baseline. Except for the treatment, all other aspects of the baseline and test sessions are held constant, including response requirements, drug delivery, and cue presentations. Typically, baseline behavior is recorded for at least a week, then the treatment is given for five consecutive sessions, followed by a return to baseline. This procedure is designed to assess the potential of a treatment for decreasing ongoing cannabis use, and to determine whether the effects are consistent over time (ie, whether an effect is immediately apparent or builds over time, and whether the effect continues to occur with extended treatment).

Drug-induced reinstatement and *cue-induced reinstatement* procedures are used to model the effects of a treatment on relapse to drug use after a period of abstinence (Bossert, Marchant, Calu, & Shaham, 2013). In humans, relapse can be triggered by reexposure to the drug, to cues that signal the availability of the drug, or to cues that have been paired with effects of the drug. In both the drug-induced and cue-induced reinstatement procedures used with squirrel monkeys, extinction is used to impose abstinence; that is, drug delivery is discontinued. However, the two procedures differ with respect to the effect of lever responding during the extinction phase. When imposing abstinence prior to a drug-induced reinstatement test, vehicle is simply substituted for THC (ie, the usual cue signals the availability of an injection, and responding produces the usual visual cues as well as any interoceptive cues from intravenous injection). In contrast, when imposing abstinence prior to a *cue*-induced reinstatement test, the visual cue previously associated with drug availability is not presented, and responding has no programmed effect (ie, it does not produce cues or injections). Under both of these extinction procedures, responding decreases to a very low level after a few sessions, at which point a reinstatement test can be conducted. For drug-induced reinstatement, the only procedural difference between the extinction phase and the test session is that the monkey is given an automatic intravenous injection of THC just before the test session. This reexposure to THC typically has a priming effect, causing lever responding to increase substantially. For cue-induced reinstatement, no THC injection is given before the test, but the normal cues are presented and responding produces vehicle injections during the session. Like drug priming, reexposure to the cues that were previously associated with THC typically causes a relapse-like increase in the drug-seeking response. To screen potential medications to treat CUD, the test drug is given before the drug-induced or cue-induced reinstatement session to see if it will prevent the effects of drug or cue reexposure. Potential medications are also tested alone to determine whether they might have effects of their own that are liable to induce relapse.

Second-order schedules of drug self-administration are used to model the long sequences of behavior that are involved in obtaining, preparing, and ingesting drugs (Schindler, Katz, & Goldberg, 1988). In the second-order schedule that has been used with THC self-administration (Justinova et al., 2008), squirrel monkeys press a lever that produces brief (2 s) presentations of a cue light for every 10th response. After 30 min has elapsed since the beginning of the session, the next cue light that is presented is extended to 90 s, during which 10 injections of THC are delivered intravenously. Thus, all THC is delivered at the end of the session, and responding during the session is not directly influenced by the pharmacological effects of THC, and therefore responding during the session can be described specifically as THC seeking. This drug seeking is maintained at high rates because it produces the brief stimulus, and the stimulus has this effect because it has been associated with the drug. The effects of this end-of-session THC delivery are clear because: (1) when the drug is not delivered, responding decreases substantially in the next session and (2) when THC delivery is resumed, responding resumes in the next session (see Fig. 1A). The powerful effects of the brief stimulus are clear from the fact that response rates are about six times higher under baseline conditions compared to when the brief stimulus is not presented (see Fig. 1B). By testing the effects of potential treatments on drug seeking in the second-order schedule, with the monkey receiving no THC until to the end of the session, it can be determined whether the treatment specifically affects drug seeking (ie, the motivation to receive the drug), as opposed to altering the effects of THC after it is received, such as in a maintenance test.

Withdrawal symptoms can occur when heavy, prolonged cannabis use is discontinued (Lee et al., 2014). DSM-5 diagnostic criteria for cannabis withdrawal include symptoms such as irritability, anxiety, insomnia, decreased appetite, and depressed mood. Avoidance of such symptoms might be one factor that contributes to the persistence of chronic cannabis use, and

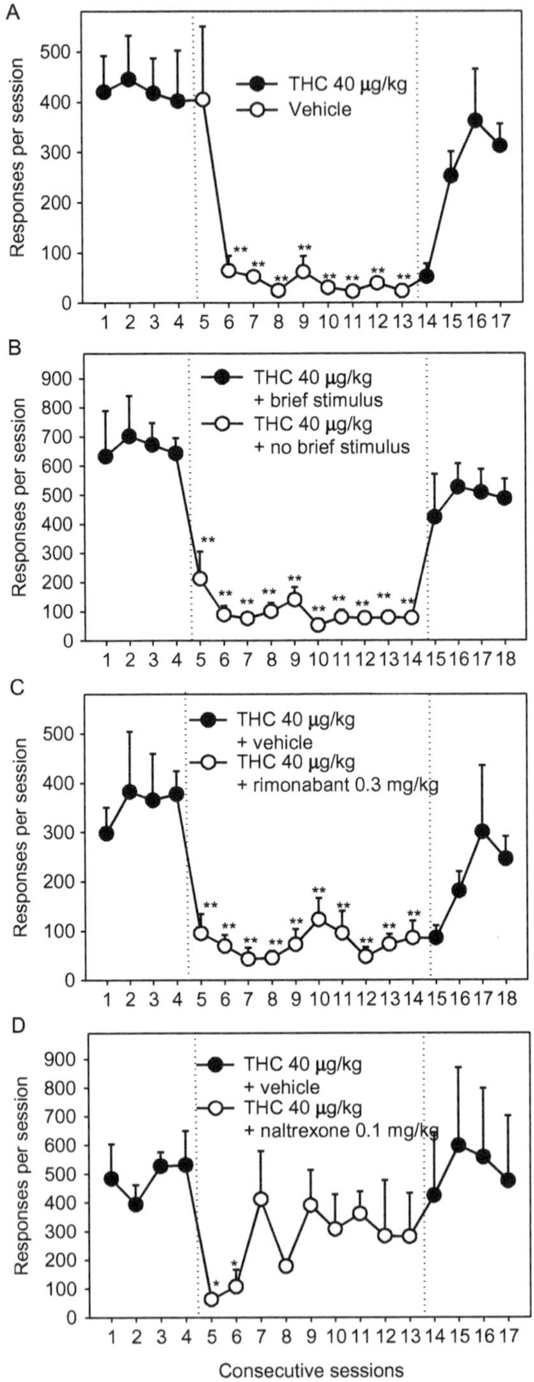

Fig. 1 See legend on opposite page.

medications that relieve such symptoms might be helpful for people attempting to quit. In animals, cannabis withdrawal is usually studied in rodents. However, simply discontinuing regular THC administration in animals does not produce prominent, easily observable symptoms. Therefore, withdrawal is usually precipitated by administering a cannabinoid antagonist, which rapidly blocks CB_1 receptors even if there is still THC in the system (Lichtman, Fisher, & Martin, 2001; Tai et al., 2015). Precipitated withdrawal in rodents produces symptoms such as scratching, face rubbing, licking, and wet-dog shakes.

Fig. 1 Effects of pharmacological treatments and contingency-based environmental manipulations on THC seeking in squirrel monkeys. A second-order schedule was used to model the environmental cues that maintain cannabis-seeking behavior in humans. Under baseline conditions (the first 4 days in each panel), lever responding produced a brief stimulus presentation (2 s of amber cue light) for every 30th response. The first stimulus presentation that occurred at least 30 min after the start of the session was accompanied by THC (40 mcg/kg, IP), after which the session ended. (A) When vehicle was substituted for THC (sessions 5–13), the rate of the THC-seeking response decreased to a low level, and when THC delivery was reinstituted (sessions 14–17), responding resumed at a high rate. Note that the effects of these manipulations were evident in the second session of each condition, because injection occurred only at the very end of the session. (B) When presentation of the brief stimulus was discontinued (sessions 5–14), there was an immediate and sustained decrease in the THC-seeking response, even though THC was still delivered at the end of the session. This demonstrates the important role of environmental cues in maintaining THC-seeking behavior when long sequences of responding are required to obtain the drug. (C) When the CB_1-receptor antagonist/inverse-agonist rimonabant (0.3 mg/kg, IM) was given prior to the daily session (with cues presented and THC delivered as under the baseline conditions), there was an immediate and sustained decrease in THC seeking (sessions 5–14). The immediacy of this effect suggests that rimonabant decreased the effectiveness of the brief stimulus in addition to blocking the rewarding effects of THC. Drug seeking gradually resumed when rimonabant treatment was discontinued (sessions 15–18). (D) When the μ-opioid antagonist naltrexone (0.1 mg/kg, IM) was given before each daily session under maintenance conditions, THC seeking was immediately decreased, but this effect was diminished after the second day. This suggests that manipulation of opioid receptors can affect THC seeking, but that the effects might be less robust than those of direct manipulation of CB_1 receptors. *Asterisks* indicate statistically significant differences between treatment and baseline (*$p < 0.05$, ** $p < 0.01$). *Figure adapted from Justinova, Z., Munzar, P., Panlilio, L. V., Yasar, S., Redhi, G. H., Tanda, G., & Goldberg, S. R. (2008). Blockade of THC-seeking behavior and relapse in monkeys by the cannabinoid CB(1)-receptor antagonist rimonabant.* Neuropsychopharmacology, *33(12), 2870–2877. doi:10.1038/npp.2008.21 with permission.*

3. FINDINGS FROM RESEARCH WITH ANIMAL MODELS

The fact that there are large numbers of cannabis users who have difficulty controlling their use and who seek treatment is not well recognized by the public. Even among researchers, the possibility of developing medications to treat CUD has received little attention compared to medications for psychostimulant, opioid, nicotine, or alcohol use disorders. This lack of attention stems at least in part from the fact that the adverse effects of these other drugs tend to be more devastating than those of cannabis, to both the user and to society. But, it also stems from the fact that cannabinoid self-administration procedures are relatively difficult to implement and have only been established in a few laboratories. Nonetheless, the studies conducted so far have identified some pharmacologic strategies that might eventually lead to approved medications.

The most straightforward approach to preventing the rewarding effects of cannabis would be to administer a CB_1 receptor antagonist. This would be analogous to opioid antagonist therapy as a treatment for heroin addiction. The CB_1 antagonist rimonabant blocks THC self-administration in squirrel monkeys in maintenance tests and second-order schedules (see Fig. 1C; Justinova et al., 2008; Tanda et al., 2000), and it also blocks the rewarding effects of WIN 55,212-2 in rodents (Fattore et al., 2001; Lefever et al., 2014; Martellotta, Cossu, Fattore, Gessa, & Fratta, 1998). CB_1 antagonist treatment also blocks the effects of THC in drug discrimination procedures with rats and monkeys (Wiley, Lowe, Balster, & Martin, 1995). Another potential advantage of CB_1 antagonists is that they decrease both drug-induced and cue-induced reinstatement in squirrel monkeys (Justinova et al., 2008), which could be a highly valuable property given that propensity to relapse is a substantial impediment to most therapeutic interventions. Cannabinoid antagonists also decrease self-administration and cue-induced reinstatement of drug seeking for some noncannabinoid drugs (De Vries & Schoffelmeer, 2005; Maccioni, Colombo, & Carai, 2010; Schindler et al., 2010) and might therefore be beneficial for the treatment of polydrug abuse. One drawback to this treatment strategy is that CB_1 antagonists could precipitate withdrawal if taken while a person is cannabinoid dependent. However, it should be noted that rimonabant and similar drugs have not only antagonist effects (blocking the effects of THC at the CB_1 receptor) but also inverse-agonist effects (producing effects opposite to those of THC). These inverse-agonist properties might be responsible for the depression-like side effects that have been associated with rimonabant (Le Foll, Gorelick, & Goldberg, 2009). For this reason, neutral antagonists of CB_1 are being developed that can presumably block the effects of THC effectively and precipitate withdrawal (Tai et al., 2015), but might not produce depression-like effects in humans (Bergman et al., 2008).

Another general approach to treating CUD involves manipulating noncannabinoid transmitter systems that interact with the endocannabinoid system. This strategy has been applied with squirrel monkey THC self-administration procedures to investigate drugs that affect opioid, adenosine, and acetylcholine receptors. It is well established that the cannabinoid and opioid systems of the brain interact, and there is evidence from animal models suggesting that μ-opioid receptor antagonists might decrease the rewarding effects of cannabis. In squirrel monkeys, the opioid antagonist naltrexone decreased THC self-administration in maintenance tests with a simple fixed-ratio schedule, in which THC was delivered for every 30th response (Justinova, Tanda, Munzar, & Goldberg, 2004), and under the second-order schedule described in the section earlier (Justinova et al., 2008). However, even though this effect of naltrexone was consistent when tested for 5 days under the fixed-ratio schedule, the effect only lasted for about 2 days under the second-order schedule, in which high rates of drug seeking are maintained by both THC delivery and the presentation of drug-related environmental cues (see Fig. 1D). These findings suggest that an opioid antagonist would probably not be as effective as a cannabinoid antagonist at decreasing cannabinoid use, perhaps because they decrease the rewarding effects of THC but do not alter the effects of THC-associated cues.

The endogenous cannabinoid system is also influenced by adenosine A_{2A} receptors, which are known to form heteromeric complexes with CB_1 receptors. Two studies were conducted looking at the effects of A_{2A} antagonists on THC self-administration in squirrel monkeys (Justinova, Ferre, et al., 2011; Justinova, Redhi, Goldberg, & Ferre, 2014). These studies showed that selective antagonism of presynaptic A_{2A} receptors can block the rewarding effects of THC, while selective antagonism of postsynaptic receptors (or nonselective antagonism) has the opposite effect, increasing the rewarding effects of THC. These findings suggest that selective presynaptic A_{2A} receptor antagonists should be further developed as candidate medications to treat CUD.

Like A_{2A} antagonists, α-7 nicotinic acetylcholine receptor antagonists such as methyllycaconitine can decrease corticostriatal glutamate signaling, which might be an important component of cannabinoid reward circuitry (Solinas et al., 2007). Methyllycaconitine blocked the interoceptive effects of THC in a drug discrimination procedure and blocked THC's ability to increase dopamine levels in the nucleus accumbens shell of rats (Solinas et al., 2007). Since direct orthosteric α7 antagonists such as methyllycaconitine tend to have adverse side effects; Justinova and colleagues (2013) studied the effects of Ro 61-8048, a negative allosteric modulator of α7 receptors, by increasing endogenous levels of kynurenic acid. Negative allosteric modulators differ from orthosteric antagonists in that they do not directly affect the receptor, but alter the effects of receptor ligands when they bind to the receptor. Thus, kynurenic acid decreases the effects of acetylcholine on the neuron. Ro 61-8048 showed promising effects in several models of cannabinoid reward, including maintenance tests with THC self-administration in squirrel monkeys (Fig. 2A), cue-induced reinstatement of WIN 55,212-2 seeking in rats, drug-induced and cue-induced reinstatement of THC seeking in squirrel monkeys (Fig. 2B and C), and THC-induced elevation of dopamine levels in the nucleus accumbens of rats. Drug discrimination procedures indicated that Ro 61-8048 blocked the subjective effects of THC in monkeys but not in rats, perhaps consistent with THC discrimination not being indicative of nonreward-related effects in rats. Overall, Ro 61-8048 was highly effective in behavioral, neurochemical, and electrophysiological models of cannabinoid reward and reward-related effects. These findings suggest that enhancing endogenous levels of kynurenic acid might be an effective and safe strategy for treating CUD.

Depression is known to be associated with heavy use of cannabis and other drugs, and cannabis might serve a self-medicating function in depressed users (Degenhardt, Hall, & Lynskey, 2003). Consistent with this hypothesis, THC shows positive effects in some animal models of depression, including olfactory bulbectomy in rats (Rodriguez-Gaztelumendi, Rojo, Pazos, & Diaz, 2009). Amchova et al. (2014) tested the effects of bulbectomy on self-administration of a synthetic cannabinoid CB_1 agonist (WIN55-212) in rats, finding that it caused rats to self-administer about twice as much drug as control rats. Because serotonin 5-HT_{1B} receptors have been implicated in depression and modulation of its activity decreases the self-administration of some drugs, Amchova et al. (2014) also tested the effects of a 5-HT_1 agonist on WIN55-212 self-administration in bulbectomized and control rats. However, the 5-HT_1 agonist did not decrease cannabinoid self-administration in either group of rats.

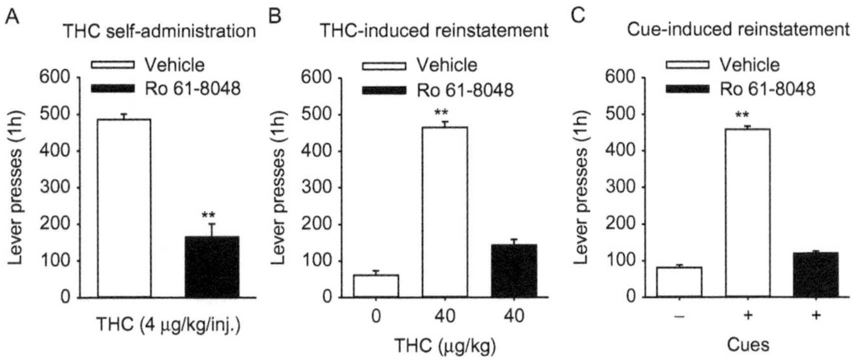

Fig. 2 Effects of treatment with Ro 61-8048 in squirrel monkey models of relapse to cannabis seeking after a period of abstinence. Ro 61-8048 increases endogenous levels of kynurenic acid, a negative allosteric modulator of the α-7 nicotinic receptor. Monkeys had extensive experience self-administering THC under a schedule in which a green light indicated that THC (4 mcg/kg, IV) could be obtained by pressing a lever 10 times (fixed-ratio 10). Under these training conditions (maintenance), sessions were 1 h long, and each THC delivery was accompanied by a brief stimulus (amber light), followed by a 60-s timeout period (lights off) before the green light signaled that another injection could be obtained. (A) Treatment with Ro 61-8048 (20 mg/kg, IM) decreased THC self-administration under these maintenance conditions, with all cues and THC delivery maintained. (B) To model relapse induced by reexposure to THC, abstinence was first imposed by simply substituting vehicle for THC (but continuing to present cues) for several sessions prior to and including the test session. Responding decreased to low levels, but increased substantially when THC (40 mcg/kg, IV) was given automatically prior to the test session. Ro 61-8048 blocked this THC-induced reinstatement effect. (C) To model relapse induced by reexposure to cannabis-associated cues, abstinence was first induced by discontinuing all intravenous injections and all visual cue presentations. Then, to test for cue-induced reinstatement, the green light was presented, and every 10th response produced the amber light and an intravenous vehicle injection (to provide interoceptive injection-related cues), but no THC was given. These cues reinstated THC seeking, but their effect was blocked when monkeys were pretreated with with Ro 61-8048. Thus, Ro 61-8048 blocked the direct reinforcing effects of THC, and it also blocked reinstatement by reexposure to THC or THC-related cues. These findings suggest that enhancing kynurenic acid levels might be a useful strategy for decreasing cannabis smoking and also for preventing relapse when users who have achieved abstinence are reexposed to the drug or to environmental cues associated with the drug. Double *asterisks* indicate statistically significant differences between treatment and baseline ($p < 0.01$). *Figure adapted from Justinova, Z., Mascia, P., Wu, H. Q., Secci, M. E., Redhi, G. H., Panlilio, L. V., … Goldberg, S. R. (2013). Reducing cannabinoid abuse and preventing relapse by enhancing endogenous brain levels of kynurenic acid.* Nature Neuroscience, 16*(11), 1652–1661. doi:10.1038/nn.3540 with permission.*

Administration of THC or synthetic CB_1 agonists in rodents increases the synthesis of pregnenolone, the precursor of all steroid hormones, including neurosteroids synthesized directly in the brain (Vallee et al., 2014). Using the tetrad test (THC-induced hypolocomotion, hypothermia, catalepsy, and analgesia) and neurochemical and electrophysiological measures of dopamine activity in the nucleus accumbens, Vallee et al. (2014) found that pregnenolone decreased the effects of THC through negative feedback, acting as an allosteric modulator of the CB_1 receptor (ie, altering the effects of THC when it binds to the receptor). When pregnenolone was given to mice trained to self-administer the CB_1 agonist WIN55-212, it decreased cannabinoid intake, and it decreased the number of responses mice performed to obtain the cannabinoid when the requirement was progressively increased during the session. These results suggest that pregnenolone decreases the motivation to receive cannabinoid reward. Allosteric modulators also have two advantages over orthosteric antagonists such as rimonabant: they should not inhibit all CB_1 receptor activities (and might therefore be less prone to side effects), and their effects should not be overridden even if cannabis intake is increased. Thus, treatment with pregnenolone or other allosteric modulators of CB_1 receptor represent another new strategy that should be further developed and assessed for treating CUD.

Noting that some of the symptoms of cannabis withdrawal resemble bipolar mood disorder, which is treated with the mood stabilizer lithium, Cui et al. (2001) studied the effects of lithium on precipitated withdrawal from a synthetic cannabinoid agonist in rats. They found that lithium prevented withdrawal symptoms, that this effect was blocked when lithium was combined with an oxytocin antagonist, and that administration of oxytocin could also prevent precipitated withdrawal. However, when lithium was subsequently tested during withdrawal in treatment-seeking cannabis users, it prevented certain symptoms (loss of appetite, stomach aches, and nightmares/strange dreams), but it did not reduce overall withdrawal scores (Johnston et al., 2014).

Another potential strategy for treating cannabinoid withdrawal is to increase levels of endogenous cannabinoid ligands, such as anandamide and 2-arachidonoylglycerol. For example, levels of anandamide can be increased by administering an inhibitor of fatty acid amide hydrolase (FAAH), the main enzyme responsible for the inactivation of anandamide. The FAAH inhibitor URB597 increases anandamide levels in the brain, which could potentially prevent withdrawal symptoms when cannabis use is discontinued. Importantly, URB597 does not produce adverse THC-like effects in animal models, suggesting that it would not be liable to be abused, to reinstate cannabis use, or have cannabis-like side effects. Stewart and McMahon (2011) tested URB597 in a withdrawal-related drug discrimination procedure. Rhesus monkeys were chronically treated with the CB_1 agonist WIN55-212 and trained to detect the effects of the CB_1 antagonist rimonabant, essentially combining drug discrimination and precipitated withdrawal procedures. They found that URB597 did not block the interoceptive effects of rimonabant in this model, which suggests that it would not block all the subjective effects of cannabis withdrawal in humans. However, the strategy of relieving withdrawal by increasing endogenous cannabinoid levels should be further investigated, because other FAAH inhibitors and monoacylglycerol lipase inhibitors (which increase levels of 2-arachidonoylglycerol) can have different effects than URB597, and because the results might also be different with precipitated withdrawal vs simple discontinuation of cannabinoid administration. Other experiments with this discrimination procedure showed that treatment with the CB_1 agonist WIN55-212 or the alpha(2)-adrenergic agonist clonidine partially blocked withdrawal symptoms (Stewart & McMahon, 2010; see also Lichtman et al., 2001).

4. HUMAN LABORATORY APPROACHES

Recent systematic reviews indicate a growing interest in exploring different therapeutic options for the treatment of CUD (Laprevote, Schwan, Schwitzer, Rolland, & Thome, 2015; Marshall, Gowing, Ali, & Le Foll, 2014). In this section, we will describe two different approaches that have been used to study a variety of potentially therapeutic options for the treatment of CUD: human laboratory studies and randomized clinical trials. Results obtained using different pharmacotherapies on human studies will be described in more detail in Section 5.

Human laboratory studies have advanced drug abuse research in a variety of ways. They are an invaluable tool for the study of potential pharmacotherapies for drug dependence, and they also provide a means of investigating the mechanisms underlying drugs' addictive effects and the ways that experimental treatments alter these effects. Human laboratory studies can be used to test the effects of different doses of the same medication, to compare the effectiveness of different medications, and even to assess preference for medication vs placebo (eg, self-administration studies, see later), often using a single sample of participants (eg, within-subject designs), and short-term treatments (eg, 5 or 7 days). Laboratory studies using within-subject designs usually control for possible carryover effects of the different doses or medications by administering the different treatments in a counter-balanced order and by including intertreatment clearance phases in which participants can often consume drugs as usual (ie, smoke cannabis). Due to the intrinsic nature of these studies, only those participants not seeking treatment are included.

The number of human laboratory studies for CUD has increased over the past decade. Several studies have investigated the effects of pharmacological interventions involving diverse neurotransmitter systems, including gamma-aminobutyric acid (GABA) (agonist: baclofen), norepinephrine/serotonin (eg, antidepressants: mirtazapine, nefazodone), norepinephrine/dopamine (eg, bupropion), α2A-adrenergic receptors (eg, agonist: lofexidine), CB_1 agonists (eg, THC, dronabinol, nabilone), and histamine (eg, antagonist: quetiapine). However, so far, the majority of human laboratory studies have focused on studying the effects of CB_1 agonists on cannabis withdrawal symptoms (Budney, Vandrey, Hughes, Moore, & Bahrenburg, 2007; Haney et al., 2008, 2004, 2013). Subsequent studies in laboratory settings exploring the effects of Sativex (a ~1:1 combination of THC and cannabidiol) on withdrawal are expected to provide further input in this respect (Trigo et al., 2016). Several human laboratory studies by Haney et al. have also explored the ability of a variety of compounds including THC, baclofen, mirtazapine, and quetiapine (Cooper et al., 2013; Haney et al., 2013, 2010) to prevent marijuana-ingestion relapse. Therefore, laboratory studies can be used to study the effects of pharmacotherapy, in participants who are not seeking treatment, on different aspects of the addictive process including drug taking, withdrawal, and relapse.

Drug self-administration studies are a subtype of human laboratory studies that allows participants who are not seeking treatment to consume drugs under controlled conditions. Self-administration procedures have been used to study preferences for different concentrations of addictive substances (eg, nicotine or THC), or whether certain treatments are able to modify drug self-administration (Haney, Ramesh, et al., 2015; Hart, Haney, Ward, Fischman, & Foltin, 2002). Drug self-administration studies have also been useful for evaluating motivation for different types of reinforcers, such as assessing whether the decision to take a drug can be affected by monetary contingencies (Haney, Comer, Ward, Foltin, & Fischman, 1997; Ward, Comer, Haney, Foltin, & Fischman, 1997), or for determining whether a treatment affects how much participants are willing to pay (ie, using study earnings) to self-administer marijuana (Haney et al., 2008). Laboratory conditions on drug self-administration studies have allowed assessment of marijuana dose–response functions for both subjective ratings (ie, "high" or "good effect") and relevant functioning parameters such as psychomotor activity, attention, and cardiovascular effects. (Ramesh, Haney, & Cooper, 2013). This type of experimental design can also be used to compare the effectiveness of fixed vs self-administered doses of a particular pharmacotherapy on CUD or to assess preferences in medication dosage (eg, whether, when given the opportunity, participants choose lower doses, or a high fixed dose) (Trigo et al., 2016).

Cannabis *withdrawal* syndrome has been included, after years of debate, in DSM-5. The relevance of including cannabis withdrawal syndrome as a mental disorder is underlined by the fact that relapse to cannabis use is associated with subsequent withdrawal symptoms (Allsop et al., 2012) and that 65% of treatment seekers report using marijuana to alleviate withdrawal symptoms (Budney, Novy, & Hughes, 1999; Vandrey, Budney, Kamon, & Stanger, 2005). Before this inclusion, several laboratory studies modeled cannabis withdrawal under controlled conditions (Haney, Ward, Comer, Foltin, & Fischman, 1999a, 1999b; Haney et al., 2004; Hart, Ward, et al., 2002; Jones, Benowitz, & Bachman, 1976; Jones, Benowitz, & Herning, 1981). These studies showed the existence of "negative" effects (eg, anxiety, sleep difficulty, and feeling "irritable," "depressed," or "miserable") during abstinence (eg, 4–5 days with no marijuana or only placebo preparations). These effects were evident even following low daily doses of THC (Haney et al., 1999a; Hart, Haney, et al., 2002) or short time administration of THC (eg, 3–7 days). In a typical laboratory study modeling cannabis withdrawal, marijuana smokers are alternatively exposed to active marijuana (oral or smoked preparations) followed by placebo. The effects of both marijuana and its absence are then evaluated using subjective/objective measures (eg, feeling "high," "irritable," food intake, psychomotor performance, etc.).

Randomized clinical trials can provide the strongest evidence for the efficacy of preventive and therapeutic procedures and are an essential step toward approval for clinical use. Clinical trials for CUD typically test the effectiveness of a particular medication, in combination with counseling (see Section 5), in reducing craving/withdrawal and cannabis use as compared with a placebo control group of treatment seekers.

Recent systematic reviews have shown an increasing incidence of double-blind randomized placebo-controlled trials evaluating treatments for CUD (vs the total number of relevant studies in the literature, ~1:3 ratio) (Laprevote et al., 2015; Marshall et al., 2014). Clinical trials have evaluated the effectiveness of both psychotherapy and pharmacotherapy (alone or in combination) for CUD. Therefore, several clinical trials have evaluated the effects of brief/extended cognitive-behavioral interventions, motivational enhancement therapy plus cognitive-behavioral therapy and adolescent community reinforcement approach, on CUD with, in general, better outcomes for extended (vs brief) treatments (Copeland, Swift, Roffman, & Stephens, 2001; Dennis et al., 2004; Marijuana Treatment Project Research Group, 2004; Stephens, Roffman, & Curtin, 2000). On the other hand, randomized clinical trials have also been used to test the effects of different pharmacotherapies including antidepressants (eg, escitalopram, nefazodone, and bupropion), anticonvulsants (eg, divalproex sodium and gabapentin), and CB_1 agonists (eg, dronabinol and Sativex), usually in combination with cognitive-behavior therapy or other type of counseling to increase coping skills (Allsop et al., 2014; Carpenter, McDowell, Brooks, Cheng, & Levin, 2009; Levin et al., 2011, 2004; Mason et al., 2012; Weinstein et al., 2014).

5. FINDINGS FROM CLINICAL TRIALS AND LABORATORY STUDIES IN HUMAN VOLUNTEERS

Currently, medications with a wide variety of mechanisms have been assessed for the treatment of CUD in humans. However, most of these have failed to decrease cannabis use, withdrawal symptoms, or laboratory measures of relapse (Balter, Cooper, & Haney, 2014; Marshall et al., 2014). The ineffective medications were antidepressants, anxiolytics, and antipsychotics, including fluoxetine, venlafaxine, escitalopram, atomoxetine, buspirone, bupropion, nefazodone, mirtazapine, and quetiapine (cf. Balter et al., 2014). Here, we will concentrate on medications that showed some effectiveness in human studies or are currently in clinical testing, particularly cannabinoid CB_1 agonists and antagonists and opioid ligands.

The CB_1 agonists that were studied as a replacement therapy for CUD are dronabinol and nabilone. Dronabinol (Marinol) is a synthetic form of THC that has been approved for treatment of chemotherapy-associated nausea and for appetite stimulation in patients with AIDS or cancer. Nabilone is a synthetic, highly bioavailable THC analogue with clinical indications similar to those of dronabinol. Dronabinol was tested in several small within-subject, placebo-controlled human laboratory studies for its effect on withdrawal symptoms. Several studies found a dose-dependent reduction of withdrawal symptoms and reversal of the anorexia and weight loss associated with withdrawal while observing minimal adverse effects (Budney, Vandrey, et al., 2007; Haney et al., 2008, 2004; Vandrey et al., 2013; Fig. 3). However, dronabinol did not reduce cannabis self-administration or relapse (Haney et al., 2008; Hart, Haney, et al., 2002). Levin et al. (2011) conducted the first randomized, double-blind, placebo-controlled clinical trial that evaluated the safety and efficacy of dronabinol in combination with behavioral therapies in treating cannabis dependence. Both placebo and dronabinol groups showed a reduction in marijuana

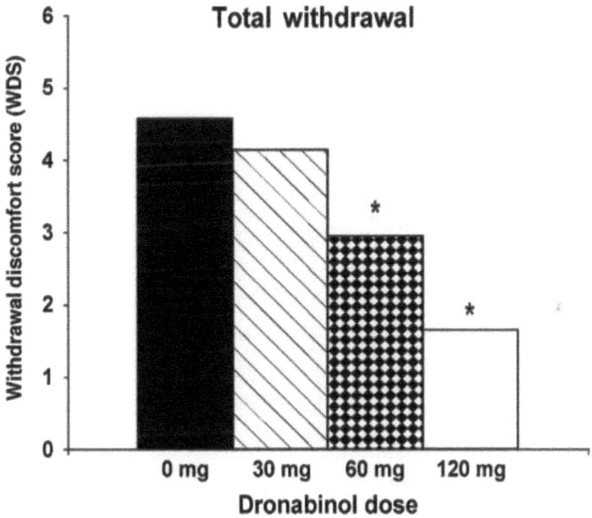

Fig. 3 Effects of treatment with dronabinol on cannabis withdrawal symptoms in daily cannabis users. Significant withdrawal symptoms were observed during the placebo maintenance phase for subjective ratings of decreased appetite, diarrhea, nausea, stomach pain, irritability, sleep difficulty, total sleep time, subjective sleep quality, mood at morning awakening, alertness at morning awakening, restlessness, nervousness/anxiety, chills, increased aggression, increased anger, headaches, difficulty concentrating, and total withdrawal discomfort score (WDS). Significant dose-dependent effects of dronabinol were observed for each withdrawal item except "sleep difficulty" and "nervousness/anxiety." Mean subjective withdrawal ratings (WDS) are presented as a function of dronabinol maintenance dose. *Asterisks* indicate statistically significant differences between dronabinol dose conditions and placebo. These results suggest that replacement therapy with a CB_1 receptor agonist can decrease withdrawal symptoms, analogous to nicotine replacement therapy for tobacco smokers. *Figure from Vandrey, R., Stitzer, M. L., Mintzer, M. Z., Huestis, M. A., Murray, J. A., & Lee, D. (2013). The dose effects of short-term dronabinol (oral THC) maintenance in daily cannabis users.* Drug and Alcohol Dependence, *128(1–2), 64–70. doi:10.1016/j.drugalcdep.2012.08.001 with permission.*

use over time, but there were no differences between the groups in cannabis use outcomes. Although it failed to improve abstinence, agonist substitution pharmacotherapy with dronabinol reduced withdrawal symptoms and improved retention in treatment with few adverse events. Nabilone, which has higher bioavailability, clearer dose–response linearity, and longer duration of action than dronabinol, also showed promise as a potential treatment medication for marijuana dependence (Bedi, Cooper, & Haney, 2013; Haney et al., 2013). Nabilone maintenance produced a robust attenuation of marijuana withdrawal symptoms and a laboratory measure of relapse in a small within-subject study (Haney et al., 2013). Another clinical study is currently evaluating nabilone as a treatment for cannabis dependence and aims to assess the correlation of neuropsychological performance to brain changes using functional MRI brain scans (ClinicalTrials.gov).

The activation of CB_1 receptors can also be achieved by pharmacological inhibition of the enzymes that degrade the endogenous CB_1 agonists anandamide and 2-arachidonoylglycerol. Inhibitors of FAAH or monoacylglycerol lipase enhance and prolong the effects of endocannabinoids at the sites of their release in the brain, and might therefore be effective as a replacement therapy or for attenuation of withdrawal symptoms. The FAAH inhibitor PF-04457845 was well tolerated in previous clinical trials in healthy subjects and patients with osteoarthritis of the knee (Huggins, Smart, Langman, Taylor, & Young, 2012; Li et al., 2012), in which no cannabis-like adverse effects were reported. Currently, PF-04457845 is being tested in a clinical trial as a treatment for marijuana withdrawal (ClinicalTrials.gov).

Cannabidiol is a major nonintoxicating ingredient in marijuana that has been shown to exert multiple pharmacological effects, mediated by multiple mechanisms (Devinsky et al., 2014; Izzo, Borrelli, Capasso, Di Marzo, & Mechoulam, 2009). For example, cannabidiol has a low affinity for CB_1 and CB_2 receptors, increases anandamide levels via FAAH inhibition, acts as an agonist at TRPV1,2 and $5HT_{1A}$ receptors, and acts as an antagonist at the orphan GPR55 receptor and nicotinic α-7 receptors (Grotenhermen, 2005; Izzo et al., 2009; Mahgoub et al., 2013; Mechoulam, Parker, & Gallily, 2002; Pertwee et al., 2010; Ryberg et al., 2007). Cannabidiol has been shown to have anticonvulsant, neuroprotective, and antiinflammatory effects in animal models and anxiolytic and antipsychotic effects in humans (Bergamaschi, Queiroz, Zuardi, & Crippa, 2011; Crippa et al., 2011; Devinsky et al., 2014; Leweke et al., 2012). Cannabidiol alone or in combination with THC (nabiximols, Sativex) has been investigated in clinical trials for management of CUD (Allsop et al., 2014; Haney, Malcolm, et al., 2015).

It is hypothesized that when administered in combination with THC, cannabidiol likely adds anxiolytic, antidepressant, and antipsychotic effects to the simple agonist THC substitution approach, and facilitates the safe delivery of the THC doses needed to control cannabis cravings (Allsop, Lintzeris, Copeland, Dunlop, & McGregor, 2015). In a double-blind, randomized, placebo-controlled, inpatient clinical trial (Allsop et al., 2014), 6-day treatment with nabiximols attenuated cannabis withdrawal symptoms and improved patient retention in treatment, but was no more effective than placebo in encouraging long-term reductions in cannabis use. Recently, Haney, Malcolm, et al. (2015) conducted a double-blind; randomized, within-subject laboratory study and showed that acute oral administration of cannabidiol alone does not reduce the reinforcing, physiological, or positive subjective effects of smoked cannabis. There are several clinical trials currently underway or have recently been completed that will hopefully shed more light on effectiveness of cannabidiol or nabiximols on cannabis use outcomes.

Drugs that prevent activation of cannabinoid CB_1 receptors represent an alternative approach to agonist substitution treatments of cannabis addiction, by directly blocking the subjective and reinforcing effects of THC. Huestis and colleagues showed that acute and repeated administration of the CB_1 antagonists/inverse-agonist rimonabant attenuated the subjective and physiological effects of smoked marijuana (Gorelick et al., 2006; Huestis et al., 2001, 2007). CB_1 antagonists/inverse agonists (like rimonabant, taranabant, and surinabant) showed effectiveness in treatment of obesity and in smoking cessation, but were eventually pulled from clinical use and further development due to reports of psychiatric side effects (anxiety, depression, suicidality) (Le Foll et al., 2009). It is possible that CB_1 antagonists that lack inverse-agonist effects (neutral antagonists) would effectively block the abuse-related effects of THC, but without producing effects opposite to those of the endogenous cannabinoids, and might therefore be safer than rimonabant (Janero & Makriyannis, 2009). The prototypical and best-characterized neutral CB_1 antagonists (eg, AM4113) have not yet been tested clinically, but a natural cannabinoid tetrahydrocannabivarin was shown to act as a neutral CB_1 antagonist at lower doses (McPartland, Duncan, Di Marzo, & Pertwee, 2015). In healthy subjects, tetrahydrocannabivarin increased neural responding to rewarding and aversive stimuli (Tudge, Williams, Cowen, & McCabe, 2015), and when administered in combination with THC, it significantly inhibited THC-induced impairment of delayed recall as well as THC-induced increase of heart rate (Englund et al., 2016). Although tetrahydrocannabivarin was well tolerated among the participants in these studies, and no serious adverse effects were observed, a larger study is needed to confirm both the safety profile and the potentially beneficial effects of tetrahydrocannabivarin.

So far in Section 5, we have described compounds that bind to the orthosteric site of the CB_1 receptor, where endogenous cannabinoid ligands also bind. Cannabinoid receptors can also contain allosteric binding sites that are discrete from the orthosteric site (Ross, 2007). Allosteric modulators are ligands that bind to the allosteric site and induce conformational changes that alter the affinity and/or efficacy of orthosteric ligands (Bosier, Muccioli, Hermans, & Lambert, 2010). Newly developed CB_1 allosteric modulators can have complex effects and unclear mechanism of action (Keov, Sexton, & Christopoulos, 2011), but they have the potential to be signaling-specific and to regulate only some of the functions of the receptor, which could lead to a more targeted action compared to orthosteric compounds (Ross, 2007). Thus, allosteric modulators of CB_1 receptors might be able to modulate specific functions of endogenous cannabinoids while avoiding the adverse effects associated with orthosteric agonism or antagonism. Recently, the precursor of steroid hormones, pregnenolone that acts as negative allosteric modulator of CB_1 receptors, was shown to block behavioral and neurobiological effects of cannabinoid drugs in rodents (Vallee et al., 2014). Based on these and other preclinical results, a clinical study has been initiated that will investigate the effect of acute administration of pregnenolone on cue-related craving in individuals with CUD (ClinicalTrials.gov).

A wealth of preclinical evidence indicates an interplay between the opioid (μ-opioid) and the cannabinoid receptor systems in modulating addictive behavior, and it has been suggested that opioid antagonists can block the reinforcing effects of cannabinoids (Fattore et al., 2004). Treatment with μ-opioid antagonists such as naltrexone decreased THC self-administration in squirrel monkeys in some conditions (Justinova et al., 2004), but was less effective in others (Justinova et al., 2008; see Section 3). Results in humans have also been mixed. In some human studies, acute naltrexone administration actually increased the positive subjective effects of oral THC or cannabis in marijuana smokers (Cooper & Haney, 2010; Haney, Bisaga, & Foltin, 2003). In another study, acute naltrexone administration did not attenuate the behavioral or cognitive effects of intravenous THC in humans who were prescreened to be free from medical or psychiatric disorders (Ranganathan et al., 2012). There is evidence that the drug history of the person might be an important determinant of the effects of naltrexone, as THC intoxication was blunted by naltrexone in marijuana users but not in naïve subjects (Haney, 2007).

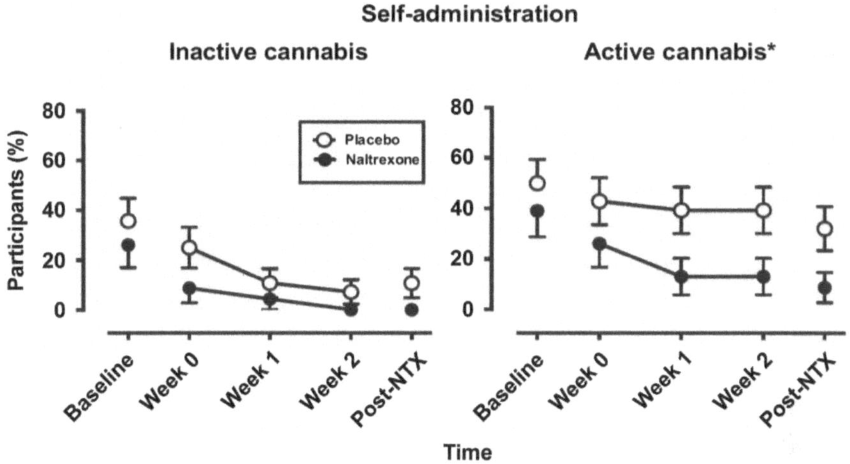

Fig. 4 Effects of naltrexone (NTX) or placebo treatment on cannabis self-administration by chronic cannabis users in a laboratory study. Points show the percentage of participants self-administering inactive (0.0% THC) or active (5.5% THC) smoked cannabis as a function of time and naltrexone dose. Baseline refers to the session before naltrexone administration commenced. Post-NTX refers to the session that occurred at least 1 week following termination of naltrexone administration. All participants received placebo capsules during baseline and post-NTX sessions. Self-administration of inactive cannabis was low and not significantly influenced by naltrexone (left panel), but naltrexone significantly decreased self-administration of active cannabis relative to placebo (right panel). *Asterisks* indicate significant difference between the naltrexone and placebo groups. These findings suggest that opioid antagonists might be useful as medications for decreasing cannabis smoking. *Figure from Haney, M., Malcolm, R. J., Babalonis, S., Nuzzo, P. A., Cooper, Z. D., Bedi, G., … Walsh, S. L. (2015). Oral cannabidiol does not alter the subjective, reinforcing or cardiovascular effects of smoked Cannabis.* Neuropsychopharmacology. *doi:10.1038/npp.2015.367; Haney, M., Ramesh, D., Glass, A., Pavlicova, M., Bedi, G., & Cooper, Z. D. (2015). Naltrexone maintenance decreases Cannabis self-administration and subjective effects in daily Cannabis smokers.* Neuropsychopharmacology, 40*(11), 2489–2498. doi:10.1038/npp.2015.108 with permission.*

A recent human laboratory study showed that chronic administration of naltrexone attenuates the reinforcing and positive subjective effects of cannabis in daily smokers (Haney, Ramesh, et al., 2015), which encourages further testing in clinical trials (Fig. 4). In fact, a clinical trial studying effectiveness of long-acting injectable naltrexone (Vivitrol) in patients with cannabis dependence is currently recruiting participants (ClinicalTrials.gov).

There are also several drugs from different pharmacological groups that showed some promise in human studies. A study by Haney et al. (2008) showed that combined treatment with the α2-adrenergic receptor agonist lofexidine and dronabinol decreased marijuana withdrawal, craving, and relapse in a small group of daily marijuana smokers. However, a recently concluded, randomized, double-blind, placebo-controlled trial did not find the combination of dronabinol and lofexidine effective in promoting abstinence among a larger sample of cannabis-dependent patients (Levin et al., 2016). Nevertheless, there is currently another clinical study underway that evaluates effectiveness of combining dronabinol and another α2-adrenergic receptor agonist, clonidine, in cannabis-dependent schizophrenics (ClinicalTrials.gov). Based on the hypothesis of the involvement of glutamatergic pathways in the pathophysiology of addiction, an over-the-counter supplement *N*-acetylcysteine (NAC) has been suggested for the treatment of addiction. NAC stabilizes extracellular glutamate levels and increases tonic activation of type 2 and 3 metabotropic glutamate receptors (mGluR2/3), with the latter inhibiting excessive glutamate transmission and excitotoxicity (Asevedo, Mendes, Berk, & Brietzke, 2014). In young marijuana users, an open-label pilot study showed decreases in self-report measures of marijuana use and craving during 4 weeks of daily administration of NAC (Gray, Watson, Carpenter, & Larowe, 2010). This study was followed by a double-blind, randomized placebo-controlled trial in cannabis-dependent adolescents. Treatment with NAC, when added to contingency management and brief cessation counseling, yielded improved cannabis abstinence during treatment compared to placebo (Gray et al., 2012). Currently, the National Institute on Drug Abuse Clinical Trials Network (NIDA CTN) is conducting a study to test the efficacy of NAC vs placebo, added to contingency management, for cannabis cessation in adults (McClure et al., 2014). Chronic cannabis use and withdrawal is associated with neural dysregulation in stress systems, and the calcium channel/GABA-modulating antiepileptic gabapentin can restore homeostasis in brain stress systems. In a proof-of-concept study in adults, gabapentin significantly decreased withdrawal severity and reduced cannabis use (Mason et al., 2012). Another randomized, double-blind, placebo-controlled clinical trial with a larger sample is currently underway (ClinicalTrials.gov).

6. CONCLUSION

The current prevalence of cannabis use worldwide, together with the estimated prospect for developing dependence in occasional and daily cannabis users (7–9% and 10–20%, respectively), underscore the necessity of efficacious tools to treat CUD. This chapter summarizes and provides an update on the different preclinical and clinical models for the screening and evaluation of medications for treating CUD.

Preclinical research is an early phase in pharmacotherapy development, and many of the aspects involved in drug dependence can be studied by using specific animal models. Among these models, the self-administration paradigm provides the most direct evidence of a drug's reinforcing properties. However, the infeasibility of modeling THC self-administration in rodents has (so far) constituted a drawback in this area. Encouragingly, nonhuman primates readily self-administer THC, and do so reliably over long periods of time, providing an animal model of chronic use in humans. Therefore, THC self-administration in monkeys, together with other paradigms in rodents (eg, drug discrimination), have allowed testing of the efficacy of several compounds for modifying the reinforcing/perceived effects of THC. In this respect, it is interesting that findings in squirrel monkeys showing the ability of naltrexone to reduce THC's reinforcing properties have been confirmed in human self-administration studies, which supports the validity of this animal model. In addition to nonhuman primate research, it is also possible to take advantage of the fact that rodents will self-administer synthetic cannabinoids such as WIN55-212. In fact, synthetic cannabinoids and treatments that affect endogenous cannabinoids (eg, agonists, FAAH inhibitors, antagonists) are useful tools in animal models of CUD and withdrawal, allowing manipulation of many different components of the endocannabinoid system.

Clinical research on CUD has been relatively scarce compared to other addictions, but there is a growing interest in this area, and human studies have become increasingly prevalent. Importantly, these human studies have modeled and established the existence of the cannabis withdrawal syndrome which is now included in the DSM-5. Human laboratory studies and clinical trials have also produced relevant information regarding the effectiveness of different pharmacotherapies (and counseling). In this regard, THC analogues such as dronabinol and nabilone have been shown to be effective in reducing withdrawal symptoms, but have no apparent effect on cannabis self-administration or relapse. Similarly, studies using a combination of THC/CBD (Sativex) showed reduced withdrawal with no changes in cannabis use. On the other hand, new pharmacological tools such as FAAH inhibitors, neutral CB_1 antagonists, allosteric modulators of CB_1 receptors, and drugs from different pharmacological groups such as α2-adrenergic receptor agonists, NMDA/AMPA, or GABA ligands have shown promising results in small or proof-of-concept studies and might yield relevant findings in ongoing clinical trials.

ACKNOWLEDGMENTS

The preparation of this manuscript was supported in part by Intramural Research Program of National Institute on Drug Abuse, National Institutes of Health, and CAMH.

REFERENCES

Allsop, D. J., Copeland, J., Lintzeris, N., Dunlop, A. J., Montebello, M., Sadler, C., … McGregor, I. S. (2014). Nabiximols as an agonist replacement therapy during cannabis withdrawal: A randomized clinical trial. *JAMA Psychiatry*, *71*(3), 281–291. http://dx.doi.org/10.1001/jamapsychiatry.2013.3947.

Allsop, D. J., Copeland, J., Norberg, M. M., Fu, S., Molnar, A., Lewis, J., & Budney, A. J. (2012). Quantifying the clinical significance of cannabis withdrawal. *PLoS One*, 7(9), e44864. http://dx.doi.org/10.1371/journal.pone.0044864.

Allsop, D. J., Lintzeris, N., Copeland, J., Dunlop, A., & McGregor, I. S. (2015). Cannabinoid replacement therapy (CRT): Nabiximols (Sativex) as a novel treatment for cannabis withdrawal. *Clinical Pharmacology and Therapeutics*, *97*(6), 571–574. http://dx.doi.org/10.1002/cpt.109.

Amchova, P., Kucerova, J., Giugliano, V., Babinska, Z., Zanda, M. T., Scherma, M., … Fattore, L. (2014). Enhanced self-administration of the CB1 receptor agonist WIN55,212-2 in olfactory bulbectomized rats: Evaluation of possible serotonergic and dopaminergic underlying mechanisms. *Frontiers in Pharmacology*, *5*, 44. http://dx.doi.org/10.3389/fphar.2014.00044.

American Psychiatric Association. (2013). *Diagnostic and statistical manual of mental disorders* (5th ed.). Arlington, VA: American Psychiatric Publishing.

Asevedo, E., Mendes, A. C., Berk, M., & Brietzke, E. (2014). Systematic review of N-acetylcysteine in the treatment of addictions. *Revista Brasileira de Psiquiatria*, *36*(2), 168–175. Retrieved from, http://www.ncbi.nlm.nih.gov/pubmed/24676047.

Balter, R. E., Cooper, Z. D., & Haney, M. (2014). Novel pharmacologic approaches to treating Cannabis use disorder. *Current Addiction Reports*, *1*(2), 137–143. http://dx.doi.org/10.1007/s40429-014-0011-1.

Batts, K., Pemberton, M., Bose, J., Weimer, B., Henderson, L., Penne, M., … Strashny, H. (2014). Comparing and evaluating substance use treatment utilization estimates from the national survey on drug use and health and other data sources. *CBHSQ Data Review*, 1–120.

Bedi, G., Cooper, Z. D., & Haney, M. (2013). Subjective, cognitive and cardiovascular dose-effect profile of nabilone and dronabinol in marijuana smokers. *Addiction Biology*, *18*(5), 872–881. http://dx.doi.org/10.1111/j.1369-1600.2011.00427.x.

Bergamaschi, M. M., Queiroz, R. H., Zuardi, A. W., & Crippa, J. A. (2011). Safety and side effects of cannabidiol, a Cannabis sativa constituent. *Current Drug Safety*, *6*(4), 237–249. Retrieved from, http://www.ncbi.nlm.nih.gov/pubmed/22129319.

Bergman, J., Delatte, M. S., Paronis, C. A., Vemuri, K., Thakur, G. A., & Makriyannis, A. (2008). Some effects of CB1 antagonists with inverse agonist and neutral biochemical properties. *Physiology & Behavior*, *93*(4–5), 666–670. http://dx.doi.org/10.1016/j.physbeh.2007.11.007.

Bosier, B., Muccioli, G. G., Hermans, E., & Lambert, D. M. (2010). Functionally selective cannabinoid receptor signalling: Therapeutic implications and opportunities. *Biochemical Pharmacology*, *80*(1), 1–12. http://dx.doi.org/10.1016/j.bcp.2010.02.013.

Bossert, J. M., Marchant, N. J., Calu, D. J., & Shaham, Y. (2013). The reinstatement model of drug relapse: Recent neurobiological findings, emerging research topics, and translational research. *Psychopharmacology*, *229*(3), 453–476. http://dx.doi.org/10.1007/s00213-013-3120-y.

Budney, A. J., Novy, P. L., & Hughes, J. R. (1999). Marijuana withdrawal among adults seeking treatment for marijuana dependence. *Addiction*, *94*(9), 1311–1322. Retrieved from, http://www.ncbi.nlm.nih.gov/pubmed/10615717.

Budney, A. J., Roffman, R., Stephens, R. S., & Walker, D. (2007). Marijuana dependence and its treatment. *Addiction Science & Clinical Practice*, *4*(1), 4–16. Retrieved from, http://www.ncbi.nlm.nih.gov/pmc/articles/PMC2797098/pdf/ascp-04-1-4.pdf.

Budney, A. J., Vandrey, R. G., Hughes, J. R., Moore, B. A., & Bahrenburg, B. (2007). Oral delta-9-tetrahydrocannabinol suppresses cannabis withdrawal symptoms. *Drug and Alcohol Dependence*, *86*(1), 22–29. http://dx.doi.org/10.1016/j.drugalcdep.2006.04.014.

Carpenter, K. M., McDowell, D., Brooks, D. J., Cheng, W. Y., & Levin, F. R. (2009). A preliminary trial: Double-blind comparison of nefazodone, bupropion-SR, and placebo in the treatment of cannabis dependence. *American Journal on Addictions*, *18*(1), 53–64. http://dx.doi.org/10.1080/10550490802408936.

Cheer, J. F., Kendall, D. A., & Marsden, C. A. (2000). Cannabinoid receptors and reward in the rat: A conditioned place preference study. *Psychopharmacology*, *151*(1), 25–30.

Cooper, Z. D., Foltin, R. W., Hart, C. L., Vosburg, S. K., Comer, S. D., & Haney, M. (2013). A human laboratory study investigating the effects of quetiapine on marijuana withdrawal and relapse in daily marijuana smokers. *Addiction Biology*, *18*(6), 993–1002. http://dx.doi.org/10.1111/j.1369-1600.2012.00461.x.

Cooper, Z. D., & Haney, M. (2010). Opioid antagonism enhances marijuana's effects in heavy marijuana smokers. *Psychopharmacology*, *211*(2), 141–148. http://dx.doi.org/10.1007/s00213-010-1875-y.

Copeland, J., Swift, W., Roffman, R., & Stephens, R. (2001). A randomized controlled trial of brief cognitive-behavioral interventions for cannabis use disorder. *Journal of Substance Abuse Treatment*, *21*(2), 55–64. discussion 65–56. Retrieved from, http://www.ncbi.nlm.nih.gov/pubmed/11551733.

Crippa, J. A., Derenusson, G. N., Ferrari, T. B., Wichert-Ana, L., Duran, F. L., Martin-Santos, R., … Hallak, J. E. (2011). Neural basis of anxiolytic effects of cannabidiol (CBD) in generalized social anxiety disorder: A preliminary report. *Journal of Psychopharmacology*, *25*(1), 121–130. http://dx.doi.org/10.1177/0269881110379283.

Cui, S. S., Bowen, R. C., Gu, G. B., Hannesson, D. K., Yu, P. H., & Zhang, X. (2001). Prevention of cannabinoid withdrawal syndrome by lithium: Involvement of oxytocinergic neuronal activation. *Journal of Neuroscience*, *21*(24), 9867–9876. Retrieved from, http://www.jneurosci.org/content/21/24/9867.full.pdf.

De Vries, T. J., & Schoffelmeer, A. N. (2005). Cannabinoid CB1 receptors control conditioned drug seeking. *Trends in Pharmacological Sciences*, *26*(8), 420–426. http://dx.doi.org/10.1016/j.tips.2005.06.002.

Degenhardt, L., Hall, W., & Lynskey, M. (2003). Exploring the association between cannabis use and depression. *Addiction*, *98*(11), 1493–1504. Retrieved from, http://onlinelibrary.wiley.com/store/10.1046/j.1360-0443.2003.00437.x/asset/j.1360-0443.2003.00437.x.pdf?v=1&t=ijyh6b60&s=d1560f9fd2425d9c2d709d5f6e828478a126ab64.

Dennis, M., Godley, S. H., Diamond, G., Tims, F. M., Babor, T., Donaldson, J., … Funk, R. (2004). The Cannabis youth treatment (CYT) study: Main findings from two randomized trials. *Journal of Substance Abuse Treatment*, *27*(3), 197–213. http://dx.doi.org/10.1016/j.jsat.2003.09.005.

Devinsky, O., Cilio, M. R., Cross, H., Fernandez-Ruiz, J., French, J., Hill, C., … Friedman, D. (2014). Cannabidiol: Pharmacology and potential therapeutic role in epilepsy and other neuropsychiatric disorders. *Epilepsia*, *55*(6), 791–802. http://dx.doi.org/10.1111/epi.12631.

Englund, A., Atakan, Z., Kralj, A., Tunstall, N., Murray, R., & Morrison, P. (2016). The effect of five day dosing with THCV on THC-induced cognitive, psychological and physiological effects in healthy male human volunteers: A placebo-controlled, double-blind, crossover pilot trial. *Journal of Psychopharmacology*, *30*(2), 140–151. http://dx.doi.org/10.1177/0269881115615104.

Fantegrossi, W. E., Moran, J. H., Radominska-Pandya, A., & Prather, P. L. (2014). Distinct pharmacology and metabolism of K2 synthetic cannabinoids compared to Delta(9)-THC: Mechanism underlying greater toxicity? *Life Sciences*, *97*(1), 45–54. http://dx.doi.org/10.1016/j.lfs.2013.09.017.

Fattore, L., Cossu, G., Martellotta, C. M., & Fratta, W. (2001). Intravenous self-administration of the cannabinoid CB1 receptor agonist WIN 55,212-2 in rats. *Psychopharmacology*, *156*(4), 410–416.

Fattore, L., Cossu, G., Spano, M. S., Deiana, S., Fadda, P., Scherma, M., & Fratta, W. (2004). Cannabinoids and reward: Interactions with the opioid system. *Critical Reviews in Neurobiology*, *16*(1–2), 147–158. Retrieved from, http://www.ncbi.nlm.nih.gov/pubmed/15581410.

Gates, P., Jaffe, A., & Copeland, J. (2014). Cannabis smoking and respiratory health: Consideration of the literature. *Respirology*, *19*(5), 655–662. http://dx.doi.org/10.1111/resp.12298.

Gorelick, D. A., Heishman, S. J., Preston, K. L., Nelson, R. A., Moolchan, E. T., & Huestis, M. A. (2006). The cannabinoid CB1 receptor antagonist rimonabant attenuates the hypotensive effect of smoked marijuana in male smokers. *American Heart Journal*, *151*(3), 754.e1–754.e5. http://dx.doi.org/10.1016/j.ahj.2005.11.006.

Gray, K. M., Carpenter, M. J., Baker, N. L., DeSantis, S. M., Kryway, E., Hartwell, K. J., … Brady, K. T. (2012). A double-blind randomized controlled trial of N-acetylcysteine in cannabis-dependent adolescents. *The American Journal of Psychiatry*, *169*(8), 805–812. http://dx.doi.org/10.1176/appi.ajp.2012.12010055.

Gray, K. M., Watson, N. L., Carpenter, M. J., & Larowe, S. D. (2010). N-acetylcysteine (NAC) in young marijuana users: An open-label pilot study. *American Journal on Addictions*, *19*(2), 187–189. http://dx.doi.org/10.1111/j.1521-0391.2009.00027.x.

Grotenhermen, F. (2005). Cannabinoids. *Current Drug Targets. CNS and Neurological Disorders*, *4*(5), 507–530.

Hall, W. (2015). What has research over the past two decades revealed about the adverse health effects of recreational cannabis use? *Addiction*, *110*(1), 19–35. http://dx.doi.org/10.1111/add.12703.

Haney, M. (2007). Opioid antagonism of cannabinoid effects: Differences between marijuana smokers and nonmarijuana smokers. *Neuropsychopharmacology*, *32*(6), 1391–1403. http://dx.doi.org/10.1038/sj.npp.1301243.

Haney, M., Bisaga, A., & Foltin, R. W. (2003). Interaction between naltrexone and oral THC in heavy marijuana smokers. *Psychopharmacology*, *166*(1), 77–85. http://dx.doi.org/10.1007/s00213-002-1279-8.

Haney, M., Comer, S. D., Ward, A. S., Foltin, R. W., & Fischman, M. W. (1997). Factors influencing marijuana self-administration by humans. *Behavioural Pharmacology*, *8*(2–3), 101–112. Retrieved from, http://www.ncbi.nlm.nih.gov/pubmed/9833006.

Haney, M., Cooper, Z. D., Bedi, G., Vosburg, S. K., Comer, S. D., & Foltin, R. W. (2013). Nabilone decreases marijuana withdrawal and a laboratory measure of marijuana relapse. *Neuropsychopharmacology*, *38*(8), 1557–1565. http://dx.doi.org/10.1038/npp.2013.54.

Haney, M., Hart, C. L., Vosburg, S. K., Comer, S. D., Reed, S. C., Cooper, Z. D., & Foltin, R. W. (2010). Effects of baclofen and mirtazapine on a laboratory model of marijuana withdrawal and relapse. *Psychopharmacology*, *211*(2), 233–244. http://dx.doi.org/10.1007/s00213-010-1888-6.

Haney, M., Hart, C. L., Vosburg, S. K., Comer, S. D., Reed, S. C., & Foltin, R. W. (2008). Effects of THC and lofexidine in a human laboratory model of marijuana withdrawal and relapse. *Psychopharmacology*, *197*(1), 157–168. http://dx.doi.org/10.1007/s00213-007-1020-8.

Haney, M., Hart, C. L., Vosburg, S. K., Nasser, J., Bennett, A., Zubaran, C., & Foltin, R. W. (2004). Marijuana withdrawal in humans: Effects of oral THC or divalproex. *Neuropsychopharmacology*, *29*(1), 158–170. http://dx.doi.org/10.1038/sj.npp.1300310.

Haney, M., Malcolm, R. J., Babalonis, S., Nuzzo, P. A., Cooper, Z. D., Bedi, G., ... Walsh, S. L. (2015). Oral cannabidiol does not alter the subjective, reinforcing or cardiovascular effects of smoked Cannabis. *Neuropsychopharmacology*. http://dx.doi.org/10.1038/npp.2015.367.

Haney, M., Ramesh, D., Glass, A., Pavlicova, M., Bedi, G., & Cooper, Z. D. (2015). Naltrexone maintenance decreases Cannabis self-administration and subjective effects in daily Cannabis smokers. *Neuropsychopharmacology*, *40*(11), 2489–2498. http://dx.doi.org/10.1038/npp.2015.108.

Haney, M., Ward, A. S., Comer, S. D., Foltin, R. W., & Fischman, M. W. (1999a). Abstinence symptoms following oral THC administration to humans. *Psychopharmacology*, *141*(4), 385–394. Retrieved from, http://www.ncbi.nlm.nih.gov/pubmed/10090646.

Haney, M., Ward, A. S., Comer, S. D., Foltin, R. W., & Fischman, M. W. (1999b). Abstinence symptoms following smoked marijuana in humans. *Psychopharmacology*, *141*(4), 395–404. Retrieved from, http://www.ncbi.nlm.nih.gov/pubmed/10090647.

Hart, C. L., Haney, M., Ward, A. S., Fischman, M. W., & Foltin, R. W. (2002). Effects of oral THC maintenance on smoked marijuana self-administration. *Drug and Alcohol Dependence*, *67*(3), 301–309. Retrieved from, http://www.ncbi.nlm.nih.gov/pubmed/12127201.

Hart, C. L., Ward, A. S., Haney, M., Comer, S. D., Foltin, R. W., & Fischman, M. W. (2002). Comparison of smoked marijuana and oral delta(9)-tetrahydrocannabinol in humans. *Psychopharmacology*, *164*(4), 407–415. http://dx.doi.org/10.1007/s00213-002-1231-y.

Heyman, G. M. (2009). *Addiction: A disorder of choice*. Cambridge, MA: Harvard University Press.

Huestis, M. A., Boyd, S. J., Heishman, S. J., Preston, K. L., Bonnet, D., Le Fur, G., & Gorelick, D. A. (2007). Single and multiple doses of rimonabant antagonize acute effects of smoked cannabis in male cannabis users. *Psychopharmacology*, *194*(4), 505–515. http://dx.doi.org/10.1007/s00213-007-0861-5.

Huestis, M. A., Gorelick, D. A., Heishman, S. J., Preston, K. L., Nelson, R. A., Moolchan, E. T., & Frank, R. A. (2001). Blockade of effects of smoked marijuana by the CB1-selective cannabinoid receptor antagonist SR141716. *Archives of General Psychiatry*, *58*(4), 322–328.

Huggins, J. P., Smart, T. S., Langman, S., Taylor, L., & Young, T. (2012). An efficient randomised, placebo-controlled clinical trial with the irreversible fatty acid amide hydrolase-1 inhibitor PF-04457845, which modulates endocannabinoids but fails to induce effective analgesia in patients with pain due to osteoarthritis of the knee. *Pain*, *153*(9), 1837–1846. http://dx.doi.org/10.1016/j.pain.2012.04.020.

Izzo, A. A., Borrelli, F., Capasso, R., Di Marzo, V., & Mechoulam, R. (2009). Non-psychotropic plant cannabinoids: New therapeutic opportunities from an ancient herb. *Trends in Pharmacological Sciences*, *30*(10), 515–527. http://dx.doi.org/10.1016/j.tips.2009.07.006.

Janero, D. R., & Makriyannis, A. (2009). Cannabinoid receptor antagonists: Pharmacological opportunities, clinical experience, and translational prognosis. *Expert Opinion on Emerging Drugs*, *14*(1), 43–65. http://dx.doi.org/10.1517/14728210902736568.

Johnston, J., Lintzeris, N., Allsop, D. J., Suraev, A., Booth, J., Carson, D. S., ... McGregor, I. S. (2014). Lithium carbonate in the management of cannabis withdrawal: A randomized placebo-controlled trial in an inpatient setting. *Psychopharmacology*, *231*(24), 4623–4636. http://dx.doi.org/10.1007/s00213-014-3611-5.

Jones, R. T., Benowitz, N., & Bachman, J. (1976). Clinical studies of cannabis tolerance and dependence. *Annals of the New York Academy of Sciences*, *282*, 221–239. Retrieved from, http://www.ncbi.nlm.nih.gov/pubmed/798533.

Jones, R. T., Benowitz, N. L., & Herning, R. I. (1981). Clinical relevance of cannabis tolerance and dependence. *Journal of Clinical Pharmacology*, *21*(8–9 Suppl.), 143S–152S. Retrieved from, http://www.ncbi.nlm.nih.gov/pubmed/6271820.

Justinova, Z., Ferre, S., Redhi, G. H., Mascia, P., Stroik, J., Quarta, D., … Goldberg, S. R. (2011). Reinforcing and neurochemical effects of cannabinoid CB1 receptor agonists, but not cocaine, are altered by an adenosine A2A receptor antagonist. *Addiction Biology*, *16*(3), 405–415. http://dx.doi.org/10.1111/j.1369-1600.2010.00258.x.

Justinova, Z., Mascia, P., Wu, H. Q., Secci, M. E., Redhi, G. H., Panlilio, L. V., … Goldberg, S. R. (2013). Reducing cannabinoid abuse and preventing relapse by enhancing endogenous brain levels of kynurenic acid. *Nature Neuroscience*, *16*(11), 1652–1661. http://dx.doi.org/10.1038/nn.3540.

Justinova, Z., Munzar, P., Panlilio, L. V., Yasar, S., Redhi, G. H., Tanda, G., & Goldberg, S. R. (2008). Blockade of THC-seeking behavior and relapse in monkeys by the cannabinoid CB(1)-receptor antagonist rimonabant. *Neuropsychopharmacology*, *33*(12), 2870–2877. http://dx.doi.org/10.1038/npp.2008.21.

Justinova, Z., Panlilio, L. V., Moreno-Sanz, G., Redhi, G. H., Auber, A., Secci, M. E., … Goldberg, S. R. (2015). Effects of fatty acid amide hydrolase (FAAH) inhibitors in non-human primate models of nicotine reward and relapse. *Neuropsychopharmacology*, *40*(9), 2185–2197. http://dx.doi.org/10.1038/npp.2015.62.

Justinova, Z., Redhi, G. H., Goldberg, S. R., & Ferre, S. (2014). Differential effects of presynaptic versus postsynaptic adenosine A2A receptor blockade on delta9-tetrahydrocannabinol (THC) self-administration in squirrel monkeys. *Journal of Neuroscience*, *34*(19), 6480–6484. http://dx.doi.org/10.1523/jneurosci.5073-13.2014.

Justinova, Z., Solinas, M., Tanda, G., Redhi, G. H., & Goldberg, S. R. (2005). The endogenous cannabinoid anandamide and its synthetic analog R(+)-methanandamide are intravenously self-administered by squirrel monkeys. *Journal of Neuroscience*, *25*(23), 5645–5650. http://dx.doi.org/10.1523/jneurosci.0951-05.2005.

Justinova, Z., Tanda, G., Munzar, P., & Goldberg, S. R. (2004). The opioid antagonist naltrexone reduces the reinforcing effects of delta 9 tetrahydrocannabinol (THC) in squirrel monkeys. *Psychopharmacology*, *173*(1–2), 186–194. http://dx.doi.org/10.1007/s00213-003-1693-6.

Justinova, Z., Tanda, G., Redhi, G. H., & Goldberg, S. R. (2003). Self-administration of delta9-tetrahydrocannabinol (THC) by drug naive squirrel monkeys. *Psychopharmacology*, *169*(2), 135–140. http://dx.doi.org/10.1007/s00213-003-1484-0.

Justinova, Z., Yasar, S., Redhi, G. H., & Goldberg, S. R. (2011). The endogenous cannabinoid 2-arachidonoylglycerol is intravenously self-administered by squirrel monkeys. *Journal of Neuroscience*, *31*(19), 7043–7048. http://dx.doi.org/10.1523/jneurosci.6058-10.2011.

Karila, L., Roux, P., Rolland, B., Benyamina, A., Reynaud, M., Aubin, H. J., & Lancon, C. (2014). Acute and long-term effects of cannabis use: A review. *Current Pharmaceutical Design*, *20*(25), 4112–4118.

Keov, P., Sexton, P. M., & Christopoulos, A. (2011). Allosteric modulation of G protein-coupled receptors: A pharmacological perspective. *Neuropharmacology*, *60*(1), 24–35. http://dx.doi.org/10.1016/j.neuropharm.2010.07.010.

Klein, C., Karanges, E., Spiro, A., Wong, A., Spencer, J., Huynh, T., … McGregor, I. S. (2011). Cannabidiol potentiates delta(9)-tetrahydrocannabinol (THC) behavioural effects and alters THC pharmacokinetics during acute and chronic treatment in adolescent rats. *Psychopharmacology*, *218*(2), 443–457. http://dx.doi.org/10.1007/s00213-011-2342-0.

Laprevote, V., Schwan, R., Schwitzer, T., Rolland, B., & Thome, J. (2015). Is there a place for off-label pharmacotherapy in Cannabis use disorder? A review on efficacy and safety. *Current Pharmaceutical Design*, *21*(23), 3298–3305. Retrieved from, http://www.ncbi.nlm.nih.gov/pubmed/26088109.

Le Foll, B., & Goldberg, S. R. (2005). Cannabinoid CB1 receptor antagonists as promising new medications for drug dependence. *Journal of Pharmacology and Experimental Therapeutics*, *312*(3), 875–883. http://dx.doi.org/10.1124/jpet.104.077974.

Le Foll, B., Gorelick, D. A., & Goldberg, S. R. (2009). The future of endocannabinoid-oriented clinical research after CB1 antagonists. *Psychopharmacology*, *205*(1), 171–174. http://dx.doi.org/10.1007/s00213-009-1506-7.

Lee, D., Schroeder, J. R., Karschner, E. L., Goodwin, R. S., Hirvonen, J., Gorelick, D. A., & Huestis, M. A. (2014). Cannabis withdrawal in chronic, frequent cannabis smokers during sustained abstinence within a closed residential environment. *American Journal on Addictions*, *23*(3), 234–242. http://dx.doi.org/10.1111/j.1521-0391.2014.12088.x.

Lefever, T. W., Marusich, J. A., Antonazzo, K. R., & Wiley, J. L. (2014). Evaluation of WIN 55,212-2 self-administration in rats as a potential cannabinoid abuse liability model. *Pharmacology, Biochemistry and Behavior*, *118*, 30–35. http://dx.doi.org/10.1016/j.pbb.2014.01.002.

Levin, F. R., Mariani, J. J., Brooks, D. J., Pavlicova, M., Cheng, W., & Nunes, E. V. (2011). Dronabinol for the treatment of cannabis dependence: A randomized, double-blind, placebo-controlled trial. *Drug and Alcohol Dependence*, *116*(1–3), 142–150. http://dx.doi.org/10.1016/j.drugalcdep.2010.12.010.

Levin, F. R., Mariani, J. J., Pavlicova, M., Brooks, D., Glass, A., Mahony, A., … Choi, J. C. (2016). Dronabinol and lofexidine for cannabis use disorder: A randomized, double-blind, placebo-controlled trial. *Drug and Alcohol Dependence*, *159*, 53–60. http://dx.doi.org/10.1016/j.drugalcdep.2015.11.025.

Levin, F. R., McDowell, D., Evans, S. M., Nunes, E., Akerele, E., Donovan, S., & Vosburg, S. K. (2004). Pharmacotherapy for marijuana dependence: A double-blind, placebo-controlled pilot study of divalproex sodium. *American Journal on Addictions*, *13*(1), 21–32. Retrieved from, http://www.ncbi.nlm.nih.gov/pubmed/14766435.

Leweke, F. M., Piomelli, D., Pahlisch, F., Muhl, D., Gerth, C. W., Hoyer, C., … Koethe, D. (2012). Cannabidiol enhances anandamide signaling and alleviates psychotic symptoms of schizophrenia. *Translational Psychiatry*, *2*, e94. http://dx.doi.org/10.1038/tp.2012.15.

Li, G. L., Winter, H., Arends, R., Jay, G. W., Le, V., Young, T., & Huggins, J. P. (2012). Assessment of the pharmacology and tolerability of PF-04457845, an irreversible inhibitor of fatty acid amide hydrolase-1, in healthy subjects. *British Journal of Clinical Pharmacology*, *73*(5), 706–716. http://dx.doi.org/10.1111/j.1365-2125.2011.04137.x.

Lichtman, A. H., Fisher, J., & Martin, B. R. (2001). Precipitated cannabinoid withdrawal is reversed by Delta(9)-tetrahydrocannabinol or clonidine. *Pharmacology, Biochemistry and Behavior*, *69*(1–2), 181–188. Retrieved from, http://ac.els-cdn.com/S0091305701005147/1-s2.0-S0091305701005147-main.pdf?_tid=ff38e0b2-c5d2-11e5-a683-00000aab0f6c&acdnat=1453994752_1d124e1e739c2e05d8db3cf5152650f0.

Maccioni, P., Colombo, G., & Carai, M. A. (2010). Blockade of the cannabinoid CB1 receptor and alcohol dependence: Preclinical evidence and preliminary clinical data. *CNS & Neurological Disorders Drug Targets*, *9*(1), 55–59.

Mahgoub, M., Keun-Hang, S. Y., Sydorenko, V., Ashoor, A., Kabbani, N., Al Kury, L., … Oz, M. (2013). Effects of cannabidiol on the function of alpha7-nicotinic acetylcholine receptors. *European Journal of Pharmacology*, *720*(1–3), 310–319. http://dx.doi.org/10.1016/j.ejphar.2013.10.011.

Marijuana Treatment Project Research Group. (2004). Brief treatments for cannabis dependence: Findings from a randomized multisite trial. *Journal of Consulting and Clinical Psychology*, *72*(3), 455–466. http://dx.doi.org/10.1037/0022-006X.72.3.455.

Marshall, K., Gowing, L., Ali, R., & Le Foll, B. (2014). Pharmacotherapies for cannabis dependence. *Cochrane Database of Systematic Reviews, 12*, Cd008940. http://dx.doi.org/10.1002/14651858.CD008940.pub2.

Martellotta, M. C., Cossu, G., Fattore, L., Gessa, G. L., & Fratta, W. (1998). Self-administration of the cannabinoid receptor agonist WIN 55,212-2 in drug-naive mice. *Neuroscience, 85*(2), 327–330.

Mason, B. J., Crean, R., Goodell, V., Light, J. M., Quello, S., Shadan, F., … Rao, S. (2012). A proof-of-concept randomized controlled study of gabapentin: Effects on cannabis use, withdrawal and executive function deficits in cannabis-dependent adults. *Neuropsychopharmacology, 37*(7), 1689–1698. http://dx.doi.org/10.1038/npp.2012.14.

McClure, E. A., Sonne, S. C., Winhusen, T., Carroll, K. M., Ghitza, U. E., McRae-Clark, A. L., … Gray, K. M. (2014). Achieving cannabis cessation—Evaluating N-acetylcysteine treatment (ACCENT): Design and implementation of a multi-site, randomized controlled study in the National Institute on Drug Abuse Clinical Trials Network. *Contemporary Clinical Trials, 39*(2), 211–223. http://dx.doi.org/10.1016/j.cct.2014.08.011.

McPartland, J. M., Duncan, M., Di Marzo, V., & Pertwee, R. G. (2015). Are cannabidiol and delta(9)-tetrahydrocannabivarin negative modulators of the endocannabinoid system? A systematic review. *British Journal of Pharmacology, 172*(3), 737–753. http://dx.doi.org/10.1111/bph.12944.

Mechoulam, R., Parker, L. A., & Gallily, R. (2002). Cannabidiol: An overview of some pharmacological aspects. *Journal of Clinical Pharmacology, 42*(11 Suppl.), 11S–19S. Retrieved from, http://www.ncbi.nlm.nih.gov/pubmed/12412831.

Mendizabal, V., Zimmer, A., & Maldonado, R. (2006). Involvement of kappa/dynorphin system in WIN 55,212-2 self-administration in mice. *Neuropsychopharmacology, 31*(9), 1957–1966. http://dx.doi.org/10.1038/sj.npp.1300957.

Oleson, E. B., & Cheer, J. F. (2012). A brain on cannabinoids: The role of dopamine release in reward seeking. *Cold Spring Harbor Perspectives in Medicine, 2*(8), a012229. http://dx.doi.org/10.1101/cshperspect.a012229.

Panlilio, L. V., & Goldberg, S. R. (2007). Self-administration of drugs in animals and humans as a model and an investigative tool. *Addiction, 102*(12), 1863–1870. http://dx.doi.org/10.1111/j.1360-0443.2007.02011.x.

Panlilio, L. V., Goldberg, S. R., & Justinova, Z. (2015). Cannabinoid abuse and addiction: Clinical and preclinical findings. *Clinical Pharmacology and Therapeutics, 97*(6), 616–627. http://dx.doi.org/10.1002/cpt.118.

Panlilio, L. V., Justinova, Z., & Goldberg, S. R. (2010). Animal models of cannabinoid reward. *British Journal of Pharmacology, 160*(3), 499–510. http://dx.doi.org/10.1111/j.1476-5381.2010.00775.x.

Pertwee, R. G., Howlett, A. C., Abood, M. E., Alexander, S. P., Di Marzo, V., Elphick, M. R., … Ross, R. A. (2010). International Union of Basic and Clinical Pharmacology. LXXIX. Cannabinoid receptors and their ligands: Beyond CB(1) and CB(2). *Pharmacological Reviews, 62*(4), 588–631. http://dx.doi.org/10.1124/pr.110. 003004.

Polissidis, A., Chouliara, O., Galanopoulos, A., Marselos, M., Papadopoulou-Daifoti, Z., & Antoniou, K. (2009). Behavioural and dopaminergic alterations induced by a low dose of WIN 55,212-2 in a conditioned place preference procedure. *Life Sciences, 85*(5–6), 248–254. http://dx.doi.org/10.1016/j.lfs.2009.05.015.

Ramesh, D., Haney, M., & Cooper, Z. D. (2013). Marijuana's dose-dependent effects in daily marijuana smokers. *Experimental and Clinical Psychopharmacology, 21*(4), 287–293. http://dx.doi.org/10.1037/a0033661.

Ranganathan, M., Carbuto, M., Braley, G., Elander, J., Perry, E., Pittman, B., … D'Souza, D. C. (2012). Naltrexone does not attenuate the effects of intravenous delta9-tetrahydrocannabinol in healthy humans. *International Journal of Neuropsychopharmacology, 15*(9), 1251–1264. http://dx.doi.org/10.1017/s1461145711001830.

Robinson, L., Hinder, L., Pertwee, R. G., & Riedel, G. (2003). Effects of delta9-THC and WIN-55,212-2 on place preference in the water maze in rats. *Psychopharmacology*, *166*(1), 40–50. http://dx.doi.org/10.1007/s00213-002-1302-0.

Rodriguez-Gaztelumendi, A., Rojo, M. L., Pazos, A., & Diaz, A. (2009). Altered CB receptor-signaling in prefrontal cortex from an animal model of depression is reversed by chronic fluoxetine. *Journal of Neurochemistry*, *108*(6), 1423–1433. http://dx.doi.org/10.1111/j.1471-4159.2009.05898.x.

Ross, R. A. (2007). Allosterism and cannabinoid CB(1) receptors: The shape of things to come. *Trends in Pharmacological Sciences*, *28*(11), 567–572. http://dx.doi.org/10.1016/j.tips.2007.10.006.

Ryberg, E., Larsson, N., Sjogren, S., Hjorth, S., Hermansson, N. O., Leonova, J., … Greasley, P. J. (2007). The orphan receptor GPR55 is a novel cannabinoid receptor. *British Journal of Pharmacology*, *152*(7), 1092–1101. http://dx.doi.org/10.1038/sj.bjp.0707460.

Schindler, C. W., Katz, J. L., & Goldberg, S. R. (1988). The use of second-order schedules to study the influence of environmental stimuli on drug-seeking behavior. *NIDA Research Monograph*, *84*, 180–195.

Schindler, C. W., Panlilio, L. V., Gilman, J. P., Justinova, Z., Vemuri, V. K., Makriyannis, A., & Goldberg, S. R. (2010). Effects of cannabinoid receptor antagonists on maintenance and reinstatement of methamphetamine self-administration in rhesus monkeys. *European Journal of Pharmacology*, *633*(1–3), 44–49. http://dx.doi.org/10.1016/j.ejphar.2010.02.005.

Solinas, M., Panlilio, L. V., Justinova, Z., Yasar, S., & Goldberg, S. R. (2006). Using drug-discrimination techniques to study the abuse-related effects of psychoactive drugs in rats. *Nature Protocols*, *1*(3), 1194–1206. http://dx.doi.org/10.1038/nprot.2006.167.

Solinas, M., Scherma, M., Fattore, L., Stroik, J., Wertheim, C., Tanda, G., … Goldberg, S. R. (2007). Nicotinic alpha 7 receptors as a new target for treatment of cannabis abuse. *Journal of Neuroscience*, *27*(21), 5615–5620. http://dx.doi.org/10.1523/JNEUROSCI.0027-07.2007.

Stephens, R. S., Roffman, R. A., & Curtin, L. (2000). Comparison of extended versus brief treatments for marijuana use. *Journal of Consulting and Clinical Psychology*, *68*(5), 898–908. Retrieved from, http://www.ncbi.nlm.nih.gov/pubmed/11068976.

Stewart, J. L., & McMahon, L. R. (2010). Rimonabant-induced delta9-tetrahydrocannabinol withdrawal in rhesus monkeys: Discriminative stimulus effects and other withdrawal signs. *Journal of Pharmacology and Experimental Therapeutics*, *334*(1), 347–356. http://dx.doi.org/10.1124/jpet.110.168435.

Stewart, J. L., & McMahon, L. R. (2011). The fatty acid amide hydrolase inhibitor URB 597: Interactions with anandamide in rhesus monkeys. *British Journal of Pharmacology*, *164*(2b), 655–666. http://dx.doi.org/10.1111/j.1476-5381.2011.01388.x.

Tai, S., Nikas, S. P., Shukla, V. G., Vemuri, K., Makriyannis, A., & Jarbe, T. U. (2015). Cannabinoid withdrawal in mice: Inverse agonist vs neutral antagonist. *Psychopharmacology*, *232*(15), 2751–2761. http://dx.doi.org/10.1007/s00213-015-3907-0.

Tanda, G., Munzar, P., & Goldberg, S. R. (2000). Self-administration behavior is maintained by the psychoactive ingredient of marijuana in squirrel monkeys. *Nature Neuroscience*, *3*(11), 1073–1074. http://dx.doi.org/10.1038/80577.

Trigo, J. M., Lagzdins, D., Rehm, J., Selby, P., Gamaleddin, I., Fischer, B., … Le Foll, B. (2016). Effects of fixed or self-titrated dosages of Sativex on cannabis withdrawal and cravings. *Drug and Alcohol Dependence*, *156*, e224.

Tudge, L., Williams, C., Cowen, P. J., & McCabe, C. (2015). Neural effects of cannabinoid CB1 neutral antagonist tetrahydrocannabivarin on food reward and aversion in healthy volunteers. *International Journal of Neuropsychopharmacology*. *18*(6), pii: pyu094. http://dx.doi.org/10.1093/ijnp/pyu094.

Vallee, M., Vitiello, S., Bellocchio, L., Hebert-Chatelain, E., Monlezun, S., Martin-Garcia, E., … Piazza, P. V. (2014). Pregnenolone can protect the brain from cannabis intoxication. *Science, 343*(6166), 94–98. http://dx.doi.org/10.1126/science.1243985.

Vandrey, R., Budney, A. J., Kamon, J. L., & Stanger, C. (2005). Cannabis withdrawal in adolescent treatment seekers. *Drug and Alcohol Dependence, 78*(2), 205–210. http://dx.doi.org/10.1016/j.drugalcdep.2004.11.001.

Vandrey, R., Stitzer, M. L., Mintzer, M. Z., Huestis, M. A., Murray, J. A., & Lee, D. (2013). The dose effects of short-term dronabinol (oral THC) maintenance in daily cannabis users. *Drug and Alcohol Dependence, 128*(1–2), 64–70. http://dx.doi.org/10.1016/j.drugalcdep.2012.08.001.

Vlachou, S., Nomikos, G. G., Stephens, D. N., & Panagis, G. (2007). Lack of evidence for appetitive effects of delta 9-tetrahydrocannabinol in the intracranial self-stimulation and conditioned place preference procedures in rodents. *Behavioural Pharmacology, 18*(4), 311–319. http://dx.doi.org/10.1097/FBP.0b013e3282186cf2.

Ward, A. S., Comer, S. D., Haney, M., Foltin, R. W., & Fischman, M. W. (1997). The effects of a monetary alternative on marijuana self-administration. *Behavioural Pharmacology, 8*(4), 275–286. Retrieved from, http://www.ncbi.nlm.nih.gov/pubmed/9832987.

Weinstein, A. M., Miller, H., Bluvstein, I., Rapoport, E., Schreiber, S., Bar-Hamburger, R., & Bloch, M. (2014). Treatment of cannabis dependence using escitalopram in combination with cognitive-behavior therapy: A double-blind placebo-controlled study. *American Journal of Drug and Alcohol Abuse, 40*(1), 16–22. http://dx.doi.org/10.3109/00952990.2013.819362.

Wiley, J. L., Lowe, J. A., Balster, R. L., & Martin, B. R. (1995). Antagonism of the discriminative stimulus effects of delta 9-tetrahydrocannabinol in rats and rhesus monkeys. *Journal of Pharmacology and Experimental Therapeutics, 275*(1), 1–6. Retrieved from, http://jpet.aspetjournals.org/content/275/1/1.full.pdf.

Williams, C. M., & Kirkham, T. C. (2002). Reversal of delta 9-THC hyperphagia by SR141716 and naloxone but not dexfenfluramine. *Pharmacology, Biochemistry and Behavior, 71*(1–2), 333–340. Retrieved from, http://ac.els-cdn.com/S0091305701006943/1-s2.0-S0091305701006943-main.pdf?_tid=a490694c-c5c2-11e5-a09e-00000aab0f01&acdnat=1453987728_0eb2493988e2293683f8fcf8a02067db.

CHAPTER FIVE

How can we Improve on Modeling Nicotine Addiction to Develop Better Smoking Cessation Treatments?

M. Shoaib[1], Y. Buhidma

Institute of Neuroscience, Newcastle University, Newcastle upon Tyne, United Kingdom
[1]Corresponding author: e-mail address: mohammed.shoaib@newcastle.ac.uk

Contents

International Review of Neurobiology, Volume 126
ISSN 0074-7742
http://dx.doi.org/10.1016/bs.irn.2016.02.008

Abstract

Clinically effective smoking cessation treatments are few in number, mainly varenicline, bupropion, and nicotine replacement therapy being prescribed by health organizations. Of the many compounds tested for smoking cessation, a good proportion fail in human trials despite positive findings in rodents. This chapter aims to cover the uses and some pit falls of current methodologies employed to discover clinical treatments in the laboratory. Complicating factors include the complex nature of genetics in tobacco smoking and the comorbidity associated with other psychiatric disorders, which has not been addressed fully in the rodent laboratory. This chapter reviews the evidence from intravenous nicotine self-administration studies and proposes modifications on how we can improve the validity of the animal models by incorporating clinically relevant factors considered to be critical in tobacco smoking. For example, choice procedures that incorporate alternative reinforcers, use of reinstatement models, and second-order schedules of reinforcement are proposed to have better scientific validity that may lead to better clinical outcomes. Furthermore, improved experimental methods will also improve our chances of discovering effective treatments that ultimately may mitigate the effects of tobacco smoking with regard to health worldwide.

1. INTRODUCTION

Tobacco smoking, globally, is one of the leading preventable causes of morbidity and mortality (Wald & Hackshaw, 1996). Many who attempt to quit relapse within 6 months, with a statistic of 40% of smokers trying to quit each year (Shiffman, Brockwell, Pillitteri, & Gitchell, 2008), and more than 80% of ex-smokers relapse (Hughes, Gulliver, Fenwick, Valliere, et al., 1992). This is believed to be due to the addictive properties of nicotine (Benowitz, 2010). Continuation of tobacco smoking greatly increases the risk of developing cancers as well as cardiovascular and pulmonary diseases (Adams, Lee, & Hoffmann, 1984; Anderson, Wilson, & Hiller, 1989). In order to mitigate the long-term effects of tobacco smoke treating nicotine addiction is believed to be the only effective measure.

Numerous strategies have been employed spanning from nicotine replacement therapy (NRT) to targeting nicotinic/dopaminergic receptors to identifying novel receptor targets that have not previously been associated with nicotine dependence. It is believed that nicotine's ability to activate the mesolimbic dopamine pathway in the brain via heteromeric nicotinic acetylcholine receptors (nAChRs) which are made up of different subunits ($\alpha2$–$\alpha10$ and $\beta2$–$\beta4$) is the primary determinant behind its dependence-producing effects in laboratory animals and humans (Benowitz, 2010).

Nicotinic receptor activation results in burst firing of the main dopamine (DA) pathway originating from the ventral tegmental area to the nucleus accumbens. Self-administration of nicotine by rodents has been shown to increase dopamine levels in the nucleus accumbens shell compared to that of the core (Lecca et al., 2006). This pathway has several other inputs notably from the limbic system which include projections from the amygdala, hippocampus, hypothalamus, striatum, orbitofrontal cortex, and the prefrontal cortex. It is well accepted that the DA-releasing properties of nicotine in the nucleus accumbens underlie its primary reinforcing effects, which are qualitatively similar to that observed for other strongly reinforcing drugs such as cocaine, amphetamine, and morphine (Di Chiara & Bassareo, 2007; Pontieri, Tanda, Orzi, & Di Chiara, 1996).

The current behavioral models employed to develop new treatments and interventions have largely been restricted to studies on primary reinforcing effects using a simple intravenous self-administration (IVSA) procedure based on operant conditioning with compounds that target either nAChRs or DA receptors. It has served well in distinguishing neural substrates relevant to the reinforcing effects of nicotine. However, the approach to develop novel effective smoking cessation treatments has not been so forthcoming, thus highlighting the need for a rethink on how best the animal models are utilized. One important factor may be the limited funding available to develop new "clinically relevant" animal models and a priority is placed on discovering new targets and developing molecules using the existing well-established primary reinforcing models (Pouletty, 2002). The approaches taken to model nicotine dependence in rodents attempt to capture elements of the DSM-V criteria, but it is believed that the main issue holding back novel therapies is the lack of predictive validity of these models which translates poorly into clinical trials (Perkins, Stitzer, & Lerman, 2006).

In this chapter, different models will be reviewed, critiqued and proposals made on how the field of smoking cessation models can advance. The hypothesis is that by considering the factors surrounding nicotine intake and the clinical context in which medications are prescribed, there may be a greater probability of developing more effective smoking cessation treatments by evaluating in "clinically appropriate" rodent models.

2. ANIMAL MODELS

The majority of published reports evaluating potential smoking cessation compounds have chosen to utilize the IVSA procedure with parameters established in the seminal publication from the Corrigall laboratory (Corrigall & Coen, 1989). While for many, this identified optimal conditions that favored reliable IVSA of nicotine and served to provide a platform to evaluate novel compounds, there still were key issues. First, in this study, once rats acquired the self-administration, they were resistant to the conventional form of extinction; at least 10 sessions were required before significant decreases were observed following saline substitution. Second, the dose–response curves generated were shallow, not dependent on unit dose of nicotine. Third, rats were previously shaped to respond on the levers with food reinforcement before having access to intravenous nicotine. As a result of this prior shaping, the rats were also food restricted which also varied the overall motivation to self-administer nicotine. Although the Corrigall paper represented a significant contribution in the field, over the past 15 years we have learned that these conditions do not favor any systematic analyses with pharmacological compounds. Even today, many researchers are still utilizing these parameters and as a consequence, the scientific validity of the nicotine IVSA procedure is waning since the compounds do not necessarily reduce tobacco smoking in humans. This "black box" approach is not unique to nicotine as it is commonly practiced for various other treatments on self-administration baselines, for example, with cocaine dependence. Despite these issues, there are certain factors that control nicotine self-administration which should be considered before experiments are designed and executed.

3. MODELING PRIMARY REINFORCING EFFECTS OF NICOTINE

The most commonly used method to investigate if a drug has reinforcing properties is by testing drug–naïve rats which have been surgically catheterized with an indwelling intravenous catheter within an operant conditioning chamber. The IVSA procedure has allowed the study of numerous psychoactive substances in relation to their primary effects on the brain (Caille, Clemens, Stinus, & Cador, 2012). With nicotine and other reinforcing substances, the IVSA procedure captures the initiation process of the dependence as well as providing an insight into the transition from controlled use to compulsive and uncontrolled drug use when using a fixed ratio schedule (Buhler, Vollstadt-Klein, Kobiella, Budde, et al., 2010). The versatility of this method has been useful in distinguishing the neural substrates and mechanisms underlying dependence for various classes of substances and their ability to support lever-press responding during unlimited access conditions. The schedule controlled behavior can be altered to incorporate fixed interval, progressive ratios and also to explicitly pair discrete and/or contextual cues; which, with repeated associations, can lead to significantly higher levels of responses over prolonged periods in the absence of drug availability, maintained largely by presentation of the conditioned reinforcers, under a second-order schedule (Schindler, Panlilio, & Goldberg, 2002).

3.1 Use of Nicotine SA Models

The IVSA procedure shows greater face validity among other models of dependence, but the construct and predictive validity still require attention. Specifically, animal models are designed to examine the reinforcing effects based on pure nicotine solutions (Rose & Corrigall, 1997), while in humans, these methods have been rarely implemented. The main dependent measure in humans has been to examine the number of tobacco cigarettes smoked by the participant, which is not the same. Despite the different experimental parameters/end-point measures between laboratory animals and humans, both exhibit regulation of nicotine intake as evident from the inverted "U-shaped" dose–response curves (Rose & Corrigall, 1997). However, as stated by Rose and Corrigall (1997), these different routes of administration alter the primary effects of nicotine itself. This limitation in the model can be overcome by measuring the plasma concentrations of nicotine/cotinine in the subjects to compare the total amounts of the active compound in human and rodents, with consideration given to confounding factors such as differences in pharmacokinetics and possible drug-drug interactions in human subjects that may alter nicotine metabolism. In rats, there was good concordance between plasma levels of nicotine and cotinine levels following nicotine self-administration across two unit doses (Shoaib & Stolerman, 1999).

It is well accepted that nicotine's actions rely on binding to central nAChRs to transduce the reinforcing properties. Comparative pharmacological tests have been carried out, with central and peripheral nAChR antagonists which confirm that the reinforcing effects are mediated centrally (Corrigall & Coen, 1989). Furthermore, local administration of dihydro-β-erythroidine confirmed that centrally located nAChRs are responsible for mediating the reinforcing effects (Corrigall, Coen, & Adamson, 1994). Similarly, newly synthesized nicotinic antagonists have been developed that offer greater selectivity for particular nAChR receptor subtypes. For example, Madsen et al. (2015) demonstrated that an acute injection of the selective α6β2 nAChR antagonist, *N,N*-decane-1, 10-diyl-bis-3-picolinium diiodide (bPiDI) attenuated nicotine self-administration just like mecamylamine in mice (Stolerman, Naylor, Elmer, & Goldberg, 1999). However, it should be noted that this compound was tested in nicotine naïve mice prior to any self-administration sessions. Daily dosing with 5 mg/kg of the antagonist caused delayed acquisition of nicotine self-administration compared to vehicle-treated control mice. However, after the self-administration task commenced a higher dose of bPiDI (10 mg/kg) was required to reduce nicotine intake (Madsen et al., 2015). In addition to this study, there have been numerous studies with nAChR agonists and antagonists along with compounds from other neurotransmitter classes for their effects to curb nicotine-taking behavior. Since it is not the scope to list all of these, see Lerman et al. (2007) for a review article that provides a comprehensive list.

Additionally, when using this paradigm, the duration of exposure with potential treatments needs to be reconsidered. While it is known that tests with mecamylamine, a nonselective nAChR antagonist can attenuate nicotine self-administration in rodents (Shoaib, Schindler, & Goldberg, 1997), in tobacco smokers, an acute injection of mecamylamine paradoxically increases tobacco smoking behavior (Stolerman, Goldfarb, Fink, & Jarvik, 1973), and reductions in consumption are only apparent following chronic exposure over 4 weeks (5 mg twice daily) (Rose, Sampson, Levin, & Henningfield, 1989). The disparity in the mechanism behind this species difference was proposed by Rose and Corrigall to be related to mecamylamine saturating the nAChRs that are required for activating the DA-releasing effects in the nucleus accumbens and thus blocking the reinforcing effects of nicotine (Rose & Corrigall, 1997).

The IVSA procedure is sensitive to acute tests with clinically effective drugs such as varenicline, bupropion, or nicotine replacement therapies, which all have the common ability to reduce nicotine intake (Brose, West, & Stapleton, 2013; Stapleton, West, Hajek, Wheeler, et al., 2013). However, it is unclear if these reductions bear any significance to a clinical outcome and if these findings build on confidence toward the predictive validity of the animal model. Extensive research has been conducted with IVSA and conditioned place preference (CPP) procedures, but there are only a few examples of successful translation from preclinical into clinical such as varenicline. This partial agonist was marketed as a selective α4β2 nAChR agonist and observed to dose-dependently reduce nicotine-taking behavior in rats without significantly modifying responses maintained by food reinforcement (O'Connor, Parker, Rollema, & Mead, 2010). Whereas some studies demonstrate drugs such as bupropion, which is commonly prescribed for smoking cessation, can even increase nicotine intake (Rauhut, Neugebauer, Dwoskin, & Bardo, 2003; Shoaib, Sidhpura, & Shafait, 2003). If there is any progress with animal models, then some methodological changes are required to ensure that there is better translation when testing potential treatments which should reflect "clinically appropriate" regimens.

3.2 Limitations of SA Studies of Nicotine

There will always be limitations with any animal model used. However, a number of issues need to be addressed in relation to the predictive aspect of evaluating compounds for clinical potential, as these animal models have not been all that pivotal in identifying "clinically effective" smoking cessation agents (Corrigall, 1999). This can be best exemplified with bupropion, a commonly used smoking cessation treatment for tobacco dependence which does not necessarily attenuate nicotine-taking behaviors within the IVSA procedure (Shoaib et al., 2003). Furthermore, in rodents dopamine receptor antagonists SCH23390, a D1 selective antagonist, spiperone and haloperidol, both relatively selective as D2 receptor antagonists, all attenuate nicotine IVSA, which confirms the role of dopamine in nicotine reinforcement (Corrigall & Coen, 1991), but these compounds did not successfully go into clinical trials, with haloperidol being shown to be ineffective (Dawe, Gerada, Russell, & Gray, 1995). However, in tobacco smokers compensatory effects are observed following administration of haloperidol (Dawe et al., 1995). This may be due to the cognitive impairments produced by DA blockade as commonly observed in patients with schizophrenia prescribed with haloperidol (Allison & Shoaib, 2013; Babin et al., 2011). These few examples highlight some of the issues related to translation of findings between human and rodent models of nicotine and tobacco dependence. The absence of any compensatory increases following pharmacological blockade is a major limitation suggesting the model may not be amenable to regulation of nicotine intake (Dani & De Biasi, 2001; Nestler, 2005). It is proposed that rodent models need to be optimized to better demonstrate the ability of rodents to display pharmacological treatment effects. This will help improve the reliability of predicting clinical efficacy.

Another weakness in the methodology can be highlighted by the chosen route of nicotine administration. Peripherally acting nAChR antagonists, such as hexamethonium and trimetaphan when given systemically, have no appreciable effects on nicotine IVSA. However, when nicotine is inhaled, hexamethonium, trimetaphan, and also mecamylamine can alter nicotine SA (Corrigall & Coen, 1989; DeNoble & Mele, 2006; Lee, Gerhardstein, Wang, & Burki, 1993; Rose, Westman, Behm, Johnson, & Goldberg, 1999), suggesting that inhalation models offer improved face validity.

An additional limitation is based on the length of exposure to nicotine, since IVSA models reflect initiation of nicotine self-administration and a period of maintenance upon which compounds are tested. This is in contrast to the many years of nicotine exposure often seen in humans. Improved models should consider including longer periods of nicotine exposure and interventions should be given after prolonged periods of self-administration in which subjects are considered to be nicotine dependent. Sensory cues in consumption of tobacco are also considered to be important which cannot be easily implemented in animal models (Rose, 1988). Sensory cues and repeated pairings with these cues should be incorporated into animal models and perhaps this will provide greater face validity of the models which may raise the bar for detecting novel pharmacological interventions.

Issues relating to the differential tolerance of treatments may also contribute toward the confines of animal models. For instance, experiments evaluating mecamylamine, known to reliably decrease nicotine IVSA in rodents following relatively high doses cannot be conducted in humans since these doses cannot be tolerated due to side effects reported after the treatment. A strategy combining both nicotine and mecamylamine was observed to show greater success in relation to reducing smoking satisfaction in humans (Rose, Behm, & Westman, 1998). This suggests that future IVSA models may need to look into coadministration studies for potential synergistic effects with nicotinic receptor antagonists and blocking other targets to have maximal efficacy in humans (Engleman, Rodd, Bell, & Murphy, 2008; Kuwabara, Heishman, Brasic, Contoreggi, et al., 2014).

4. SECOND-ORDER SCHEDULES

It has been well documented that sensory cues associated with tobacco smoking enhance the primary reinforcing effects of nicotine in smokers (Rose, Behm, & Levin, 1993). This has been shown to include the sensory effects in the larynx caused by particles in the tobacco smoke as well as paraphernalia associated with tobacco smoking such as the visual cues of cigarettes or the act of rolling up a cigarette before smoking. IVSA procedures have been further developed to incorporate conditioned cues within IVSA models. Second-order schedules were established with nicotine in nonhuman primates to capture aspects of conditioned reinforcement within the IVSA procedure (Goldberg, Kelleher, & Goldberg, 1981). Greater success was attained in nonhuman primates rather than rodents mainly due to limited catheter patency in the latter. With developments in catheter materials and surgical implantation techniques, it is possible to train rats for prolonged periods in order to achieve second-order schedules, which include repeated pairings of nicotine with discrete stimuli such as brief period of light oscillations or presentation of an auditory stimulus such as a buzzer (Wing & Shoaib, 2010). Likewise, studies with other psychostimulants have shown that cocaine can also support second-order schedules in rats (Kantak, Black, Valencia, Green-Jordan, & Eichenbaum, 2002; Pilla, Perachon, Sautel, Garrido, et al., 1999) and the basis for the enhanced rates of responding over prolonged periods may be due to the self-administered drug enhancing the conditioned reinforcing effects of the nonpharmacological cues (Meyer, Cogan, & Robinson, 2014). This has been demonstrated using either contingent or noncontingent presentations of nicotine which can enhance responses for a visual stimulus within the IVSA milieu, with the nicotine-paired stimulus combination increasing responding for nicotine in a synergistic manner (Donny, Chaudhri, Caggiula, Evans-Martin, et al., 2003; Palmatier, Liu, Caggiula, Donny, & Sved, 2007).

4.1 Use of Second-Order Schedules

Second-order schedules have allowed the distinction to be made between drug-seeking and drug-taking behaviors especially with regard to the underlying neural mechanisms with drugs such as cocaine (Arroyo, Markou, Robbins, & Everitt, 1998; Everitt & Robbins, 2000; Kalivas & McFarland, 2003). As a consequence, these studies have encouraged further investigation with nicotine in rats (Caggiula, Donny, White, Chaudhri, et al., 2001, 2002; Wing & Shoaib, 2010). This has subsequently lead to the identification of novel targets for potential therapies. For example, Fig. 1 illustrates the effect of the cannabinoid receptor, CB1 subtype antagonist, AM251, which attenuates responses under a second-order schedule maintained by IV nicotine (Wing & Shoaib, 2010). Second-order schedules have also been a method in which animal models have been able to capture relapse-like behavior.

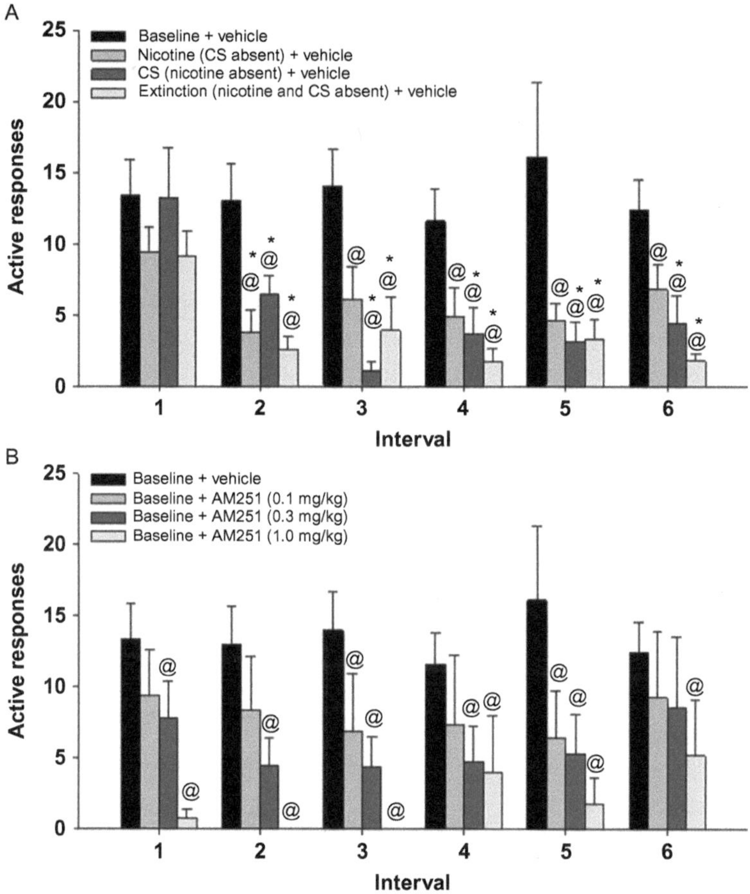

Fig. 1 Tests with the CB1 receptor antagonist AM251 in rats responding under second-order schedules for nicotine reinforcement. The total number of active responses during the six intervals of test sessions is illustrated. (A) The effect of tests highlighting the role of the conditioned stimulus (CS) within the second-order schedule; * denotes a significant difference in responding made in the same test in latter intervals compared to the first. @ denotes a significant difference in responding made in a particular test (vehicle: nicotine, CS, or extinction) compared to the baseline vehicle condition at a specific interval. (B) The effect of AM251 on baseline responding; * denotes a significant difference in responding made in the same test in latter intervals compared to the first. @ denotes a significant difference in responding made in a particular test (baseline: 0, 0.1, 0.3, and 1 mg/kg) compared to the baseline vehicle condition at a specific interval. The baseline vehicle condition shown in (A) and (B) are the same. *Figure taken from Wing, V. C., & Shoaib, M. (2010). Second-order schedules of nicotine reinforcement in rats: Effect of AM251.* Addiction Biology, 15*(4), 393–402. http://dx.doi.org/10.1111/j.1369-1600.2009.00204.x.*

4.2 Modeling Relapse of Nicotine use

It is believed by many that addiction itself is a chronic relapsing disorder and that the only clinically effective way to treat addiction in humans, is to prevent the onset of relapse (Koob & Volkow, 2010). This is modeled in animals via a negative-reinforcement variant of the cue associated SA paradigm. The reinstatement paradigm was first used to model relapse of drug use with other substances of abuse, such as cocaine and heroin (Shaham, Erb, & Stewart, 2000). The animal is trained to self-administer the drug intravenously, with their operant responses for the substance being extinguished subsequently. Following this extinction, the animal is exposed to a subtherapeutic priming dose of the drug or stress/sensory stimuli (Beardsley & Shelton, 2012). The first demonstration was reported in 1981 by de Wit and Stewart, showing how a priming dose could reinstate lever-press responding for cocaine (de Wit & Stewart, 1981). This method was then subsequently coined the "reinstatement model of drug seeking," which led to further investigation on what type of stimuli induce relapse and cravings for certain substances of abuse in humans. Such stimuli include reexposure to drug or drug-associated cues, as mentioned before, and exposure to certain stressful stimuli, such as emotional or physical stress, which can be replicated in animal models using electric foot shock or mild food deprivation.

4.3 Reinstatement Paradigms

The findings emerging from these reinstatement paradigms have shown that relapsing behavior is under control of different neuronal substrates depending on the nature of the reinstating stimulus, ie, stress or conditioned sensory stimuli associated with drug taking. The paradigm has allowed for the discovery of multiple neural mechanisms responsible for nicotine relapse and therefore has the potential to provide plausible targets for pharmacological intervention which is not restricted to just nAChRs. Studies on reinstatement of nicotine-seeking behavior have recently implicated kappa-opioid receptors (KOR) in relation to stress-induced reinstatement (Grella, Funk, Coen, Li, & Le, 2014). Nor-binaltorphimine, a relatively selective kappa-opioid receptor antagonist, when given to rats 1 and 24 h prior to testing showed no effect on cue-induced reinstatement of nicotine seeking, but was able to block yohimbine-induced reinstatement, yohimbine being a pharmacological stressor that antagonizes α-2 adrenergic receptors. This relation of KORs with reinstatement is further reinforced by the activation of opioid receptors via the novel, highly selective KOR agonist U50, 488 which successfully reinstates nicotine seeking (Grella et al., 2014). Furthermore, recent studies have shown that pharmacological inactivation, via combination of γ-aminobutyric acid (GABA) agonist, baclofen, and muscimol or by electrical inactivation of the granular insula, decreases nicotine-induced priming of reinstatement (Forget, Pushparaj, & Le Foll, 2010; Pushparaj et al., 2013). This, in conjunction with studies carried out by Naqvi, Rudrauf, Damasio, and Bechara (2007), implicate the insula as a brain region that can be targeted with future pharmacological interventions to prevent relapse of nicotine. However, when varenicline was tested within the cue-induced reinstatement model, the partial agonist was surprisingly ineffective. Part of this discrepancy may lie in the partial agonist effects of the drug since this may produce an agonist-like reinstatement effect to counter any cue-induced responses (O'Connor et al., 2010). Studies conducted following a longer latency between pretreatment and test of varenicline (30 min) revealed an effect of reducing cue-induced responses but this was not statistically significant (Wouda, Riga, De Vries, Stegeman, et al., 2011). This raises the question on the role of cues within self-administration behaviors and whether second-order schedules provide an improved baseline to test potential therapeutic treatments.

5. MODELING DRUG WITHDRAWAL

As mentioned above when discussing relevant nicotine exposure levels and how these may not be achieved using IVSA procedures, studies have been conducted in which subjects are exposed to significant levels of nicotine passively with the intent of evaluating both physical and motivational measures of dependence in the form of a withdrawal response. Nicotine withdrawal is one of the major factors behind relapse and is therefore another route to manage drug dependence (West, Hajek, & Belcher, 1989). In humans it is normally associated with anxiety, irritability, depression, cognitive/attentional disturbances, weight and appetite gain, and sleep disturbances (Breslau, Kilbey, & Andreski, 1992). In rats withdrawn from nicotine, there is an increased release of corticotrophin releasing factor, a hormone associated with emotional stress, in limbic nuclei, which may be associated with the "anxiety-like" behavior reported in nicotine-withdrawn subjects (Parrott, 1995; Torres, Pipkin, Ferree, Carcoba, & O'Dell, 2015).

In standard withdrawal experiments, the animal would be exposed to the drug chronically via repeated systemic injections for a prolonged period of time or delivery is via a minipump device that releases nicotine chronically via osmosis, which after a prescribed period the treatment is halted or the minipump device is surgically removed. The constellation of symptoms mainly somatic in nature can be measured by direct observation and the various somatic signs can be used to quantify the intensity of nicotine dependence (Malin et al., 1994; Wing & Shoaib, 2007).

5.1 Affective Measures of Nicotine Withdrawal

Withdrawal signs have been reported primarily between 1 and 3 days after abstinence. It is common that upon withdrawal due to alterations of central systems, rodents would express signs of irritability and agitation in the form of spontaneously emitted behaviors such as writhes, gasps, shakes, tremors, teeth chattering, vacuous chewing, ptosis, and scratching (Malin et al., 1994).

One observation, as shown using intracranial self-stimulation behavior, is brain reward thresholds in rodents are increased during withdrawal from several psychoactive drugs, including nicotine (Epping-Jordan, Watkins, Koob, & Markou, 1998), suggesting a reward deficit or anhedonia analogous to the affective symptoms of withdrawal in humans (Johnson, Hollander, & Kenny, 2008). In order to measure nicotine abstinence in rodent models, standard measures of anxiety and depression have been implemented using the elevated plus maze, avoidance of the light side of the light/dark apparatus, keeping to the side of the open field and reduction of social interaction (Irvine, Cheeta, & File, 2001). Another possible cause of these affective aspects of withdrawal may be the interaction between nicotine and its ability to modulate regions associated with stress regulation and hypothalamic-pituitary-adrenal (HPA) axis activation during withdrawal. Nicotine IVSA has been shown to increase levels of corticosterone, adrenocorticotropic hormone and prolactin, making the rodents hypersensitive to stress-induced reinstatement of nicotine seeking (Chen, Fu, & Sharp, 2008; Yu & Sharp, 2010). From this it was reasonable to suggest that nicotine withdrawal may be treated using antidepressant or anxiolytic compounds that have been clinically effective.

As seen with antidepressants, fluoxetine can reduce the ICSS thresholds in animals when given in combination with the 5-HT1A antagonist p-MPPI (Harrison, Liem, & Markou, 2001). However, when tested in humans, it showed no therapeutic effects (Saules et al., 2004), thus highlighting issues on translation. These findings lead to the testing of acute administration of clinically effective antidepressants such as bupropion and nortriptyline, to treat nicotine withdrawal (Biala & Kruk, 2009; Wing & Shoaib, 2007). However, targeting the anxiety-like symptoms of withdrawal can be done with the GABA-B agonists, baclofen, which has clinical efficacy in relation to alcohol dependence at high doses (Addolorato, Leggio, Abenavoli, Agabio, et al., 2006; de Beaurepaire, 2014). Preliminary trials have shown that baclofen may also have an effect on withdrawal from nicotine manifestations (Varani, Aso, Moutinho, Maldonado, & Balerio, 2014), suggesting that GABA(B)-ergic activation may be a successful way to prevent the somatic and affective aspects of nicotine withdrawal in humans.

5.2 Neurocognitive Models of Withdrawal

Nicotine withdrawal, as mentioned before, causes cognitive/attentional disturbances in humans. It is believed that this reduction in cognitive flexibility in humans from nicotine withdrawal has an impact on decision making capabilities and therefore may account for the difficulty in quitting tobacco smoking resulting in high relapse rates (Hare, Camerer, & Rangel, 2009; Kouneiher, Charron, & Koechlin, 2009). nAChRs are putatively involved in this process, as they are widely distributed throughout the brain. More specifically, there is evidence that the main nAChR subtypes in rats are highly expressed in brain regions subserving cognitive functions, such as the prefrontal cortex (PFC), a brain region known to play a critical role in the modulation of executive function (Braida et al., 2014). Withdrawal from nicotine administered via osmotic minipumps impairs attentional processes in a time-dependent manner, as measured with the 5-choice serial reaction time task (5-CSRTT) (Shoaib & Bizarro, 2005). This is similar in nature to the disturbances in attention reported by smokers during withdrawal (Foulds et al., 2006). Furthermore, nicotine and other subtype nAChR agonists have been demonstrated to improve sustained and divided aspects of attention using the 5-CSRTT (Hahn, Shoaib, & Stolerman, 2002). Nicotine administered both acutely and subacutely in normal noncompromised rats improves performance in the attentional set shifting test as well (Allison & Shoaib, 2013).

The neurocognitive model proposes that targeting impaired cognition by pharmacological means may be more effective than simply attenuating the reinforcing properties of nicotine. As a result, a number of researchers are repositioning the use of commonly available cognitive enhancers as potential treatment for tobacco withdrawal.

6. CONSIDERATIONS FOR FUTURE MODELS

In order to find new interventions for nicotine dependence, it is important to revise the current models. The following section will cover proposed revisions to methods.

6.1 Improved Experimental Design

The majority of the studies examining compounds on nicotine IVSA have adopted a "black box" strategy and have simply tested different doses of a compound in a randomized order, as an acute pretreatment over a single session. In most cases, the compound produces the desired effect; an immediate suppression of nicotine-taking behavior that is dose-related and similar doses tested in another cohort of subjects trained to respond for food pellets have a lesser effect. The relative degree of selectivity for nicotine-taking over responding for food is taken as a hallmark indicative of potential as a smoking cessation treatment. These studies are conducted without much information on the nature of the reinforcers supporting the lever-press behavior. Most publications do not show the levels of behavior following extinction. In most cases when conducted, significant decreases are observed following a minimum of three to five extinction sessions in which only the nicotine is removed from the syringe while keeping all the visual and auditory cues in place (Donny et al., 1998; O'Connor et al., 2010; Panlilio, Hogarth, & Shoaib, 2015; Shoaib et al., 1997). Therefore, it should be a requirement for all doses of each compound are repeatedly tested for as many sessions as necessary to demonstrate extinction in these groups of rats. By conducting tests with repeated treatments a more accurate reflection on how individuals will respond and adapt to a change in reinforcing efficacy and thus will be more informative and provide better translation to the clinic.

6.2 Tests on Concurrent Schedules

Given the efforts to demonstrate selectivity of potential treatments across two cohorts of rats, an improved design would be to implement both within the same subject. Concurrent schedules are beginning to be adopted since these schedules might be more naturalistic and have a higher predictive validity than procedures in which only one reinforcer is available.

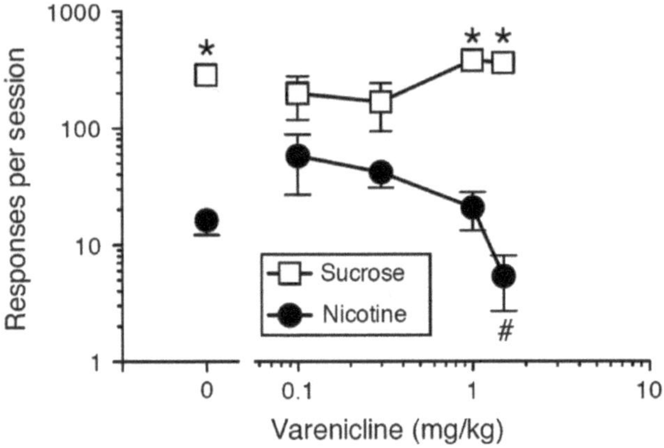

Fig. 2 Effects of varenicline treatment on nicotine and sucrose responding under a concurrent schedule. Asterisks indicate significant differences between sucrose and nicotine lever responding within specific treatment levels. Pound sign indicates a significant decrease in nicotine responding at the highest dose of varenicline compared to nicotine responding under baseline conditions (the zero dose of varenicline). A significant familywise significance level of $p < 0.05$ was maintained for all paired comparisons. All points represent mean ± s.e.m. *Taken from Panlilio, L. V., Hogarth, L., & Shoaib, M. (2015). Concurrent access to nicotine and sucrose in rats.* Psychopharmacology (Berl), *232(8), 1451–1460. http://dx.doi.org/10.1007/s00213-014-3787-8.*

Within a free-choice operant procedure, in which both reinforcers were available, varenicline was shown to dose-dependently attenuate responding on the nicotine lever without significantly modifying responses on the sucrose lever (Fig. 2). Under concurrent conditions, rats tend to prefer sucrose over nicotine and this preference for the highly palatable food is also observed with other drugs such as cocaine and heroin (Cantin, Lenoir, Augier, Vanhille, et al., 2010; Lenoir, Guillem, Koob, & Ahmed, 2012). Concurrent schedules also provide another level of translation since it can provide an insight into how individuals will respond in the presence of an alternative reinforcer. It is known that the alternative reinforcers can decrease drug use in animals (Lesage, 2009; Norman & Tsibulsky, 2001) and in humans (Heishman & Henningfield, 2000; Higgins et al., 1994) and therefore the concurrent schedule might provide an animal model of tobacco use that models the human environment, where tobacco use is persistent despite the availability of non-drug reinforcers.

6.3 Genetic Aspect of Nicotine Dependence

Genome-wide association studies (GWAS) carried out on populations of tobacco smokers, comparing cigarettes smoked per day to genotype has allowed the identification of a number of genes associated with tobacco smoking behavior. Nicotine itself is metabolized in humans and rodents via the drug metabolizing enzyme, cytochrome P450 subtype 2A6 (CYP2A6) (Messina, Tyndale, & Sellers, 1997). Altered activity of CYP2A6 would clearly affect the bioavailability of nicotine in the systemic circulation and therefore would be a key factor involved with the addictive properties of nicotine itself. Several genetic single nucleotide polymorphisms (SNPs) have been identified that reduce function of CYP2A6*2, *4, and *5, which would lead to a prolonged effect of nicotine in the body, as well as increased gene replication of CYP2A6 in some individuals which would lead to lower sensitivity to the effects of nicotine (Ariyoshi, Sawamura, & Kamataki, 2001; Benowitz, Tyndale, Jacob, & Swan, 2002; Hadidi, Zahlsen, Idle, & Cholerton, 1997; Tyndale & Sellers, 2001).

SNPs in the choline acetyltransferase (ChAT) gene (rs1880676, rs3810950, and rs868750) were also significantly associated with level of nicotine dependence (allele p-values were 0.01, 0.02, and 0.04, respectively). With multiple haplotype variants being associated with a predisposition to relapse and nicotine dependence phenotypes (Ray, Mitra, Baldwin, Guo, et al., 2010).

6.4 Impulsivity as a Behavioral Trait

There is also evidence to suggest that impulsivity in human or animal behavior is a major risk factor as to whether a subject exposed to a substance of abuse will develop dependence to that certain compound (McGue & Bouchard, 1984; Tarter, Kirisci, Feske, & Vanyukov, 2007). In rats, the levels of impulsivity measured within an attention task can predict vulnerability to nicotine seeking within a relapse model (Diergaarde et al., 2008). It has been shown that serotonin transporter (SERT) function in the orbitofrontal cortex plays a prominent role in impulsive choice in rodents, and that increased SERT function with a short allele variant correlates with high impulsive choice behavior as measured in delay-discounting tasks (Darna et al., 2015). Animal models should take these different behavioral traits and genetic factors into consideration as they are associated with addiction in humans and effect the addictive properties of the compound for each individual.

6.5 Transgenic Models for Nicotine use

Another recent GWAS has identified a significant association between the genetic variant, rs1051730, in the nAChR gene cluster, *CHRNA5-CHRNA3-CHRNA4*, and having a reduced ability to quit smoking for women in pregnancy, with a 1.27-fold higher odds (95% CI 1.11–1.45) of continuing smoking (Freathy, Ring, Shields, Galobardes, et al., 2009). This data replicated a previous study investigating the gene variant in relation to dependence rather than smoking cessation (Freathy et al., 2009; Thorgeirsson, Geller, Sulem, Rafnar, et al., 2008). From these gene association studies, extensive preclinical investigations have been carried out on the association between specific nAChR subunit variants and their influence nicotine dependent behaviors (Fowler, Arends, & Kenny, 2008; Markou, 2008). This is summarized in Table 1.

Table 1 The Effect of Pharmacologically Blocking or Genetically Knocking Out Specific Subunits of the Nicotinic Acetylcholine Receptor in Rodent Nicotine Self-Administration Tasks

Subunit of nAChR	Effect of KO/Blocking Pharmacologically	References
B2*	Attenuates SA and reduces sensitivity to nicotine	Picciotto, Zoli, Rimondini, Lena, et al. (1998), Champtiaux, Gotti, Cordero-Erausquin, David, et al. (2003), and Gotti et al. (2008)
B3	Modulates locomotor activity, prepulse inhibition of acoustic startle, and dysregulation of α6 assembly	Cui, Booker, Allen, Grady, et al. (2003), Gotti, Moretti, Clementi, Riganti, et al. (2005), and Gotti, Fornasari, and Clementi (1997)
B4	No distinguishable effect	Salminen et al. (2004)
A3	Activate DA release into striatum	Azam and McIntosh (2005)
A4*	Increases sensitivity to nicotine in SA protocol	Gotti et al. (2008)
A5	Regulatory subunit; indirectly increases SA and reduces nicotine-induced behaviors	Fowler, Lu, Johnson, Marks, and Kenny (2011), Gotti et al. (1997), and Shiffman (2008)
A6	Activate DA release into striatum and reduces SA dose dependently	Azam and McIntosh (2005) and Drenan, Grady, Whiteaker, McClure-Begley, et al. (2008)
A7	Reduced SA of nicotine	Markou and Paterson (2001), Salminen et al. (2004), and Brunzell and McIntosh (2012)

*Denotes that the stated effect is only seen when the noted subtypes are expressed on the same receptor.

From the genetic information, transgenic mouse models can address some important questions. For example, using the α4-S248F mice, it was shown that mecamylamine was not able to attenuate nicotine taking behavior (Madsen et al., 2015). This is due to this specific mouse having a point mutation within the α4 nAChR subunit that increases nicotine sensitivity and is directly implicated in nicotine addiction (Cahir, Pillidge, Drago, & Lawrence, 2011; Tapper, McKinney, Nashmi, Schwarz, et al., 2004). Similarly, Madsen et al. were able to target α6β2* or α4α6β2* nAChRs using the novel compound, bPiDI, to attenuate nicotine SA (Jones & Wonnacott, 2004). KO of α5 nAChRs have been shown to have a substantial effect on nicotine sensitivity and short-term effects of nicotine (Salas et al., 2003; Shiffman, 2008); A5 KO mice show unregulated self-administration rates. Even when nicotine dose has been increased, there is no sign of compensatory decrease in intake (Fowler et al., 2011). This implicates the α5 nAChR subunit in the role of controlling nicotine intake, which is also consistent with human data gathered from GWAS (MacQueen et al., 2014). For a deeper understanding of the genetics behind nicotine addiction, future models may need to incorporate polygenic knockin/knockout paradigms and specific drug targets that focus on specific aspects of nicotine dependence.

Further to this point, different strains of isogenic rats have been shown to have different responses to nicotine SA. A study, carried out by Chen, Hiler, Tolley, Matta, and Sharp (2012), showed there was no perceived correlation between high food SA and high nicotine SA, showing that there is a different mechanism involved in the motivation for either food or nicotine. However, the different rat strains could be split into high, intermediate and low responders of nicotine. Upon subsequent transcriptome analysis of the nucleus accumbens shell (AcbS) it was shown that there are genes present that correlate negatively to nicotine SA, and may act in a protective manner (Chen, Luo, Gong, Matta, & Sharp, 2014). This might be a new way of isolating animals that may be at risk of developing addictive behavior toward nicotine and thus making more effective drug models or alternatively, upregulate specific genes in the AcbS via medication to possibly protect against potential nicotine dependence.

6.6 Environment Factors of Nicotine Addiction

It is well known that in humans, there are a multitude of factors that influence tobacco smoking on a social and environmental level. However, this has been rarely touched upon in animal models due to the difficulty in successfully capturing this behavior or replicating these conditions. A model, established by Chen, Sharp, Matta, and Wu (2011), provides the sensory input of tobacco smoking in humans, replicated by an olfactory and gustatory (OG) cue accompanied by a dose of nicotine to the rat. This OG cue prevents the animals from SA nicotine. However, peri-adolescent rats that were presented with visual presentation and orofacial contact with other rats established SA, which implicates adolescent social interactions in promoting nicotine seeking behavior by surmounting the aversive properties associated with the drug.

Furthermore, fostering has also been shown to increase nicotine SA in drug–naïve offspring but does not change SA behavior of prenatally nicotine-exposed offspring (Roguski, Chen, Sharp, & Matta, 2013). Nicotine exposure during the full gestational period, along with ethanol, augments nicotine SA by possibly altering DA transmission via NMDA receptors in the VTA (Roguski, Sharp, Chen, & Matta, 2014). These models effectively capture effects that can reduce nicotine-seeking behavior based on how nicotine intake interacts with environmental or social factors.

6.7 Effect of Nonnicotine Constituents of Tobacco

It has been shown that smoking cigarettes is more rewarding than receiving nicotine by itself via nicotine gum, patches, or nasal spray (Henningfield & Goldberg, 1983; Pomerleau, Marks, & Pomerleau, 2000; Sutherland et al., 1992). This gestures toward the belief that there is something in the tobacco that modulates smoking behavior in humans (Rose, 2006).

More currently, self-administration studies have been undertaken to investigate the differential effects of the nonnicotine constituents of tobacco and how it modifies nicotine self-administration in rats. It was discovered that some compounds modulate nicotine intake by increasing or decreasing its reinforcing properties, as shown in Table 2.

These chemicals have clear effects related to nicotine intake and investigation of these compounds may lead to novel nicotine IVSA models in rodents that include these different constituents and thus result in a more reliable method to treat nicotine dependence vis-à-vis tobacco smoke.

The observation that monoamine oxidase B (MAO-B) activity is 40% lower in the brain of smokers compared to nonsmokers, with the decrease in MAO-B being reversible after long-term smoking abstinence (Fowler et al., 1996a, 1996b) provides another insight as a potential target for

Table 2 The Effect of Nonnicotinic Tobacco Compounds On the Primary Effects, Intake, Absorption, and Metabolism of Nicotine

Nonnicotinic Tobacco Compound	Effects	References
Acetaldehyde	Increases reinforcing effects of nicotine	Belluzzi, Wang, and Leslie (2005)
Pyridine	Decreases reinforcing effects of nicotine	Farone (1996) and Flinchum (1993)
Beta-carboline	Increases reinforcing effects of nicotine	Guillem et al. (2006)
Nornicotine	Increases DA release from nucleus accumbens and contributes to nicotine dependence	Green, Crooks, Bardo, and Dwoskin (2001)
Ammonia/ammonia derivatives	Increase the amount of nicotine volatilised and increasing nicotine intake	Henningfield, Pankow, and Garrett (2004)
Levulinic acid	Enhances binding of nicotine to nAChRs	Keithly, Ferris Wayne, Cullen, and Connolly (2005) and Rabinoff, Caskey, Rissling, and Park (2007)
Propylene glycol and pyrazine	Correlated to the favorable sensory effects of cigarettes in humans	Glantz, Bero, Hanauer, and Barnes (1998)
Gamma-valerolactone	Inhibitor of CYP2A6 and increases systemic nicotine levels	Juvonen, Gynther, Pasanen, Alhava, and Poso (2000)
Tranylcypromine	Inhibits MAO and increases rewarding effect of low doses of nicotine	Villegier, Lotfipour, McQuown, Belluzzi, and Leslie (2007)

intervention. Inhibition of MAO-B by tobacco smoke may contribute to the rewarding effects of smoking and is believed to be due to compounds found in tobacco, like tranylcypromine (Villegier et al., 2006). With this knowledge, the selective and irreversible MAO-B inhibitor, selegiline, has the potential to reduce withdrawal symptoms and increase abstinence rates in patients. Using the IVSA paradigm with morphine, it has been shown that selegiline modifies the rates of extinction but does not alter cue-induced reinstatement, reacquisition or precipitated withdrawal in rats (Grasing, He, & Li, 2005). However, in human clinical trials, selegiline has been shown to significantly reduce smoking satisfaction during smoking and decrease cravings during abstinence in combination with a nicotine skin patch (Biberman, Neumann, Katzir, & Gerber, 2003). This discrepancy between observations from animal models compared with those observed in human trials raises issues on translation between preclinical models and clinical trials. However, studies have found that MAO-A inhibition and not MAO-B may play a more important role in sustaining nicotine seeking behavior (Fowler et al., 1996a, 1996b; Villegier et al., 2006), which has lead to the evaluation of moclobemide (a reversible MAO-A inhibitor) in smoking sessation in an ongoing phase II trial adjunct to NRT.

6.8 Importance of Food Restriction

For IVSA models, it is important whilst training, the animals are maintained at a certain level of food restriction. It has been shown both clinically and preclinically that caloric restrictions or overeating affect neural processes associated with reinforcement (Carr, 2002). Lower body weight and food restriction increases the self-administration and motor activating effects of reinforcing drugs. Specifically, there is increased nicotine SA in rats that have been weight restricted and food deprived (Donny et al., 1998). Further studies have shown that dopamine D1 receptors in the accumbens shell, dorsal medial prefrontal cortex, and basolateral amygdala are associated with food-deprivation stress-induced reinstatement of drug seeking (Tobin, Sedki, Abbas, & Shalev, 2013). In a study carried out in human, there were greater dopamine model reward-learning signals in the ventral striatum, insula, and prefrontal cortex in patients suffering from anorexia nervosa (Frank et al., 2012). These studies show that via food deprivation, the brain becomes more sensitive to the activation of the DA reward pathway via illicit drugs.

6.9 Effect of Nicotine on Appetite

Nicotine is known to exert anorectic effects in humans and animals (Filozof, Fernandez Pinilla, & Fernandez-Cruz, 2004; Grunberg, 2007; Mineur, Abizaid, Rao, Salas, et al., 2011) via tobacco smoke or self-administered nicotine. However, the method by which it produces this effect is not fully understood. It is believed that it suppresses appetite and therefore alters feeding patterns which ultimately results in reduction of body mass (Bray, 2000;

Grunberg, 1986; Miyata, Meguid, Fetissov, Torelli, & Kim, 1999; Zhang, Meguid, Miyata, Varma, & Fetissov, 2001). It is also believed that neurons in the hypothalamus play a key integrative role in feeding-linked information by connecting brain areas associated with motivation and emotion-related behavior (medial PFC and cingulate cortex, basal forebrain, and medial septal nuclei) with areas associated with habitual behavior and aversive responses (nucleus accumbens, anterior cingulate, amygdala, ventral tegmental area, substantia nigra, and raphe nuclei) (Bittencourt & Elias, 1998; Broberger, De Lecea, Sutcliffe, & Hokfelt, 1998; Fadel & Deutch, 2002; Pajolla et al., 2001; Peyron et al., 1998; Stratford & Kelley, 1999). Appetite suppression may prove to be a potential target for drug therapy for smoking cessation, as weight gain upon cessation is another reason for high relapse rate in humans, as well as a common reason why some people initiate tobacco smoking (Audrain-McGovern & Benowitz, 2011; Meyers et al., 1997). This approach was validated indirectly via partial clinical success of the CB-1 receptor antagonist, SR141716A, to reduce both conditioned nicotine and sucrose seeking in a long-term extinction–reinstatement model (De Vries, de Vries, Janssen, & Schoffelmeer, 2005).

6.10 Studies in Comorbid Populations

The aforementioned models that are used to characterize the different aspects of drug addiction use naïve rats from an outbred colony. However, the main issue with subjects is that in many cases within the human population only a minor proportion of these subjects will become dependent on nicotine after short or intermediate exposure (up to 40 days with 2 h sessions per day). Studies carried out using cocaine revealed that about 20% of rats showed three addiction-like behaviors after three months of self-administration, to a point where the behavior is classified as being pathological in accordance with the DSM-V (Deroche-Gamonet, Belin, & Piazza, 2004). These findings show that addiction is not restricted to just humans but also exists in laboratory animals. Despite the use of large amounts of a drug over a prolonged period of time, most individuals are resistant to addiction and ultimately some of the nosographic constructions used to classify psychiatric diseases in humans correspond to unitary behavioral dimensions in laboratory animals.

Alternative models have been proposed in relation to nicotine addiction to enhance the efficacy of tested pharmacotherapies, as it is well established that addiction is often comorbid with other psychiatric disorders. Studies regarding eating disorders have gathered substantial evidence suggesting that substance abuse can be found to occur in up to 49% of people with eating disorders and there is cooccurence in up to 32% of substance use cases (Courbasson & Brunshaw, 2009; Dunn, Neighbors, Fossos, & Larimer, 2009; Piran & Robinson, 2006). Patients who undergo withdrawal from nicotine have been shown to exhibit symptoms of anxiety and depression as stated previous, and schizophrenic patients (60–90%) are chronic tobacco smokers (Chapman, Ragg, & McGeechan, 2009; de Leon & Diaz, 2005; de Leon et al., 1995).

This has led to the use of animal models of nicotine dependence within subjects exhibiting some of the symptoms of schizophrenia or depression using models of nicotine addiction (Chambers & Self, 2002; Paterson & Markou, 2007). Studies have been carried out using varenicline to treat tobacco addiction in schizophrenia, with rats, in order to reverse the cognitive deficits associated with schizophrenia, as there has been emerging evidence of cognitive benefits in both patients and controls using the drug, as stated previously. It was found effective in patients that had heavier smoking behavior and tobacco abstinence impaired visuospatial working memory, which was not found in controls (George, Vessicchio, Termine, Sahady, et al., 2002; Tidey, Rohsenow, Kaplan, & Swift, 2005). Varenicline reduced craving induced by overnight abstinence in patients and controls. Additionally, varenicline attenuated neurobiological and cognitive decline biomarkers in smokers and nonsmokers with schizophrenia or schizoaffective disorder (Hong, Thaker, McMahon, Summerfelt, et al., 2011), as well as being able to attenuate tobacco-withdrawal-induced visuospatial working memory deficits in patients with schizophrenia (Wing, Wass, Soh, & George, 2012). These findings suggest that the tobacco withdrawal syndrome has specific features of schizophrenia, and that the mechanism by which varenicline aids smoking cessation may differ in schizophrenic patients. This may lead to proposed use of rodents that incorporate developmental aspects of schizophrenia, bipolar disorder, anxiety disorders and other psychiatric conditions with nicotine dependence. Comparison between these groups will help to distinguish underlying mechanisms for treatment of nicotine dependence and the respective disorder concomitantly.

7. SUMMARY

Traditional operant conditioning procedures have shown without a doubt that intravenous nicotine can support lever-press responding in rodents. However, the primary reinforcing effect is not a good representation of tobacco smoking in humans since there are many factors that control nicotine intake. This review postulates that more clinically relevant rodents models for smoking cessation may yield more effective smoking cessation treatments. Better experimental design with more appropriate treatment regimens will ultimately lead to better translation of the models and lead to more successful identification of potential treatments for tobacco smoking.

REFERENCES

Adams, J. D., Lee, S. J., & Hoffmann, D. (1984). Carcinogenic agents in cigarette smoke and the influence of nitrate on their formation. *Carcinogenesis*, *5*(2), 221–223.

Addolorato, G., Leggio, L., Abenavoli, L., Agabio, R., et al. (2006). Baclofen in the treatment of alcohol withdrawal syndrome: A comparative study vs diazepam. *American Journal of Medicine*, *119*(3). http://dx.doi.org/10.1016/j.amjmed.2005.08.042. 276. e213-278.

Allison, C., & Shoaib, M. (2013). Nicotine improves performance in an attentional set shifting task in rats. *Neuropharmacology*, *64*, 314–320. http://dx.doi.org/10.1016/j.neuropharm.2012.06.055.

Anderson, P. J., Wilson, J. D., & Hiller, F. C. (1989). Particle size distribution of mainstream tobacco and marijuana smoke. Analysis using the electrical aerosol analyzer. *The American Review of Respiratory Disease*, *140*(1), 202–205. http://dx.doi.org/10.1164/ajrccm/140.1.202.

Ariyoshi, N., Sawamura, Y., & Kamataki, T. (2001). A novel single nucleotide polymorphism altering stability and activity of CYP2a6. *Biochemical and Biophysical Research Communications*, *281*(3), 810–814. http://dx.doi.org/10.1006/bbrc.2001.4422.

Arroyo, M., Markou, A., Robbins, T. W., & Everitt, B. J. (1998). Acquisition, maintenance and reinstatement of intravenous cocaine self-administration under a second-order schedule of reinforcement in rats: Effects of conditioned cues and continuous access to cocaine. *Psychopharmacology*, *140*(3), 331–344.

Audrain-McGovern, J., & Benowitz, N. L. (2011). Cigarette smoking, nicotine, and body weight. *Clinical Pharmacology and Therapeutics*, *90*(1), 164–168. http://dx.doi.org/10.1038/clpt.2011.105.

Azam, L., & McIntosh, J. M. (2005). Effect of novel α-conotoxins on nicotine-stimulated [3H]dopamine release from rat striatal synaptosomes. *Journal of Pharmacology and Experimental Therapeutics*, *312*(1), 231–237. http://dx.doi.org/10.1124/jpet.104.071456.

Babin, S. L., Hood, A. J., Wassef, A. A., Williams, N. G., Patel, S. S., & Sereno, A. B. (2011). Effects of haloperidol on cognition in schizophrenia patients depend on baseline performance: A saccadic eye movement study. *Progress in Neuro-Psychopharmacology & Biological Psychiatry*, *35*(7), 1753 1764. http://dx.doi.org/10.1016/j.pnpbp.2011.06.004.

Beardsley, P. M., & Shelton, K. L. (2012). Prime-, stress-, and cue-induced reinstatement of extinguished drug-reinforced responding in rats: Cocaine as the prototypical drug of abuse. *Current Protocols in Neuroscience*. http://dx.doi.org/10.1002/0471142301.ns0939s61. Chapter 9, Unit 9.39.

Belluzzi, J. D., Wang, R., & Leslie, F. M. (2005). Acetaldehyde enhances acquisition of nicotine self-administration in adolescent rats. *Neuropsychopharmacology*, *30*(4), 705–712. http://dx.doi.org/10.1038/sj.npp.1300586.

Benowitz, N. L. (2010). Nicotine addiction. *New England Journal of Medicine*, *362*(24), 2295–2303. http://dx.doi.org/10.1056/NEJMra0809890.

Benowitz, N. L., Tyndale, R. F., Jacob, P., III, & Swan, G. E. (2002). CYP2A6 polymorphisms and nicotine metabolism [abstract]. *Clinical Pharmacology and Therapeutics*, *71*, P41.

Biala, G., & Kruk, M. (2009). Effects of co-administration of bupropion and nicotine or D-amphetamine on the elevated plus maze test in mice. *Journal of Pharmacy and Pharmacology*, *61*(4), 493–502. http://dx.doi.org/10.1211/jpp/61.04.0012.

Biberman, R., Neumann, R., Katzir, I., & Gerber, Y. (2003). A randomized controlled trial of oral selegiline plus nicotine skin patch compared with placebo plus nicotine skin patch for smoking cessation. *Addiction*, *98*(10), 1403–1407.

Bittencourt, J. C., & Elias, C. F. (1998). Melanin-concentrating hormone and neuropeptide EI projections from the lateral hypothalamic area and zona incerta to the medial septal nucleus and spinal cord: A study using multiple neuronal tracers. *Brain Research*, *805*(1–2), 1–19.

Braida, D., Ponzoni, L., Martucci, R., Sparatore, F., Gotti, C., & Sala, M. (2014). Role of neuronal nicotinic acetylcholine receptors (nAChRs) on learning and memory in zebrafish. *Psychopharmacology*, *231*(9), 1975–1985. http://dx.doi.org/10.1007/s00213-013-3340-1.

Bray, G. A. (2000). Reciprocal relation of food intake and sympathetic activity: Experimental observations and clinical implications. *International Journal of Obesity and Related Metabolic Disorders*, *24*(Suppl. 2), S8–S17.

Breslau, N., Kilbey, M. M., & Andreski, P. (1992). Nicotine withdrawal symptoms and psychiatric disorders: Findings from an epidemiologic study of young adults. *The American Journal of Psychiatry*, *149*(4), 464–469.

Broberger, C., De Lecea, L., Sutcliffe, J. G., & Hokfelt, T. (1998). Hypocretin/orexin- and melanin-concentrating hormone-expressing cells form distinct populations in the rodent lateral hypothalamus: Relationship to the neuropeptide Y and agouti gene-related protein systems. *Journal of Comparative Neurology*, *402*(4), 460–474.

Brose, L. S., West, R., & Stapleton, J. A. (2013). Comparison of the effectiveness of varenicline and combination nicotine replacement therapy for smoking cessation in clinical practice. *Mayo Clinic Proceedings*, *88*(3), 226–233. http://dx.doi.org/10.1016/j.mayocp.2012.11.013.

Brunzell, D. H., & McIntosh, J. M. (2012). Alpha7 nicotinic acetylcholine receptors modulate motivation to self-administer nicotine: Implications for smoking and schizophrenia. *Neuropsychopharmacology*, *37*(5), 1134–1143. http://dx.doi.org/10.1038/npp.2011.299.

Buhler, M., Vollstadt-Klein, S., Kobiella, A., Budde, H., et al. (2010). Nicotine dependence is characterized by disordered reward processing in a network driving motivation. *Biological Psychiatry*, *67*(8), 745–752. http://dx.doi.org/10.1016/j.biopsych.2009.10.029.

Caggiula, A. R., Donny, E. C., White, A. R., Chaudhri, N., et al. (2001). Cue dependency of nicotine self-administration and smoking. *Pharmacology Biochemistry and Behavior*, *70*(4), 515–530.

Caggiula, A. R., Donny, E. C., White, A. R., Chaudhri, N., et al. (2002). Environmental stimuli promote the acquisition of nicotine self-administration in rats. *Psychopharmacology*, *163*(2), 230–237. http://dx.doi.org/10.1007/s00213-002-1156-5.

Cahir, E., Pillidge, K., Drago, J., & Lawrence, A. J. (2011). The necessity of alpha4* nicotinic receptors in nicotine-driven behaviors: Dissociation between reinforcing and motor effects of nicotine. *Neuropsychopharmacology*, *36*(7), 1505–1517. http://dx.doi.org/10.1038/npp.2011.35.

Caille, S., Clemens, K., Stinus, L., & Cador, M. (2012). Modeling nicotine addiction in rats. *Methods in Molecular Biology*, *829*, 243–256. http://dx.doi.org/10.1007/978-1-61779-458-2_15.

Cantin, L., Lenoir, M., Augier, E., Vanhille, N., et al. (2010). Cocaine is low on the value ladder of rats: Possible evidence for resilience to addiction. *PLoS One*, *5*(7), e11592. http://dx.doi.org/10.1371/journal.pone.0011592.

Carr, K. D. (2002). Augmentation of drug reward by chronic food restriction: Behavioral evidence and underlying mechanisms. *Physiology and Behavior, 76*(3), 353–364.

Chambers, R. A., & Self, D. W. (2002). Motivational responses to natural and drug rewards in rats with neonatal ventral hippocampal lesions: An animal model of dual diagnosis schizophrenia. *Neuropsychopharmacology, 27*(6), 889–905. http://dx.doi.org/10.1016/s0893-133x(02)00365-2.

Champtiaux, N., Gotti, C., Cordero-Erausquin, M., David, D. J., et al. (2003). Subunit composition of functional nicotinic receptors in dopaminergic neurons investigated with knock-out mice. *Journal of Neuroscience, 23*(21), 7820–7829.

Chapman, S., Ragg, M., & McGeechan, K. (2009). Citation bias in reported smoking prevalence in people with schizophrenia. *The Australian and New Zealand Journal of Psychiatry, 43*(3), 277–282. http://dx.doi.org/10.1080/00048670802653372.

Chen, H., Fu, Y., & Sharp, B. M. (2008). Chronic nicotine self-administration augments hypothalamic-pituitary-adrenal responses to mild acute stress. *Neuropsychopharmacology, 33*(4), 721–730. http://dx.doi.org/10.1038/sj.npp.1301466.

Chen, H., Hiler, K. A., Tolley, E. A., Matta, S. G., & Sharp, B. M. (2012). Genetic factors control nicotine self-administration in isogenic adolescent rat strains. *PLoS One*, 7(8), e44234. http://dx.doi.org/10.1371/journal.pone.0044234.

Chen, H., Luo, R., Gong, S., Matta, S. G., & Sharp, B. M. (2014). Protection genes in nucleus accumbens shell affect vulnerability to nicotine self-administration across isogenic strains of adolescent rat. *PLoS One, 9*(1), e86214. http://dx.doi.org/10.1371/journal.pone.0086214.

Chen, H., Sharp, B. M., Matta, S. G., & Wu, Q. (2011). Social interaction promotes nicotine self-administration with olfactogustatory cues in adolescent rats. *Neuropsychopharmacology, 36*(13), 2629–2638. http://dx.doi.org/10.1038/npp.2011.149.

Corrigall, W. A. (1999). Nicotine self-administration in animals as a dependence model. *Nicotine & Tobacco Research, 1*(1), 11–20.

Corrigall, W. A., & Coen, K. M. (1989). Nicotine maintains robust self-administration in rats on a limited-access schedule. *Psychopharmacology, 99*(4), 473–478.

Corrigall, W. A., & Coen, K. M. (1991). Selective dopamine antagonists reduce nicotine self-administration. *Psychopharmacology, 104*(2), 171–176.

Corrigall, W. A., Coen, K. M., & Adamson, K. L. (1994). Self-administered nicotine activates the mesolimbic dopamine system through the ventral tegmental area. *Brain Research, 653*(1–2), 278–284.

Courbasson, C., & Brunshaw, J. M. (2009). The relationship between concurrent substance use disorders and eating disorders with personality disorders. *International Journal of Environmental Research and Public Health, 6*(7), 2076–2089. http://dx.doi.org/10.3390/ijerph6072076.

Cui, C., Booker, T. K., Allen, R. S., Grady, S. R., et al. (2003). The β3 nicotinic receptor subunit: A component of α-conotoxin MII-binding nicotinic acetylcholine receptors that modulate dopamine release and related behaviors. *Journal of Neuroscience, 23*(35), 11045–11053.

Dani, J. A., & De Biasi, M. (2001). Cellular mechanisms of nicotine addiction. *Pharmacology Biochemistry and Behavior, 70*(4), 439–446.

Darna, M., Chow, J. J., Yates, J. R., Charnigo, R. J., Beckmann, J. S., Bardo, M. T., et al. (2015). Role of serotonin transporter function in rat orbitofrontal cortex in impulsive choice. *Behavioural Brain Research, 293*, 134–142. http://dx.doi.org/10.1016/j.bbr.2015.07.025.

Dawe, S., Gerada, C., Russell, M. A., & Gray, J. A. (1995). Nicotine intake in smokers increases following a single dose of haloperidol. *Psychopharmacology, 117*(1), 110–115.

de Beaurepaire, R. (2014). The use of very high-doses of baclofen for the treatment of alcohol-dependence: A case series. *Frontiers in Psychiatry, 5*, 143. http://dx.doi.org/10.3389/fpsyt.2014.00143.

de Leon, J., Dadvand, M., Canuso, C., White, A. O., Stanilla, J. K., & Simpson, G. M. (1995). Schizophrenia and smoking: An epidemiological survey in a state hospital. *The American Journal of Psychiatry*, *152*(3), 453–455.

de Leon, J., & Diaz, F. J. (2005). A meta-analysis of worldwide studies demonstrates an association between schizophrenia and tobacco smoking behaviors. *Schizophrenia Research*, *76*(2–3), 135–157. http://dx.doi.org/10.1016/j.schres.2005.02.010.

De Vries, T. J., de Vries, W., Janssen, M. C., & Schoffelmeer, A. N. (2005). Suppression of conditioned nicotine and sucrose seeking by the cannabinoid-1 receptor antagonist SR141716A. *Behavioural Brain Research*, *161*(1), 164–168. http://dx.doi.org/10.1016/j.bbr.2005.02.021.

de Wit, H., & Stewart, J. (1981). Reinstatement of cocaine-reinforced responding in the rat. *Psychopharmacology*, *75*(2), 134–143.

DeNoble, V., & Mele, P. (2006). Intravenous nicotine self-administration in rats: Effects of mecamylamine, hexamethonium and naloxone. *Psychopharmacology*, *184*(3–4), 266–272. http://dx.doi.org/10.1007/s00213-005-0054-z.

Deroche-Gamonet, V., Belin, D., & Piazza, P. V. (2004). Evidence for addiction-like behavior in the rat. *Science*, *305*(5686), 1014–1017. http://dx.doi.org/10.1126/science.1099020.

Di Chiara, G., & Bassareo, V. (2007). Reward system and addiction: What dopamine does and doesn't do. *Current Opinion in Pharmacology*, 7(1), 69–76. http://dx.doi.org/10.1016/j.coph.2006.11.003.

Diergaarde, L., Pattij, T., Poortvliet, I., Hogenboom, F., de Vries, W., Schoffelmeer, A. N., et al. (2008). Impulsive choice and impulsive action predict vulnerability to distinct stages of nicotine seeking in rats. *Biological Psychiatry*, *63*(3), 301–308. http://dx.doi.org/10.1016/j.biopsych.2007.07.011.

Donny, E. C., Caggiula, A. R., Mielke, M. M., Jacobs, K. S., Rose, C., & Sved, A. F. (1998). Acquisition of nicotine self-administration in rats: The effects of dose, feeding schedule, and drug contingency. *Psychopharmacology*, *136*(1), 83–90.

Donny, E. C., Chaudhri, N., Caggiula, A. R., Evans-Martin, F. F., et al. (2003). Operant responding for a visual reinforcer in rats is enhanced by noncontingent nicotine: Implications for nicotine self-administration and reinforcement. *Psychopharmacology*, *169*(1), 68–76. http://dx.doi.org/10.1007/s00213-003-1473-3.

Drenan, R. M., Grady, S. R., Whiteaker, P., McClure-Begley, T., et al. (2008). In vivo activation of midbrain dopamine neurons via sensitized, high-affinity α6* nicotinic acetylcholine receptors. *Neuron*, *60*(1), 123–136. http://dx.doi.org/10.1016/j.neuron.2008.09.009.

Dunn, E. C., Neighbors, C., Fossos, N., & Larimer, M. E. (2009). A cross-lagged evaluation of eating disorder symptomatology and substance-Use problems. *Journal of Studies on Alcohol and Drugs*, *70*(1), 106–116. http://dx.doi.org/10.15288/jsad.2009.70.106.

Engleman, E. A., Rodd, Z. A., Bell, R. L., & Murphy, J. M. (2008). The role of 5-HT3 receptors in drug abuse and as a target for pharmacotherapy. *CNS & Neurological Disorders Drug Targets*, 7(5), 454–467. http://dx.doi.org/10.2174/187152708786927886.

Epping-Jordan, M. P., Watkins, S. S., Koob, G. F., & Markou, A. (1998). Dramatic decreases in brain reward function during nicotine withdrawal. *Nature*, *393*(6680), 76–79. http://dx.doi.org/10.1038/30001.

Everitt, B. J., & Robbins, T. W. (2000). Second-order schedules of drug reinforcement in rats and monkeys: Measurement of reinforcing efficacy and drug-seeking behaviour. *Psychopharmacology*, *153*(1), 17–30.

Fadel, J., & Deutch, A. Y. (2002). Anatomical substrates of orexin-dopamine interactions: Lateral hypothalamic projections to the ventral tegmental area. *Neuroscience*, *111*(2), 379–387.

Farone, W. (1996). *Pyridine studies*. Retrieved 23 November 2015, from, http://legacy.library.ucsf.edu/tid/kgn93c00.

Filozof, C., Fernandez Pinilla, M. C., & Fernandez-Cruz, A. (2004). Smoking cessation and weight gain. *Obesity Reviews*, *5*(2), 95–103. http://dx.doi.org/10.1111/j.1467-789X.2004.00131.x.

Flinchum, G. (1993). *German high nicotine extract*. Bates no. 510783956/3956. Retrieved 23 November 2015, from, http://legacy.library.ucsf.edu/tid/bju53d00.

Forget, B., Pushparaj, A., & Le Foll, B. (2010). Granular insular cortex inactivation as a novel therapeutic strategy for nicotine addiction. *Biological Psychiatry*, *68*(3), 265–271. http://dx.doi.org/10.1016/j.biopsych.2010.01.029.

Foulds, J., Gandhi, K. K., Steinberg, M. B., Richardson, D. L., Williams, J. M., Burke, M. V., et al. (2006). Factors associated with quitting smoking at a tobacco dependence treatment clinic. *American Journal of Health Behavior*, *30*(4), 400–412. http://dx.doi.org/10.5555/ajhb.2006.30.4.400.

Fowler, C. D., Arends, M. A., & Kenny, P. J. (2008). Subtypes of nicotinic acetylcholine receptors in nicotine reward, dependence, and withdrawal: Evidence from genetically modified mice. *Behavioural Pharmacology*, *19*(5–6), 461–484. http://dx.doi.org/10.1097/FBP.0b013e32830c360e.

Fowler, C. D., Lu, Q., Johnson, P. M., Marks, M. J., & Kenny, P. J. (2011). Habenular alpha5 nicotinic receptor subunit signalling controls nicotine intake. *Nature*, *471*(7340), 597–601. http://dx.doi.org/10.1038/nature09797.

Fowler, J. S., Volkow, N. D., Wang, G. J., Pappas, N., et al. (1996a). Inhibition of monoamine oxidase B in the brains of smokers. *Nature*, *379*(6567), 733–736. http://dx.doi.org/10.1038/379733a0.

Fowler, J. S., Volkow, N. D., Wang, G.-J., Pappas, N., et al. (1996b). Brain monoamine oxidase A inhibition in cigarette smokers. *Proceedings of the National Academy of Sciences of the United States of America*, *93*(24), 14065–14069.

Frank, G. K., Reynolds, J. R., Shott, M. E., Jappe, L., Yang, T. T., Tregellas, J. R., et al. (2012). Anorexia nervosa and obesity are associated with opposite brain reward response. *Neuropsychopharmacology*, *37*(9), 2031–2046. http://dx.doi.org/10.1038/npp.2012.51.

Freathy, R. M., Ring, S. M., Shields, B., Galobardes, B., et al. (2009). A common genetic variant in the 15q24 nicotinic acetylcholine receptor gene cluster (CHRNA5–CHRNA3–CHRNB4) is associated with a reduced ability of women to quit smoking in pregnancy. *Human Molecular Genetics*, *18*(15), 2922–2927. http://dx.doi.org/10.1093/hmg/ddp216.

George, T. P., Vessicchio, J. C., Termine, A., Sahady, D. M., et al. (2002). Effects of smoking abstinence on visuospatial working memory function in schizophrenia. *Neuropsychopharmacology*, *26*(1), 75–85. http://dx.doi.org/10.1016/s0893-133x(01)00296-2.

Glantz, S. A., Bero, L., Hanauer, P., & Barnes, D. E. (1998). *The cigarette papers*. Berkeley and Los Angeles: University of California Press.

Goldberg, S. R., Kelleher, R. T., & Goldberg, D. M. (1981). Fixed-ratio responding under second-order schedules of food presentation or cocaine injection. *Journal of Pharmacol and Experimental Therapeutics*, *218*(1), 271–281.

Gotti, C., Fornasari, D., & Clementi, F. (1997). Human neuronal nicotinic receptors. *Progress in Neurobiology*, *53*(2), 199–237. http://dx.doi.org/10.1016/S0301-0082(97)00034-8.

Gotti, C., Moretti, M., Clementi, F., Riganti, L., et al. (2005). Expression of nigrostriatal α6-containing nicotinic acetylcholine receptors is selectively reduced, but not eliminated, by β3 subunit gene deletion. *Molecular Pharmacology*, *67*(6), 2007–2015. http://dx.doi.org/10.1124/mol.105.011940.

Gotti, C., Moretti, M., Meinerz, N. M., Clementi, F., Gaimarri, A., Collins, A. C., et al. (2008). Partial deletion of the nicotinic cholinergic receptor α4 or β2 subunit genes changes the acetylcholine sensitivity of receptor-mediated 86Rb+ efflux in cortex and thalamus and alters relative expression of α4 and β2 subunits. *Molecular Pharmacology*, *73*(6), 1796–1807. http://dx.doi.org/10.1124/mol.108.045203.

Grasing, K., He, S., & Li, N. (2005). Selegiline modifies the extinction of responding following morphine self-administration, but does not alter cue-induced reinstatement, reacquisition of morphine reinforcement, or precipitated withdrawal. *Pharmacological Research*, *51*(1), 69–78. http://dx.doi.org/10.1016/j.phrs.2004.07.004.

Green, T. A., Crooks, P. A., Bardo, M. T., & Dwoskin, L. P. (2001). Contributory role for nornicotine in nicotine neuropharmacology: Nornicotine-evoked [3H]dopamine overflow from rat nucleus accumbens slices. *Biochemical Pharmacology*, *62*(12), 1597–1603.

Grella, S. L., Funk, D., Coen, K., Li, Z., & Le, A. D. (2014). Role of the kappa-opioid receptor system in stress-induced reinstatement of nicotine seeking in rats. *Behavioural Brain Research*, *265*, 188–197. http://dx.doi.org/10.1016/j.bbr.2014.02.029.

Grunberg, N. E. (1986). Nicotine as a psychoactive drug: Appetite regulation. *Psychopharmacology Bulletin*, *22*(3), 875–881.

Grunberg, N. E. (2007). A neurobiological basis for nicotine withdrawal. *Proceedings of the National Academy of Sciences of the United States of America*, *104*(46), 17901–17902. http://dx.doi.org/10.1073/pnas.0708964104.

Guillem, K., Vouillac, C., Azar, M. R., Parsons, L. H., Koob, G. F., Cador, M., et al. (2006). Monoamine oxidase A rather than monoamine oxidase B inhibition increases nicotine reinforcement in rats. *European Journal of Neuroscience*, *24*(12), 3532–3540. http://dx.doi.org/10.1111/j.1460-9568.2006.05217.x.

Hadidi, H., Zahlsen, K., Idle, J. R., & Cholerton, S. (1997). A single amino acid substitution (Leu160His) in cytochrome P450 CYP2A6 causes switching from 7-hydroxylation to 3-hydroxylation of coumarin. *Food and Chemical Toxicology*, *35*(9), 903–907.

Hahn, B., Shoaib, M., & Stolerman, I. P. (2002). Effects of dopamine receptor antagonists on nicotine-induced attentional enhancement. *Behavioural Pharmacology*, *13*(8), 621–632. http://dx.doi.org/10.1097/01.fbp.0000046506.84018.28.

Hare, T. A., Camerer, C. F., & Rangel, A. (2009). Self-control in decision-making involves modulation of the vmPFC valuation system. *Science*, *324*(5927), 646–648. http://dx.doi.org/10.1126/science.1168450.

Harrison, A. A., Liem, Y. T., & Markou, A. (2001). Fluoxetine combined with a serotonin-1A receptor antagonist reversed reward deficits observed during nicotine and amphetamine withdrawal in rats. *Neuropsychopharmacology*, *25*(1), 55–71. http://dx.doi.org/10.1016/s0893-133x(00)00237-2.

Heishman, S. J., & Henningfield, J. E. (2000). Tolerance to repeated nicotine administration on performance, subjective, and physiological responses in nonsmokers. *Psychopharmacology*, *152*(3), 321–333.

Henningfield, J. E., & Goldberg, S. R. (1983). Control of behavior by intravenous nicotine injections in human subjects. *Pharmacology Biochemistry and Behavior*, *19*(6), 1021–1026.

Henningfield, J., Pankow, J., & Garrett, B. (2004). Ammonia and other chemical base tobacco additives and cigarette nicotine delivery: Issues and research needs. *Nicotine & Tobacco Research*, *6*(2), 199–205. http://dx.doi.org/10.1080/1462220042000202472.

Higgins, S. T., Budney, A. J., Bickel, W. K., Foerg, F. E., Donham, R., & Badger, G. J. (1994). Incentives improve outcome in outpatient behavioral treatment of cocaine dependence. *Archives of General Psychiatry*, *51*(7), 568–576.

Hong, L. E., Thaker, G. K., McMahon, R. P., Summerfelt, A., et al. (2011). Effects of moderate-dose treatment with varenicline on neurobiological and cognitive biomarkers in smokers and nonsmokers with schizophrenia or schizoaffective disorder. *Archives of General Psychiatry*, *68*(12), 1195–1206. http://dx.doi.org/10.1001/archgenpsychiatry.2011.83.

Hughes, J. R., Gulliver, S. B., Fenwick, J. W., Valliere, W. A., et al. (1992). Smoking cessation among self-quitters. *Health Psychology*, *11*(5), 331–334. http://dx.doi.org/10.1037//0278-6133.11.5.331.

Irvine, E. E., Cheeta, S., & File, S. E. (2001). Tolerance to nicotine's effects in the elevated plus-maze and increased anxiety during withdrawal. *Pharmacology Biochemistry and Behavior*, *68*(2), 319–325.

Johnson, P. M., Hollander, J. A., & Kenny, P. J. (2008). Decreased brain reward function during nicotine withdrawal in C57BL6 mice: Evidence from intracranial self-stimulation (ICSS) studies. *Pharmacology Biochemistry and Behavior*, *90*(3), 409–415. http://dx.doi.org/10.1016/j.pbb.2008.03.024.

Jones, I. W., & Wonnacott, S. (2004). Precise localization of alpha7 nicotinic acetylcholine receptors on glutamatergic axon terminals in the rat ventral tegmental area. *Journal of Neuroscience*, *24*(50), 11244–11252. http://dx.doi.org/10.1523/jneurosci.3009-04.2004.

Juvonen, R. O., Gynther, J., Pasanen, M., Alhava, E., & Poso, A. (2000). Pronounced differences in inhibition potency of lactone and non-lactone compounds for mouse and human coumarin 7-hydroxylases (CYP2A5 and CYP2A6). *Xenobiotica*, *30*(1), 81–92.

Kalivas, P. W., & McFarland, K. (2003). Brain circuitry and the reinstatement of cocaine-seeking behavior. *Psychopharmacology*, *168*(1–2), 44–56. http://dx.doi.org/10.1007/s00213-003-1393-2.

Kantak, K. M., Black, Y., Valencia, E., Green-Jordan, K., & Eichenbaum, H. B. (2002). Dissociable effects of lidocaine inactivation of the rostral and caudal basolateral amygdala on the maintenance and reinstatement of cocaine-seeking behavior in rats. *Journal of Neuroscience*, *22*(3), 1126–1136.

Keithly, L., Ferris Wayne, G., Cullen, D. M., & Connolly, G. N. (2005). Industry research on the use and effects of levulinic acid: A case study in cigarette additives. *Nicotine & Tobacco Research*, 7(5), 761–771. http://dx.doi.org/10.1080/14622200500259820.

Koob, G. F., & Volkow, N. D. (2010). Neurocircuitry of addiction. *Neuropsychopharmacology*, *35*(1), 217–238. http://dx.doi.org/10.1038/npp.2009.110.

Kouneiher, F., Charron, S., & Koechlin, E. (2009). Motivation and cognitive control in the human prefrontal cortex. *Nature Neuroscience*, *12*(7), 939–945. http://dx.doi.org/10.1038/nn.2321.

Kuwabara, H., Heishman, S. J., Brasic, J. R., Contoreggi, C., et al. (2014). Mu opioid receptor binding correlates with nicotine dependence and reward in smokers. *PLoS One*, *9*(12), e113694. http://dx.doi.org/10.1371/journal.pone.0113694.

Lecca, D., Cacciapaglia, F., Valentini, V., Gronli, J., Spiga, S., & Di Chiara, G. (2006). Preferential increase of extracellular dopamine in the rat nucleus accumbens shell as compared to that in the core during acquisition and maintenance of intravenous nicotine self-administration. *Psychopharmacology*, *184*(3–4), 435–446. http://dx.doi.org/10.1007/s00213-005-0280-4.

Lee, L. Y., Gerhardstein, D. C., Wang, A. L., & Burki, N. K. (1993). Nicotine is responsible for airway irritation evoked by cigarette smoke inhalation in men. *Journal of Applied Physiology (Bethesda, Md: 1985)*, *75*(5), 1955–1961.

Lenoir, M., Guillem, K., Koob, G. F., & Ahmed, S. H. (2012). Drug specificity in extended access cocaine and heroin self-administration. *Addiction Biology*, *17*(6), 964–976. http://dx.doi.org/10.1111/j.1369-1600.2011.00385.x.

Lerman, C., LeSage, M. G., Perkins, K. A., O'Malley, S. S., Siegel, S. J., Benowitz, N. L., et al. (2007). Translational research in medication development for nicotine dependence. *Nature Reviews Drug Discovery*, *6*(9), 746–762.

Lesage, M. G. (2009). Toward a nonhuman model of contingency management: Effects of reinforcing abstinence from nicotine self-administration in rats with an alternative nondrug reinforcer. *Psychopharmacology*, *203*(1), 13–22. http://dx.doi.org/10.1007/s00213-008-1362-x.

MacQueen, D. A., Heckman, B. W., Blank, M. D., Van Rensburg, K. J., Park, J. Y., Drobes, D. J., et al. (2014). Variation in the alpha 5 nicotinic acetylcholine receptor subunit gene predicts cigarette smoking intensity as a function of nicotine content. *The Pharmacogenomics Journal, 14*(1), 70–76. http://dx.doi.org/10.1038/tpj.2012.50.

Madsen, H. B., Koghar, H. S., Pooters, T., Massalas, J. S., Drago, J., & Lawrence, A. J. (2015). Role of α4- and α6-containing nicotinic receptors in the acquisition and maintenance of nicotine self-administration. *Addiction Biology, 20*(3), 500–512. http://dx.doi.org/10.1111/adb.12148.

Malin, D. H., Lake, J. R., Carter, V. A., Cunningham, J. S., Hebert, K. M., Conrad, D. L., et al. (1994). The nicotinic antagonist mecamylamine precipitates nicotine abstinence syndrome in the rat. *Psychopharmacology, 115*(1–2), 180–184.

Markou, A. (2008). Review. Neurobiology of nicotine dependence. *Philosophical Transactions of the Royal Society of London, Series B: Biological Sciences, 363*(1507), 3159–3168. http://dx.doi.org/10.1098/rstb.2008.0095.

Markou, A., & Paterson, N. E. (2001). The nicotinic antagonist methyllycaconitine has differential effects on nicotine self-administration and nicotine withdrawal in the rat. *Nicotine & Tobacco Research, 3*(4), 361–373. http://dx.doi.org/10.1080/14622200110073380.

McGue, M., & Bouchard, T. J., Jr. (1984). Adjustment of twin data for the effects of age and sex. *Behavior Genetics, 14*(4), 325–343.

Messina, E. S., Tyndale, R. F., & Sellers, E. M. (1997). A major role for CYP2A6 in nicotine C-oxidation by human liver microsomes. *Journal of Pharmacol and Experimental Therapeutics, 282*(3), 1608–1614.

Meyer, P. J., Cogan, E. S., & Robinson, T. E. (2014). The form of a conditioned stimulus Can influence the degree to which it acquires incentive motivational properties. *PLoS One, 9*(8)e107194. http://dx.doi.org/10.1371/journal.pone.0098163.

Meyers, A. W., Klesges, R. C., Winders, S. E., Ward, K. D., Peterson, B. A., & Eck, L. H. (1997). Are weight concerns predictive of smoking cessation? A prospective analysis. *Journal of Consulting and Clinical Psychology, 65*(3), 448–452.

Mineur, Y. S., Abizaid, A., Rao, Y., Salas, R., et al. (2011). Nicotine decreases food intake through activation of POMC neurons. *Science, 332*(6035), 1330–1332. http://dx.doi.org/10.1126/science.1201889.

Miyata, G., Meguid, M. M., Fetissov, S. O., Torelli, G. F., & Kim, H. J. (1999). Nicotine's effect on hypothalamic neurotransmitters and appetite regulation. *Surgery, 126*(2), 255–263.

Naqvi, N. H., Rudrauf, D., Damasio, H., & Bechara, A. (2007). Damage to the insula disrupts addiction to cigarette smoking. *Science, 315*(5811), 531–534. http://dx.doi.org/10.1126/science.1135926.

Nestler, E. J. (2005). Is there a common molecular pathway for addiction? *Nature Neuroscience, 8*(11), 1445–1449. http://dx.doi.org/10.1038/nn1578.

Norman, A. B., & Tsibulsky, V. L. (2001). Satiety threshold regulates maintained self-administration: Comment on Lynch and Carroll (2001). *Experimental and Clinical Psychopharmacology, 9*(2), 151–154. discussion 160–152.

O'Connor, E. C., Parker, D., Rollema, H., & Mead, A. N. (2010). The alpha4beta2 nicotinic acetylcholine-receptor partial agonist varenicline inhibits both nicotine self-administration following repeated dosing and reinstatement of nicotine seeking in rats. *Psychopharmacology, 208*(3), 365–376. http://dx.doi.org/10.1007/s00213-009-1739-5.

Pajolla, G. P., Crippa, G. E., Correa, S. A., Moreira, K. B., Tavares, R. F., & Correa, F. M. (2001). The lateral hypothalamus is involved in the pathway mediating the hypotensive response to cingulate cortex-cholinergic stimulation. *Cellular and Molecular Neurobiology, 21*(4), 341–356.

Palmatier, M. I., Liu, X., Caggiula, A. R., Donny, E. C., & Sved, A. F. (2007). The role of nicotinic acetylcholine receptors in the primary reinforcing and reinforcement-enhancing effects of nicotine. *Neuropsychopharmacology, 32*(5), 1098–1108. http://dx.doi.org/10.1038/sj.npp.1301228.

Panlilio, L. V., Hogarth, L., & Shoaib, M. (2015). Concurrent access to nicotine and sucrose in rats. *Psychopharmacology, 232*(8), 1451–1460. http://dx.doi.org/10.1007/s00213-014-3787-8.

Parrott, A. C. (1995). Stress modulation over the day in cigarette smokers. *Addiction, 90*(2), 233–244.

Paterson, N. E., & Markou, A. (2007). Animal models and treatments for addiction and depression co-morbidity. *Neurotoxicity Research, 11*(1), 1–32.

Perkins, K., Stitzer, M., & Lerman, C. (2006). Medication screening for smoking cessation: A proposal for new methodologies. *Psychopharmacology, 184*(3-4), 628–636. http://dx.doi.org/10.1007/s00213-005-0105-5.

Peyron, C., Tighe, D. K., van den Pol, A. N., de Lecea, L., Heller, H. C., Sutcliffe, J. G., et al. (1998). Neurons containing hypocretin (orexin) project to multiple neuronal systems. *Journal of Neuroscience, 18*(23), 9996–10015.

Picciotto, M. R., Zoli, M., Rimondini, R., Lena, C., et al. (1998). Acetylcholine receptors containing the beta2 subunit are involved in the reinforcing properties of nicotine. *Nature, 391*(6663), 173–177. http://dx.doi.org/10.1038/34413.

Pilla, M., Perachon, S., Sautel, F., Garrido, F., et al. (1999). Selective inhibition of cocaine-seeking behaviour by a partial dopamine D3 receptor agonist. *Nature, 400*(6742), 371–375. http://dx.doi.org/10.1038/22560.

Piran, N., & Robinson, S. R. (2006). Associations between disordered eating behaviors and licit and illicit substance use and abuse in a university sample. *Addictive Behaviors, 31*(10), 1761–1775. http://dx.doi.org/10.1016/j.addbeh.2005.12.021.

Pomerleau, C. S., Marks, J. L., & Pomerleau, O. F. (2000). Who gets what symptom? Effects of psychiatric cofactors and nicotine dependence on patterns of smoking withdrawal symptomatology. *Nicotine & Tobacco Research, 2*(3), 275–280.

Pontieri, F. E., Tanda, G., Orzi, F., & Di Chiara, G. (1996). Effects of nicotine on the nucleus accumbens and similarity to those of addictive drugs. *Nature, 382*(6588), 255–257. http://dx.doi.org/10.1038/382255a0.

Pouletty, P. (2002). Drug addictions: Towards socially accepted and medically treatable diseases. *Nature Reviews Drug Discovery, 1*(9), 731–736.

Pushparaj, A., Hamani, C., Yu, W., Shin, D. S., Kang, B., Nobrega, J. N., et al. (2013). Electrical stimulation of the insular region attenuates nicotine-taking and nicotine-seeking behaviors. *Neuropsychopharmacology, 38*(4), 690–698. http://dx.doi.org/10.1038/npp.2012.235.

Rabinoff, M., Caskey, N., Rissling, A., & Park, C. (2007). Pharmacological and chemical effects of cigarette additives. *American Journal of Public Health, 97*(11), 1981–1991. http://dx.doi.org/10.2105/AJPH.2005.078014.

Rauhut, A. S., Neugebauer, N., Dwoskin, L. P., & Bardo, M. T. (2003). Effect of bupropion on nicotine self-administration in rats. *Psychopharmacology, 169*(1), 1–9. http://dx.doi.org/10.1007/s00213-003-1450-x.

Ray, R., Mitra, N., Baldwin, D., Guo, M., et al. (2010). Convergent evidence that choline acetyltransferase gene variation is associated with prospective smoking cessation and nicotine dependence. *Neuropsychopharmacology, 35*(6), 1374–1382. http://dx.doi.org/10.1038/npp.2010.7.

Roguski, E. E., Chen, H., Sharp, B. M., & Matta, S. G. (2013). Fostering itself increases nicotine self-administration in young adult male rats. *Psychopharmacology, 229*(2), 227–234. http://dx.doi.org/10.1007/s00213-013-3093-x.

Roguski, E. E., Sharp, B. M., Chen, H., & Matta, S. G. (2014). Full-gestational exposure to nicotine and ethanol augments nicotine self-administration by altering ventral tegmental dopaminergic function due to NMDA receptors in adolescent rats. *Journal of Neurochemistry, 128*(5), 701–712. http://dx.doi.org/10.1111/jnc.12504.

Rose, J. E. (1988). The role of upper airway stimulation in smoking. *Progress in Clinical and Biological Research, 261*, 95–106.

Rose, J. E. (2006). Nicotine and nonnicotine factors in cigarette addiction. *Psychopharmacology, 184*(3–4), 274–285. http://dx.doi.org/10.1007/s00213-005-0250-x.

Rose, J. E., Behm, F. M., & Levin, E. D. (1993). Role of nicotine dose and sensory cues in the regulation of smoke intake. *Pharmacology Biochemistry and Behavior, 44*(4), 891–900.

Rose, J. E., Behm, F. M., & Westman, E. C. (1998). Nicotine-mecamylamine treatment for smoking cessation: The role of pre-cessation therapy. *Experimental and Clinical Psychopharmacology, 6*(3), 331–343.

Rose, J. E., & Corrigall, W. A. (1997). Nicotine self-administration in animals and humans: Similarities and differences. *Psychopharmacology, 130*(1), 28–40.

Rose, J. E., Sampson, A., Levin, E. D., & Henningfield, J. E. (1989). Mecamylamine increases nicotine preference and attenuates nicotine discrimination. *Pharmacology Biochemistry and Behavior, 32*(4), 933–938.

Rose, J. E., Westman, E. C., Behm, F. M., Johnson, M. P., & Goldberg, J. S. (1999). Blockade of smoking satisfaction using the peripheral nicotinic antagonist trimethaphan. *Pharmacology Biochemistry and Behavior, 62*(1), 165–172.

Salas, R., Orr-Urtreger, A., Broide, R. S., Beaudet, A., Paylor, R., & De Biasi, M. (2003). The nicotinic acetylcholine receptor subunit α5 mediates short-term effects of nicotine in vivo. *Molecular Pharmacology, 63*(5), 1059–1066. http://dx.doi.org/10.1124/mol.63.5.1059.

Salminen, O., Murphy, K. L., McIntosh, J. M., Drago, J., Marks, M. J., Collins, A. C., et al. (2004). Subunit composition and pharmacology of two classes of striatal presynaptic nicotinic acetylcholine receptors mediating dopamine release in mice. *Molecular Pharmacology, 65*(6), 1526–1535. http://dx.doi.org/10.1124/mol.65.6.1526.

Saules, K. K., Schuh, L. M., Arfken, C. L., Reed, K., Kilbey, M. M., & Schuster, C. R. (2004). Double-blind placebo-controlled trial of fluoxetine in smoking cessation treatment including nicotine patch and cognitive-behavioral group therapy. *The American Journal on Addictions, 13*(5), 438–446. http://dx.doi.org/10.1080/10550490490512762.

Schindler, C. W., Panlilio, L. V., & Goldberg, S. R. (2002). Second-order schedules of drug self-administration in animals. *Psychopharmacology, 163*(3–4), 327–344. http://dx.doi.org/10.1007/s00213-002-1157-4.

Shaham, Y., Erb, S., & Stewart, J. (2000). Stress-induced relapse to heroin and cocaine seeking in rats: A review. *Brain Research. Brain Research Reviews, 33*(1), 13–33.

Shiffman, S. (2008). Effect of nicotine lozenges on affective smoking withdrawal symptoms: Secondary analysis of a randomized, double-blind, placebo-controlled clinical trial. *Clinical Therapeutics, 30*(8), 1461–1475. http://dx.doi.org/10.1016/j.clinthera.2008.07.019.

Shiffman, S., Brockwell, S. E., Pillitteri, J. L., & Gitchell, J. G. (2008). Use of smoking-cessation treatments in the United States. *American Journal of Preventive Medicine, 34*(2), 102–111. http://dx.doi.org/10.1016/j.amepre.2007.09.033.

Shoaib, M., & Bizarro, L. (2005). Deficits in a sustained attention task following nicotine withdrawal in rats. *Psychopharmacology, 178*(2–3), 211–222. http://dx.doi.org/10.1007/s00213-004-2004-6.

Shoaib, M., Schindler, C. W., & Goldberg, S. R. (1997). Nicotine self-administration in rats: Strain and nicotine pre-exposure effects on acquisition. *Psychopharmacology, 129*(1), 35–43.

Shoaib, M., Sidhpura, N., & Shafait, S. (2003). Investigating the actions of bupropion on dependence-related effects of nicotine in rats. *Psychopharmacology, 165*(4), 405–412. http://dx.doi.org/10.1007/s00213-002-1277-x.

Shoaib, M., & Stolerman, I. P. (1999). Plasma nicotine and cotinine levels following intravenous nicotine self-administration in rats. *Psychopharmacology, 143*(3), 318–321.

Stapleton, J., West, R., Hajek, P., Wheeler, J., et al. (2013). Randomized trial of nicotine replacement therapy (NRT), bupropion and NRT plus bupropion for smoking cessation: Effectiveness in clinical practice. *Addiction, 108*(12), 2193–2201. http://dx.doi.org/10.1111/add.12304.

Stolerman, I. P., Goldfarb, T., Fink, R., & Jarvik, M. E. (1973). Influencing cigarette smoking with nicotine antagonists. *Psychopharmacologia, 28*(3), 247–259. http://dx.doi.org/10.1007/BF00429305.

Stolerman, I. P., Naylor, C., Elmer, G. I., & Goldberg, S. R. (1999). Discriminsation and self-administration of nicotine by inbred strains of mice. *Psychopharmacology, 14*(3), 297–306. http://dx.doi.org/10.1007/s002130050837.

Stratford, T. R., & Kelley, A. E. (1999). Evidence of a functional relationship between the nucleus accumbens shell and lateral hypothalamus subserving the control of feeding behavior. *Journal of Neuroscience, 19*(24), 11040–11048.

Sutherland, G., Stapleton, J. A., Russell, M. A., Jarvis, M. J., Hajek, P., Belcher, M., et al. (1992). Randomised controlled trial of nasal nicotine spray in smoking cessation. *Lancet, 340*(8815), 324–329.

Tapper, A. R., McKinney, S. L., Nashmi, R., Schwarz, J., et al. (2004). Nicotine activation of alpha4* receptors: Sufficient for reward, tolerance, and sensitization. *Science, 306*(5698), 1029–1032. http://dx.doi.org/10.1126/science.1099420.

Tarter, R. E., Kirisci, L., Feske, U., & Vanyukov, M. (2007). Modeling the pathways linking childhood hyperactivity and substance use disorder in young adulthood. *Psychology of Addictive Behaviors, 21*(2), 266–271. http://dx.doi.org/10.1037/0893-164x.21.2.266.

Thorgeirsson, T. E., Geller, F., Sulem, P., Rafnar, T., et al. (2008). A variant associated with nicotine dependence, lung cancer and peripheral arterial disease. *Nature, 452*(7187), 638–642. http://dx.doi.org/10.1038/nature06846.

Tidey, J. W., Rohsenow, D. J., Kaplan, G. B., & Swift, R. M. (2005). Cigarette smoking topography in smokers with schizophrenia and matched non-psychiatric controls. *Drug and Alcohol Dependence, 80*(2), 259–265. http://dx.doi.org/10.1016/j.drugalcdep.2005.04.002.

Tobin, S., Sedki, F., Abbas, Z., & Shalev, U. (2013). Antagonism of the dopamine D1-like receptor in mesocorticolimbic nuclei attenuates acute food deprivation-induced reinstatement of heroin seeking in rats. *European Journal of Neuroscience, 37*(6), 972–981. http://dx.doi.org/10.1111/ejn.12112.

Torres, O. V., Pipkin, J. A., Ferree, P., Carcoba, L. M., & O'Dell, L. E. (2015). Nicotine withdrawal increases stress-associated genes in the nucleus accumbens of female rats in a hormone-dependent manner. *Nicotine & Tobacco Research, 17*(4), 422–430. http://dx.doi.org/10.1093/ntr/ntu278.

Tyndale, R. F., & Sellers, E. M. (2001). Variable CYP2A6-mediated nicotine metabolism alters smoking behavior and risk. *Drug Metabolism and Disposition, 29*(4 Pt. 2), 548–552.

Varani, A. P., Aso, E., Moutinho, L. M., Maldonado, R., & Balerio, G. N. (2014). Attenuation by baclofen of nicotine rewarding properties and nicotine withdrawal manifestations. *Psychopharmacology, 231*(15), 3031–3040. http://dx.doi.org/10.1007/s00213-014-3469-6.

Villegier, A. S., Lotfipour, S., McQuown, S. C., Belluzzi, J. D., & Leslie, F. M. (2007). Tranylcypromine enhancement of nicotine self-administration. *Neuropharmacology, 52*(6), 1415–1425. http://dx.doi.org/10.1016/j.neuropharm.2007.02.001.

Villegier, A. S., Salomon, L., Granon, S., Changeux, J. P., Belluzzi, J. D., Leslie, F. M., et al. (2006). Monoamine oxidase inhibitors allow locomotor and rewarding responses to nicotine. *Neuropsychopharmacology, 31*(8), 1704–1713. http://dx.doi.org/10.1038/sj.npp.1300987.

Wald, N. J., & Hackshaw, A. K. (1996). Cigarette smoking: An epidemiological overview. *British Medical Bulletin*, *52*(1), 3–11.

West, R. J., Hajek, P., & Belcher, M. (1989). Severity of withdrawal symptoms as a predictor of outcome of an attempt to quit smoking. *Psychological Medicine*, *19*(4), 981–985.

Wing, V. C., & Shoaib, M. (2007). Examining the clinical efficacy of bupropion and nortriptyline as smoking cessation agents in a rodent model of nicotine withdrawal. *Psychopharmacology*, *195*(3), 303–313. http://dx.doi.org/10.1007/s00213-007-0902-0.

Wing, V. C., & Shoaib, M. (2010). Second-order schedules of nicotine reinforcement in rats: Effect of AM251. *Addiction Biology*, *15*(4), 393–402. http://dx.doi.org/10.1111/j.1369-1600.2009.00204.x.

Wing, V. C., Wass, C. E., Soh, D. W., & George, T. P. (2012). A review of neurobiological vulnerability factors and treatment implications for comorbid tobacco dependence in schizophrenia. *Annals of the New York Academy of Sciences*, *1248*, 89–106. http://dx.doi.org/10.1111/j.1749-6632.2011.

Wouda, J. A., Riga, D., De Vries, W., Stegeman, M., et al. (2011). Varenicline attenuates cue-induced relapse to alcohol, but not nicotine seeking, while reducing inhibitory response control. *Psychopharmacology*, *216*(2), 267–277. http://dx.doi.org/10.1007/s00213-011-2213-8.

Yu, G., & Sharp, B. M. (2010). Nicotine self-administration diminishes stress-induced norepinephrine secretion but augments adrenergic-responsiveness in the hypothalamic paraventricular nucleus and enhances adrenocorticotropic hormone and corticosterone release. *Journal of Neurochemistry*, *112*(5), 1327–1337. http://dx.doi.org/10.1111/j.1471-4159.2009.06551.x.

Zhang, L., Meguid, M. M., Miyata, G., Varma, M., & Fetissov, S. O. (2001). Role of hypothalamic monoamines in nicotine-induced anorexia in menopausal rats. *Surgery*, *130*(2), 133–142. http://dx.doi.org/10.1067/msy.2001.115513.

CHAPTER SIX

An Animal Model of Alcohol Dependence to Screen Medications for Treating Alcoholism

H.C. Becker[*,†,‡,1], **M.F. Lopez**[*]
[*]Charleston Alcohol Research Center, Charleston, SC, United States
[†]Medical University of South Carolina, Charleston, SC, United States
[‡]RHJ Department of Veterans Affairs Medical Center, Charleston, SC, United States
[1]Corresponding author: e-mail address: beckerh@musc.edu

Contents

Abstract

Despite the high prevalence of alcohol use disorders in the United States, only a relatively small percentage of those afflicted seek treatment. This is further compounded by the fact that there are too few medications available to effectively treat this significant public health problem. The need for identifying and evaluating more effective treatments that aid in preventing relapse and/or tempering risky and harmful alcohol consumption cannot be overstated. Use of animal models represents a critical step in the process of screening, identifying, and informing plans for prioritizing the most promising candidate medications that can be advanced to the next stage of evaluation (clinical laboratory paradigms and controlled clinical trials). Numerous animal models have been developed to study excessive levels of alcohol self-administration. In recent years, a large literature has amassed of studies in which rodent models of dependence have been linked with alcohol self-administration procedures. This chapter focuses on studies employing a dependence model that involves chronic exposure to alcohol vapor by inhalation, which yields in both mice and rats significant escalation of voluntary alcohol

International Review of Neurobiology, Volume 126
ISSN 0074-7742
http://dx.doi.org/10.1016/bs.irn.2016.02.006

consumption. These animal models of dependence and alcohol self-administration have revealed valuable insights about underlying mechanisms that drive excessive drinking. Additionally, this preclinical approach is useful in evaluating the effects of medications on escalated drinking associated with dependence vs more stable levels displayed by nondependent animals.

Alcohol abuse and dependence are serious medical and social problems in the United States and, thus, constitute a major public health concern. In the past decade, alcohol use and abuse has significantly increased in the United States (Dawson, Goldstein, Saha, & Grant, 2015). In a recent epidemiological study, the estimated 12-month prevalence for alcohol use disorder (AUD) indicated that over 32 million Americans met DSM-5 diagnostic criteria spanning the full spectrum of mild to moderate to severe AUD (Grant et al., 2015). Excessive alcohol use has devastating medical consequences, levying a tremendous toll on the health-care industry. Heavy drinking can cause or increase risk for many medical illnesses and accounts for a substantial number of emergency medical admissions related to accidents and violent crime. The economic costs of alcohol abuse and dependence are estimated at 235 billion dollars per year due to health-care expenditures, lost productivity, and damage/loss of property (Rehm et al., 2009). In addition to these medical and economic burdens, the cost to society in terms of personal tragedy and loss is enormous. Unfortunately, only a relatively small percentage of those suffering with AUD that would benefit from treatment actually seek treatment. A recent study estimated that less than 10% of those with a history of alcohol abuse seek treatment (Grant et al., 2015). This indicates the need to raise public awareness of the disease and available treatment options and, at the same time, invest resources in developing new and more effective treatments for this major public health problem.

Alcoholism is a chronic relapsing disease, and relapse represents a major challenge to treatment efforts. To date, there is no therapeutic intervention that has proven to be fully satisfactory in preventing relapse, sustaining abstinence, or tempering amount of drinking when a "slip" occurs. For some individuals, heavy and frequent binge-like drinking can lead to general loss of control over regulation of alcohol consumption. In many cases, such drinking patterns can lead to the development of alcohol dependence, rendering these individuals more susceptible to relapse as well as more vulnerable to engaging in drinking behavior that often spirals out of control. Many such individuals make numerous attempts at curtailing their alcohol use, only to find themselves reverting back to patterns of excessive consumption again. The need to provide effective treatments for these individuals suffering from AUDs cannot be overstated.

Significant advancements have been made in our understanding of neurobiological underpinnings and environmental factors that influence motivation to drink as well as consequences of excessive alcohol use. The development of preclinical models has played a central role in expanding our knowledge of the myriad biological, genetic, and environmental forces that influence excessive alcohol consumption. However, despite significant advancements in our understanding of the complexities of the addiction process as well as factors that influence motivation to engage in risky unhealthy drinking, development of new medications for treating AUDs has lagged behind. Thus, a major challenge for the field is to employ established preclinical models to identify and evaluate new therapeutics, which may be added to the armament of treatment strategies that can be advanced to clinical trials and, ultimately, delivered to those suffering from the ravages of alcohol abuse and dependence.

1. ANIMAL MODELS OF EXCESSIVE ALCOHOL DRINKING

Heavy (excessive) levels of drinking and increased vulnerability to relapse represent hallmark features of AUDs and alcoholism. As noted earlier, the development of animal models that incorporate these key behavioral characteristics has played a key role in advancing our knowledge about biological underpinnings and environmental circumstances that engender such maladaptive behavior. These preclinical models also are crucial in providing testing procedures for identifying new potential therapeutic targets and evaluating efficacy and safety of various treatment strategies.

Numerous experimental approaches have been employed in developing rodent models of excessive alcohol self-administration. One of the major obstacles in this work is that rodents typically do not self-administer alcohol in sufficient amounts to produce overt signs of intoxication. Further, when given the opportunity to voluntarily drink alcohol, even under circumstances when access is unlimited, rodents rarely will consume alcohol in a manner that results in significant elevation in blood alcohol levels (above legal limits). Thus, a major challenge for the field has been to overcome these critical problems so that animal models developed for studying alcohol consumption have greater clinical relevance and, thereby, greater potential for use in elucidating underlying mechanisms as well as identifying and evaluating new and more effective treatment approaches.

In recent years, several new models have been developed, and some older ones have been resurrected and refined that demonstrate excessive and physiologically relevant levels of alcohol consumption (Becker, 2013). In general, these models have incorporated genetic manipulations (eg, selective breeding for high alcohol drinking and preference) or environmental manipulations that involve modifying scheduled access to alcohol, scheduled periods of alcohol deprivation, and linking drinking procedures with dependence models (Becker, 2013). Thus, a variety of experimental approaches have been used to approximate clinical aspects and stages of alcohol addiction. This includes modeling genetic predisposition, binge-like patterns of drinking, relapse vulnerability, and alcohol consumption that reflects motivation related to acquired negative reinforcing effects of alcohol. Not surprisingly, each of these models possesses distinct advantages and disadvantages. Further, all of these models have been used to examine effects of various pharmacological agents (eg, Albrechet-Souza et al., 2015; Anderson, Becker, Adams, Jesudason, & Rorick-Kehn, 2014; Bell et al., 2012; Carnicella, Ron, & Barak, 2014; Crabbe, Harris, & Koob, 2011; Simms, Nielsen, Li, & Bartlett, 2014; Vengeliene, Bilbao, & Spanagel, 2014). This chapter will focus on models of dependence that engender high levels of alcohol self-administration, as well as use of these models for evaluation of medications that may temper excessive drinking associated with dependence.

2. ANIMAL MODELS OF DEPENDENCE AND EXCESSIVE ALCOHOL DRINKING

Alcohol dependence has long been viewed as playing a significant role in promoting and sustaining excessive levels of alcohol consumption, as well as driving increased susceptibility to relapse after periods of attempted abstinence (Cappell & LeBlanc, 1981; Grant, 1995). Indeed, a large literature has amassed of studies in which rodent models of dependence have been linked with alcohol self-administration procedures. Key features of these models that have contributed to their success include first establishing the positive reinforcing effects of alcohol and then presenting alcohol in the context of dependence, but in a manner that facilitates associating consumption of alcohol with the drug's ability to alleviate dysphoric aspects of dependence (withdrawal). This latter feature defines acquisition of the negative reinforcing effects of alcohol, which is thought to play an important role in increasing vulnerability to relapse and then maintaining heavy drinking once a slip occurs (Becker, 2013, 2014).

Over the past decade or so, numerous studies have demonstrated escalation of alcohol consumption using home-cage free-choice drinking models in dependent mice (Becker & Lopez, 2004; Dhaher, Finn, Snelling, & Hitzemann, 2008; Finn et al., 2007; Griffin, Lopez, & Becker, 2009; Lopez & Becker, 2005) and rats (Rimondini, Arlinde, Sommer, & Heilig, 2002; Rimondini, Sommer, & Heilig, 2003; Sommer et al., 2008). Studies have also employed operant conditioning procedures to demonstrate elevated alcohol self-administration in dependent mice (Chu, Koob, Cole, Zorrilla, & Roberts, 2007; Lopez, Anderson, & Becker, 2008) and rats (Funk & Koob, 2007; Funk, O'Dell, Crawford, & Koob, 2006; Funk, Zorrilla, Lee, Rice, & Koob, 2007; Gilpin, Misra, & Koob, 2008; Gilpin, Richardson, & Koob, 2008; Gilpin, Richardson, Lumeng, & Koob, 2008; Gilpin et al., 2009; O'Dell, Roberts, Smith, & Koob, 2004; Richardson et al., 2008; Rimondini, Thorsell, & Heilig, 2005; Roberts, Cole, & Koob, 1996; Roberts, Heyser, Cole, Griffin, & Koob, 2000). In these models, several procedures have been used to induce a state of alcohol dependence. For example, increased alcohol self-administration has been demonstrated in studies where dependence was induced by chronic administration of alcohol in a nutritionally fortified liquid diet (that served as the animals' sole source of calories and fluid) (Brown, Jackson, & Stephens, 1998; Chu et al., 2007; Gilpin et al., 2009; Schulteis, Hyytia, Heinrichs, & Koob, 1996), via intragastric infusions (Cunningham, Fidler, Murphy, Mulgrew, & Smitasin, 2013; Fidler et al., 2011, 2012) and inhalation of alcohol vapors (eg, Becker & Lopez, 2004; Rimondini et al., 2002; Roberts et al., 2000).

Additionally, in many instances these models have provided opportunities to study various withdrawal symptoms. Of particular relevance are symptoms that contribute to a negative emotional and dysphoric state associated with dependence, such as anxiety, heightened stress, anhedonia, and sleep disturbances (Becker, 2014; Heilig, Egli, Crabbe, & Becker, 2010; Koob, 2013). In as much as symptoms related to negative affect that often linger into protracted phases of abstinence contribute to relapse and potential for self-medication with alcohol, these models have proven useful for studying mechanisms underlying motivation to drink in the context of alcohol dependence. Nevertheless, focus on alcohol consumption in these dependence models has been the primary target for evaluation of potential medications.

3. AN ANIMAL MODEL OF DEPENDENCE-RELATED EXCESSIVE ALCOHOL DRINKING

Animal studies linking alcohol dependence and self-administration procedures have predominantly involved administering alcohol vapor in inhalation chambers, with the chronic alcohol exposure delivered in an intermittent fashion such that animals experience multiple withdrawal episodes. Using this experimental approach, escalation of alcohol self-administration has been demonstrated in dependent mice and rats compared to independent groups of animals that display relatively stable and modest levels of alcohol intake (Becker, 2013, 2014; Griffin, 2014; Lopez & Becker, 2014; Vendruscolo & Roberts, 2014). For example, rats exposed to chronic alcohol treatment interspersed with repeated episodes of withdrawal subsequently consumed significantly more alcohol than controls under free-choice unlimited (24 h/day) access conditions (Rimondini et al., 2002, 2003; Sommer et al., 2008). Similar results have been reported in mice using a dependence model involving repeated cycles of chronic intermittent ethanol vapor exposure and with voluntary alcohol consumption assessed using a limited access (2 h/day) schedule (Becker & Lopez, 2004; Dhaher et al., 2008; Finn et al., 2007; Lopez & Becker, 2005). The intensity of repeated chronic alcohol exposure (producing high and sustained blood alcohol levels) was shown to be critical in favoring escalation of alcohol consumption in the model (Griffin, Lopez, & Becker, 2009). Analysis of the temporal pattern of alcohol consumption revealed that dependent mice not only consumed more alcohol than nondependent animals over the entire 2-h access period, but the rate of consumption was faster and progressively increased over successive withdrawal test periods (Griffin, Lopez, Yanke, Middaugh, & Becker, 2009). Further, the effect appears specific to alcohol because repeated cycles of chronic intermittent alcohol exposure did not produce alterations in water intake or consumption of highly palatable fluids such as sucrose and saccharin (Becker & Lopez, 2004; Lopez, Griffin, Melendez, & Becker, 2012). This suggests that the increase in alcohol consumption is not a nonspecific effect related to a general need to hydrate with fluids or increase caloric intake.

Studies in rodents involving chronic alcohol vapor exposure to induce dependence also have employed operant conditioning procedures to demonstrate increased alcohol self-administration. These studies have been conducted with mice (Chu et al., 2007; Lopez et al., 2008) and rats (Funk & Koob, 2007; Funk et al., 2006, 2007; Gilpin, Misra, et al., 2008; Gilpin, Richardson, & Koob, 2008; Gilpin, Richardson, Lumeng, et al., 2008; Gilpin et al., 2009; O'Dell et al., 2004; Richardson et al., 2008; Rimondini et al., 2005; Roberts et al., 1996, 2000). Use of operant conditioning procedures has enabled the demonstration that chronic intermittent alcohol delivered by inhalation increases both alcohol seeking (responding) as well as consumption components of alcohol self-administration behavior. Further, employing progressive ratio schedules has demonstrated that the amount of work mice (Lopez et al., 2008) and rats (Vendruscolo et al., 2012; Walker & Koob, 2007) are willing to expend in order to gain access to alcohol is significantly increased following repeated cycles of chronic alcohol exposure and withdrawal experience. Recent studies also have shown that dependent animals display compulsive-like responding for alcohol, at least as indexed by relative insensitivity to devaluation of alcohol's rewarding effects (ie, persistent responding for alcohol even after its rewarding value is reduced by adulteration with quinine, or it is associated with lithium chloride-induced aversion; Lopez, Becker, & Chandler, 2014; Vendruscolo et al., 2012).

Regardless of whether free-choice drinking or operant conditioning procedures are used, enhanced alcohol responding/intake in dependent animals has been demonstrated at time points well beyond acute withdrawal (Lopez & Becker, 2005; Rimondini et al., 2003; Roberts et al., 2000; Valdez et al., 2002). For example, increasing the number of chronic alcohol exposure/withdrawal cycles was shown to not only further augment alcohol consumption but also sustain elevated levels of intake for a longer period of time (several weeks) following final withdrawal compared to intake in a separate group of nondependent mice (Lopez & Becker, 2005). Thus, increased alcohol self-administration in these dependence models is not a transient effect, and this provides opportunities for evaluation of medication effects on excessive drinking that occurs at temporally more distal time points relative to acute withdrawal.

Additionally, studies have shown that escalation of alcohol self-administration in these models is especially facilitated when dependence induction involved delivery of chronic alcohol in an intermittent rather than continuous fashion (Lopez & Becker, 2005; O'Dell et al., 2004). Thus, while elevated alcohol self-administration may eventually develop following long-term alcohol exposure, a pattern of chronic intermittent alcohol exposure that involves repeated withdrawal experience (which more closely models typical patterns of alcohol consumption in humans) accelerates the rate at which escalation of drinking is displayed. This suggests that providing an opportunity to consume alcohol in the context of repeated withdrawal experience plays a significant role in promoting enhanced motivation for alcohol.

One procedural difference between studies in rats and mice regards the timing of when animals are given the opportunity to self-administer alcohol in relation to withdrawal from chronic alcohol vapor exposure. Studies in rats have shown that alcohol responding and intake progressively increases when the rats are allowed to self-administer alcohol during repeated acute withdrawal episodes (1–12 h postinhalation exposure) (Roberts et al., 1996). In contrast, mice develop an aversion to alcohol when it is presented for consumption early after vapor exposure. Providing a forced abstinence period of at least 48 h following chronic alcohol vapor exposure enables escalation of self-administration in dependent mice (Lopez & Becker, 2005). However, this effect in mice may be dependent on genotype and the model of chronic alcohol exposure (Cunningham et al., 2013).

Studies in both mice and rats have shown that the escalation of alcohol self-administration following repeated cycles of chronic intermittent alcohol exposure produced significantly higher blood alcohol levels compared to that achieved by more modest and stable levels of intake by nondependent animals (Becker & Lopez, 2004; Roberts et al., 2000). Similar results were obtained in a study involving continuous alcohol vapor exposure and operant oral alcohol self-administration in rats (Gilpin et al., 2009). In another study, microdialysis procedures were used to demonstrate that the faster rate of alcohol intake and greater overall amount consumed by dependent mice produced significantly higher peak and more sustained alcohol concentrations measured in brain compared to levels achieved from consumption of alcohol in nondependent animals (Griffin, Lopez, Yanke, et al., 2009). Moreover, greater voluntary alcohol consumption in dependent mice produced brain alcohol concentrations that approximated those levels experienced during chronic intermittent alcohol exposure that rendered the subjects dependent in the first place. The extent to which achieving a threshold brain alcohol concentration plays a role in promoting as well as perpetuating enhanced alcohol drinking in dependent animals remains to be determined.

In summary, a large number of studies have successfully linked procedures for inducing alcohol dependence with self-administration protocols.

Most commonly, induction of dependence has been accomplished by delivery of chronic alcohol exposure via the inhalation route. This route of administration has many advantages (eg, ability to exert rigorous experimental control over variables such as duration, frequency, and intensity of exposure while minimizing compromised health), but a detraction relates to the fact that the inhalation procedure departs from the manner in which humans normally consume alcohol (orally). It is important to note, however, that these models of dependence and relapse drinking are not designed to examine how dependence develops, but, rather, the focus is on how a history of chronic alcohol exposure that renders subjects dependent alters motivational processes that engender excessive levels of consumption. As previously indicated, with few exceptions, rodents even when given free access to alcohol will not consume sufficient amounts to produce dependence. Thus, in order to study the impact of dependence on continued and sustained alcohol drinking, the dependence state must be experimentally induced. This approach has effectively been adopted in mouse and rat models. Further, incorporating alternating cycles of chronic alcohol exposure interspersed with periods when subjects have the opportunity to self-administer the drug enables evaluation of drinking in the context of both alcohol's positive and negative reinforcing effects. Indeed, a positive feature of this model relates to the ability to contrast relatively stable alcohol consumption in nondependent subjects with escalated drinking exhibited by dependent subjects. This powerful attribute of dependence and relapse drinking models has not only been exploited in studies aimed at elucidating underlying neuroadaptations and motivational mechanisms but also in studies focused on evaluation of pharmacotherapeutics.

4. EVALUATION OF PHARMACOTHERAPEUTICS IN MODELS OF ALCOHOL DEPENDENCE AND RELAPSE DRINKING

In recent years, a large number of studies have employed dependence and relapse drinking models to evaluate the capacity of various pharmacological agents to modulate alcohol consumption (Bell et al., 2015; Meinhardt & Sommer, 2015; Vendruscolo & Roberts, 2014). Table 1 summarizes work conducted evaluating the effects of various pharmacological treatments using these models. All of these studies have been conducted with rats or mice using home-cage drinking or operant self-administration procedures. Of note, all of these studies examined effects of drug treatments on

Table 1 Effect of Medication Treatments Evaluated in Rodent Models of Alcohol Dependence and Drinking

Target System	Medication	Mode of Action	Effect		Reference
			Nondep.	Dep.	
GABA	Baclofen	GABA-B agonist	→	↓	Walker and Koob (2007)
	Muscimol	GABA-A agonist	→	↓	Roberts et al. (1996)
	Gabapentin	GABA modulator	→	↓	Roberto et al. (2008)
Glutamate	Acamprosate	Modulator	→	↓	Rimondini et al. (2002)
	LY379268	mGluR2/3 agonist	↓	↓	Sidhpura, Weiss, and Martin-Fardon (2010)
	MTEP	mGluR1/5 antagonist	↓	↓	Sidhpura et al. (2010)
Opiates	Naltrexone	Nonselect antagonist	↓	↓	Gilpin, Richardson, and Koob (2008) and Walker and Koob (2008)
	Nalmefene	MOR/KOR antagonist	↓	↓	Kissler et al. (2014) and Nealey, Smith, Davis, Smith, and Walker (2011)
	BD-1063	DOR antagonist	→	↓	Sabino et al. (2009)
	CTOP + naltrindole	MOR/DOR antagonist	↓	→	Kissler et al. (2014) and Nealey et al. (2011)
	Nor-BNI	KOR antagonist	→	↓	Kissler et al. (2014), Nealey et al. (2011), Walker, Rasmussen, Raskind, and Koob (2008), and Walker, Zorrilla, and Koob (2011)

Table 1 Effect of Medication Treatments Evaluated in Rodent Models of Alcohol Dependence and Drinking—cont'd

Target System	Medication	Mode of Action	Effect Nondep.	Effect Dep.	Reference
Monoamines	Desipramine	SNRI	↓	↓	Simon O'Brien et al. (2011)
	Fluoxetine	SSRI	↓	↓	Simon O'Brien et al. (2011)
	Milnacipran	SNRI	→	↓	Simon O'Brien et al. (2011)
	Prazosin	α1-NE antagonist	↓	↓	Walker et al. (2008)
	Propranolol	β-NE antagonist	↓	↓	Gilpin and Koob (2010)
Neuropeptides	Antalarmin	CRF1R antagonist	→	↓	Funk et al. (2007)
	D-Phe-CRF_{12-41}	CRF1R antagonist	→	↓	Funk et al. (2006) and Valdez et al. (2002)
	MJL-1-109-2	CRF1R antagonist	→	↓	Funk et al. (2007)
	MTIP	CRF1R antagonist	→	↓	Gehlert et al. (2007)
	MPZP	CRF1R antagonist	→	↓	Gilpin, Richardson, and Koob (2008) and Richardson et al. (2008)
	R121919	CRF1R antagonist	→	↓	Funk et al. (2007) and Roberto et al. (2010)
	Urocortin3	CRF2R agonist	→	↓	Funk and Koob (2007)
	NPY	NPYR agonist	↓	↓	Gilpin et al. (2011) and Thorsell, Slawecki, and Ehlers (2005)

Continued

Table 1 Effect of Medication Treatments Evaluated in Rodent Models of Alcohol Dependence and Drinking—cont'd

Target System	Medication	Mode of Action	Effect		Reference
			Nondep.	Dep.	
	BIIE0246	NPY2R antagonist	→	→	Kallupi et al. (2014)
	JNJ-31020028	NPY2R antagonist	→	→	Kallupi et al. (2014)
	SSR149415	AVP V_{1b} antagonist	→	↓	Edwards, Guerrero, Ghoneim, Roberts, and Koob (2012)
Others	AC3174	Glucagon-1 agonist	→	↓	Suchankova et al. (2015)
	CGS 21680	Adenosine A2a agonist	→	↓	Houchi, Persyn, Legastelois, and Naassila (2013)
	FN-439	MP protease	–	↓	Smith, Nealey, Wright, and Walker (2011)
	Ibudilast	Nonselect PDE inhibitor	↓	↓	Bell et al. (2015)
	Mifepristone	GR antagonist	→	↓	Vendruscolo et al. (2012)
	SR141716A	CB1 antagonist	→	↓	Rodriguez de Fonseca, Roberts, Bilbao, Koob, and Navarro (1999)
	Spirapril	ACE inhibitor	↓	↓	Sommer et al. (2007)

alcohol consumption in dependent and, in most cases, nondependent animals. Thus, studies that evaluated drug effects on operant reinstatement models where effects on alcohol responding were examined under extinction conditions (ie, no alcohol consumption) are not included in Table 1.

In many instances, pharmacological agents were used to probe mechanisms underlying escalated drinking associated with dependence.

The development of alcohol dependence is a dynamic process that reflects a complex interplay among many neurobiological, genetic, experiential, and environmental factors. AUD is thought to encompass a composite of brain adaptations that manifest as dysfunctional reward processing, persistent negative affect/emotion along with sensitization of stress systems, and impaired executive function that compromises behavioral control (George & Koob, 2013; Koob, 2013). These adaptations are thought to reflect an allostatic state that is fueled by continued alcohol use (Koob, 2003), and marked by progressive dysregulation of brain mechanisms that mediate motivated behavior (reward circuitry), emotional stability (limbic and stress circuitry), and executive function (cortical circuitry). Thus, it is not surprising that many studies involving alcohol dependence and self-administration models have focused on drug treatments that target neurochemical systems that mediate pharmacological effects of alcohol (eg, GABA, glutamate, opioids, monoamines), as well as neurotransmitter and neuropeptide systems (eg, CRF, NPY) that contribute to adaptations in reward and stress circuitry in the brain associated with dependence. Such changes in brain function are thought to underlie expression of withdrawal-related symptoms that contribute to enhanced relapse vulnerability as well as promote transition to excessive, uncontrolled drinking (Becker, 2014; Hansson, Rimondini, Neznanova, Sommer, & Heilig, 2008; Koob, 2013).

As previously mentioned, a positive feature of alcohol dependence and self-administration models is that treatment effects may be evaluated in both dependent and nondependent animals. This offers an excellent platform for evaluating efficacy of medications in reducing escalated drinking associated with dependence, more modest and stable levels of intake exhibited by nondependent animals, or alcohol consumption in both conditions. As shown in Table 1, a large number of drug treatments were shown to be effective in selectively reducing elevated alcohol self-administration in dependent subjects while not altering intake in nondependent animals. This outcome suggests that such drug treatments may be influencing adaptive changes associated with dependence that significantly contribute to enhanced motivation to imbibe. This profile of results is best exemplified by drugs that target CRF1 receptors (CRF1R antagonists).

In several instances, pharmacological agents were shown to reduce alcohol self-administration in both dependent and nondependent animals. Drugs that target monoamine systems are a good example of this scenario. In many of these studies, however, higher doses of the drug treatments were required to reduce alcohol consumption in nondependent animals in comparison to dependent subjects. This was the case for studies testing the effects of the serotonin–norepinephrine reuptake inhibitor milnacipran (Simon O'Brien et al., 2011) and the β-adrenergic receptor antagonist propranolol (Gilpin & Koob, 2010), as well as other drugs such as the GABA-B receptor antagonist baclofen (Walker & Koob, 2007). The greater sensitivity to drug-induced modulation of alcohol consumption in dependent animals would also suggest that the treatment might be targeting brain systems/processes that have changed as a function of chronic alcohol exposure in these models.

5. CHALLENGES IN DEVELOPING PHARMACOTHERAPIES FOR ALCOHOL DEPENDENCE AND RELAPSE

When one considers the diverse and widespread neuroadaptive changes that are set in motion as a consequence of prolonged excessive alcohol drinking, perhaps it is not surprising that no single pharmacological agent has proven to be fully successful in the treatment of alcoholism. Further, it is plausible that various pharmacological treatments may differ in effectiveness depending on, among several variables, the stage of addiction and motivation for drinking (Koob & Mason, 2016). The challenge of using pharmacotherapies and, perhaps more accurately, choosing the most appropriate drug for treatment of alcoholism is no doubt complicated by the complexity and heterogeneity of this relapsing disease, along with a host of other variables (eg, genotype, comorbidities, treatment regimens, compliance) that must be considered in the context of treatment interventions (eg, Litten et al., 2015; McLellan, Lewis, O'Brien, & Kleber, 2000). This is further compounded by the difficulty and time-consuming, not to mention costly, exercise of moving medications along the path toward FDA approval (Litten et al., 2012; Litten, Falk, Ryan, & Fertig, 2014).

Nevertheless, the alcohol field is at an exciting junction, poised to make inroads in the discovery and development of new medications. In the past two decades, an explosion of new discoveries from studies using molecular biology, neurobiology, neuroimaging, and behavioral neuroscience research has provided new insights about potential novel targets for developing new medications and treatment strategies for alcohol dependence. The burgeoning areas of pharmacogenetics and genomics have highlighted the important role of genotype and endophenotypes in defining differential responsiveness to medications, holding promise for the development of new treatment approaches that are more specifically tailored to subpopulations of individuals battling relapse as well as attempting to moderate their alcohol use and/or sustain abstinence. Unfortunately, despite these advancements, there are too few pharmacotherapeutic agents that have proven to be effective in treating alcohol dependence.

Currently, there are four medications approved by the FDA for treatment of AUD: disulfiram, oral naltrexone, acamprosate, and a long-acting injectable form of naltrexone (O'Malley & O'Connor, 2011). None of these medications have proven to be fully satisfactory with regard to magnitude of effects and uniformity of clinical outcomes. The need for identifying and developing new medications that will be more effective in treating AUDs is of paramount importance from a public health standpoint. No doubt, animal models of dependence and drinking will continue to play a pivotal role in not only the discovery of new and novel targets but also providing a valuable platform for evaluating efficacy of medications in reducing excessive, harmful levels of alcohol consumption. Such preclinical work is a critical step in the process of screening, identifying, and informing plans for prioritizing the most promising candidate medications that will be advanced to the next stage of evaluation—investigation in human laboratory paradigms and controlled clinical trials.

6. SUMMARY

Alcoholism is a complex disease that represents a highly significant public health concern. However, despite remarkable progress in elucidating biological and environmental mechanisms that drive excessive alcohol consumption along with its devastating consequences, few treatments are available for tackling this problem and providing relief for those suffering with the disease. Use of preclinical models that closely resemble the clinical situation plays a critical role in not only screening new test compounds but also identifying the most promising agents to be advanced for further clinical investigation. Numerous animal models have been developed to study underlying mechanisms as well as evaluate potential pharmacotherapeutic effects on excessive levels of drinking. In recent years, a great deal of attention has been focused on models that link alcohol dependence with self-administration procedures. These studies have been conducted in both mice and rats, and they have largely involved chronic intermittent exposure to alcohol vapor by inhalation. After establishing the positive reinforcing effects of alcohol (stable baseline levels of consumption), rodents exposed to chronic intermittent alcohol vapor display an escalation of self-administration behavior in comparison to nondependent animals that maintain a stable level of intake. This dependence-related increase in alcohol consumption has been demonstrated with different drinking models, including home-cage drinking as well as self-administration involving operant conditioning procedures. Further, this alcohol dependence and drinking model has been widely used to evaluate the effects of medications on self-administration of alcohol in dependent compared to nondependent animals. Thus, studies employing alcohol dependence and self-administration procedures have not only provided critical insights about neuroadaptations that increase susceptibility to relapse and promote excessive drinking, but these models also have served as a valuable platform for identifying novel targets and evaluating new therapeutic strategies for treating AUDs.

ACKNOWLEDGMENTS

This work was supported by NIH grants P50 AA010761, U01 AA014095, and U01 AA020929 from the National Institute on Alcohol Abuse and Alcoholism and grant 2101BX000813 from the Department of Veterans Affairs Medical Research. The views and opinions expressed herein reflect those of the authors and do not necessarily reflect those of the funding agencies. The authors declare no implicit or explicit conflicts of interest with this material.

REFERENCES

Albrechet-Souza, L., Hwa, L. S., Han, X., Zhang, E. Y., DeBold, J. F., & Miczek, K. A. (2015). Corticotropin releasing factor binding protein and CRF2 receptors in the ventral tegmental area: Modulation of ethanol binge drinking in C57BL/6J mice. *Alcoholism, Clinical and Experimental Research*, *39*, 1609–1618.

Anderson, R. I., Becker, H. C., Adams, B. L., Jesudason, C. D., & Rorick-Kehn, L. M. (2014). Orexin-1 and orexin-2 receptor antagonists reduce ethanol self-administration in high-drinking rodent models. *Frontiers in Neuroscience*, *8*, 33.

Becker, H. C. (2013). Animal models of excessive alcohol consumption in rodents. *Current Topics in Behavioral Neurosciences*, *13*, 355–377.

Becker, H. C. (2014). Alcohol dependence, withdrawal, and relapse. In A. B. C. Noronha, C. Cui, R. A. Harris, & J. C. Crabbe (Eds.), *Neurobiology of alcohol dependence* (pp. 377–410). London: Academic Press; Elsevier.

Becker, H. C., & Lopez, M. F. (2004). Increased ethanol drinking after repeated chronic ethanol exposure and withdrawal experience in C57BL/6 mice. *Alcoholism: Clinical and Experimental Research*, *28*, 1829–1838.

Bell, R. L., Lopez, M. F., Cui, C., Egli, M., Johnson, K. W., Franklin, K. M., et al. (2015). Ibudilast reduces alcohol drinking in multiple animal models of alcohol dependence. *Addiction Biology*, *20*, 38–42.

Bell, R. L., Sable, H. J. K., Colombo, G., Hyytia, P., Rodd, Z. A., & Lumeng, L. (2012). Animal models for medications development targeting alcohol abuse using selectively bred rat lines: Neurobiological and pharmacological validity. *Pharmacology, Biochemistry, and Behavior*, *103*, 119–155.

Brown, G., Jackson, A., & Stephens, D. N. (1998). Effects of repeated withdrawal from chronic ethanol on oral self-administration of ethanol on a progressive ratio schedule. *Behavioural Pharmacology*, *9*, 149–161.

Cappell, H., & LeBlanc, A. (1981). Tolerance and physical dependence: Do they play a role in alcohol and drug self-administration? In Y. Israel, F. B. Glaser, H. Kalant, R. E. Popham, W. Schmidt, & R. G. Smart (Eds.), *Research advances in alcohol and drug problems* (pp. 159–196). New York: Plenum Press.

Carnicella, S., Ron, D., & Barak, S. (2014). Intermittent ethanol access schedule in rats as a preclinical model of alcohol abuse. *Alcohol*, *48*, 243–252.

Chu, K., Koob, G. F., Cole, M., Zorrilla, E. P., & Roberts, A. J. (2007). Dependence-induced increases in ethanol self-administration in mice are blocked by the CRF1 receptor antagonist antalarmin and by CRF1 receptor knockout. *Pharmacology, Biochemistry, and Behavior*, *86*, 813–821.

Crabbe, J. C., Harris, R. A., & Koob, G. F. (2011). Preclinical studies of alcohol binge drinking. *Annals of the New York Academy of Sciences*, *1216*, 24–40.

Cunningham, C. L., Fidler, T. L., Murphy, K. V., Mulgrew, J. A., & Smitasin, P. J. (2013). Time-dependent negative reinforcement of ethanol intake by alleviation of acute withdrawal. *Biological Psychiatry*, *73*, 249–255.

Dawson, D. A., Goldstein, R. B., Saha, T. D., & Grant, B. F. (2015). Changes in alcohol consumption: United States, 2001–2002 to 2012–2013. *Drug and Alcohol Dependence*, *148*, 56–61.

Dhaher, R., Finn, D., Snelling, C., & Hitzemann, R. (2008). Lesions of the extended amygdala in C57BL/6J mice do not block the intermittent ethanol vapor-induced increase in ethanol consumption. *Alcoholism, Clinical and Experimental Research*, *32*, 197–208.

Edwards, S., Guerrero, M., Ghoneim, O. M., Roberts, E., & Koob, G. F. (2012). Evidence that vasopressin V1b receptors mediate the transition to excessive drinking in ethanol-dependent rats. *Addiction Biology*, *17*, 76–85.

Fidler, T. L., Dion, A. M., Powers, M. S., Ramirez, J. J., Mulgrew, J. A., Smitasin, P. J., et al. (2011). Intragastric self-infusion of ethanol in high- and low-drinking mouse genotypes after passive ethanol exposure. *Genes, Brain, and Behavior*, *10*, 264–275.

Fidler, T. L., Powers, M. S., Ramirez, J. J., Crane, A., Mulgrew, J., Smitasin, P., et al. (2012). Dependence induced increases in intragastric alcohol consumption in mice. *Addiction Biology*, *17*, 13–32.

Finn, D. A., Snelling, C., Fretwell, A. M., Tanchuck, M. A., Underwood, L., Cole, M., et al. (2007). Increased drinking during withdrawal from intermittent ethanol exposure is blocked by the CRF receptor antagonist D-Phe-CRF(12–41). *Alcoholism, Clinical and Experimental Research*, *31*, 939–949.

Funk, C. K., & Koob, G. F. (2007). A CRF(2) agonist administered into the central nucleus of the amygdala decreases ethanol self-administration in ethanol-dependent rats. *Brain Research*, *1155*, 172–178.

Funk, C. K., O'Dell, L. E., Crawford, E. F., & Koob, G. F. (2006). Corticotropin-releasing factor within the central nucleus of the amygdala mediates enhanced ethanol self-administration in withdrawn, ethanol-dependent rats. *The Journal of Neuroscience*, *26*, 11324–11332.

Funk, C. K., Zorrilla, E. P., Lee, M. J., Rice, K. C., & Koob, G. F. (2007). Corticotropin-releasing factor 1 antagonists selectively reduce ethanol self-administration in ethanol-dependent rats. *Biological Psychiatry*, *61*, 78–86.

Gehlert, D. R., Cippitelli, A., Thorsell, A., Le, A. D., Hipskind, P. A., Hamdouchi, C., et al. (2007). 3-(4-Chloro-2-morpholin-4-yl-thiazol-5-yl)-8-(1-ethylpropyl)-2,6-dimethyl-imidazo [1,2-b]pyridazine: A novel brain-penetrant, orally available corticotropin-releasing factor receptor 1 antagonist with efficacy in animal models of alcoholism. *The Journal of Neuroscience*, *27*, 2718–2726.

George, O., & Koob, G. F. (2013). Control of craving by the prefrontal cortex. *Proceedings of the National Academy of Sciences of the United States of America, 110*, 4165–4166.
Gilpin, N. W., & Koob, G. F. (2010). Effects of beta-adrenoceptor antagonists on alcohol drinking by alcohol-dependent rats. *Psychopharmacology, 212*, 431–439.
Gilpin, N. W., Misra, K., Herman, M. A., Cruz, M. T., Koob, G. F., & Roberto, M. (2011). Neuropeptide Y opposes alcohol effects on gamma-aminobutyric acid release in amygdala and blocks the transition to alcohol dependence. *Biological Psychiatry, 69*, 1091–1099.
Gilpin, N. W., Misra, K., & Koob, G. F. (2008). Neuropeptide Y in the central nucleus of the amygdala suppresses dependence-induced increases in alcohol drinking. *Pharmacology, Biochemistry, and Behavior, 90*, 475–480.
Gilpin, N. W., Richardson, H. N., & Koob, G. F. (2008). Effects of CRF1-receptor and opioid-receptor antagonists on dependence-induced increases in alcohol drinking by alcohol-preferring (P) rats. *Alcoholism, Clinical and Experimental Research, 32*, 1535–1542.
Gilpin, N. W., Richardson, H. N., Lumeng, L., & Koob, G. F. (2008). Dependence-induced alcohol drinking by alcohol-preferring (P) rats and outbred Wistar rats. *Alcoholism, Clinical and Experimental Research, 32*, 1688–1696.
Gilpin, N. W., Smith, A. D., Cole, M., Weiss, F., Koob, G. F., & Richardson, H. N. (2009). Operant behavior and alcohol levels in blood and brain of alcohol-dependent rats. *Alcoholism, Clinical and Experimental Research, 33*, 2113–2123.
Grant, K. (1995). Animal models of the alcohol addiction process. In H. Kranzler (Ed.), *The pharmacology of alcohol abuse* (pp. 185–230). Berlin: Springer-Verlag.
Grant, B. F., Goldstein, R. B., Saha, T. D., Chou, S. P., Jung, J., Zhang, H., et al. (2015). Epidemiology of DSM-5 alcohol use disorder: Results from the national epidemiologic survey on alcohol and related conditions III. *JAMA Psychiatry, 72*, 757–766.
Griffin, W. C., 3rd (2014). Alcohol dependence and free-choice drinking in mice. *Alcohol, 48*, 287–293.
Griffin, W. C., 3rd, Lopez, M. F., & Becker, H. C. (2009a). Intensity and duration of chronic ethanol exposure is critical for subsequent escalation of voluntary ethanol drinking in mice. *Alcoholism, Clinical and Experimental Research, 33*, 1893–1900.
Griffin, W. C., 3rd, Lopez, M. F., Yanke, A. B., Middaugh, L. D., & Becker, H. C. (2009b). Repeated cycles of chronic intermittent ethanol exposure in mice increases voluntary ethanol drinking and ethanol concentrations in the nucleus accumbens. *Psychopharmacology, 201*, 569–580.
Hansson, A. C., Rimondini, R., Neznanova, O., Sommer, W. H., & Heilig, M. (2008). Neuroplasticity in brain reward circuitry following a history of ethanol dependence. *The European Journal of Neuroscience, 27*, 1912–1922.
Heilig, M., Egli, M., Crabbe, J. C., & Becker, H. C. (2010). Acute withdrawal, protracted abstinence and negative affect in alcoholism: Are they linked? *Addiction Biology, 15*, 169–184.
Houchi, H., Persyn, W., Legastelois, R., & Naassila, M. (2013). The adenosine A2A receptor agonist CGS 21680 decreases ethanol self-administration in both non-dependent and dependent animals. *Addiction Biology, 18*, 812–825.
Kallupi, M., Vendruscolo, L. F., Carmichael, C. Y., George, O., Koob, G. F., & Gilpin, N. W. (2014). Neuropeptide YY(2)R blockade in the central amygdala reduces anxiety-like behavior but not alcohol drinking in alcohol-dependent rats. *Addiction Biology, 19*, 755–757.
Kissler, J. L., Sirohi, S., Reis, D. J., Jansen, H. T., Quock, R. M., Smith, D. G., et al. (2014). The one-two punch of alcoholism: Role of central amygdala dynorphins/kappa-opioid receptors. *Biological Psychiatry, 75*, 774–782.
Koob, G. F. (2003). Alcoholism: Allostasis and beyond. *Alcoholism, Clinical and Experimental Research, 27*, 232–243.

Koob, G. F. (2013). Addiction is a reward deficit and stress surfeit disorder. *Frontiers in Psychiatry*, *4*, 72.

Koob, G. F., & Mason, B. J. (2016). Existing and future drugs for the treatment of the dark side of addiction. *Annual Review of Pharmacology and Toxicology*, *56*, 299–322.

Litten, R. Z., Egli, M., Heilig, M., Cui, C., Fertig, J. B., Ryan, M. L., et al. (2012). Medications development to treat alcohol dependence: A vision for the next decade. *Addiction Biology*, *17*, 513–527.

Litten, R. Z., Falk, D., Ryan, M., & Fertig, J. (2014). Research opportunities for medications to treat alcohol dependence: Addressing stakeholders' needs. *Alcoholism, Clinical and Experimental Research*, *38*, 27–32.

Litten, R. Z., Ryan, M. L., Falk, D. E., Reilly, M., Fertig, J. B., & Koob, G. F. (2015). Heterogeneity of alcohol use disorder: Understanding mechanisms to advance personalized treatment. *Alcoholism, Clinical and Experimental Research*, *39*, 579–584.

Lopez, M. F., Anderson, R. I., & Becker, H. C. (2008). Repeated cycles of chronic intermittent ethanol exposure increase both self-administration and the reinforcing value of ethanol in C57BL/6J mice. *Alcoholism, Clinical and Experimental Research*, *32*, 163A.

Lopez, M. F., & Becker, H. C. (2005). Effect of pattern and number of chronic ethanol exposures on subsequent voluntary ethanol intake in C57BL/6J mice. *Psychopharmacology*, *181*, 688–696.

Lopez, M. F., & Becker, H. C. (2014). Operant ethanol self-administration in ethanol dependent mice. *Alcohol*, *48*, 295–299.

Lopez, M. F., Becker, H. C., & Chandler, L. J. (2014). Repeated episodes of chronic intermittent ethanol promote insensitivity to devaluation of the reinforcing effect of ethanol. *Alcohol*, *48*, 639–645.

Lopez, M. F., Griffin, W. C., Melendez, R. I., & Becker, H. C. (2012). Repeated cycles of chronic intermittent ethanol exposure leads to the development of tolerance to aversive effects of ethanol in C57BL/6J mice. *Alcoholism: Clinical and Experimental Research*, *36*, 1180–1187.

McLellan, A. T., Lewis, D. C., O'Brien, C. P., & Kleber, H. D. (2000). Drug dependence, a chronic medical illness: Implications for treatment, insurance, and outcomes evaluation. *JAMA*, *284*, 1689–1695.

Meinhardt, M. W., & Sommer, W. H. (2015). Postdependent state in rats as a model for medication development in alcoholism. *Addiction Biology*, *20*, 1–21.

Nealey, K. A., Smith, A. W., Davis, S. M., Smith, D. G., & Walker, B. M. (2011). kappa-opioid receptors are implicated in the increased potency of intra-accumbens nalmefene in ethanol-dependent rats. *Neuropharmacology*, *61*, 35–42.

O'Dell, L. E., Roberts, A. J., Smith, R. T., & Koob, G. F. (2004). Enhanced alcohol self-administration after intermittent versus continuous alcohol vapor exposure. *Alcoholism, Clinical and Experimental Research*, *28*, 1676–1682.

O'Malley, S. S., & O'Connor, P. G. (2011). Medications for unhealthy alcohol use: Across the spectrum. *Alcohol Research & Health*, *33*, 300–312.

Rehm, J., Mathers, C., Popova, S., Thavorncharoensap, M., Teerawattananon, Y., & Patra, J. (2009). Global burden of disease and injury and economic cost attributable to alcohol use and alcohol-use disorders. *Lancet*, *373*, 2223–2233.

Richardson, H. N., Zhao, Y., Fekete, E. M., Funk, C. K., Wirsching, P., Janda, K. D., et al. (2008). MPZP: A novel small molecule corticotropin-releasing factor type 1 receptor (CRF1) antagonist. *Pharmacology, Biochemistry, and Behavior*, *88*, 497–510.

Rimondini, R., Arlinde, C., Sommer, W., & Heilig, M. (2002). Long-lasting increase in voluntary ethanol consumption and transcriptional regulation in the rat brain after intermittent exposure to alcohol. *The FASEB Journal*, *16*, 27–35.

Rimondini, R., Sommer, W., & Heilig, M. (2003). A temporal threshold for induction of persistent alcohol preference: Behavioral evidence in a rat model of intermittent intoxication. *Journal of Studies on Alcohol, 64*, 445–449.

Rimondini, R., Thorsell, A., & Heilig, M. (2005). Suppression of ethanol self-administration by the neuropeptide Y (NPY) Y2 receptor antagonist BIIE0246: Evidence for sensitization in rats with a history of dependence. *Neuroscience Letters, 375*, 129–133.

Roberto, M., Cruz, M. T., Gilpin, N. W., Sabino, V., Schweitzer, P., Bajo, M., et al. (2010). Corticotropin releasing factor-induced amygdala gamma-aminobutyric acid release plays a key role in alcohol dependence. *Biological Psychiatry, 67*, 831–839.

Roberto, M., Gilpin, N. W., O'Dell, L. E., Cruz, M. T., Morse, A. C., Siggins, G. R., et al. (2008). Cellular and behavioral interactions of gabapentin with alcohol dependence. *The Journal of Neuroscience, 28*, 5762–5771.

Roberts, A. J., Cole, M., & Koob, G. F. (1996). Intra-amygdala muscimol decreases operant ethanol self-administration in dependent rats. *Alcoholism, Clinical and Experimental Research, 20*, 1289–1298.

Roberts, A. J., Heyser, C. J., Cole, M., Griffin, P., & Koob, G. F. (2000). Excessive ethanol drinking following a history of dependence: Animal model of allostasis. *Neuropsychopharmacology, 22*, 581–594.

Rodriguez de Fonseca, F., Roberts, A. J., Bilbao, A., Koob, G. F., & Navarro, M. (1999). Cannabinoid receptor antagonist SR141716A decreases operant ethanol self administration in rats exposed to ethanol-vapor chambers. *Zhongguo Yao Li Xue Bao, 20*, 1109–1114.

Sabino, V., Cottone, P., Zhao, Y., Iyer, M. R., Steardo, L., Jr., Steardo, L., et al. (2009). The sigma-receptor antagonist BD-1063 decreases ethanol intake and reinforcement in animal models of excessive drinking. *Neuropsychopharmacology, 34*, 1482–1493.

Schulteis, G., Hyytia, P., Heinrichs, S. C., & Koob, G. F. (1996). Effects of chronic ethanol exposure on oral self-administration of ethanol or saccharin by Wistar rats. *Alcoholism, Clinical and Experimental Research, 20*, 164–171.

Sidhpura, N., Weiss, F., & Martin-Fardon, R. (2010). Effects of the mGlu2/3 agonist LY379268 and the mGlu5 antagonist MTEP on ethanol seeking and reinforcement are differentially altered in rats with a history of ethanol dependence. *Biological Psychiatry, 67*, 804–811.

Simms, J. A., Nielsen, C. K., Li, R., & Bartlett, S. E. (2014). Intermittent access ethanol consumption dysregulates CRF function in the hypothalamus and is attenuated by the CRF-R1 antagonist, CP-376395. *Addiction Biology, 19*, 606–611.

Simon O'Brien, E., Legastelois, R., Houchi, H., Vilpoux, C., Alaux-Cantin, S., Pierrefiche, O., et al. (2011). Fluoxetine, desipramine, and the dual antidepressant milnacipran reduce alcohol self-administration and/or relapse in dependent rats. *Neuropsychopharmacology, 36*, 1518–1530.

Smith, A. W., Nealey, K. A., Wright, J. W., & Walker, B. M. (2011). Plasticity associated with escalated operant ethanol self-administration during acute withdrawal in ethanol-dependent rats requires intact matrix metalloproteinase systems. *Neurobiology of Learning and Memory, 96*, 199–206.

Sommer, W. H., Rimondini, R., Hansson, A. C., Hipskind, P. A., Gehlert, D. R., Barr, C. S., et al. (2008). Upregulation of voluntary alcohol intake, behavioral sensitivity to stress, and amygdala crhr1 expression following a history of dependence. *Biological Psychiatry, 63*, 139–145.

Sommer, W. H., Rimondini, R., Marquitz, M., Lidstrom, J., Siems, W. E., Bader, M., et al. (2007). Plasticity and impact of the central renin-angiotensin system during development of ethanol dependence. *Journal of Molecular Medicine (Berlin, Germany), 85*, 1089–1097.

Suchankova, P., Yan, J., Schwandt, M. L., Stangl, B. L., Caparelli, E. C., Momenan, R., et al. (2015). The glucagon-like peptide-1 receptor as a potential treatment target in alcohol

use disorder: Evidence from human genetic association studies and a mouse model of alcohol dependence. *Translational Psychiatry*, *5*, e583.
Thorsell, A., Slawecki, C. J., & Ehlers, C. L. (2005). Effects of neuropeptide Y on appetitive and consummatory behaviors associated with alcohol drinking in wistar rats with a history of ethanol exposure. *Alcoholism, Clinical and Experimental Research*, *29*, 584–590.
Valdez, G. R., Roberts, A. J., Chan, K., Davis, H., Brennan, M., Zorrilla, E. P., et al. (2002). Increased ethanol self-administration and anxiety-like behavior during acute ethanol withdrawal and protracted abstinence: Regulation by corticotropin-releasing factor. *Alcoholism, Clinical and Experimental Research*, *26*, 1494–1501.
Vendruscolo, L. F., Barbier, E., Schlosburg, J. E., Misra, K. K., Whitfield, T. W., Jr., Logrip, M. L., et al. (2012). Corticosteroid-dependent plasticity mediates compulsive alcohol drinking in rats. *The Journal of Neuroscience*, *32*, 7563–7571.
Vendruscolo, L. F., & Roberts, A. J. (2014). Operant alcohol self-administration in dependent rats: Focus on the vapor model. *Alcohol*, *48*, 277–286.
Vengeliene, V., Bilbao, A., & Spanagel, R. (2014). The alcohol deprivation effect model for studying relapse behavior: A comparison between rats and mice. *Alcohol*, *48*, 313–320.
Walker, B. M., & Koob, G. F. (2007). The gamma-aminobutyric acid-B receptor agonist baclofen attenuates responding for ethanol in ethanol-dependent rats. *Alcoholism, Clinical and Experimental Research*, *31*, 11–18.
Walker, B. M., & Koob, G. F. (2008). Pharmacological evidence for a motivational role of kappa-opioid systems in ethanol dependence. *Neuropsychopharmacology*, *33*, 643–652.
Walker, B. M., Rasmussen, D. D., Raskind, M. A., & Koob, G. F. (2008). alpha1-Noradrenergic receptor antagonism blocks dependence-induced increases in responding for ethanol. *Alcohol*, *42*, 91–97.
Walker, B. M., Zorrilla, E. P., & Koob, G. F. (2011). Systemic kappa-opioid receptor antagonism by nor-binaltorphimine reduces dependence-induced excessive alcohol self-administration in rats. *Addiction Biology*, *16*, 116–119.

CHAPTER SEVEN

A Genetic Animal Model of Alcoholism for Screening Medications to Treat Addiction

R.L. Bell*,[1], S. Hauser*, Z.A. Rodd*, T. Liang†, Y. Sari‡, J. McClintick§, S. Rahman¶, E.A. Engleman*

*Institute of Psychiatric Research, Indiana University School of Medicine, Indianapolis, IN, United States
†Indiana University School of Medicine, Indianapolis, IN, United States
‡University of Toledo, Toledo, OH, United States
§Center for Medical Genomics, Indiana University School of Medicine, Indianapolis, IN, United States
¶Department of Pharmaceutical Sciences, South Dakota State University, Brookings, SD, United States
[1]Corresponding author: e-mail address: ribell@iupui.edu

Contents

Abstract

The purpose of this review is to present up-to-date pharmacological, genetic, and behavioral findings from the alcohol-preferring P rat and summarize similar past work.

International Review of Neurobiology, Volume 126
ISSN 0074-7742
http://dx.doi.org/10.1016/bs.irn.2016.02.017

Behaviorally, the focus will be on how the P rat meets criteria put forth for a valid animal model of alcoholism with a highlight on its use as an animal model of polysubstance abuse, including alcohol, nicotine, and psychostimulants. Pharmacologically and genetically, the focus will be on the neurotransmitter and neuropeptide systems that have received the most attention: cholinergic, dopaminergic, GABAergic, glutamatergic, serotonergic, noradrenergic, corticotrophin releasing hormone, opioid, and neuropeptide Y. Herein, we sought to place the P rat's behavioral and neurochemical phenotypes, and to some extent its genotype, in the context of the clinical literature. After reviewing the findings thus far, this chapter discusses future directions for expanding the use of this genetic animal model of alcoholism to identify molecular targets for treating drug addiction in general.

1. ALCOHOL ABUSE AND DEPENDENCE

Over half of adult Americans have a family member with an alcohol abuse or dependence disorder [alcohol use disorder (AUD)], although only a subset of these individuals have this across multiple generations (Research Society on Alcoholism, 2009, 2015). Moreover, approximately one in four Americans have had an AUD during their lifetime (Research Society on Alcoholism, 2009, 2015). The repercussions of AUDs cost the U.S. economy nearly a quarter of a trillion dollars per year (Research Society on Alcoholism, 2015). Much of this is in health care costs, with AUDs being the third leading cause of preventable death according to the Centers for Disease Control and Prevention (Mokdad, Marks, Stroup, & Gerberding, 2004; also see Johnson, 2010) and a causal relationship has been established between AUDs and at least 50 different medical conditions (Reed, Page, Viken, & Christian, 1996; Rehm, 2011; Rehm et al., 2003).

Alcohol dependence is a chronic, progressive, relapsing disorder that advances in stages from experimentation to dependence (Heilig & Egli, 2006; Jupp & Lawrence, 2010; Koob, 2009; Koob & Le Moal, 2008; Koob & Volkow, 2010; Spanagel, 2009; Volkow & Li, 2005). During the experimentation stage, the individual experiences the rewarding, euphoric, and positive-reinforcing effects of alcohol consumption. These positive reinforcing effects are often associated with acute increases in motor and autonomic (eg, heart rate) activity as well as prosocial behavior and are generally perceived, by the individual, as euphoric (ie, pleasant). This learning process results in positive reinforcement; which increases the probability, frequency, and magnitude of subsequent drinking behavior. However, with continued usage the individual experiences an increase in the magnitude, duration, and/or frequency of dysphoria (as opposed to euphoria), such as anxiety, during periods without access to alcohol. These dysphoric effects can be physiological in nature (eg, hangover, hyperthermia, and tachycardia) or associated with behavioral sequelae, such as getting arrested for driving while intoxicated. Moreover, given this increase in dysphoria, the individual often seeks to relieve this state by returning to drinking alcohol, often to excess. Therefore, during the early stages of the disease positive reinforcement generally predominates, whereas during the later stages of the disease negative reinforcement predominates (cf, Koob, Arends, & Le Moal, 2014; Koob, Buck, et al., 2014).

Fundamentally, reinforcement results in an increase in behavior and its associated cognitive processes. Thus, increases in approach behavior are associated with positive reinforcement during initial stages of the disease, whereas increases in retreat or avoidance behavior (eg, consuming alcohol upon waking to counter hangover effects) are associated with negative reinforcement during later stages of the disease. The roles of positive reinforcement vs negative reinforcement can also be characterized in terms of impulsive vs compulsive alcohol drinking (Koob, Arends, & Le Moal, 2014; Koob, Buck, et al., 2014; Koob & Le Moal, 2006, 2008). Within this construct, impulsive drinking leads to, and is associated with, binge drinking and intoxication (eg, Gray & MacKillop, 2014; Hamilton, Felton, Risco, Lejuez, & MacPherson, 2014; but see Irimia et al., 2013). However, after chronic usage, impulsive drinking, during which an individual will putatively have some volitional control, will be replaced by compulsive drinking to mitigate physical and behavioral withdrawal from alcohol. This, in turn, leads to a preoccupation with, and an anticipation of, future alcohol consumption during alcohol withdrawal (Koob & Le Moal, 2008). Therefore, impulsive drinking and positive reinforcement predominate in the early stages of alcohol dependence, whereas compulsive drinking and negative reinforcement predominate in later stages of addiction (Koob, Arends, & Le Moal, 2014; Koob, Buck, et al., 2014; Koob & Le Moal, 2006, 2008). However, despite this general trend of cycles of active drinking and relapse, with a concomitant increase in alcohol, or drugs of abuse, intake and the development of tolerance to the effects of alcohol, or drugs of abuse; the progression of the disease is not necessarily linear, such that the frequency and/or duration a person experiences these cycles differs across individuals (eg, Barker & Taylor, 2014; Mackenzie, El-Gabalaway, Chou, & Sareen, 2014; Sartor, Kranzler, & Gelernter, 2014; Van Rizen & Dishion, 2014). Moreover, not all individuals who abuse alcohol, or other drugs of abuse, need formal treatment to reduce or stop their excessive intake. For instance, many individuals who abused alcohol during their adolescence and emerging adulthood do not develop alcohol dependence.

Nevertheless, many individuals who abused alcohol during adolescence and emerging adulthood do develop alcohol dependence either during this developmental stage or later in life. Because of this, there is a strong emphasis to study the acute and long-term effects of alcohol and/or drug abuse during the periadolescent developmental window (ie, juvenile to emerging adulthood) (eg, Bell, Franklin, Hauser, & Engleman, 2013; Gulley & Juraska, 2013; Spear, 2010, 2014; Witt, 1994, 2006, 2010). According to the National Institute on Alcohol Abuse and Alcoholism (NIAAA, 2012), 11% of all alcohol consumed in the United States is done so by 12–20 year olds and 90% of this drinking is in the form of binges. Additionally, about a third of high school seniors report binge drinking during high school and ~75% of college students report binge drinking (Johnston, O'Malley, & Bachman, 1999). For ~30% of male college students (Wechsler, Lee, Kuo, & Lee, 2000; White, Kraus, & Swartzwelder, 2006), this behavior continues into college and the magnitude of these binges often exceeds, by to two- to threefold, threshold consumption levels [five drinks in one sitting resulting in blood alcohol concentrations (BACs) of 80 mg% or higher] put forth in NIAAA's definition of binge drinking (NIAAA National Advisory Council, 2004). Thus, binge drinking during periadolescence has become a serious public health concern, with research indicating it is a strong predictor of future alcohol-related problems in North America (Dawson, Grant, Stinson, & Chou, 2004; Johnston, O'Malley, Bachman, & Schulenberg, 2008; Kuntsche, Rehm, & Gmel, 2004; Presley, Meilman, & Lyerla, 1994; Wechsler et al., 2000; White et al., 2006).

Other predictors of AUDs as well as their epidemiological antecedents and trajectory are the pattern of drinking (eg, social vs binge vs continuous) and amount consumed (Flory, Lynam, Milich, Leukefeld, & Clayton, 2004; Heather, Tebbutt, Mattick, & Zamir, 1993; Lancaster, 1994; Shield, Rehm, Gmel, Rehm, & Allamani, 2013; Zucker, 1995). Characterization of these predictors and their antecedents has led to the development of different typologies and/or drinking profiles as well as subcategories of severity associated with a diagnosis of an AUD in the Diagnostic and Statistical Manual of Mental Disorders, 4th and 5th Editions (DSM-4, DSM-TR-4, and DSM-5; American Psychiatric Association, 1994, 2000, 2013; Babor et al., 1992; Cloninger, 1987; Conrod, Pihl, Stewart, & Dongier, 2000; Epstein, Kahler, McCrady, Lewis, & Lewis, 1995; Lesch & Walter, 1996;

Moss, Chen, & Yi, 2007; Prelipceanu & Mihailescu, 2005; Preuss, Watzke, & Wurst, 2014; Windle & Scheidt, 2004; Zucker, 1987). It has also been shown that an individual's ranking within a particular typology predict's the efficacy of certain pharmacotherapies (Cherpitel, Moskalewicz, & Swiatkiewicz, 2004; Dundon, Lynch, Pettinati, & Lipkin, 2004; Epstein et al., 1995; Forray & Sofuoglu, 2014; Hulse, 2012; Johnson, 2005, 2010; Johnson, Ait-Daoud, Ma, & Wang, 2003; Keating, 2013). Therefore, age-of-onset and pattern of drinking, recognizing that these are often correlated, have significant predictive validity for a life-time diagnosis of alcohol abuse or dependence and, in some cases, the effectiveness of pharmacotherapies to treat alcohol dependence. In this chapter, alcohol and ethanol are used interchangeably; such that in the clinical setting the term alcohol is preferentially used over ethanol, whereas the term ethanol (scientific name for the two-carbon chain alcohol) is preferentially used over alcohol in the preclinical setting.

2. NEUROBEHAVIORAL CORRELATES WITH ALCOHOL ABUSE AND DEPENDENCE

Clinical and basic research indicate that (1) lower responsivity to ethanol's effects is directly associated with alcohol abuse and dependence (eg, Crabbe, Bell, & Ehlers, 2010; Draski & Deitrich, 1996; Heit et al., 2013; Morean & Corbin, 2010; Morozova, Mackay, & Anholt, 2014; Schuckit & Gold, 1988; Silveri, 2012, 2014; Spear, 2010, 2014); (2) the ability to display greater levels and quicker development of tolerance (a reduction in ethanol's effects after prior treatment with ethanol) to ethanol's effects is also associated with alcohol abuse and dependence (eg, Lê & Mayer, 1996); (3) additionally, the expression of anxiety-like behavior under basal and/or withdrawal conditions is associated with a propensity to abuse alcohol (eg, Heilig, Egli, Crabbe, & Becker, 2010; Heilig, Thorsell, et al., 2010; Kirby, Zeeb, & Winstanley, 2011; Pautassi, Camarini, Quadros, Miczek, & Israel, 2010; Thorsell, 2010); and (4) moreover, the expression of low- to moderate-dose ethanol-induced stimulation [which is modeled in rodents by increased motor activity/approach behavior (Chappell & Weiner, 2008; Faria et al., 2008; Wise & Bozarth, 1987), aggression (Chiavegatto, Quadros, Ambar, & Miczak, 2010), and social facilitation (Varlinskaya & Spear, 2009, 2010)] is associated with excessive alcohol consumption. This behavioral phenotype may have pharmacological validity as well, such that the histaminergic (cf, Panula & Nuutinen, 2011 and references therein) and ghrelin (cf, Jerlhag, Landgren, Egecioglu, Dickson, & Engel, 2011 and references therein) systems have been implicated in ethanol-induced motor activation, ethanol-induced conditioned place preference, alcohol preference, and high alcohol consumption behavior. However, there remain concerns with the translatability of ethanol-induced stimulation in rodents vs humans (eg, Crabbe et al., 2010). For instance, other than low- to moderate-dose effects on (a) self-report (Morzorati, Ramchandani, Flury, Li, & O'Connor, 2002; Viken, Rose, Morzorati, Christian, & Li, 2003), (b) heart rate (Finn & Justus, 1997; Peterson et al., 1996), and (c) brain activity (Lukas, Mendelson, Benedikt, & Jones, 1986; Sorbel, Morzorati, O'Connor, Li, & Christian, 1996; Trim et al., 2010), the stimulating effects of ethanol are less apparent in humans compared with rodents.

3. NEUROCHEMICAL CORRELATES WITH ALCOHOL ABUSE AND DEPENDENCE

Clinical and basic research indicate that alcohol abuse and dependence are mediated in part by a number of neurobiological systems (cf, Koob, Arends, & Le Moal, 2014; Koob, Buck, et al., 2014; Noronha, Cui, Harris, & Crabbe, 2014; Pierce & Kenny, 2013; Robbins, Everitt, & Nutt, 2010; Self & Staley, 2010; Sommer & Spanagel, 2013; Spanagel, 2009) including: acetylcholine (ACh; Chatterjee & Bartlett, 2010; Davis & de Fiebre, 2006; Rahman, Engleman, & Bell, 2015, 2016; Soderpalm, Ericson, Olausson, Blomqvist, & Engel, 2000), adenosine (Filip, Zaniewska, Frankowska, Wydra, & Fuxe, 2012; Nam, Bruner, & Choi, 2013), dopamine (DA; Bhaskar & Kumar, 2014; Engel & Jerlhag, 2014; Heinz, 2002; Nutt, Lingfor-Hughes, Erritzoe, & Stokes, 2015; Soderpalm & Ericson, 2013), endocannabinoid (Moreira, Jupp, Belin, & Dalley, 2015), gamma-aminobutyric-acid (GABA; Agabio & Colombo, 2014; Kumar et al., 2009; Liang & Olsen, 2014; Maccioni & Colombo, 2009), glutamate (Barron et al., 2012; Bell et al., 2016; Davis & Wu, 2001; Gass & Olive, 2008; Rao, Bell, Engleman, & Sari, 2015), purinergic (Franklin et al., 2014), serotonin (5-HT; Engleman, Rodd, Bell, & Murphy, 2008; Hauser et al., 2014; Lovinger, 1999), melanocortin (Olney, Navarro, & Thiele, 2014), opiate (Charbogne, Kieffer, & Befort, 2014; Drews & Zimmer, 1997), orexin (Baimel et al., 2015), oxytocin (Buisman-Pijlman et al., 2014), neuropeptide Y (NPY; Heilig & Thorsell, 2002), corticotropin releasing factor (CRF; Burke & Miczek, 2014;

Koob, 2010), substance P (George et al., 2008), nociceptin/orphanin FQ (NOP, N/OFQ; Economidou et al., 2008; Witkin et al., 2014), ghrelin (Jerlhag, Egecioglu, Dickson, & Engel, 2011; Jerlhag et al., 2009; Jerlhag, Landgren, et al., 2011), neurotrophic factors such as brain-derived neurotrophic factor (BDNF; Logrip, Janak, & Ron, 2009), and hypothalamic–pituitary–adrenal (HPA) activity including corticosteroids, etc. (Gianoulakis, Guillaume, De Waele, & Angelogianni, 1995; Keith, Roberts, Wisen, & Crabbe, 1995; Kiefer, Jahn, Otte, Nakovics, & Wiedemann, 2006; Rasmussen, Boldt, Wilkinson, & Mitton, 2002; Richardson, Lee, O'Dell, Koob, & Rivier, 2008) systems within the brain. More recently there has been a significant increase in research on the neuroimmune system, which can modulate these neurochemical/neuropeptide systems and is an important contributor to the development of addiction (Crews, Qin, Sheedy, Ventreno, & Zou, 2013; Cui, Shurtleff, & Harris, 2014; Kane et al., 2014; Robinson et al., 2014; Ward, Lallemand, & de Witte, 2014). This is one area where basic and clinical research have informed each other, such that innate differences or ethanol-induced changes in neurotransmitter, neuropeptide, and neuroimmune systems of subjects genetically predisposed for excessive ethanol consumption strongly suggests these systems play a significant role in the development of alcoholism, at least in predisposed individuals (cf, Bell et al., 2012; Robinson et al., 2014). By extension, demonstration of these innate neurotransmitter and/or neuropeptide differences in an animal model of alcoholism underscore its utility in screening the efficacy of compounds to treat alcohol dependence.

4. THE GENETICS OF ALCOHOLISM

Family History Positive (FHP) for alcoholism individuals are persons who have relatives that meet diagnostic criteria for AUDs. The strongest expression of correlative phenotypes (discussed later) is observed in FHP individuals with this characteristic across multiple generations, starting with the immediate family. These findings provide strong support for genetics as a mediator in the development of alcoholism (cf, Schuckit, 2014). Epidemiologically, twin studies have yielded data indicating a strong genetic component to the development and expression of alcohol dependence. Heath (1995) provides an excellent overview of the early Australian (eg, Eysenck & Eysenck, 1975; Martin et al., 1985), Finnish (eg, Kaprio, Rose, Romanov, & Koskenvuo, 1991; Partanen, Bruun, & Markkanen,

1966), London Twin Family Survey (eg, Clifford, Hopper, Fulker, & Murray, 1984), Swedish (eg, Cederlof, Friberg, & Lundman, 1977), U.S. National Academy of Science/National Research Council twin registry (eg, Hrubec & Neel, 1978; Jablon, Neel, Gershowitz, & Atkinson, 1967), the U.S. Vietnam Era Twin Survey (eg, Goldberg, Eisen, True, & Henderson, 1990), and his own work with the U.S. Virginia 30,000 Survey. What is most striking about his review of these twin registries and AUDs is that, while the U.S. data tends to suggest that the development of AUDs in men is genetically influenced to a greater extent than that observed in women, multiple European studies suggest that the development of AUDs by women is genetically influenced to the same degree or more so than men (Heath, 1995). For further reading on the genetics of AUDs, see Hesselbrock (1995b) for an overview of early adoption studies, see Hesselbrock (1995a) for an overview of early work looking at Alcoholic Subtypes, see Cadoret (1995) for an overview of early studies on genetic correlates with other psychiatric disorders and see Schuckit (2014) for a recent review on the history of research investigating the genetics of alcohol and drug dependence.

5. CRITERIA FOR AN ANIMAL MODEL OF ALCOHOLISM

Different animal models have had different levels of success in research to develop treatments for both medical and psychiatric disorders (Gobrogge, 2014; Golbidi, Frisbee, & Laher, 2015; Griffin, 2002; McCairn & Isoda, 2013; McGonigle & Ruggeri, 2014; McKinney, 2001; McLarnon, 2014; Nestler & Hyman, 2010; Reser, 2014; Whiteside, Pomonis, & Kennedy, 2013). An animal model allows an experimenter to control the subject's genetic background, environmental factors, and prior drug experience. In addition, it allows for the examination of neurobehavioral, neurochemical, and neurophysiological correlates associated with the behavioral, physiological, and neurological states being modeled, in the present case alcohol abuse and dependence. These correlates can, in turn, facilitate the development of pharmacological and/or behavioral treatments for these disorders. There have been reservations as to whether a valid animal model of alcoholism could be developed (Cicero, 1979). These concerns stemmed from the fact that, in general, heterogeneous stock rats consume only modest levels of ethanol, such that BACs achieved are modest (cf, Bell, Rodd, Engleman, Toalston, & McBride, 2014 for a comparison of 22 different rat lines and strains). Nevertheless, certain criteria for an animal model of alcoholism have been put forth (Cicero, 1979; Lester & Freed, 1973). Briefly, these criteria are (1) the animal should readily consume ethanol under free-choice access conditions; (2) the amount of ethanol consumed should result in pharmacologically relevant BACs; (3) ethanol should be consumed for its pharmacological effects; (4) ethanol should be reinforcing, usually demonstrated through operant procedures; (5) chronic ethanol consumption should lead to the expression of metabolic and functional/neuronal tolerance to alcohol's effects; and (6) chronic consumption should lead to dependence, as indicated by withdrawal signs after access to ethanol is terminated. Other proposed criteria for a valid animal model of alcoholism include displaying characteristics associated with relapse-like drinking, generally demonstrated by an alcohol deprivation effect (ADE; McBride & Li, 1998); as well as excessive ethanol drinking during adolescence, such that it exceeds adult intake levels and binge-like access results in BACs > 80 mg% as well as motor impairment (Bell et al., 2011, 2013, 2014; McBride, Bell, Rodd, Strother, & Murphy, 2005; McBride, Rodd, Bell, Lumeng, & Li, 2014).

6. A RAT GENETIC ANIMAL MODEL OF ALCOHOLISM

The well-documented familial incidence of alcoholism, including multiple international twin studies, indicates a strong (40–70%) genetic component (discussed earlier) mediates a predisposition for and the development of AUDs (eg, Cloninger, 1987; Cotton, 1979; Hesselbrock, 1995a, 1995b; Schuckit, 1986). Given heterogeneous stock rats display a wide-range of ethanol consumption levels (Richter & Campbell, 1940) and very early work on selective breeding for alcohol consumption, Williams and associates (Williams, Berry, & Beerstecher, 1949) as well as Mardones and colleagues (Mardones, Segovia, & Hederra, 1953; Mardones & Segovia-Riquelme, 1983) proposed a genetic link to ethanol intake in rodents. From their work (eg, Mardones et al., 1953; Mardones & Segovia-Riquelme, 1983) and that of four other international sites, bidirectional selective breeding has resulted in at least six high alcohol-consuming vs their respective low alcohol-consuming rat lines (cf, Bell et al., 2012).

Bidirectional selective breeding is a powerful genetic tool for studying alcohol-associated phenotypes (eg, Crabbe, 2008). Compared to pure association studies such as genome-wide association studies and recombinant inbred lines, selective breeding from a heterogeneous outbred stock can make low frequency/rare alleles (minor allele frequency <0.05) more common by segregating these genetic polymorphisms into the opposite extremes of the overall population. This bidirectional selection is accomplished through systematic mating of animals with similar traits (alcohol preferring on the one hand vs alcohol avoiding on the other) over successive generations. Thus, the high and low lines will exhibit extreme phenotypes exceeding the range found in the original parent population. Additionally, selective breeding for any phenotypes (such as alcohol preference) is hypothesis driven and genetically correlated traits of the primary selected phenotype (presumably due to pleiotropic actions of genes; Crabbe, Phillips, Kosobud, & Belknap, 1990) can be identified for further study. A major advantage of this selection process is that the ethanol-drinking phenotype is observed without the stress of environmental manipulations.

7. THE P RAT AS A GENETIC ANIMAL MODEL OF ALCOHOLISM

The alcohol preferring, P, and alcohol nonpreferring, NP, rat lines were developed by bidirectional mass selection from a closed-colony of Wistar rats at the Walter Reed Army Hospital and subsequently transferred to the Indiana University School of Medicine, Indianapolis, Indiana, USA (Lumeng, Hawkins, & Li, 1977). Two metrics were used for the selection of an alcohol preference. First, the animals had to prefer an unadulterated 10% ethanol solution over water by a ratio of at least 2:1. Second, the animals had to consume at least 5 g of ethanol/kg body weight/day. The first criterion was used to prevent selecting for an artifact of body weight when using the second criterion (selection for lower body weight will appear to be equivalent to a selection for high ethanol consumption only). To place this 5 g/kg/day selection criterion in a clinical perspective, using 0.793 as the specific gravity of ethanol, 90-proof whiskey as the preferred alcoholic beverage of choice for a 70 kg male, and the fact that rats metabolize ethanol close to 1.4 × the rate of humans, 5 g/kg/day of ethanol intake by a rat would be approximately a fifth of 90-proof whiskey being consumed each day.

Regarding criteria for an animal model of alcoholism, under free-choice conditions (water and food available) P rats readily consume greater than 5 g/kg/day of ethanol, whereas NP rats consume less than 1 g/kg/day (Li, Lumeng, McBride, & Murphy, 1987). P rats readily attain pharmacologically relevant BACs (80–250 mg%: this would approximate 0.08–0.25 in forensic terms, the latter under operant binge-like conditions)

(Bell et al., 2013, 2014, 2011; Bell, Rodd, Lumeng, Murphy, & McBride, 2006; Bell, Rodd, Sable, et al., 2006; Bell, Rodd,Schultz, et al., 2008; McBride, Kimpel, McClintick, Ding, Hauser, et al., 2013; McBride, Kimpel, et al., 2014; McBride, Rodd, et al., 2014; Murphy et al., 2002). Ethanol drinking in the home cage also results in intoxication including motor impairment, as measured with an oscillating bar apparatus (Bell, McKinzie, Murphy, & McBride, 2000; Bell et al., 2001). Moreover, P rats will display tolerance to this effect after chronic binge drinking (Bell et al., 2011). P rats operantly self-administer ethanol intragastrically for its postingestive effects (Murphy et al., 1988; Waller, McBride, Gatto, Lumeng, & Li, 1984). P rats will operantly self-administer ethanol using a dipper model, indicating these rats will work for access to ethanol (Files, Samson, Denning, & Marvin, 1998; Murphy, Gatto, McBride, Lumeng, & Li, 1989; Rodd et al., 2003; Rodd-Henricks et al., 2002a,2002b; Samson, Files, Denning, & Marvin, 1998; Toalston et al., 2008) or sipper tube model (Beckwith & Czachowski, 2014; Bertholomey, Verplaetse, & Czachowski, 2013; Czachowski & Samson, 2002; Samson & Czachowski, 2003; Verplaetse, Rasmussen, Froehlich, & Czachowski, 2012; Verplaetse & Czachowski, 2015) indicating these rats will work for access to ethanol.

Whereas ethanol-naïve P and NP rats display similar levels of ethanol clearance (Li & Lumeng, 1977; Lumeng, Waller, McBride, & Li, 1982), after chronic ethanol-drinking (6–8 weeks) P rats display metabolic as well as functional tolerance to the motor impairing and aversive effects of ethanol (Gatto et al., 1987; Lumeng & Li, 1986; Stewart, McBride, Lumeng, Li, & Murphy, 1991). Moreover, similarly treated P rats also display dependence-associated signs when ethanol access is terminated (Kampov-Polevoy, Matthews, Gause, Morrow, & Overstreet, 2000; Waller, McBride, Lumeng, & Li, 1982). In addition, P rats display relapse-like drinking by exhibiting a robust ADE (Rodd et al., 2003; Rodd-Henricks et al., 2001; Rodd-Henricks, McKinzie, Shaikh, et al., 2000). The ADE is a transient increase in ethanol intake after a period of ethanol withdrawal (cf, Rodd, Bell, McKinzie, et al., 2004; Rodd, Bell, Sable, Murphy, & McBride, 2004; Vengeliene, Bilbao, & Spanagel, 2014). Regarding initial sensitivity, compared with NP rats, P rats are less sensitive to the ataxic (Bell et al., 2001) and hypothermic (Stewart, Kurtz, Zweifel, Li, & Froehlich, 1992) effects of ethanol; and P rats develop tolerance quicker to the ataxic (Bell et al., 2001) and hypnotic (Kurtz, Stewart, Zweifel, Li, & Froehlich, 1996) effects of ethanol as well. During ethanol withdrawal, P, but not NP, rats display greater acoustic startle reactivity compared with basal conditions (Chester, Blose, & Froehlich, 2004). P rats display greater low dose ethanol-induced locomotor activity, compared with NP rats (Rodd, Bell, McKinzie, et al., 2004; Waller, Murphy, McBride, Lumeng, & Li, 1986), and display locomotor activation during ethanol drinking or self-administration (Bell, Rodd, Toalston, et al., 2008; Bell et al., 2002; Melendez et al., 2002). The latter provides additional support to the view that P rats find ethanol rewarding. Thus, the P line of rats satisfies criteria proposed by multiple authors for a valid animal model of alcoholism. The key behavioral features of this genetic rat animal model of alcoholism are outlined below:

Key features of the P genetic rat animal model of alcoholism

1. >7 g ethanol/kg body weight/day is orally consumed under home-cage, free-choice, 24 h conditions (Bell et al., 2011, 2013, 2014; Bell, Rodd, Schultz, et al., 2008; Li et al., 1987)
2. >1 g/kg ethanol is orally consumed during the first 15-min of home-cage, limited access conditions (Bell et al., 2014; Bell, Rodd, Lumeng, et al., 2006; Murphy et al., 1986; Russell, McBride, Lumeng, Li, & Murphy, 1996)
3. Pharmacologically relevant BACs (80–250 mg%), which parallel those observed in alcoholics, are achieved during ethanol drinking, and self-administration (Bell et al., 2011, 2014; Bell, Rodd, Lumeng, et al., 2006; Bell, Rodd, Sable, et al., 2006; Bell, Rodd,Schultz, et al., 2008; Murphy et al., 1986, 2002; Rodd et al., 2003)
4. Ethanol is self-administered intragastically (Murphy et al., 1988; Waller et al., 1984) and intracranially (Engleman et al., 2009; Toalston et al., 2014) indicating taste and calories are not the primary motivators for this behavior
5. Ethanol is consumed and self-administered despite ethanol-induced motor impairment (Bell et al., 2011; McBride, Kimpel, McClintick, Ding, Hauser, et al., 2013)
6. Ethanol drinking and self-administration induce autonomic (heart rate) and/or behavioral (motor) activation (Bell, Rodd, Toalston, et al., 2008; Bell et al., 2002; Melendez et al., 2002)
7. Ethanol is operantly self-administered, without using fading/adaptation techniques (ie, P rats work for access to ethanol) (Files et al., 1998; Murphy et al., 1989; Rodd et al., 2003; Rodd-Henricks et al., 2002a,2002b; Rodd-Henricks, McKinzie, Shaikh, et al., 2000; Samson et al., 1998; Toalston et al., 2008)
8. Chronic home-cage, free-choice ethanol consumption leads to metabolic (Lumeng & Li, 1986) and functional (Gatto et al., 1987; Stewart et al., 1991, 1996) tolerance

9. Chronic home-cage, ethanol consumption leads to dependence (eg, decreased seizure thresholds) (Kampov-Polevoy et al., 2000; Waller et al., 1982)
10. Relapse behavior is displayed under home-cage and operant conditions [eg, expression of an ADE is observed] (Rodd et al., 2003; Rodd-Henricks et al., 2001; Rodd-Henricks, McKinzie, Shaikh, et al., 2000)
11. Serve as an animal model of adolescent binge alcohol abuse by exceeding ethanol intakes seen during adulthood, with motor impairment and BACs exceeding 80 mg%, under both continuous and limited access conditions (Bell et al., 2011, 2014, 2006a; McBride et al., 2005; McBride, Rodd, et al., 2014)
12. Self-administer or consume drugs of abuse by (a) orally consuming nicotine-adulterated solutions (Hauser et al., 2012); (b) operantly self-administering nicotine both orally (Hauser et al., 2014) and intravenously (Le et al., 2006); and (c) intracranially self-administering cocaine (Katner et al., 2011), nicotine (Deehan et al., 2015; Hauser et al., 2014), and nicotine + ethanol (Truitt et al., 2015).

8. SOME FHP CORRELATES

The above survey of the literature concerning neurobehavioral and neurobiological phenotypes found in these selectively bred high vs low alcohol-consuming rat lines indicates that many phenotypes present in alcohol abusing or dependent individuals are also present in these lines. For example, similar to the animal literature, clinical studies of FHP subjects (ie, with a family history of alcoholism) report an inverse relationship between low level responsivity to ethanol and risk for the development of AUDs (cf, Crabbe et al., 2010; Schuckit, 1994, 2009, 2014). Thus, after an ethanol challenge young adult FHP females (Eng, Schuckit, & Smith, 2005; Lex, Lukas, Greenwald, & Mendelson, 1988) and males (Schuckit, 1985; Schuckit & Gold, 1988) display less body sway than family history negative (FHN) controls. An individual's level of response (low vs high) to ethanol also influences brain regional activation following an acute ethanol challenge (Trim et al., 2010). Moreover, an individual's level of response to ethanol is associated with the long vs short allele for the 5-HT transporter (*5htt*) gene and this association has significant predictive validity for alcohol intake by adolescents (Hinckers et al., 2006). As discussed earlier, the FHP, P, rat displays lower levels of ethanol-induced behavioral and physiological changes compared with FHN, NP, rats (Bell et al., 2001; Stewart et al., 1992).

Alcohol dependence is a chronic relapsing disorder, with craving and relapse often precipitated by physiological and behavioral responses (ie, cue reactivity) to environmental and interoceptive cues associated with alcohol and/or drugs of abuse consumption (Childress et al., 1993; Drummond, Cooper, & Glautier, 1990; Greeley, Swift, Prescott, & Heather, 1993; Kaplan, Meyer, & Stroebel, 1983; Kaplan et al., 1985; O'Brien, Childress, McLellan, & Ehrman, 1992; Rajan, Naga Venkatesha Murthy, Ramakrishnan, & Gangadhar, 1998; Stormark, Laberg, Nordby, & Hugdahl, 1998). In addition, ethanol's effects on heart rate reactivity are associated with level of genetic density in FHP individuals as well dissociating FHP from FHN subjects, such that persons that are FHP display greater sensitivity to ethanol reward and display sensation-seeking behavior (Assaad et al., 2003; Conrod, Pihl, & Vassileva, 1998; Finn, Earleywine, & Pihl, 1992; Peterson, Pihl, Seguin, Finn, & Stewart, 1993). This has also been shown in cross-sensitivity by stimulant users displaying a characteristic alcohol-reward heart rate response (Brunelle, Barrett, & Pihl, 2006). It is noteworthy that both male (Bell et al., 2002) and female (Bell, Rodd, Toalston, et al., 2008) P rats display increased heart rate during ethanol drinking. In addition, this autonomic reactivity can be conditioned to the environment associated with ethanol consumption (Bell, Rodd, Toalston, et al., 2008; Bell et al., 2002). Thus, monitoring autonomic reactivity in P rats may be an important model system for testing compounds targeting craving, especially in the context of cue reactivity.

9. SOME NEUROCHEMICAL, NEUROPHARMACOLOGICAL AS WELL AS NEUROGENETIC CORRELATES

9.1 The Cholinergic System

Acetylcholine is released from neurons projecting to a broad range of cortical and subcortical structures and influences cellular physiology and neuronal function throughout the brain (Newman, Gupta, Climer, Monaghan, & Hasselmo, 2012). There are two classifications of cholinergic receptors: nicotinic and muscarinic. The neuronal nicotinic acetylcholine receptors (nAChRs) belong to the family of ligand-gated ion channel receptors (Albuquerque, Pereira, Alkondon, & Rogers, 2009; Gotti et al., 2009). nAChRs consist of 11 neuronal subunits, which are divided into 8 alpha subunits (α2–α10) and 3 beta subunits (β2–β4). nAChR subtypes with diverse subunit combinations are distributed across multiple brain regions, including the mesocorticolimbic and extended amygdala reward circuitry, where they regulate dopaminergic function. A notable difference among the receptor subtypes is that the homomeric α7 nAChR does not desensitize to nicotine stimulation as the heteromeric nAChRs (eg, α4β2 nAChR) do. The α7 nAChRs are located presynaptically on glutamatergic projections from the mPFC to the NAcbSh. Therefore, activation α7 nAChRs may enhance glutamatergic excitatory drive and promote DA release in the NAcb after α4β2 receptors are desensitized. Muscarinic acetylcholine receptors (mAChRs) are G-protein coupled receptors that are widely distributed in the brain. There are five mAChRs neuronal subunits (M1–M5) (Bymaster, McKinzie, Felder, & Wess, 2003; Wess, 2003; Wess et al., 2003; Yamada et al., 2003). Studies examining the striatum indicated that M1 receptors are expressed on spiny projection neurons (Wang et al., 2006), whereas M2/M4 receptors are primarily presynaptic autoreceptors (Yan & Surmeier, 1996; Zhang et al., 2002). As autoreceptors, M2/M4 receptors inhibit ACh release and subsequent nAChR-dependent DA release in the striatum (Shin, Adrover, Wess, & Alvarez, 2015). In addition, M2 receptors are located on glutamatergic terminals which inhibit its release (Hersch, Gutekunst, Rees, Heilman, & Levey, 1994). The M5 receptor is the only mAChR subtype found on midbrain DA neurons (Vilaró, Palacios, & Mengod, 1990; Weiner, Levey, & Brann, 1990) and these receptors modulate DA and DA/glutamate projections from the midbrain (Shin et al., 2015).

Substantial research indicates that nAChR activity mediates, in part, the rewarding effects of drugs of abuse (Chatterjee & Bartlett, 2010; Corrigall & Coen, 1994; Ericson, Blomqvist, Engel, & Söderpalm, 1998; Hendrickson, Guildford, & Tapper, 2013; McGehee & Role, 1995; Nisell, Nomikos, & Svensson, 1994; Rahman, 2013; Rahman et al., 2015, 2016; Rahman & Prendergast, 2012; Sajja, Dwivedi, & Rahman, 2010). For example, both alcohol (Brodie, Pesold, & Appel, 1999; Brodie, Shefner, & Dunwiddie, 1990) and nicotine (Calabresi, Lacey, & North, 1989; Nisell et al., 1994) activate ventral tegmental area (VTA) DA projection neurons and stimulation of nAChRs within the VTA modulate, at least in part, the reinforcing effects of nicotine (Corrigall & Coen, 1994; Nisell et al., 1994) and alcohol (Blomqvist, Ericson, Johnson, Engel, & Söderpalm, 1996; Ericson, Molander, Lof, Engel, & Soderpalm, 2003; Soderpalm et al., 2000). Additionally, a number of reports indicate that preexposure to nicotine significantly increases operant or free-choice ethanol self-administration and reinstates ethanol-seeking behavior in animal models (Bito-Onon, Simms, Chatterjee, Holgate, & Bartlett, 2011; Hauser et al., 2012; Lê et al., 2003). Moreover, coadministration of ethanol and nicotine produces an additive effect on their reinforcing effects and associated dopamine release in the nucleus accumbens (NAcb, Ericson, Löf, Stomberg, & Söderpalm, 2009; Tizabi, Bai, Copeland, & Taylor, 2007; Sajja et al., 2010).

Emerging preclinical evidence indicates that a number of ligands targeting nAChRs modulate ethanol-drinking behavior. For example, mecamylamine, a nonselective nAChR antagonist or varenicline, a partial agonist at $\alpha4\beta2^*$ nAChRs, reduces ethanol-drinking behavior by targeting nAChRs in the mesocorticolimbic dopamine system (Ericson et al., 1998; Lê et al., 2000; Steensland, Simms, Holgate, Richards, & Bartlett, 2007). In addition, nAChR ligands such as cytisine, a partial agonist at $\alpha4\beta2^*$, and lobeline, a nonselective antagonist, were found to reduce ethanol self-administration or nicotine-induced ethanol drinking in rodents (Bell, Eiler, Cook, & Rahman, 2009; Chatterjee, Steensland, Rollema, Bartlett, et al., 2011; Hendrickson, Zhao-Shea, & Tapper, 2009; Sajja & Rahman, 2011, 2012, 2013). Overall, brain nAChRs have emerged as important therapeutic targets for the rewarding effects of ethanol in numerous animal models. Regarding P rats, sazetidine-A, a novel ligand that desensitizes $\alpha4\beta2$ nAChRs with partial agonistic activity reduces ethanol drinking by P rats (Rezvani et al., 2010). In addition, pretreatment with nicotine increases operant ethanol self-administration and relapse behavior in P rats (Hauser et al., 2012). Furthermore, P rats have higher sensitivity to the reinforcing effects of nicotine in the pVTA compared with outbred Wistar rats (Hauser et al., 2014). These reinforcing effects of nicotine can be blocked by mecamylamine, an nAChR antagonist, when injected simultaneously into the pVTA of P rats (Hauser et al., 2014), indicating that mesolimbic cholinergic activity modulates the reinforcing effects of ethanol and nicotine. More recent findings indicate that P rats will also coself-administer ethanol + nicotine into the pVTA (Truitt et al., 2015) and self-administer nicotine into the NAcbSh (Deehan et al., 2015). In addition, oral binge ethanol + nicotine self-administration by P rats induces greater sensitivity to the reinforcing effects of nicotine in the NAcbSh by shifting the self-administration dose–response curve to the left (Deehan et al., 2015).

Receptor binding studies have shown that P rats have lower striatal $\alpha7$ nAChR expression than NP rats (Tizabi, Getachew, Davila-Garcia, & Taylor, 2001). Additionally, there is higher expression of *Chat* (choline acetyltransferase), *Chrm3* (mACh3R), *Slc5a7* (transporter uptake for acetylcholine synthesis), *Slc18a3* (vesicular amine transport into secretory vesicles) in the NAcbSh of adult P rats compared with NP rats, whereas adult P rats have lower expression of *Chrm4* (mACh4R) in the NAcbSh than NP rats (McBride, Kimpel, McClintick, Ding, Hauser, et al., 2013; McBride, Kimpel, McClintick, Ding, Hyytia, et al., 2013). These findings suggest that the accumbal cholinergic system may be more active in P rats than NP rats.

Clinical research indicates individuals with AUDs have lower ChAT activity and mAChR density in the hippocampus (Antuono, Sorbi, Bracco, Fusco, & Amaducci, 1980; Nordberg, Larsson, Perdahl, & Winblad, 1983) as well as ChAT protein expression in the basal forebrain (Vetreno, Broadwater, Liu, Spear, & Crews, 2014). However, these clinical findings do not necessarily indicate that basal levels are different from control subjects. Our microarray findings indicate that ethanol drinking by adult P rats increased *Chrn7a* expression in the NAcb (Bell, Kimpel, et al., 2009), but reduced its expression in the CeA (McBride et al., 2010) and the DRN of ethanol-drinking adolescent P rats (McClintick et al., 2015). Overall, a modest number of differences in ACh-associated gene expression levels have been detected between P and NP rats (data for the pVTA came from McBride et al., 2012; data for the CeA and NAcbSh came from McBride, Kimpel, McClintick, Ding, Hyytia, et al., 2013). Similarly, few changes in ACh-associated gene expression levels have been detected following binge drinking in adult or adolescent P rats (adult data for the whole NAcb came from Rodd et al., 2008; adult data for the NAcbSh came from Bell, Kimpel, et al., 2009; McBride et al., 2010; adult data for the CeA came from McBride et al., 2010; adult data for the pVTA came from McBride, Kimpel, McClintick, Ding, Hyytia, et al., 2013; adolescent data for the CeA came from McBride, Kimpel, et al., 2014; McBride, Rodd, et al., 2014; adolescent data for the DRN came from McClintick et al., 2015) thus far (Fig. 1). Taken together, these data indicate that brain cholinergic activity regulates behaviors associated with alcohol abuse and alcoholism and the P rat displays some treatment characteristics seen in the clinical population. For instance, varenicline has modest effects on reducing ethanol intake by P rats. However, the tests done in P rats thus far have not incorporated ethanol and nicotine coabuse, which appears to be necessary to observe efficacy with varenicline in the clinical setting.

9.2 The Dopaminergic System

DAergic projections (Fig. 2) emanating from the tegmentum (especially the VTA) and terminating in limbic, forebrain and cortical regions (eg, the mesocorticolimbic and extended amygdala reward circuitry), are involved in the appetitive and consummatory behaviors associated with, as well as the positive and negative reinforcing properties of, addictive drugs. DA activates metabotropic receptors (D_{1-5}) that are generally classified as D_1-like (D_1 and D_5) and D_2-like (D_{2-4}) (Cooper, Bloom, & Roth, 2002).

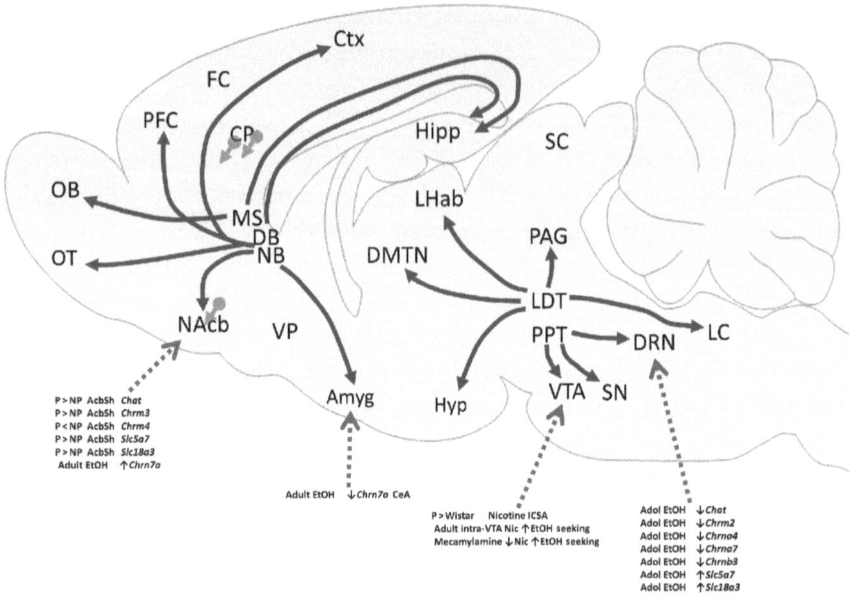

Fig. 1 Innate differences in gene expression between P and NP rats, ethanol-drinking-induced changes in gene expression of P rats as well as pharmacological changes in cholinergic activity and/or effects. *AcbSh*, nucleus accumbens shell; *Amyg*, amygdala; *CeA*, central amygdala; *Ctx*, cortex; *CP*, caudate-putamen; *DB*, diagonal band of Broca; *DMTN*, dorsal medial thalamic nucleus; *DRN*, dorsal raphe nucleus; *FC*, frontal cortex; *Hipp*, hippocampus; *Hyp*, hypothalamus; *LC*, locus coereleus; *LDT*, lateral dorsal tegmentum; *LHab*, lateral habenula; *MS*, medial septum; *NAcb*, nucleus accumbens; *NB*, nucleus basalis; *OB*, olfactory bulb; *OT*, olfactory tubercle; *PAG*, periaqueductal gray; *PFC*, prefrontal cortex; *PPT*, pedunculopontine tegmentum; *SC*, superior colliculus; *SN*, substantia nigra; *VP*, ventral pallidum; *VTA*, ventral tegmental area. *Dark arrows* indicate cholinergic projections. *Up-arrow* indicates an increase, whereas *down-arrow* indicates a decrease in expression levels, activity, or consumption. *Short filled arrow* indicates interneuron.

D_1-like receptors are typically coupled to the activation of adenylyl cyclase whereas D_2-like receptors are coupled to the inhibition of adenylyl cyclase (Cooper et al., 2002). DA neurotransmission is terminated via clearance by uptake through the high affinity DA transporter (DAT; Cooper et al., 2002). The DAergic system plays a central role in the processing of natural-, alcohol-, and drug of abuse-associated reward and reinforcement (eg, Di Chiara & Imperato, 1988; McBride & Li, 1998; Melendez et al., 2002; Nogueira, Kalivas, & Lavin, 2006; Palmer, Low, Grandy, & Phillips, 2003; Volkow & Morales, 2015), with imaging studies indicating $D_{2/3}$ receptor dysfunction in subjects addicted to several different drugs of abuse (Cosgrove, 2010). When an individual or animal ingests alcohol and/or

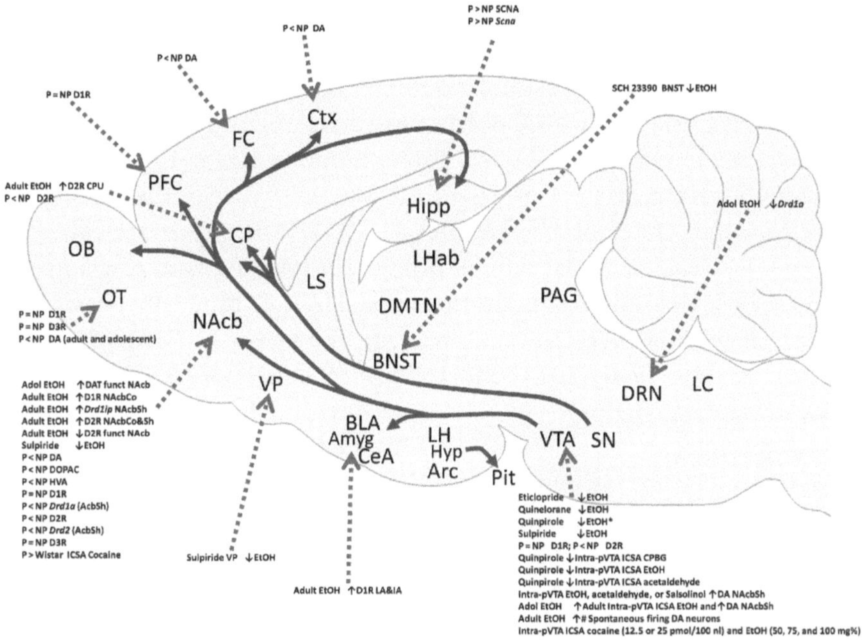

Fig. 2 Innate differences in gene (*italics*) and/or protein expression between P and NP rats, ethanol-drinking-induced changes in gene expression of P rats as well as pharmacological changes in dopaminergic activity and/or effects. *Arc*, arcuate nucleus of the hypothalamus; *BLA*, basolateral amygdala; *BNST*, bednucleus of the stria terminalis; *ICSA*, intracranial self-administration; *IA*, intercalated amygdala; *LA*, lateral amygdala; *LH*, lateral hypothalamus; *NAcbCo*, nucleus accumbens core; *Pit*, pituitary; *pVTA*, posterior ventral tegmental area. For other abbreviations see Fig. 1 legend. *Dark arrows* indicate dopaminergic projections. *Up-arrow* indicates an increase, whereas a *down-arrow* indicates a decrease in expression levels, activity, or consumption/self-administration. *, indicates multiple studies.

other drugs of abuse, DA efflux is increased in several key mesocorticolimbic brain reward centers (eg, Brodie et al., 1990; Franklin et al., 2009; Gessa, Muntoni, Collu, Vargiu, & Mereu, 1985; Imperato & Di Chiara, 1986; Smith & Weiss, 1999; Yoshimoto & McBride, 1992). This change in DAergic activity and other associated neuroplastic changes promote further alcohol and drug taking behavior. Following prolonged alcohol and/or drug abuse, individuals may display tolerance to these DA-elevating properties. Thus, the requirement for more alcohol and/or drug taking to get the same level of initial response moves the individual further through the addiction cycle, which starts out as impulsive use for intoxication and progresses to compulsive use to avoid or reduce the negative consequences of alcohol/drug withdrawal.

Regarding P rats, these animals exhibit reduced basal NAcb tissue DA levels and/or enhanced ethanol-induced extracellular DA efflux, compared to outbred rats or their alcohol nonpreferring counterparts (Engleman, Ingraham, McBride, Lumeng, & Murphy, 2006; McBride, Chernet, Dyr, Lumeng, & Li, 1993; Murphy, McBride, Lumeng, & Li, 1982; Murphy, McBride, Lumeng, & Li, 1987; Smith & Weiss, 1999; Strother, Lumeng, Li, & McBride, 2005; also see Bell et al., 2012; Murphy et al., 2002 for reviews). DAergic neuronal activity from the VTA appears to play a major role in these alterations of mesocorticolimbic DAergic tone either basally or under ethanol-induced conditions (eg, Engleman, Keen, Tilford, Thielen, & Morzorati, 2011; Morzorati, 1998; Morzorati & Marunde, 2006; Morzorati, Marunde, & Downey, 2010), which is modulated in part by glutamatergic activity in the pVTA (Fitzgerald, Liu, & Morzorati, 2012). In addition, chronic or binge-like alcohol drinking by P rats reduces D_2 autoreceptor function (Engleman, McBride, Li, Lumeng, & Murphy, 2003; Engleman et al., 2000), elevates extracellular DA levels (Thielen et al., 2004), and increases DA reuptake (Sahr, Thielen, Lumeng, Li, & McBride, 2004) in the NAcb. Sari, Bell, and Zhou (2006) also reported that long-term ethanol consumption by P rats increased D_1 and D_2 expression levels in the NAcb core (NAcbCo) with D_2 expression also increased in NAcb shell (NAcbSh). These authors indicated that intermittent periods of ethanol deprivation increased D_1 receptor expression in the amygdala and D_2 receptor expression in the caudate putamen as well.

Given the evidence for differences in, and ethanol-induced changes of, DAergic activity, it is not surprising that a number of DA-associated compounds have been tested for their effects on ethanol drinking or self-administration by this line. Systemic administration of the DA agonist ibogaine (Rezvani, Overstreet, & Lee, 1995) and DAT inverse modulator amphetamine (McBride, Murphy, Lumeng, & Li, 1990) reduced ethanol drinking. Ethanol drinking by P rats also was disrupted by systemic administration of the D_2 agonist bromocriptine (Mason, Rezvani, Grady, & Garbutt, 1994; McBride et al., 1990; Weiss, Mitchiner, Bloom, & Koob, 1990) and the D_3 agonist 7-OH-DPAT (Mason et al., 1997) as well as intra-VTA infusion of the D_2 agonists quinpirole (Hauser et al., 2011; Nowak, McBride, Lumeng, Li, & Murphy, 2000) and quinelorane (Nowak et al., 2000). Both systemic administration (Mason et al., 1997) and intra-bed nucleus of the stria terminalis (BNST) infusion (Eiler, Seyoum, Foster, Mailey, & June, 2003) of the D_1 antagonist SCH23390 reduced ethanol drinking and self-administration by P rats. Similarly, both systemic administration (Mason et al., 1997) and intra-BNST as well as intra-VTA infusion (Eiler et al., 2003) of the D_2

antagonist eticlopride reduced ethanol drinking and self-administration by P rats. Additionally, ethanol drinking by P rats was reduced by microinfusion of the D_2 antagonist sulpiride into the VTA (Nowak et al., 2000), NAcb (Levy, Murphy, McBride, Lumeng, & Li, 1991), and ventral pallidum (Melendez, Rodd, McBride, & Murphy, 2005). Finally, systemic administration of the D_3 antagonist SB-277011-A (Thanos et al., 2005) as well as the DAT inhibitors GBR 12909 (McBride et al., 1990) and DOV 102,677 (Yang et al., 2012) reduced ethanol intake by this line of rats.

It is noteworthy that P rats have lower levels of D_2 receptors than NP rats in the VTA, NAcb and caudate putamen (McBride, Chernet, Dyr, et al., 1993; Strother, Lumeng, Li, & McBride, 2003) and smaller subpopulations of DA projections from the VTA compared with NP rats (Zhou, Zhang, Lumeng, & Li, 1995), but no differences in D_1 or D_3 receptor levels have been detected thus far (McBride et al., 1997). Finally, differences in both mRNA and protein expression levels of alpha-synuclein (SNCA, often associated with DAergic function) have been found between P and NP rats, such that hippocampal levels were twice as high in inbred P rats compared with inbred NP rats (Liang et al., 2003). Interestingly, clinical research indicates that *Snca* expression is associated with alcohol craving, hazardous alcohol-drinking and posttraumatic stress disorder (PTSD) symptomology in hazardous drinkers (Foroud et al., 2007; Guillot, Fanning, Liang, Leventhal, & Berman, 2015; Guillot, Pang, Leventhal, Liang, & Berman, 2015). Regarding gene expression differences, similar to the cholinergic system, only a modest number of DA-associated differences have been detected between P and NP rats so far (data for the pVTA came from McBride et al., 2012; data for the CeA and NAcbSh came from McBride, Kimpel, McClintick, Ding, Hyytia, et al., 2013). This paucity of findings was also true for DA-associated changes following ethanol drinking, usually binge like, by P rats (adult data for the whole NAcb came from Rodd et al., 2008; adult data for the NAcbSh came from Bell, Kimpel, et al., 2009; McBride et al., 2010; adult data for the CeA came from McBride et al., 2010; adult data for the pVTA came from McBride, Kimpel, McClintick, Ding, Hauser, et al., 2013; adolescent data for the CeA came from McBride, Kimpel, et al., 2014; McBride, Rodd, et al., 2014; adolescent data for the DRN came from McClintick et al., 2015) (Fig. 2). Overall, whenever ethanol-induced changes were detected, the direction of change was generally an upregulation of mRNA expression. In summary, the DAergic system is the center of the mesocorticolimbic reward circuit and it appears that effective medications for alcoholism modulate the circuit, rather than impacting the DA-system directly.

9.3 The GABAergic System

GABA is the primary inhibitory neurotransmitter in the central nervous system (CNS). Thus, it and its receptors are found throughout the brain (eg, Benham, Engin, & Rudolph, 2014). In addition, there are four GABA transporters: GAT1 and GAT3 located presynaptically; GAT2 and GAT3 located postsynaptically; GAT1, GAT2, GAT3, and GAT4 on glia (cf, Clausen et al., 2006; Gonzalez-Burgos, 2010; Madsen, White, & Schousboe, 2010). The GABA receptors are classified as either $GABA_A$ or $GABA_B$. There are multiple subunit isoforms for $GABA_A$ receptors [(alpha1–6), (beta1–3), (gamma1–3), delta, (rho1–3), epsilon, pi, and theta], but the most common in the CNS is a pentamer comprising $(\alpha1)2(\beta2)2(\gamma2)$. The receptor itself is a ligand-gated chloride channel that has binding sites for GABA, benzodiazepine, picrotoxin, steroids, and anesthetics. Activation of the $GABA_A$ receptor opens the chloride channel for influx, which induces a hyperpolarized state that decreases the probability of an action potential resulting in an inhibitory state (cf, Fritschy, Panzanelli, & Tyagarajan, 2012; McCarson & Enna, 2014). $GABA_A$ receptors are located extra- and postsynaptically, whereas $GABA_B$ receptors are located both pre- and postsynaptically (Hanchar, Dodson, Olsen, Otis, & Wallner, 2005; Lovinger & Roberto, 2013). $GABA_B$ receptors are G-protein coupled, heteromer, receptors with two known subunits R1 and R2, where it appears the R1 subunit binds GABA and the R2 subunit interacts with the G-protein (Gaiarsa, Kuczewski, & Porcher, 2011; Terunuma et al., 2014).

Several research groups have reported significant associations between GABA gene variants, expression levels, and activation in brain regions such as the mesocorticolimbic system and the extended amygdala (which includes substructures of the bed nucleus of the stria terminalis, amygdala, nucleus accumbens, and prefrontal cortex), with high alcohol-consuming phenotypes and risk for developing alcohol dependence in alcoholics as well as alcohol-preferring rats (Dick & Bierut, 2006; Enoch et al., 2012; Herman et al., 2012; Korpi & Sinkkonen, 2006; McBride et al., 2010; Tabakoff et al., 2009). Thus, differential GABA signaling could reflect one mechanism that predisposes individuals to consume alcohol. In addition, GABAergic activity regulates, in part, other neuromodulator systems in the mesocorticolimbic reward circuit (Eiler & June, 2007; Melis, Camarini, Ungless, & Bonci, 2002; Rahman & McBride, 2002), supporting the role of GABA in DA-associated responses to reward. Given there are genetic differences, between P and NP rats, in these other neuromodulator systems of the mesocorticolimbic reward circuit as well (eg, Bell et al., 2012, 2016; Franklin et al., 2014; Rahman et al.,

2014, 2016 and discussed herein), this gene-by-gene interactional effect may serve as another contributing factor for the development of alcohol abuse and dependence (eg, Saba et al., 2015; Tabakoff et al., 2009). Acute alcohol experience potentiates GABA signaling and facilitates its hyperpolarizing actions (Koob, 2004). And, $GABA_A$ and $GABA_B$ receptors mediate some of the rewarding, reinforcing, and motivational effects of alcohol consumption and alcohol binge drinking (Eiler & June, 2007; Nowak, McBride, Lumeng, Li, & Murphy, 1998; Tanchuck, Yoneyama, Ford, Fretwell, & Finn, 2011; also see Agabio & Colombo, 2014). Systemically, the $GABA_A$ agonist topiramate (Breslin, Johnson, & Lynch, 2010; Lynch, Bond, Breslin, & Johnson, 2011), $GABA_B$ agonist baclofen (Liang et al., 2006; Maccioni et al., 2012), $GABA_B$ positive modulators GHB (June et al., 1995), CGP7930 (Liang et al., 2006), and GS39783 (Maccioni et al., 2012) all reduced ethanol drinking and/or self-administration by P rats. Similarly, negative modulators of the benzodiazepine-site Ro 15–4513 (McBride, Murphy, Lumeng, & Li, 1988), Ro 19–4603 (June et al., 1996; June, Murphy, Mellor-Burke, Lumeng, & Li, 1994; June, Torres, et al., 1998), Ru 34000 (June, Eggers, et al., 1998), Ro 15–1788 (June et al., 1994; June, Torres, et al., 1998), CGS 8216, and ZK 93426 (June, Devaraju, et al., 1998; June, Zuccarelli, et al., 1998) all reduced ethanol intake and/or self-administration by P rats. The partial agonist/antagonist βCCt systemically also reduced ethanol self-administration by P rats (June et al., 2003). Centrally, intra-VTA infusion of the $GABA_A$ receptor antagonists bicuculline, picrotoxin (Nowak et al., 1998), and SR95531 (Eiler & June, 2007) reduced ethanol intake or self-administration by P rats. Similarly, intra-VTA infusion of the negative modulator of the benzodiazepine-site Ru 34000 (June, Eggers, et al., 1998) as well as intra-CeA (Foster et al., 2004) and intra-VP (June et al., 2003) infusion of the partial agonist/antagonist βCCt reduced operant self-administration of ethanol by P rats. Also, shRNA-induced reductions of $GABA_A$-α2-subunit and its associated toll-like receptor 4 (*Tlr4*), in the CeA of P rats, significantly reduced alcohol self-administration (Liu et al., 2011).

Previous work indicated that P rats have more $GABA_A$ receptors in the NAcb than NP rats (Hwang, Lumeng, Wu, & Li, 1990), and P rats display a greater response to benzodiazepines in the PFC, NAcbSh, CPU, cingulate gyrus, and dorsal lateral septum than NP rats as well (Thielen, McBride, Chernet, Lumeng, & Li, 1997). Additionally, as seen in Fig. 3, gene expression for a number of $GABA_A$ and $GABA_B$ receptor subunits differ between P and NP rats (data for the pVTA came from McBride et al., 2012; data for the CeA and NAcbSh came from McBride, Kimpel, McClintick, Ding,

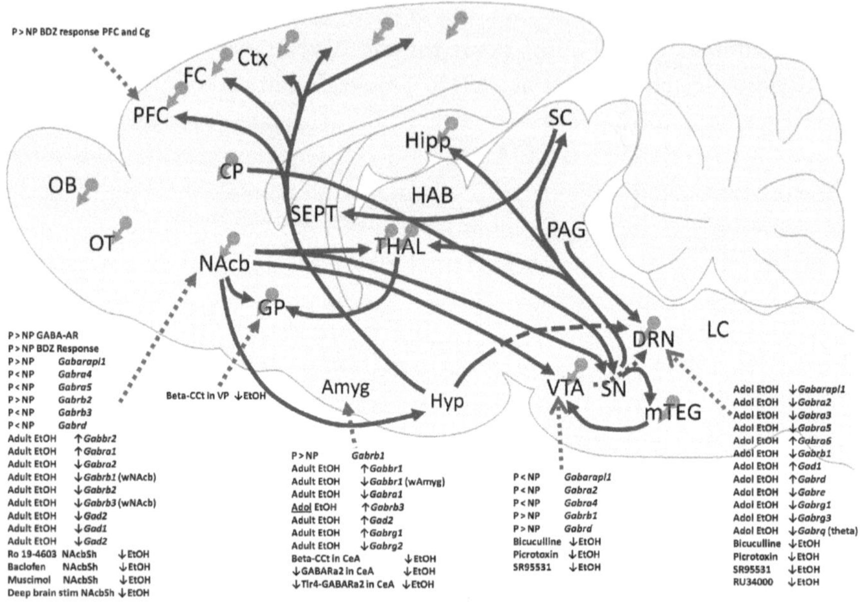

Fig. 3 Innate differences in gene (*italics*) and/or protein expression between P and NP rats, ethanol-drinking-induced changes in gene expression of P rats as well as pharmacological changes in GABAergic activity and/or effects. *BDZ*, benzodiazepine; *Cg*, cingulate cortex; *GP*, globus pallidus; *HAB*, habenula; *mTEG*, medial tegmentum; *SEPT*, septum; *THAL*, thalamus; *wAmyg*, whole amygdala; *wNAcb*, whole nucleus accumbens. For other abbreviations see Figs. 1 and 2 legend. *Dark arrows* indicate GABAergic projections. *Up-arrow* indicates an increase, whereas a *down-arrow* indicates a decrease in expression levels, activity, or consumption/self-administration. *Short filled arrow*, interneuon.

Hyytia, et al., 2013). Also seen in Fig. 3 are our findings on the role of ethanol drinking, usually in a binge-like manner, on $GABA_A$ and $GABA_B$ receptor subunit gene expression (adult data for the whole NAcb came from Rodd et al., 2008; adult data for the NAcbSh came from Bell, Kimpel, et al., 2009; McBride et al., 2010; adult data for the CeA came from McBride et al., 2010; adult data for the pVTA came from McBride, Kimpel, McClintick, Ding, Hauser, et al., 2013; adolescent data for the CeA came from McBride, Kimpel, et al., 2014; McBride, Rodd, et al., 2014; adolescent data for the DRN came from McClintick et al., 2015). In general, where differences in mRNA expression were detected in the NAcbSh and pVTA, P rats had lower GABAR-associated levels than NP rats. Following ethanol consumption by adult P rats, the majority of detected changes in the NAcbSh represented downregulated gene expression, whereas detected changes in the CeA were approximately equally down- vs upregulated gene expression. Following ethanol drinking by adolescent P rats, practically all detected changes represented downregulation of GABA-associated gene expression. These findings suggest lowered GABAergic function in the NAcbSh and DRN following excessive ethanol intake by P rats, which would support increased excitatory/glutamatergic activity as a contributor to excessive ethanol intake (see glutamatergic section directly below).

9.4 The Glutamatergic System

The amino acid glutamate is the primary excitatory neurotransmitter in the CNS. Given this, it is not surprising that glutamatergic projections, transporters, and receptors are found throughout the brain. As the primary excitatory neurotransmitter, glutamate plays a crucial role in neuroplasticity, learning, and memory (cf, Henley & Wilkinson, 2013; Morris, 2013; Warburton, Barker, & Brown, 2013). Glutamate interacts with both metabotropic (Grm1–Grm8) and ionotropic receptors, which include those that can bind to *N*-methyl-D-aspartate (NMDA) subunits [Grin1, Grin2a–Grin2d, and Grin3a–Grin3b], α-amino-3-hydroxy-5-methyl-4-isoxazole propionic acid (AMPA) subunits [Gria1–Gria4] or kainite subunits [Grik1–Grik4], for excellent reviews see (Danbolt, 2001; Niciu, Kelmendi, & Sanacora, 2012; Traynelis et al., 2010). Because of glutamate's role in excitotoxicity, extracellular glutamate must be tightly controlled (Danbolt, 2001; Sari, 2014; Wang & Qin, 2010). Multiple glutamate transporters have been implicated in this process (Anderson & Swanson, 2000; Danbolt, 2001; Gegelashvili & Schousboe, 1997; Seal & Amara, 1999). Nevertheless, the human excitatory amino acid transporter 2 (EAAT2) and its rodent analog glutamate transporter 1 (GLT1) appear to be the primary transporters performing this function (Danbolt, 2001; Mitani & Tanaka, 2003; Rothstein, Van Kammen, Levey, Martin, & Kuncl, 1995; Sari, 2014). However, the glial protein cystine/glutamate exchanger (xCT) appears to exchange extracellular cystine for intracellular (glial) glutamate (Bannai et al., 1984; Bannai & Ishii, 1982), which seems counterintuitive except for the fact that cystine is converted to cysteine, among other things, and reduces oxidative stress (cf, Ishii & Mann, 2014).

Glutamatergic activity has been shown to mediate natural as well as drug and nondrug associated reward through direct and indirect interactions with other neurotransmitter/neuromodulatory systems within the mesocorticolimbic and extended amygdala reward neurocircuitry (eg, Carlezon & Wise, 1996; Grace, Floresco, Goto, & Lodge, 2007; Kupila et al., 2013). For instance, research with the P rat has shown that free-choice ethanol drinking results in elevated levels of extracellular glutamate in the NAcb (Das, Yamamoto, Hristov, & Sari, 2015), as well as its subregion the NAcbSh, and the pVTA (Ding et al., 2013). The latter authors reported that this increase in glutamate was inversely related to decreases in glutamate clearance. Considerable research has examined the hypothesis that sensitized mesocorticolimbic and extended amygdala glutamate neurotransmission mediate in part alcohol and drug dependence. For example, it appears continued alcohol or drug intake results in a hypergluatmatergic state within mesocorticolimbic and extended amygdala reward circuits (Gass & Olive, 2008; Kryger & Wilce, 2010; Vengeliene, Bilbao, Molander, & Spanagel, 2008). Preclinical evidence supports clinical findings that alcohol acutely inhibits, and chronically sensitizes and upregulates glutamate neurotransmission, in brain reward regions of the mesocorticolimbic and extended amygdala circuits (eg, Carlezon & Wise, 1996; Chandler, Newsom, Sumners, & Crews, 1993; Cui et al., 2013; Ding, Engleman, Rodd, & McBride, 2012; Floyd, Jung, & McCool, 2003; Gass & Olive, 2008; Kapasova & Szumlinski, 2008; Nevo & Hamon, 1995; Nie, Madamba, & Siggins, 1994; Nie, Yuan, Madamba, & Siggins, 1993; Tabakoff & Hoffman, 2013; Weitlauf & Woodward, 2008), which may be due, in part, to changes in glutamate clearance (Ding et al., 2013; Kapasova & Szumlinski, 2008; Othman, Sinclair, Haughey, Geiger, & Parkinson, 2002; Parks et al., 2002; Rao, Bell, et al., 2015; Sari, Sakai, Weedman, Rebec, & Bell, 2011; Smith, 1997; Smith & Zsigo, 1996; Thoma et al., 2011). Moreover, recent data showed that 10 weeks of operant binge-like self-administration of solutions containing both ethanol and nicotine resulted in elevation of extracellular glutamate levels in the PFC (Deehan et al., 2015).

The role of glutamate in alcohol consumption is prominent in binge drinking as well (cf, Bell et al., 2016; Rao, Bell, et al., 2015). Genetic animal models of alcoholism, the P rat in this case, engaging in binge-like drinking, which results in BACs of 80 mg% and higher (cf, Bell et al., 2011, 2014), display numerous changes in glutamate receptor and/or subunits, transporters, scaffolding proteins as well as other associated gene expression levels in discrete brain regions of the mesocorticolimbic and extended amygdala circuits (Bell et al., 2016; Bell, Kimpel, et al., 2009; Coleman et al., 2011; McBride, Kimpel, et al., 2014; McBride et al., 2009, 2010; McBride, Rodd, et al., 2014; McClintick et al., 2015; Rodd et al., 2008). These changes in glutamatergic neurotransmission include enhanced receptor activation and intracellular downstream signaling cascades (Cozzoli et al., 2009; Szumlinski et al., 2007; Tabakoff et al., 2009). In support of this contention, glutamate receptor antagonists such as acamprosate and MPEP reduce binge-like drinking dose dependently (Grace et al., 2007; Gupta et al., 2008). Increases in excitatory neurotransmission may be greater during periods of acute ethanol withdrawal, which is commonly associated with binge drinking, compared to more protracted withdrawal periods (Ward et al., 2009). This may support the hypothesis that binge alcohol abuse increases susceptibility to alcohol-induced excitotoxic brain damage to a greater extent than continuous excessive drinking (eg, Hunt, 1993; see also discussion and references in Bell et al., 2013). Overall, it is likely that glutamatergic neuroadaptations following repeated binge-like drinking behavior lead to a glutamate-GABA functional imbalance (Enna, 1997; Fadda & Rossetti, 1998; Szumlinski et al., 2007) and are responsible, in part, for withdrawal symptomology when ethanol access is terminated. This withdrawal symptomology in turn increases the negative reinforcement-associated properties of continued binge drinking (Everitt & Robbins, 2005; Koob & Le Moal, 2008; Robinson & Berridge, 2008). These effects are consistent with a proposed transition from binge/impulsive alcohol drinking to habitual/compulsive drinking to dependence (cf, Koob, 2013; Koob & Volkow, 2010).

Systemically, the GRM1 antagonist CPCCOEt did not affect operant self-administration (Schroeder, Overstreet, & Hodge, 2005a,2005b), whereas the GRM1 antagonist JNJ 16259685 (Besheer, Faccidomo, Grondin, & Hodge, 2008a,2008b) did significantly reduce operant self-administration by P rats. Slight procedural and/or motor effect differences may explain the difference between the former and latter findings from the same laboratory. The GRM2/3 antagonist LY404039 significantly reduced ethanol-seeking behavior as well as relapse, but did not affect the maintenance, of ethanol self-administration by P rats (Rodd et al., 2006). The GRM5 antagonists MPEP (Schroeder et al., 2005a,2005b) and MTEP (Cowen, Djouma, & Lawrence, 2005) both reduced operant self-administration of ethanol by P rats. Intra-NAcb infusion of the GRM2/3 agonist LY379268 and GRM5 antagonist MPEP both significantly reduced operant ethanol self-administration by P rats (Besheer et al., 2010). Much more research has been done exploring the effects of manipulating GLT1 transporter levels, within subregions of the mesocorticolimbic reward circuit, on ethanol intake by P rats, to which we turn next.

The P rat has been very useful in assessing the efficacy of beta-lactam antibiotic, and similar, molecules in reducing alcohol intake, which appears to be due to a reversal of ethanol-induced downregulation of GLT1 levels and the concomitant increase in extracellular glutamate. For example, chronic free-choice drinking by P rats reduced GLT1 expression in the NAcb but not in the PFC, although xCT (the glutamate-cystine antiporter) was reduced in both regions (Alhaddad, Das, & Sari, 2014; Sari & Sreemantula, 2012; Sari, Sreemantula, Lee, & Choi, 2013). Ceftriaxone administered peripherally reversed ethanol-induced downregulation of GLT1 (both isoforms GLT1a and GLT1b) in the PFC and NAcb as well as xCT in the NAcb, PFC, and Amyg (Alhaddad, Das, et al., 2014; Rao & Sari, 2014a, 2014b). Similarly, ceftriaxone reversed ethanol-induced increases in extracellular glutamate of the NAcb (Das et al., 2015). Research with other beta-lactam antibiotics in P rats has shown that amoxicillin, augmentin, and ampicillin, which have the potential to be orally administered, reduced alcohol intake and increased GLT-1 expression in PFC and NAcb of P rats (Goodwani, Rao, Bell, & Sari, 2015; Rao, Goodwani, et al., 2015). Nonantibiotic compounds with a putative ability to upregulate GLT1 expression and/or activity have been tested in P rats as well. For example, GPI-1046, an analog of FK506, significantly reduces free-choice alcohol intake by male P rats with a concomitant increase in GLT1 expression levels in the NAcb and PFC (Sari & Sreemantula, 2012). Similar results were found for MS-153 with significant decreases in ethanol intake paralleling reversals of ethanol-induced GLT1 downregulation in the NAcb, Amyg, and hippocampus (Aal-Aaboda, Alhaddad, Osowik, Nauli, & Sari, 2015; Alhaddad, Kim, et al., 2014). Finally, it has been demonstrated that upregulation of GLT1 (including both isoforms) and xCT by ceftriaxone and MS-153 involves NF-kB and Akt signaling pathways (Alhaddad, Kim, et al., 2014; Rao, Saternos, Goodwani, & Sari, 2015).

An early study found that, globally, CNS *Grm3* and *Grm7* and the glycine-α1 subunit mRNA levels are lower in inbred P rats compared with inbred NP rats (Kimpel et al., 2007). More recently, Zhou et al. (2013) reported that P rats are homozygous for a *Grm2* stop codon (Grm2*407) that essentially renders them a functional KO of *Grm2*. At the same time, Meinhardt et al. (2013) published their own work indicating GRM2 deficits are inversely related to drug-seeking behavior. As highlighted in Bell et al. (2016), a number of site-specific differences in glutamate-associated mRNA expression levels (see Fig. 4) have been observed between P and NP rats

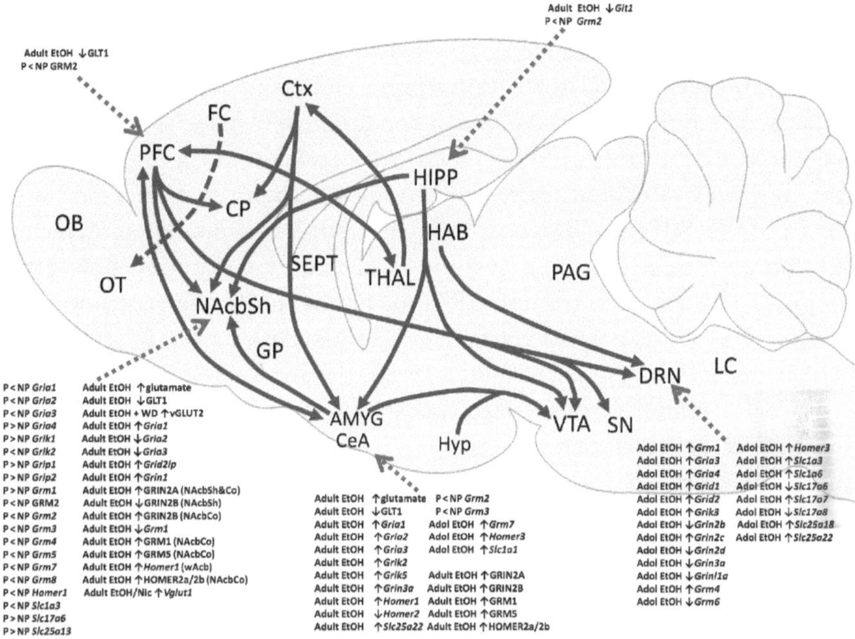

Fig. 4 Innate differences in gene (*italics*) and/or protein (*upper case*) expression between P and NP rats, ethanol-drinking-induced changes in gene expression of P rats as well as pharmacological changes in glutamatergic activity and/or effects. *WD*, withdrawal. For other abbreviations see Figs. 1–3 legends. *Dark arrows* indicate glutamatergic projections. *Up-arrow* indicates an increase, whereas a *down-arrow* indicates a decrease in expression levels, activity, or consumption/self-administration.

(data for the pVTA came from McBride et al., 2012; data for the CeA and NAcbSh came from McBride, Kimpel, McClintick, Ding, Hyytia, et al., 2013). As with differences between P and NP rats, a number of glutamate-associated gene and/or protein expression changes are observed after ethanol drinking by P rats (adult data for the whole NAcb came from Rodd et al., 2008; adult data for the NAcbSh came from Bell, Kimpel, et al., 2009; McBride et al., 2010; adult data for the CeA came from McBride et al., 2010; adult data for the pVTA came from McBride, Kimpel, McClintick, Ding, Hauser, et al., 2013; adolescent data for the CeA came from McBride, Kimpel, et al., 2014; McBride, Rodd, et al., 2014; adolescent data for the DRN came from McClintick et al., 2015) (Fig. 4). The protein changes after ethanol drinking in the NAcb and amygdala are from Obara et al. (2009). In general, differences, between P and NP rats, in accumbal ionotropic-associated glutamate receptor subunits are equally greater or lesser than that observed in the other line. However, metabotropic glutamate receptors are, for the most part, expressed lower in the P rat than in the NP within this brain region. Following ethanol consumption by adult P rats, expression levels for metabotropic glutamate receptors remained unchanged, whereas gene expression for ionotropic-associated glutamate receptor subunits are equally up- and downregulated in the NAcb. In the CeA, the only observed difference was lower expression of *Grm2* and *Grm3* in P vs NP rats. However following ethanol drinking, all detected ionotropic subunit-associated and metabotropic glutamate receptor gene expression changes reflected upregulation. This would suggest a strong neuroplastic response associated with ethanol-induced elevations in glutamate, such that substantial, enhanced glutamatergic neurotransmission is occurring in this brain region of the P rat. Similar observations have been seen in glutamate-associated protein levels of P rats (Obara et al., 2009) and electrophysiological activity in Marchigian sP rats (Herman et al., 2016). As discussed in our recent glutamate review (Bell et al., 2016), the findings from P rats support the hyperglutamatergic hypothesis of alcohol and drug dependence. Given this, the P rat serves as a genetic animal model of alcoholism with characteristics of glutamatergic function paralleling clinical observations including the efficacy of topirmate in significantly reducing ethanol intake.

9.5 The Serotonergic System

The neurotransmitter serotonin (5-HT) is associated with addictive behaviors, appetite regulation, behavioral inhibition, mood, and cognitive functions. Thus, dysregulation of the 5-HT system is implicated in the development of alcohol dependence. The serotonin transporter (SERT) clears 5-HT from the synapse through reuptake into the presynapse. There are seven families of 5-HT receptors (5-HT$_{1-7}$) and at least 14 distinct 5-HT receptor subtypes (Barnes & Sharp, 1999), which makes the task of understanding which 5-HT receptor subtypes mediate addictive behaviors a complex one. The raphe nucleus, where 5-HT neurons originate, sends 5-HT projections to numerous regions including the VTA, NAc, and PFC and studies have shown that the 5-HT system regulates DA neuronal activity in these subregions of the mesocorticolimbic system (Azmitia & Segal, 1978; Halliday & Törk, 1989; Herve, Pickel, Joh, & Beaudet, 1987; Parent, Descarries, & Beaudet, 1981; Van Bockstaele, Cestari, & Pickel, 1994). For example, 5-HT activates VTA-DA neurons (Pessia, Jiang, North, & Johnson, 1994), induces DA release in VTA slices

(Beart & McDonald, 1982), enhances DA release in NAc when locally applied to the VTA (Guan & McBride, 1989) or the dorsal raphe is activated (Yoshimoto & McBride, 1992), potentiates the excitatory actions of alcohol on VTA-DA neurons (Brodie, Trifunović, & Shefner, 1995), and increases extracellular DA release in the PFC (Iyer & Bradberry, 1996).

Acute alcohol exposure appears to increase 5-HT activity (McBride, Chernet, Rabold, Lumeng, & Li, 1993; Smith & Weiss, 1999), whereas chronic exposure to alcohol may result in the development of tolerance to this effect (Smith & Weiss, 1999). Clinical and/or preclinical studies have reported deficiencies of 5-HT and/or its major metabolite 5-HIAA in the brains of human alcoholics (Pivac, Mück-Seler, Mustapić, Nenadić-Sviglin, & Kozarić-Kovacić, 2004; Schmidt, Dufeu, Heinz, Kuhn, & Rommelspacher, 1997) and genetically selected alcohol-preferring rats (McBride, Chernet, Rabold, et al., 1993; Murphy et al., 1987; Strother, Chernet, Lumeng, Li, & McBride, 2001; Zhou, Bledsoe, Lumeng, & Li, 1991a,1991b). Moreover, treatments that reduce 5-HT neurotransmission can elevate self-administration of alcohol (Ciccocioppo, Angeletti, Colombo, Gessa, & Massi, 1999; Lyness & Smith, 1992). Drug treatments with antidepressants that affect 5-HT CNS activity have been shown to reduce craving and/or symptomatic behavior associated with alcohol dependence (cf, Goodman, 2008) and alcoholic individuals with a polymorphism of the 5-HT transporter can respond favorably to certain medication combinations (Johnson, 2010). Therefore, it has been proposed that modulation of the 5-HT system is a viable therapy for alcoholism in a subset of patients (Johnson, 2005, 2010; Wrase, Reimold, Puls, Kienast, & Heinz, 2006). Research on the involvement of 5-HT in binge alcohol drinking has been limited, with some evidence that binge drinking induces a blunted 5-HT response in the Scheduled High Alcohol Consumption mouse binge drinking model (Szumlinski et al., 2007). Additionally, acute withdrawal from alcohol after binge-like exposure leads to a wide-spread reduction in 5-HT and other neurotransmitters in several brain regions including those associated with the mesocorticolimbic system (Smith, Co, McIntosh, & Cunningham, 2008). In general, these findings indicate that serotonergic treatments may disrupt binge alcohol drinking and may interfere with the progression to alcohol dependence, in certain individuals. However, actual efficacy of 5HT manipulation to treat alcohol dependence in the clinic has been rather modest and may be relevant only in certain subpopulations of alcoholics.

Systemically, the 5HT precursor, D1L-5-HTP (McBride et al., 1990), reverse SERT modulator, fenfluramine (McBride et al., 1990); as well as agonists for the $5HT_1$, TFMPP (McBride et al., 1990), $5HT_{1A}$, 8-OH-DPAT (McBride et al., 1990), $5HT_2$, DOI (McBride et al., 1990), all reduced ethanol intake by P rats. However, most work has examined receptor antagonists, including WAY 100,635 (Zhou, McKinzie, Patel, Lumeng, & Li, 1998) which targets the $5HT_{1A}$ receptor; amperozide/FG 5606 (Lankford, Bjork, & Myers, 1996; Overstreet, McArthur, Rezvani, & Post, 1997), and FG 5974 (Lankford et al., 1996; Overstreet et al., 1997; Piercy, Bjork, & Myers, 1996) which target $5HT_2$ receptors; as well as MDL 72222 (Rodd-Henricks, McKinzie, Edmundson, et al., 2000) and ICS 205–930 (Rodd et al., 2010; Rodd-Henricks, McKinzie, Edmundson, et al., 2000) which target $5HT_3$ receptors, all of which reduce ethanol intake or the acquisition of operant ethanol self-administration by P rats. Additionally, the SERT inhibitors fluoxetine (Murphy et al., 1985, 1988; Rezvani et al., 2000; Zhou et al., 1998), fluvoxamine (Murphy et al., 1985), and DOV 102,677 (Yang et al., 2012) all reduced ethanol intake by P rats. The last compound also inhibits the norepinephrine (NE) and dopamine transporters.

A number of differences in serotonin-associated protein (Ciccocioppo, Ge, Barnes, & Cooper, 1998; McBride, Chernet, Rabold, et al., 1993; McBride et al., 1997; McBride, Guan, Chernet, Lumeng, & Li, 1994; Murphy et al., 1987; Pandey, Lumeng, & Li, 1996) and mRNA expression levels (see Fig. 5) have been reported between P and NP rats (data for the pVTA came from McBride et al., 2012; data for the CeA and NAcbSh came from McBride, Kimpel, McClintick, Ding, Hyytia, et al., 2013). Observed ethanol-drinking-induced changes in gene expression have come mainly from our work with adolescent bingeing P rats and in the DRN, from which all 5HT projections emanate (adult data for the whole NAcb came from Rodd et al., 2008; adult data for the NAcbSh came from Bell, Kimpel, et al., 2009; McBride et al., 2010; adult data for the CeA came from McBride et al., 2010; adult data for the pVTA came from McBride, Kimpel, McClintick, Ding, Hauser, et al., 2013; adolescent data for the CeA came from McBride, Kimpel, et al., 2014; McBride, Rodd, et al., 2014; adolescent data for the DRN came from McClintick et al., 2015) (Fig. 5). For the most part, P rats have lower levels of central 5HT, 5HIAA, and 5HT receptors than NP rats. Following ethanol drinking by both adult and adolescent P rats, usually in binge-like form, nearly all detected changes in 5HT-associated gene expression represented downregulation. These findings provide strong evidence that serotonergic deficits predispose an individual to abuse ethanol and, without intervention, foster continued excessive drinking.

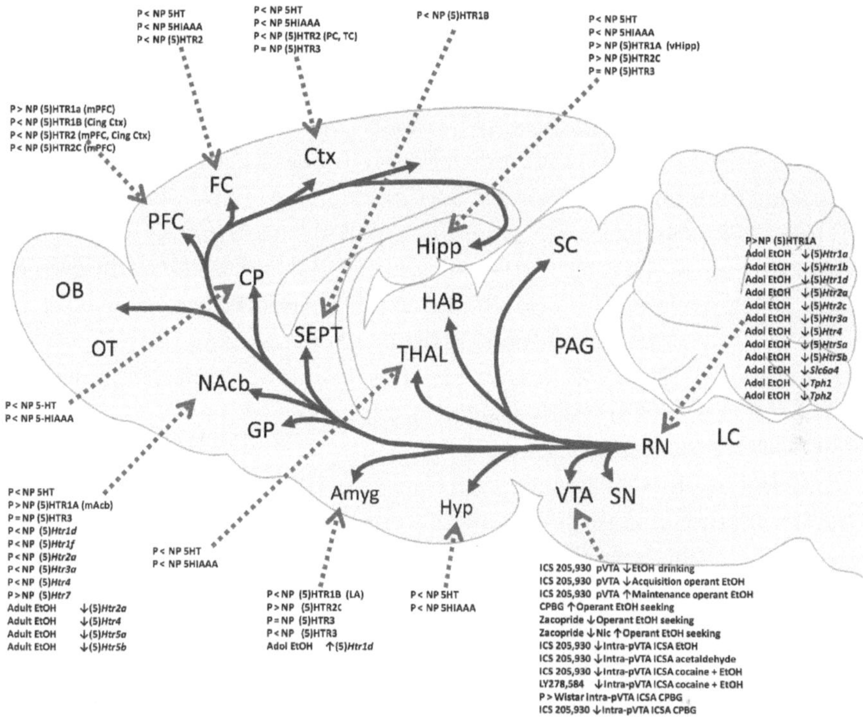

Fig. 5 Innate differences in gene (*italics*) and/or protein expression between P and NP rats, ethanol-drinking-induced changes in gene expression of P rats as well as pharmacological changes in serotonergic activity and/or effects. *mAcb*, medial nucleus accumbens; *LA*, lateral amygdala; *PC*, pyriform cortex; *RN*, raphe nuclei; *TC*, temporal cortex; *vHipp*, ventral hippocampus. For other abbreviations see Figs. 1–4 legends. *Dark arrows* indicate serotonergic projections. *Up-arrow* indicates an increase, whereas a *down-arrow* indicates a decrease in expression levels, activity, or consumption/self-administration.

10. THE NORADRENERGIC AND CORTICOTROPHIN RELEASING HORMONE SYSTEMS

It is widely established that the central noradrenergic system serves a global function in neuromodulation, controlling vigilance, attention, and the sleep–wake cycle as well as contributing to learning and memory processes. Neuroanatomical evidence indicates that noradrenergic system in the brain arise from the cell bodies in the locus coeruleus and project to different cerebral regions and to the spinal cord (Cooper et al., 2002). Moreover, there is a complex interaction between norepinephrine (NE) and CRF, a neuropeptide strongly associated with central autonomic and stress activity, receptors within the locus coeruleus (Reyes, Bangasser, Valentino, & Van Bockstaele, 2014). In addition to major projections to the frontal cortex, NE neurons project to the limbic system (Flavin & Winder, 2013), including amygdala, bed nucleus of the stria terminalis, hippocampus, and hypothalamus where it is implicated in addiction (Becker, 2012; Koob, 2013; Sofuoglu, Rosenheck, & Petrakis, 2014; see also Al' Absi, 2007), anxiety (Geiger, Neufang, Stein, & Domschke, 2014), attention (Geiger et al., 2014; Hegerl & Hensch, 2014), cognition (Chandler, Waterhouse, & Gao, 2014), memory, mood (Gold, 2015), pain (Elman, Borsook, & Volkow, 2013; Strobel, Hunt, Sullivan, Sun, & Sah, 2014), PTSD (Sofuoglu et al., 2014; Wimalawansa, 2014), sleep (Zeitzer, 2013), suicide (Elman et al., 2013), and associated physiological processes (Klimek, Rajkowska, Luker, Dilley, et al., 1999; Moret & Briley, 2011).

NE in synaptic vesicles is derived from two sources. The synthesis of NE begins with the synthesis of dopamine from tyrosine and is transported into the vesicle by the vesicular monoamine transporter (VMAT). Once dopamine is synthesized and stored in synaptic vesicles, an enzyme called dopamine-β-hydroxylase further hydroxylates dopamine into NE. The synthesis of NE is different than the other neurotransmitters as they are usually made in the cytoplasm of the terminal buttons, whereas NE's final stage of synthesis occurs in synaptic vesicles. Neurotransmission is initiated by an action potential which triggers the release of NE into the synaptic cleft. Released NE interacts with multiple adrenergic receptors, including presynaptic α2 and postsynaptic α1, β1, and β2 receptors. NE is removed from the synaptic cleft by both selective NE transporter (NET) as well nonselective transporters. NE's stimulation of α2-adrenergic receptors provide's feedback inhibition of further release. Cytoplasmic NE that is not sequestered in synaptic vesicles by VMAT is degraded into its metabolites by the enzyme monoamine oxidase, type A (MAO-A) (Cooper et al., 2002; Golan, Tashjian, Armstrong, & Armstrong, 2012; Krishnan & Nestler, 2008).

Evidence indicates that central NE activity modulates alcohol drinking behavior (Ehrenreich, Schuck, Stender, et al., 1997; Getachew, Hauser, Taylor, & Tizabi, 2010; see also Al' Absi, 2007). For example, selective NE uptake inhibition may normalize the behavioral and negative affective effects of alcohol (Getachew et al., 2010). Similarly, α1-adrenergic antagonists, α2-adrenergic agonists, and β-adrenergic antagonists modulate alcohol drinking or associated withdrawal behavior (eg, Gilpin & Koob, 2010; Riihioja, Jaatinen, Oksanen, et al., 1997; Walker, Rasmussen, Raskind, & Koob, 2008). There have been several studies investigating the adrenergic system's role in excessive ethanol drinking by P rats using peripheral routes of administration. Prazosin, an α1-adrenergic antagonist (Menkes, Baraban, & Aghajanian, 1981), reduces home-cage limited access drinking (Rasmussen, Alexander, Raskind, & Froehlich, 2009), relapse drinking (Froehlich, Hausauer, Fischer, Wise, & Rasmussen, 2015), operant self-administration (Verplaetse & Czachowski, 2015; Verplaetse et al., 2012), and operant ethanol seeking (Verplaetse & Czachowski, 2015; Verplaetse et al., 2012). It also disrupts the acquisition of excessive ethanol drinking by P rats (Froehlich, Hausauer, Federoff, Fischer, & Rasmussen, 2013). Combining prazosin and naltrexone was more effective in reducing ethanol drinking than either compound alone, at least in P rats (Froehlich, Hausauer, & Rasmussen, 2013). Additionally, tolerance did not appear to develop following repeated daily treatments (Rasmussen, Kincaid, & Froehlich, 2015). Another study found that clonidine, an α2-adrenergic receptor agonist, can also reduce ethanol drinking by P rats (Rasmussen, Alexander, Malone, Federoff, & Froehlich, 2014; Rasmussen, Beckwith, Kincaid, & Froehlich, 2014). In addition, the triple monoamine uptake inhibitor (ie, DAT, NET, and SERT) DOV 102,677 reduced ethanol intake by P rats (Yang et al., 2012). Paralleling the preclinical findings, prazosin also has shown promise in the treatment of AUDs in humans (eg, Simpson, Saxon, Meredith, et al., 2009). Overall, limited data suggest the involvement of the noradrenergic system in AUDs but substantial territory still needs to be explored regarding the role of NE and stress-associated systems in alcohol dependence.

The CRF system is localized, often colocalizing with NE, in multiple brain regions associated with addiction, anxiety, consummatory behavior, sleep, stress, learning, and memory (Reul & Holsboer, 2002; Sajdyk, Shekhar, & Gehlert, 2004). These brain regions include the raphe nucleus (Lukkes et al., 2011), multiple nuclei of the hypothalamus and the amygdala (Blume et al., 2009; Campbell, Grove, & Smith, 2003; Sajdyk et al., 2004; Smialowska, Wieronska, & Wedzony, 2002), pituitary (Stanley et al., 2004), cortex and lateral septum (Miyata, Shiota, Chaki, Okuyama, & Inagami, 2001), as well as bed nucleus of the stria terminalis and hippocampus (Van Pett et al., 2000). Moreover, CRFR1 receptors colocalize with cholinergic, noradrenergic, and DAergic neurons in many of these brain regions as well (Sauvage & Steckler, 2001). There is increasing research into the role of corticotrophin releasing hormone (CRH), NPY, glucocorticoids, and HPA activity (the endogenous opioid system is discussed next) in alcohol dependence. One important reason for this is the fact that most earlier research on pharmacological treatments was conducted in animal models mimicking early stages of the dependence cycle, rather than the later stages observed in dependence. In the P rat, CRF protein expression levels are lower in the PFC, pyriform cortex, hypothalamus, and amygdala compared with NP rats (Ehlers et al., 1992). A subsequent study revealed that both CRF protein and mRNA are lower in the CeA of P rats compared with NP rats (Hwang, Stewart, Zhang, Lumeng, & Li, 2004). More recent work has shown that P rats have a polymorphism in the promoter region of the CRFR2 gene, which is not present in NP rats (Yong et al., 2014). This polymorpshism is associated with lower *Crfr2* expression, especially in the amygdala; and, similar to innately reduced NPY levels in the amygdala, reduced *Crfr2* expression appears to be directly associated with higher ethanol intake and anxiety in P rats (Yong et al., 2014). Regarding CRF, ligands targeting the CRF1 receptor, such as antalarmin and MPZP, reduce ethanol drinking by P rats but dependence is a requirement to see these effects (Gilpin, Richardson, & Koob, 2008; Gilpin, Stewart, & Badia-Elder, 2008; Heilig & Egli, 2006).

Similar to observations for the cholinergic and DAergic systems, very few differences in adrenergic and CRF-associated mRNA expression differences (see Fig. 6) between P and NP rats (data for the pVTA came from McBride et al., 2012; data for the CeA and NAcbSh came from McBride, Kimpel, McClintick, Ding, Hyytia, et al., 2013). The same is true for ethanol-drinking-induced changes in P rats (adult data for the whole NAcb came from Rodd et al., 2008; adult data for the NAcbSh came from Bell, Kimpel, et al., 2009; McBride et al., 2010; adult data for the CeA came from McBride et al., 2010; adult data for the pVTA came from McBride, Kimpel, McClintick, Ding, Hauser, et al., 2013; adolescent data for the CeA came from McBride, Kimpel, et al., 2014; McBride, Rodd, et al., 2014; adolescent data for the DRN came from McClintick et al., 2015) (Fig. 6). However, most drinking protocols used for our recent microarray work have been binge-like and limited access in nature. Thus, the animals were probably not physically dependent on alcohol, despite being quite motivated to consume it and achieving BACs greater than 80–100 mg%.

10.1 The Opioid System(s)

There are several classes of endogenous opioids including enkephalins, endorphins, dynorphins, and endomorphins. These classes of ligands bind with some specificity to the delta, kappa, and mu-receptors, respectively.

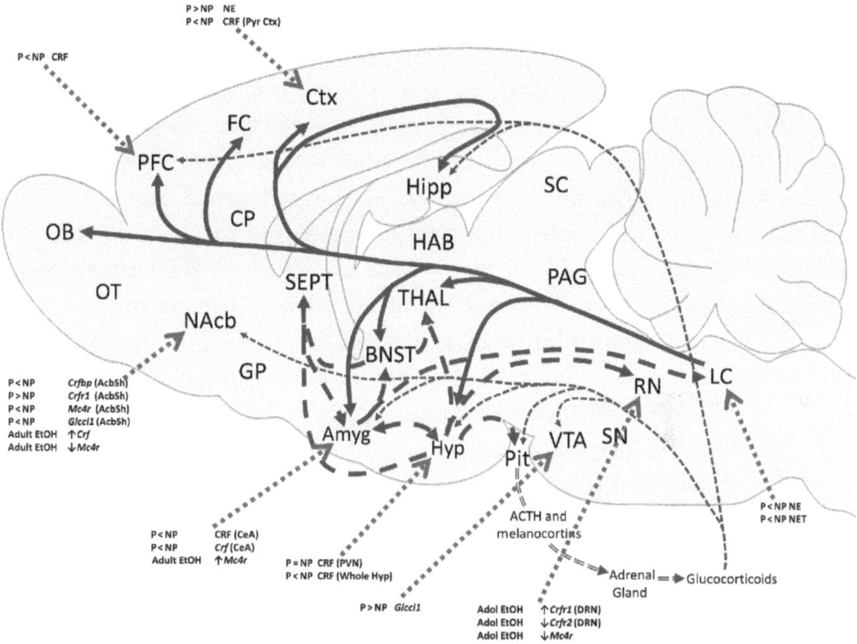

Fig. 6 Innate differences in gene (*italics*) and/or protein expression between P and NP rats, ethanol-drinking-induced changes in gene expression of P rats as well as pharmacological changes in noradrenergic and corticotrophic releasing factor-system activity and/or effects. *ACTH*, adrenal corticotrophic hormone; *MC4R*, melanocortin-4 receptor. For other abbreviations see Figs. 1–5 legends. *Dark, solid arrows* indicate noradrenergic projections. *Dark, thick, dashed arrows* indicate CRF projections. *Dark, thin, dashed arrows* point to nuclei with glucocorticoid binding sites and activity. *Up-arrow* indicates an increase, whereas a *down-arrow* indicates a decrease in expression levels, activity, or consumption/self-administration.

One role of these peptides in the brain is to process information about rewarding stimuli, including alcohol (cf, Oswald & Wand, 2004). Therefore, it is not surprising that these peptides have been shown to influence the development of alcohol abuse and dependence. Opioid receptors are found presynaptically on DAergic neurons of the mesocorticolimbic system (eg, within the NAc) where they control the release of DA. Thus, opioid activity, similar to the glutamatergic and GABAergic systems, modulates DA activity in this "reward" neurocircuit.

Variations in opioid-related gene expression and function may contribute to high levels of alcohol consumption as well (eg, Marini et al., 2013). For example, high alcohol drinking rats exhibit a greater level of mu-opioid receptor (MOR)-associated and enkephalin mRNA, compared to low alcohol drinking rats (Morganstern, Liang, Ye, Karatayev, & Leibowitz, 2012). For a review of neurobiological differences in the opioid system between selectively bred high and low alcohol-consuming rats see Bell et al. (2012). A great deal of existing evidence for the role of opioids in alcohol abuse and dependence comes from pharmacological experiments using the Food and Drug Administration (FDA)-approved treatment for alcoholism, naltrexone (ReVia) and other nonspecific opioid antagonists. Naltrexone blocks alcohol-induced changes in gene transcription in several receptor systems, including the mu-opioid system. Evidence from knockout mice lacking MORs or dynorphin suggest that MORs and kappa-opioid receptors (KORs) are involved in the rewarding or reinforcing effects of alcohol (Blednov, Walker, Martinez, & Harris, 2006; Charbogne et al., 2014; Roberts, Heyser, Cole, Griffin, & Koob, 2000).

There is a substantial literature on peripheral administration of opioid-associated ligands to reduce ethanol drinking by P rats. Essentially, peripheral naltrexone, a pan-opioid antagonist, has been tested repeatedly (Coonfield, Kiefer, Ferraro, & Sinclair, 2004; Dhaher et al., 2012; June, Grey, et al., 1998; Sable, Bell, Rodd, & McBride, 2006) along with a study testing the effects of CeA microinjections (Foster et al., 2004). These publications all reported significant reductions in ethanol intake by male and female, adolescent and adult P rats using both home-cage and operant procedures. Peripheral testing with the pan-opioid antagonists, naloxone (Badia-Elder, Mosemiller, Elder, & Froehlich, 1999; June, Lummis, Colker, Moore, & Lewis, 1991) and nalmefene (June, Grey, et al., 1998), also revealed significant reductions in ethanol intake by male and female, adolescent and adult P rats using both home-cage and operant procedures. Another study reported that nalmefene microinjections into the NAcb, Hipp and VTA significantly reduced operant alcohol self-administration by adult female P rats (June et al., 2004). A study examining the acquisition, maintenance and relapse operant self-administration of ethanol by adult female P rats found that the MOR antagonist LY255582 significantly reduced all of these behaviors (Dhaher et al., 2012). Other studies examined the role of the delta-opioid receptor (DOR) in mediating ethanol intake by adult P rats. Systemic treatment with the DOR antagonists naltriben, naltrindole and ICI 174,864 reduced home-cage ethanol drinking (Krishnan-Sarin, Jing, et al., 1995; Krishnan-Sarin, Portoghese, Li, & Froehlich, 1995) and operant self-administration (June et al., 1999) of the same.

Despite a clear role for the opioid system (including ligand-associated changes in drinking) in alcohol drinking and consummatory behavior across reinforcers, very few changes induced by ethanol and/or differences between P and NP rats have been reported to date (Fig. 7). Data of line-dependent differences for the pVTA came from McBride et al., 2012 and data for the CeA and NAcbSh came from McBride, Kimpel, McClintick, Ding, Hyytia, et al., 2013. Data for ethanol-drinking-induced changes in P rats (adult data for the whole NAcb came from Rodd et al., 2008; adult data for the NAcbSh came from Bell, Kimpel, et al., 2009; McBride et al., 2010; adult data for the CeA came from McBride et al., 2010; adult data for the pVTA came from McBride, Kimpel, McClintick, Ding, Hauser, et al., 2013; adolescent data for the CeA came from McBride, Kimpel, et al., 2014; McBride, Rodd, et al., 2014; adolescent data for the DRN came from McClintick et al., 2015) (Fig. 7). The findings thus far indicate that multiple opioid systems are involved in excessive ethanol drinking and modulation of

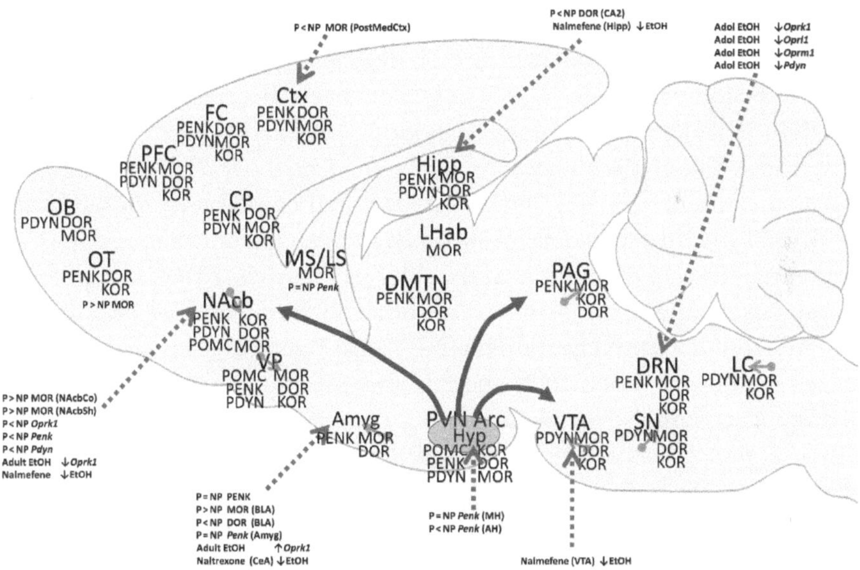

Fig. 7 Innate differences in gene (*first letter upper case others lower case* and in *italics*) and/or protein expression between P and NP rats, ethanol-drinking-induced changes in gene expression of P rats as well as pharmacological changes in opioid-system activity and/or effects. *DOR*, delta opioid receptor; *KOR*, kappa opioid receptor; *MOR*, mu opioid receptor; *Pdyn*, prodynorphin; *Penk*, proenkephalin; *POMC*, proopiomelanocortin. For other abbreviations see Figs. 1–6 legends. When multiple opioid precursors (PDYN, PENK, or POMC) and/or opioid receptors (DOR, KOR, or MOR) are localized in the brain region, the order of density are presented from highest (*top*) to lowest (*bottom*). *Dark, solid arrows* indicate opioid projections. *Short light-shaded arrows* represent opioid containing interneurons. *Up-arrow* indicates an increase, whereas a *down-arrow* indicates a decrease in expression levels, activity, or consumption/self-administration.

this system consistently reduces ethanol intake significantly. Thus, the P rat also displays the relatively robust reduction in alcohol abuse induced by the pan-opioid receptor antagonist naltrexone, let alone similar observations for more selective mu- and delta-opioid antagonists.

10.2 The NPY System

NPY is a 36 amino acid peptide abundantly expressed throughout the central and peripheral nervous systems (Allen, Adrian, Allen, et al., 1983) and acts centrally on target cells through the G-protein, coupled NPY receptors Y1, Y2, and Y5 (Dumont, Satoh, Cadieux, et al., 1993; Fetissov, Kopp, & Hokfelt, 2004; Wolak et al., 2003). NPYir cell bodies have been found in the hypothalamus, hippocampus, amygdala, brain stem nuclei, and ganglions of the sympathetic and parasympathetic nervous systems. There is evidence that NPY is locally synthesized (van den Pol, 2012), therefore its expression is not dependent upon projection neurons. For the most part, NPY neurons are interneurons and found in most brain regions except for the thalamus and the cerebellum, although Y-receptors have been detected in the thalamus (eg, Kaji, 2013). NPY is implicated in food intake and energy balance, anxiety, stress, autonomic function, learning, and memory (cf, Parker, 2013). Given the role of stress and anxiety (at least their comorbidity for the latter) in alcohol and drug dependence, it is not surprising that NPY, and for that matter the interaction of NPY and CRF (eg, Thorsell, 2010), is implicated in the development and maintenance of alcohol and/or drug dependence (cf, Al' Absi, 2007). Regarding the NPY–CRF interaction, There appears to be a confluence of activity from multiple systems in the CeA that mediates anxiety and alcohol dependence with a complex interaction of the CRF, GABAergic, glutamatergic, and NPY systems (Gilpin, Herman, & Roberto, 2015; Roberto, Gilpin, & Siggins, 2012). Additionally, cAMP response element-binding protein, BDNF, activity-regulated cytoskeleton (ARC) associated protein are involved in the modulation of CeA–NPY activity and, by extension, are involved in anxiety and alcohol dependence as well (cf, Pandey, 2003; Wand, 2005).

Centrally, site-specific injections of NPY point to the amygdala (Gilpin, Richardson, et al., 2008; Gilpin, Stewart, et al., 2008; Pandey, Chartoff, Carlezon, et al., 2005) and hypothalamus (Gilpin, Stewart, Murphy, & Badia-Elder, 2004; Kelley, Nannini, Bratt, & Hodge, 2001) as possible primary neuroanatomical loci for NPY-induced alterations in ethanol drinking. Previous studies suggest that NPY is a major factor which distinguishes the ethanol-drinking behavior of P vs NP rats (Cowen, Chen, & Lawrence, 2004; Pandey, Carr, Heilig, Ilveskoski, & Thiele, 2003; Thiele & Badia-Elder, 2003).

P rats display lower levels of NPY immunoreactivity in various regions of the brain, including the CeA, hippocampus and the FC with higher levels in the paraventricular hypothalamic nucleus and arcuate nucleus of the hypothalamus (Ehlers, Li, Lumeng, et al., 1998; Hwang, Zhang, Ehlers, Lumeng, & Li, 1999). Furthermore, decreased levels of NPY are associated with increased anxiety in P rats (Spence, Liang, Habegger, & Carr, 2005; Stewart, Gatto, Lumeng, Li, & Murphy, 1993), and ICV infusion of NPY has been shown to reduce ethanol intake in the P rat (Gilpin, Stewart, Murphy, Li, & Badia-Elder, 2003) (Fig. 8). Genomically, NPY is localized in an interval that is highly associated with alcohol preference and consumption, mapping to a quantitative trait locus (QTL) with a LOD score of 9.2 on rat chromosome 4, using an F2 population bred from iP and iNP rats (Bice, Foroud, Bo, et al., 1998; Carr, Foroud, Bice, et al., 1998).

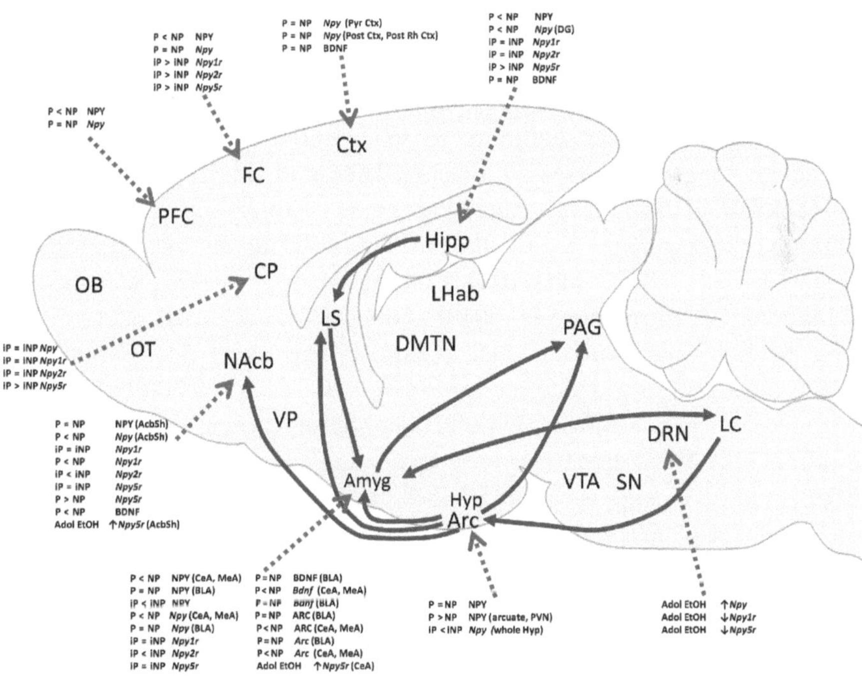

Fig. 8 Innate differences in gene (*italics*) and/or protein expression between P and NP rats, ethanol-drinking-induced changes in gene expression of P rats as well as pharmacological changes in the neuropeptide Y (NPY)-system activity and/or effects. *ARC*, activity-regulated cytoskeleton-associated protein; *BDNF*, brain-derived neurotrophic factor. For abbreviations see Figs. 1–7 legends. *Dark, solid arrows* indicate NPY projections. It is important to remember that NPY is also synthesized locally in the brain. *Up-arrow* indicates an increase, whereas a *down-arrow* indicates a decrease in expression levels, activity, or consumption/self-administration.

11. PHARMACOGENOMICS AND ALCOHOLISM/ADDICTION

The interest in using pharmacogenomic research to treat alcohol dependence spans more than a decade (eg, Anton et al., 2008; Goldman, Oroszi, O'Malley, & Anton, 2005). Numerous studies have identified a number of single nucleotide polymorphisms that are associated with alcohol dependence and/or drug codependence including (a) CHRM2 (Luo et al., 2005; Wang et al., 2004), CHRNA4 (Kim et al., 2004), CHRNA5 (Saccone et al., 2007; Wang et al., 2009) as well as the CHRNA5–CHRNA3–CHRNB4 cluster and alcohol abuse/dependence (Schlaepfer et al., 2008); (b) DAT (Heinz, Goldman, Gallinat, Schumann, & Puls, 2004; see also Bhaskar & Kumar, 2014 for this and other DA-associated polymorphisms), DA beta hydroxylase and alcohol dependence in women (Preuss et al., 2013), DRD3 and alcohol craving (Agrawal et al., 2013) as well as DA dysfunction and Cloninger Type I alcoholism (Leggio & Addolorato, 2008); (c) GABRA1, GABRA2, GABRB3, GABRG3, and alcohol dependence or sensitivity to its intoxicating effects during the ascending slope of the BAC curve (eg, Bierut et al., 2010; Dick et al., 2004; Dick et al., 2006; Edenberg et al., 2004; Enoch, Schwartz, Albaugh, Virkkunen, & Goldman, 2006; Haughey et al., 2008; Noble et al., 1998); (d) GRIK3 (Grzywacz, Małecka, Suchanecka, Bieńkowski, & Samochowiec, 2013) and GRIN2A (Domart et al., 2012) with alcohol dependence as well as GRM8 and event-related potential (ERP) theta power and alcohol dependence (Chen et al., 2009); (e) 5HT dysfunction and Cloninger Type II alcoholism (Leggio & Addolorato, 2008), HTR1A and alcohol as well as nicotine codependence (Zuo et al., 2013), HTR1B and alcohol as well as multiple drug abuse (Cao, LaRocque, & Li, 2013; Contini et al., 2012), HTR2A and alcohol as well as heroin abuse (Cao et al., 2014), HTR7 and alcohol dependence as well as electrophysiological measures (Zlojutro et al., 2010; Zuo et al., 2014), alcohol dependence and SERT (eg, Heinz et al., 2004; cf, Johnson, 2010; McHugh, Hofmann, Asnaani, Sawyer, & Otto, 2010; Plemenitas et al., 2015); (f) OPRM1 and level of response to ethanol in Native Americans (Ehlers, Lind, & Wilhelmsen, 2008), OPRM1 polymorphisms and naltrexone's efficacy for treating alcohol dependence (eg, Jonas et al., 2014), as well as PDYN and OPRK1 with alcohol dependence (Gerra et al., 2007; Williams et al., 2007; Xuei et al., 2006); (g) CRFR1 polymorphism with P3 ERP and alcohol dependence (Chen et al., 2010); and (h) NPY and its receptor's association with alcohol as well as multiple drug abuse and dependencies (Bhaskar et al., 2013; Frances et al., 2011; Okahisa et al., 2009; Sato et al., 2010; Wetherill et al., 2008). While there is overlap in the ethanol affected neurotransmitter and neuropeptide systems between the clinical alcohol dependent population and the P rat, much more research needs to be done.

12. CONCLUSIONS

The present chapter sought to present the existing neuropharmacological findings on P rats in a more holistic manner than done in the past. While previous reviews listed many differences between the P rat and its control line the NP rat, this was mostly done in tabular form or buried in the text. By using figures of the primary neurotransmitter and neuropeptide systems examined thus far in the P rat, our objective was to map the published findings in the context of the projections and/or localization of each respective neuromodulatory system. This chapter also outlines how the P rat has neurochemical, physiological and behavioral characteristics often seen in individuals with alcohol, and in some cases drug, dependence. Finally, it has been proposed that an animal model of alcoholism should display similar pharmacological efficacy as that seen in the alcohol dependent treatment population. Because excessive alcohol is under genetic control, individual differences are expected in animal models just as individual differences prevail in the clinical treatment population. Thus, an animal model of alcoholism also should display some pharmacological treatment validity (Dyr & Kostowski, 2008; Litten et al., 2012; Overstreet, Rezvani, Djouma, Parsian, & Lawrence, 2007).

For the cholinergic system, the P rat displays a modest effect of varenicline on ethanol intake. This parallels the clinical literature in that there are mixed findings for the efficacy of varenicline, especially in the context of smokers vs nonsmokers and/or comorbid psychiatric conditions. For the DAergic system, findings with the P rat do not match clinical observations; such that whereas manipulations of the DAergic system consistently affect ethanol intake in P rats, this is not true in the treatment setting. For the GABAergic system, the results are mixed. GABAergic modulators are often used during ethanol withdrawal, but cross-tolerance with the effects of ethanol and inherent abuse liability limit their usefulness beyond acute care.

One exception should be noted and that is topiramate, which has shown similar efficacy in P rats, other animal models and a tested treatment population. Topiramate, as an anticonvulsant, modulates both GABAergic and glutamatergic activity. Acamprosate, another modulator of the glutamatergic system, has demonstrated modest effects in the clinical treatment population, but only has marginal effects in P rats. Similar to the DAergic system, modulators of the serotonergic system have had limited success in the clinical setting even though robust effects are seen in P rats. Ondansetron is one exception, with mixed findings in both P rats and human laboratory subjects. In addition, it appears that variances in the SERT gene (*5htt*) can determine some pharmacotherapeutic efficacy supporting a role for pharmacogenomics. For the noradrenergic system, prazosin consistently decreases ethanol intake in P rats and promising results are being seen in human clinical studies as well. For the opioid system, naltrexone consistently reduces ethanol intake in P rats and while some clinical studies report robust effects, other clinical studies have reported modest to marginal results. Similar to observations of *5htt* polymorphisms, variances in the OPRM1, and possibly OPRK1, gene appear to have some predictive validity for naltrexone's efficacy in the treatment of alcohol dependence. A final example, which does not fall into any of the systems discussed in this chapter, is ibudilast and more selective inhibitors of phosphodiesterase 4 (eg, Bell et al., 2015; Franklin, Hauser, Lasek, Bell, & McBride, 2015). Robust findings in P rats and multiple other animal models parallel early results from clinical laboratory studies. Regarding these early findings, it should be noted that in 2015 the FDA gave ibudilast a fast-track designation for the treatment of methamphetamine dependence. While considerable progress has been made in the treatment of alcohol and drug dependence, considerable more work needs to be done. One future direction for research with the P rat is to determine if this genetic animal model will self-administer other drugs of abuse. Present work, and some previous work, from our laboratory indicates P rats will readily self-administer nicotine and cocaine into discrete regions of the mesocorticolimbic reward circuit. Intravenous work has not been published yet, but the results look promising for this route of administration as well. These findings suggest that the P rat may be a genetic animal model of polysubstance abuse/dependence.

To date, the vast majority of the genomic information about the P rat has come from microarray and RT-PCR techniques. Future research should use next-generation DNA sequencing to identify genomic signatures of selection between P and NP rats and next-generation RNA sequencing methodologies to analyze allele-specific expression of genes in F1 crosses of these lines (eg, Farris & Mayfield, 2014; Wang, Kapoor, & Goate, 2012). This will help move the field from QTL analyses to quantitative trait nucleotide (ie, SNPs) and quantitative trait gene analyses (eg, Ehlers, Walter, Dick, Buck, & Crabbe, 2010; Milner & Buck, 2010; Spence et al., 2009). By doing so, the level of genomic resolution and the power of these analyses will be exponentially increased over the existing techniques. In addition, by localizing genetic variation to genes and SNPs, research on the role of epigenomics/epigenetics (eg, Moonat, Starkman, Sakharkar, & Pandey, 2010; Renthal & Nestler, 2009) in alcohol preference can also be advanced. These advances will allow investigators to combine traditional hypothesis-driven research based on deductive reasoning with unprejudiced genome association studies. These approaches will delineate putative neuromolecular pathways (eg, intracellular cascades) mediating alcohol dependence and identify possible new drugable targets to prevent and/or treat alcohol abuse and dependence.

Another direction for future research is the use of emerging and evolving neuroscience methodologies to examine the role of second messenger systems, synaptic plasticity, protein–protein interactions, gene–gene interactions, and the role of noncoding RNAs (eg, Clerget, Bouguignon-Igel, & Rederstorff, 2015; Gedik et al., 2015; Gorini, Bell, & Mayfield, 2011; Gorini, Harris, & Mayfield, 2014; Manzardo, McGuire, & Butler, 2015; Nunez et al., 2013; Ponomarev, Wang, Zhang, Harris, & Mayfield, 2012). Emerging bioinformatic strategies would synthesize the large amounts of data obtained with high throughput gene and/or protein expression techniques (eg, Gorini et al., 2011; Gorini et al., 2014). Presently, this synthesis has started to reveal the complex neurobiology of alcoholism and the multiple roles of genetics in its development through functional and genetical genomics (eg, Spanagel et al., 2013; Zuo et al., 2014). We believe that continued research with the P rat using these more advanced genomic, proteomic, and bioinformatic techniques will yield new information on molecular substrates to target for repurposing existing FDA-approved medications, or those that are in advanced clinical trials, to treat alcohol and drug abuse/dependence.

ACKNOWLEDGMENTS

This work was supported in part by AA013522 an INIA-Neuroimmune project (R.L.B.) and AA020396 to E.A.E./R.L.B. The authors do not have any conflicts of interest with this material. The views expressed herein are solely those of the authors and do not necessarily represent the views of the National Institutes of Health (NIH) funding agencies.

REFERENCES

Aal-Aaboda, M., Alhaddad, H., Osowik, F., Nauli, S. M., & Sari, Y. (2015). Effects of (R)-(-)-5-methyl-1-nicotinoyl-2-pyrazoline on glutamate transporter 1 and cysteine/glutamate exchanger as well as ethanol drinking behavior in male, alcohol-preferring rats. *Journal of Neuroscience Research*, *93*, 930–937.

Agabio, R., & Colombo, G. (2014). GABAB receptor ligands for the treatment of alcohol use disorder: Preclinical and clinical evidence. *Frontiers in Neuroscience/Neuropharmacology*, *8*, 140. http://dx.doi.org/10.3389/fnins.2014.00140.

Agrawal, A., Wetherill, L., Bucholz, K. K., Kramer, J., Kuperman, S., Lynskey, M. T., et al. (2013). Genetic influences on craving for alcohol. *Addictive Behaviors*, *38*, 1501–1508.

Al' Absi, M. (2007). *Stress and addiction: Biological and psychological mechanisms*. New York, NY: Elsevier Academic Press.

Albuquerque, E. X., Pereira, E. F. R., Alkondon, M., & Rogers, S. W. (2009). Mammalian nicotinic acetylcholine receptors: From structure to function. *Physiological Reviews*, *89*, 73–120.

Alhaddad, H., Das, S. C., & Sari, Y. (2014). Effects of ceftriaxone on ethanol intake: A possible role for xCT and GLT-1 isoforms modulation of glutamate levels in P rats. *Psychopharmacology*, *231*, 4049–4057.

Alhaddad, H., Kim, N. T., Aal-Aaboda, M., Althobaiti, Y. S., Leighton, J., Boddu, S. H., et al. (2014). Effects of MS-153 on chronic ethanol consumption and GLT1 modulation of glutamate levels in male alcohol-preferring rats. *Frontiers in Behavioral Neuroscience/Neuropharmacology*, *8*, 366.

Allen, Y. S., Adrian, T. E., Allen, J. M., et al. (1983). Neuropeptide Y distribution in the rat brain. *Science*, *221*, 877–879.

American Psychiatric Association. (1994). *Diagnostic and statistical manual of mental disorders* (4th ed.). Washington, DC: Author.

American Psychiatric Association. (2000). *Diagnostic and statistical manual of mental disorders-text revision* (4th ed.). Washington, DC: Author.

American Psychiatric Association. (2013). *Diagnostic and statistical manual of mental disorders* (5th ed.). Washington, DC: Author.

Anderson, C. M., & Swanson, R. A. (2000). Astrocyte glutamate transport: Review of properties, regulation, and physiological functions. *Glia*, *32*, 1–14.

Anton, R. F., Oroszi, G., O'Malley, S., Couper, D., Swift, R., Pettinati, H., et al. (2008). An evaluation of mu-opioid receptor (OPRM1) as a predictor of naltrexone response in the treatment of alcohol dependence. *Archives of General Psychiatry*, *65*, 135–144.

Antuono, P., Sorbi, S., Bracco, L., Fusco, T., & Amaducci, L. (1980). A discrete sampling technique in senile dementia of the Alzheimer type and alcoholic dementia: Study of the cholinergic system. In L. Amaducci, A. N. Davison, & P. Antuono (Eds.), *Aging of brain and dementia* (pp. 151–158). New York, NY: Raven Press.

Assaad, J.-M., Pihl, R. O., Seguin, J. R., Nagin, D., Vitaro, F., Carbonneau, R., et al. (2003). Aggressiveness, family history of alcoholism, and the heart rate response to alcohol intoxication. *Experimental and Clinical Psychopharmacology*, *11*, 158–166.

Azmitia, E. C., & Segal, M. (1978). An autoradiographic analysis of the differential ascending projections of the dorsal and median raphe nuclei in the rat. *Journal of Comparative Neurology*, *179*, 641–667.

Babor, T. F., De Hofmann, M. I., Boca, F. K., Hesselbrock, V., Meyer, R. E., & Rounsaville, B. (1992). Types of alcoholics, I. Evidence for an empirically derived typology based on indicator of vulnerability and severity. *Archives of General Psychiatry*, *49*, 599–608.

Badia-Elder, N. E., Mosemiller, A. K., Elder, R. L., & Froehlich, J. C. (1999). Naloxone retards the expression of a genetic predisposition toward alcohol drinking. *Psychopharmacology*, *144*, 205–212.

Baimel, C., Bartlett, S. E., Chiou, L. C., Lawrence, A. J., Muschamp, J. W., Patkar, O., et al. (2015). Orexin/hypocretin role in reward: Implication for opioid and other addictions. *British Journal of Pharmacology*, *172*, 334–348.

Bannai, S., Christensen, H. N., Vadgama, J. V., Ellory, J. C., Englesberg, E., Guidotti, G. G., et al. (1984). Amino acid transport systems. *Nature*, *311*, 308.

Bannai, S., & Ishii, T. (1982). Transport of cystine and cysteine and cell growth in cultured human diploid fibroblasts: Effect of glutamate and homocysteate. *Journal of Cellular Physiology*, *112*, 265–272.

Barker, J. M., & Taylor, J. R. (2014). Habitual alcohol seeking: Modeling the transition from casual drinking to addiction. *Neuroscience and Biobehavioral Reviews*, *47*, 281–294. http://dx.doi.org/10.1016/j.neubiorev.2014.08.012.

Barnes, N. M., & Sharp, T. (1999). A review of central 5-HT receptors and their function. *Neuropharmacology*, *38*, 1083–1152.

Barron, S., Lewis, B., Wellman, K., Carter, M., Farook, J., Ring, J., et al. (2012). Polyamine modulation of NMDARs as a mechanism to reduce effects of alcohol dependence. *Recent Patents on CNS Drug Discovery*, *7*, 129–144. http://dx.doi.org/10.2174/157488912800673128.

Beart, P. M., & McDonald, D. (1982). 5-Hydroxytryptamine and 5-hydroxytryptaminergic dopaminergic interactions in the ventral tegmental area of rat brain. *Journal of Pharmacy and Pharmacology*, *34*, 591–593.

Becker, H. C. (2012). Effects of alcohol dependence and withdrawal on stress responsiveness and alcohol consumption. *Alcohol Research*, *34*, 448–458.

Beckwith, S. W., & Czachowski, C. L. (2014). Increased delay discounting tracks with a high ethanol-seeking phenotype and subsequent ethanol seeking but not consumption. *Alcoholism, Clinical and Experimental Research*, *38*, 2607–2614.

Bell, R. L., Eiler, W. J. A., II, Cook, J. B., & Rahman, S. (2009). Nicotinic receptor ligands reduce ethanol intake by high alcohol-drinking HAD-2 rats. *Alcohol*, *43*, 581–592.

Bell, R. L., Franklin, K. M., Hauser, S. R., & Engleman, E. A. (2013). Next stop dependence. Binge drinking on the road to alcoholism: Preclinical findings on its neurobiology from rat animal models. In S. B. Harris (Ed.), *Binge eating and binge drinking: Psychological, social and medical implications* (pp. 1–60). New York, NY: Nova Science Publishers.

Bell, R. L., Hauser, S. R., McClintick, J., Rahman, S., Edenberg, H. J., Szumlinski, K. K., et al. (2016). Ethanol-associated changes in glutamate reward neurocircutiry: A minireview of clinical and preclinical genetic findings. *Progress in Molecular Biology and Translational Science*, *137*, 41–85. http://dx.doi.org/10.1016/bs.pmbts.2015.10.018.

Bell, R. L., Kimpel, M. W., McClintick, J. N., Strother, W. N., Carr, L. G., Liang, T., et al. (2009). Gene expression changes in the nucleus accumbens of alcohol-preferring rats following chronic ethanol consumption. *Pharmacology, Biochemistry, and Behavior*, *94*, 131–147. http://dx.doi.org/10.1016/j.pbb.2009.07.019.

Bell, R. L., Lopez, M., Changhai, C., Egli, M., Johnson, K., Franklin, K., et al. (2015). Ibudilast reduces alcohol drinking in multiple animal models of alcohol-dependence. *Addiction Biology*, *20*, 38–42. http://dx.doi.org/10.1111/adb.12106.

Bell, R. L., McKinzie, D. L., Murphy, J. M., & McBride, W. J. (2000). Sensitivity and tolerance to the motor impairing effects of moderate doses of ethanol. *Pharmacology, Biochemistry, and Behavior*, *67*, 583–586. http://dx.doi.org/10.1016/S0091-3057(00)00387-7.

Bell, R. L., Rodd, Z. A., Engleman, E. A., Toalston, J. E., & McBride, W. J. (2014). Scheduled access alcohol drinking by alcohol-preferring (P) and high alcohol-drinking (HAD) rats: Modeling adolescent and adult binge-like drinking. *Alcohol*, *48*, 225–234. http://dx.doi.org/10.1016/j.alcohol.2013.10.004.

Bell, R. L., Rodd, Z. A., Lumeng, L., Murphy, J. M., & McBride, W. J. (2006a). The alcohol-preferring P rat and animal models of excessive alcohol drinking. *Addiction Biology*, *11*, 270–288. http://dx.doi.org/10.1111/j.1369-1600.2005.00029.x.

Bell, R. L., Rodd, Z. A., Sable, H. J., Schultz, J. A., Hsu, C. C., Lumeng, L., et al. (2006b). Daily patterns of ethanol drinking in peri-adolescent and adult alcohol-preferring (P) rats. *Pharmacology, Biochemistry, and Behavior*, *83*, 35–46.

Bell, R. L., Rodd, Z. A., Schultz, J. A., Peper, C. L., Lumeng, L., Murphy, J. M., et al. (2008a). Effects of short deprivation and re-exposure intervals on the ethanol drinking behavior of selectively bred high alcohol-consuming rats. *Alcohol*, *42*, 407–416. http://dx.doi.org/10.1016/j.alcohol.2008.03.130.

Bell, R. L., Rodd, Z. A., Smith, R. J., Toalston, J. E., Franklin, K. M., & McBride, W. J. (2011). Modeling binge-like ethanol drinking by peri-adolescent and adult P rats. *Pharmacology, Biochemistry, and Behavior*, *100*, 90–97. http://dx.doi.org/10.1016/j.pbb.2011.07.017.

Bell, R. L., Rodd, Z. A., Toalston, J. E., McKinzie, D. L., Lumeng, L., Li, T.-K., et al. (2008b). Autonomic activation associated with ethanol self-administration in adult female P rats. *Pharmacology, Biochemistry, and Behavior*, *91*, 223–232. http://dx.doi.org/10.1016/j.pbb.2008.07.016.

Bell, R. L., Rodd-Henricks, Z. A., Webster, A. A., Lumeng, L., Li, T.-K., McBride, W. J., et al. (2002). Heart rate and motor activating effects of orally self-administered ethanol in alcohol-preferring (P) rats. *Alcoholism, Clinical and Experimental Research*, *26*, 1162–1170. http://dx.doi.org/10.1111/j.1530-0277.2002.tb02652.x.

Bell, R. L., Sable, H. J. K., Colombo, G., Hyytia, P., Rodd, Z. A., & Lumeng, L. (2012). Animal models for medications development targeting alcohol abuse using selectively bred rat lines: Neurobiological and pharmacological validity. *Pharmacology, Biochemistry, and Behavior*, *103*, 119–155. http://dx.doi.org/10.1016/j.pbb.2012.07.007.

Bell, R. L., Stewart, R. B., Woods, J. E., II, Lumeng, L., Li, T.-K., Murphy, J. M., et al. (2001). Responsivity and development of tolerance to the motor impairing effects of moderate doses of ethanol in alcohol-preferring (P) and -nonpreferring (NP) rat lines. *Alcoholism, Clinical and Experimental Research*, *25*, 644–650. http://dx.doi.org/10.1111/j.1530-0277.2001.tb02262.x.

Benham, R. S., Engin, E., & Rudolph, U. (2014). Diversity of neuronal inhibition: A path to novel treatments for neuropsychiatric disorders. *JAMA Psychiatry*, *71*, 91–93.

Bertholomey, M. L., Verplaetse, T. L., & Czachowski, C. L. (2013). Alterations in ethanol seeking and self-administration following yohimbine in selectively bred alcohol-preferring (P) and high alcohol drinking (HAD-2) rats. *Behavioural Brain Research*, *238*, 252–258.

Besheer, J., Faccidomo, S., Grondin, J. J. M., & Hodge, C. W. (2008a). Effects of mGlu1-receptor blockade on ethanol self-administration in inbred alcohol-preferring rats. *Alcohol*, *42*, 13–20.

Besheer, J., Faccidomo, S., Grondin, J. J. M., & Hodge, C. W. (2008b). Regulation of motivation to self-administer ethanol by mGluR5 in alcohol-preferring (P) rats. *Alcoholism, Clinical and Experimental Research*, *32*, 209–221.

Besheer, J., Grondin, J. J. M., Cannady, R., Sharko, A. C., Faccidomo, S., & Hodge, C. W. (2010). Metabotropic glutamate receptor 5 activity in the nucleus accumbens is required for the maintenance of ethanol self-administration in a rat genetic model of high alcohol intake. *Biological Psychiatry*, *67*, 812–822.

Bhaskar, L. V., & Kumar, S. A. (2014). Polymorphisms in genes encoding dopamine signaling pathway and risk of alcohol dependence: A systematic review. *Acta Neuropsychiatrica*, *26*, 69–80.

Bhaskar, L. V., Thangaraj, K., Kumar, K. P., Pardhasaradhi, G., Singh, L., & Rao, V. R. (2013). Association between neuropeptide Y gene polymorphisms and alcohol dependence: A case-control study in two independent populations. *European Addiction Research*, *19*, 307–313.

Bice, P., Foroud, T., Bo, R., et al. (1998). Genomic screen for QTLs underlying alcohol consumption in the P and NP rat lines. *Mammalian Genome*, *9*, 949–955.

Bierut, L. J., Agrawal, A., Bucholz, K. K., Doheny, K. F., Laurie, C., Pugh, E., et al. (2010). A genome wide association study of alcohol dependence. *Proceedings of the National Academy of Sciences of the United States of America*, *107*, 5082–5087.

Bito-Onon, J. J., Simms, J. A., Chatterjee, S., Holgate, J., & Bartlett, S. E. (2011). Varenicline, a partial agonist at neuronal nicotinic acetylcholine receptors, reduces nicotine-induced increases in 20% ethanol operant self-administration in Sprague-Dawley rats. *Addiction Biology*, *16*, 440–449.

Blednov, Y. A., Walker, D., Martinez, M., & Harris, R. A. (2006). Reduced alcohol consumption in mice lacking preprodynorphin. *Alcohol*, *40*, 73–86.

Blomqvist, O., Ericson, M., Johnson, D. H., Engel, J. A., & Söderpalm, B. (1996). Voluntary ethanol intake in the rat: Effects of nicotinic acetylcholine receptor blockade or subchronic nicotine treatment. *European Journal of Pharmacology*, *314*, 257–267.

Blume, A., Torner, L., Liu, Y., Subburaju, S., Aguilera, G., & Neumann, I. D. (2009). Prolactin activates mitogen-activated protein kinase signaling and corticotropin releasing hormone transcription in rat hypothalamic neurons. *Endocrinology*, *150*, 1841–1849.

Breslin, F. J., Johnson, B. A., & Lynch, W. J. (2010). Effect of topiramate treatment on ethanol consumption in rats. *Psychopharmacology*, *207*, 529–534.

Brodie, M. S., Pesold, C., & Appel, S. B. (1999). Ethanol directly excites dopaminergic ventral tegmental area reward neurons. *Alcoholism, Clinical and Experimental Research*, *23*, 1848–1852.

Brodie, M. S., Shefner, S. A., & Dunwiddie, T. V. (1990). Ethanol increases the firing rate of dopamine neurons of the rat ventral tegmental area in vitro. *Brain Research*, *508*, 65–69.

Brodie, M. S., Trifunović, R. D., & Shefner, S. A. (1995). Serotonin potentiates ethanol-induced excitation of ventral tegmental area neurons in brain slices from three different rat strains. *Journal of Pharmacology and Experimental Therapeutics*, *273*, 1139–1146.

Brunelle, C., Barrett, S. P., & Pihl, R. O. (2006). Psychostimulant users are sensitive to the stimulus properties of alcohol as indexed by alcohol-induced cardiac reactivity. *Psychology of Addictive Behaviors*, *20*, 478–483.

Buisman-Pijlman, F. T., Sumracki, N. M., Gordon, J. J., Hull, P. R., Carter, C. S., & Tops, M. (2014). Individual differences underlying susceptibility to addiction: Role for the endogenous oxytocin system. *Pharmacology, Biochemistry, and Behavior*, *119*, 22–38.

Burke, A. R., & Miczek, K. A. (2014). Stress in adolescence and drugs of abuse in rodent models: Role of dopamine, CRF, and HPA axis. *Psychopharmacology*, *231*, 1557–1580.

Bymaster, F. P., McKinzie, D. L., Felder, C. C., & Wess, J. (2003). Use of M1-M5 muscarinic receptor knockout mice as novel tools to delineate the physiological roles of the muscarinic cholinergic system. *Neurochemical Research*, *28*, 437–442.

Cadoret, R. J. (1995). Familial transmission of psychiatric disorders associated with alcoholism. In H. Begleiter & B. Kissin (Eds.), *The genetics of alcoholism* (pp. 70–81). New York, NY: Oxford University Press.

Calabresi, P., Lacey, M. G., & North, R. A. (1989). Nicotinic excitation of rat ventral tegmental neurones in vitro studied by intracellular recording. *British Journal of Pharmacology*, *98*, 135–140.

Campbell, R. E., Grove, K. L., & Smith, M. S. (2003). Distribution of corticotropin releasing hormone receptor immunoreactivity in the rat hypothalamus: Coexpression in neuropeptide Y and dopamine neurons in the arcuate. *Brain Research*, *973*, 223–232.

Cao, J., LaRocque, E., & Li, D. (2013). Associations of the 5-hydroxytryptamine (serotonin) receptor 1B gene (HTR1B) with alcohol, cocaine, and heroin abuse. *American Journal of Medical Genetics. Part B, Neuropsychiatric Genetics*, *162B*, 169–176.

Cao, J., Liu, X., Han, S., Zhang, C. K., Liu, Z., & Li, D. (2014). Association of the HTR2A gene with alcohol and heroin abuse. *Human Genetics*, *133*, 357–365.

Carlezon, W. A., & Wise, R. A. (1996). Microinjections of phencyclidine (PCP) and related drugs into nucleus accumbens shell potentiate medial forebrain bundle brain stimulation reward. *Psychopharmacology*, *128*, 413–420.

Carr, L. G., Foroud, T., Bice, P., et al. (1998). A quantitative trait locus for alcohol consumption in selectively bred rat lines. *Alcoholism, Clinical and Experimental Research, 22*, 884–887.

Cederlof, R., Friberg, L., & Lundman, T. (1977). The interaction of smoking, environment and heredity and their implications for disease aetiology: A report of epidemiological studies on the Swedish twin registries. *Acta Medica Scandinavica. Supplementum. 612*(Suppl. 1), 1–128.

Chandler, L. J., Newsom, H., Sumners, C., & Crews, F. (1993). Chronic ethanol exposure potentiates NMDA excitotoxicity in cerebral cortical neurons. *Journal of Neurochemistry, 60*, 1578–1581.

Chandler, D. J., Waterhouse, B. D., & Gao, W. J. (2014). New perspectives on catecholaminergic regulation of executive circuits: Evidence for independent modulation of prefrontal functions by midbrain dopaminergic and noradrenergic neurons. *Frontiers in Neural Circuits, 8*, 53.

Chappell, A. M., & Weiner, J. L. (2008). Relationship between ethanol's acute locomotor effects and ethanol self-administration in male Long-Evans rats. *Alcoholism, Clinical and Experimental Research, 32*, 2088–2099.

Charbogne, P., Kieffer, B. L., & Befort, K. (2014). 15 years of genetic approaches in vivo for addiction research: Opioid receptor and peptide gene knockout in mouse models of drug abuse. *Neuropharmacology, 76*, 204–217.

Chatterjee, S., & Bartlett, S. E. (2010). Neuronal nicotinic acetylcholine receptors as pharmacotherapeutic targets for the treatment of alcohol use disorders. *CNS & Neurological Disorders Drug Targets, 9*, 60–76.

Chatterjee, S., Steensland, P., Rollema, H., Bartlett, S. E., et al. (2011). Partial agonists of the alpha3beta4 neuronal nicotinic acetlycholine receptor reduce ethanol consumption and seeking in rats. *Neuropsyhcopharmacology, 36*, 603–615.

Chen, A. C., Manz, N., Tang, Y., Rangaswamy, M., Almasy, L., Kuperman, S., et al. (2010). Single-nucleotide polymorphisms in corticotropin releasing hormone receptor 1 gene (CRHR1) are associated with quantitative trait of event-related potential and alcohol dependence. *Alcoholism, Clinical and Experimental Research, 34*, 988–996.

Chen, A. C., Tang, Y., Rangaswamy, M., Wang, J. C., Almasy, L., Foroud, T., et al. (2009). Association of single nucleotide polymorphisms in a glutamate receptor gene (GRM8) with theta power of event-related oscillations and alcohol dependence. *American Journal of Medical Genetics. Part B, Neuropsychiatric Genetics, 150*, 359–368.

Cherpitel, C. J., Moskalewicz, J., & Swiatkiewicz, G. (2004). Drinking patterns and problems in emergency services in Poland. *Alcohol and Alcoholism, 39*, 256–261.

Chester, J. A., Blose, A. M., & Froehlich, J. C. (2004). Acoustic startle reactivity during acute alcohol withdrawal in rats that differ in genetic predisposition toward alcohol drinking: Effect of stimulus characteristics. *Alcoholism, Clinical and Experimental Research, 28*, 677–687.

Chiavegatto, S., Quadros, I. M. H., Ambar, G., & Miczak, K. A. (2010). Individual vulnerability to escalated aggressive behavior by a low dose of alcohol: Decreased serotonin receptor mRNA in the prefrontal cortex of male mice. *Genes, Brain, and Behavior, 9*, 110–119.

Childress, A. R., Hole, A. V, Ehrman, R. N., Robbins, S. J., McLellan, A. T., & O'Brien, C. P. (1993). Cue reactivity and cue reactivity interventions in drug dependence. *NIDA Research Monograph, 137*, 73–95.

Ciccocioppo, R., Angeletti, S., Colombo, G., Gessa, G. L., & Massi, M. (1999). Autoradiographic analysis of 5-HT_{2A} binding sites in the brain of Sardinian alcohol-preferring and nonpreferring rats. *European Journal of Pharmacology, 373*, 13–19.

Ciccocioppo, R., Ge, J., Barnes, N. M., & Cooper, S. J. (1998). Central 5-HT_3 receptors in P and in AA alcohol-preferring rats: An autoradiographic study. *Brain Research Bulletin, 46*, 311–315.

Cicero, T. J. (1979). A critique of animal analogues of alcoholism. In E. Majchrowicz & E. P. Noble (Eds.), *Biochemistry and pharmacology of ethanol: Vol. 2* (pp. 533–560). New York, NY: Plenum Press.

Clausen, R. P., Madsen, K., Larsson, O. M., Frolund, B., Krogsgaard-Larsen, P., & Schousboe, A. (2006). Structure-activity relationship and pharmacology of gamma-aminobutyric acid (GABA) transport inhibitors. *Advances in Pharmacology*, *54*, 265–284.

Clerget, G., Bouguignon-Igel, V., & Rederstorff, M. (2015). Alcoholic precipitation of small non-coding RNAs. *Methods in Molecular Biology*, *1296*, 11–16.

Clifford, C. A., Hopper, J. L., Fulker, D. W., & Murray, R. M. (1984). A genetic and environmental analysis of a twin family study of alcohol use, anxiety, and depression. *Genetic Epidemiology*, *1*, 63–79.

Cloninger, C. R. (1987). Neurogenetic adaptive mechanisms in alcoholism. *Science*, *236*, 410–416.

Coleman, J., Williams, A., Phan, T. H., Mummalaneni, S., Melone, P., Ren, Z., et al. (2011). Strain differences in the neural, behavioral, and molecular correlates of sweet and salty taste in naive, ethanol- and sucrose-exposed P and NP rats. *Journal of Neurophysiology*, *106*, 2606–2621.

Conrod, P. J., Pihl, R. O., Stewart, S. H., & Dongier, M. (2000). Validation of a system of classifying female substance abusers on the basis of personality and motivational risk factors for substance abuse. *Psychology of Addictive Behaviors*, *14*, 243–256.

Conrod, P. J., Pihl, R. O., & Vassileva, J. (1998). Differential sensitivity to alcohol reinforcement in groups of men at risk for distinct alcoholism subtypes. *Alcoholism, Clinical and Experimental Research*, *22*, 585–597.

Contini, V., Bertuzzi, G. P., Polina, E. R., Hunemeier, T., Hendler, E. M., Hutz, M. H., et al. (2012). A haplotype analysis is consistent with the role of functional HTR1B variants in alcohol dependence. *Drug and Alcohol Dependence*, *122*, 100–104.

Coonfield, D. L., Kiefer, S. W., Ferraro, F. M., III, & Sinclair, J. D. (2004). Ethanol palatability and consumption by high ethanol-drinking rats: Manipulation of the opioid system with naltrexone. *Behavioral Neuroscience*, *118*, 1189–1196.

Cooper, J. R., Bloom, F. E., & Roth, R. H. (2002). *Biochemical basis of neuropharmacology* (8th ed.). Oxford University Press: Oxford.

Corrigall, W. A., & Coen, K. M. (1994). Nicotine self-administration and locomotor activity are not modified by the 5-HT3 antagonists ICS 205-930 and MDL 72222. *Pharmacology, Biochemistry, and Behavior*, *49*, 67–71.

Cosgrove, K. P. (2010). Imaging receptor changes in human drug abusers. *Current Topics in Behavioral Neurosciences*, *3*, 199–217.

Cotton, N. S. (1979). The familial incidence of alcoholism. *Journal of Studies on Alcohol*, *40*, 89–116.

Cowen, M. S., Chen, F., & Lawrence, A. J. (2004). Neuropeptides: Implications for alcoholism. *Journal of Neurochemistry*, *89*, 273–285.

Cowen, M. S., Djouma, E., & Lawrence, A. J. (2005). The metabotropic glutamate 5 receptor antagonist 3-[(2-methyl-1,3-thiazol-4-yl) ethnyl]-pyridine reduces ethanol self-administration in multiple strains of alcohol-preferring rats and regulates olfactory glutamatergic systems. *Journal of Pharmacology and Experimental Therapeutics*, *315*, 590–600.

Cozzoli, D. K., Goulding, S. P., Zhang, P.-W., Xiao, B., Hu, J.-H., Ary, A. W., et al. (2009). Binge drinking up-regulated accumbens mGluR5-Homer2-PI3K signaling: Functional implications for alcoholism. *Journal of Neuroscience*, *29*, 8655–8668.

Crabbe, J. C. (2008). Neurogenetic studies of alcohol addiction. *Philosophical Transactions of the Royal Society of London. Series B, Biological Sciences*, *363*, 3201–3211.

Crabbe, J. C., Bell, R. L., & Ehlers, C. L. (2010). Human and laboratory rodent low response to alcohol: Is better consilience possible? *Addiction Biology*, *15*, 125–144.

Crabbe, J. C., Phillips, T. J., Kosobud, A., & Belknap, J. K. (1990). Estimation of genetic correlation: Interpretation of experiments using selectively bred and inbred animals. *Alcoholism, Clinical and Experimental Research, 14*, 141–151.

Crews, F. T., Qin, L., Sheedy, D., Ventreno, R. P., & Zou, J. (2013). High mobility group box 1/Toll-like receptor danger signaling increases brain neuroimmune activation in alcohol dependence. *Biological Psychiatry, 73*, 602–612.

Cui, C., Noronha, A., Morikawa, H., Alvarez, V. A., Stuber, G. D., Szumlinski, K. K., et al. (2013). New insights on neurobiological mechanisms underlying alcohol addiction. *Neuropharmacology, 67*, 223–232.

Cui, C., Shurtleff, D., & Harris, R. A. (2014). Neuroimmune mechanisms of alcohol and drug addiction. *International Review of Neurobiology, 118*, 1–12.

Czachowski, C. L., & Samson, H. H. (2002). Ethanol- and sucrose-reinforced appetitive and consummatory responding in HAD1, HAD2, and P rats. *Alcoholism, Clinical and Experimental Research, 26*, 1653–1661.

Danbolt, N. C. (2001). Glutamate uptake. *Progress in Neurobiology, 65*, 1–105.

Das, S. C., Yamamoto, B. K., Hristov, A. M., & Sari, Y. (2015). Ceftriaxone attenuates ethanol drinking and restores extracellular glutamate concentration through normalization of GLT-1 in nucleus accumbens of male alcohol-preferring rats. *Neuropharmacology, 97*, 67–74.

Davis, T. J., & de Fiebre, C. M. (2006). Alcohol's actions on neuronal nicotinic acetylcholine receptors. *Alcohol Research & Health, 29*, 179–185.

Davis, K. M., & Wu, J. Y. (2001). Role of glutamatergic and GABAergic systems in alcoholism. *Journal of Biomedical Science, 8*, 7–19.

Dawson, D. A., Grant, B. F., Stinson, F. S., & Chou, P. S. (2004). Another look at heavy episodic drinking and alcohol use disorders among college and noncollege youth. *Journal of Studies on Alcohol, 65*, 477–488.

Deehan, G. A., Jr., Hauser, S. R., Waeiss, R. A., Knight, C. P., Toalston, J. E., Truitt, W. A., et al. (2015). Co-administration of ethanol and nicotine: The enduring alterations in the rewarding properties of nicotine and glutamate activity within the mesocorticolimbic system of female alcohol-preferring (P) rats. *Psychopharmacology, 232*, 4293–4302.

Dhaher, R., Toalston, J. E., Hauser, S. R., Bell, R. L., McKinzie, D. L., McBride, W. J., et al. (2012). Effects of naltrexone and LY255582 on ethanol maintenance, seeking, and relapse responding by alcohol-preferring (P) rats. *Alcohol, 46*, 17–27.

Di Chiara, G., & Imperato, A. (1988). Drugs abused by humans preferentially increase synaptic dopamine concentrations in the mesolimbic system of freely moving rats. *Proceedings of the National Academy of Sciences of the United States of America, 85*, 5274–5278.

Dick, D. M., & Bierut, L. J. (2006). The genetics of alcohol dependence. *Current Psychiatry Reports, 8*, 151–157.

Dick, D. M., Edenberg, H. J., Xuei, X., Goate, A., Kuperman, S., Schuckit, M., et al. (2004). Association of GABRG3 with alcohol dependence. *Alcoholism, Clinical and Experimental Research, 28*, 4–9.

Dick, D. M., Plunkett, J., Wetherill, L. F., Xuei, X., Goate, A., Hesselbrock, V., et al. (2006). Association between GABRA1 and drinking behaviors in the collaborative study on the genetics of alcoholism sample. *Alcoholism, Clinical and Experimental Research, 30*, 1101–1110.

Ding, Z. M., Engleman, E. A., Rodd, Z. A., & McBride, W. J. (2012). Ethanol increases glutamate neurotransmission in the posterior ventral tegmental area of female Wistar rats. *Alcoholism, Clinical and Experimental Research, 36*, 633–640.

Ding, Z. M., Rodd, Z. A., Engleman, E. A., Bailey, J. A., Lahiri, D. K., & McBride, W. J. (2013). Alcohol drinking and deprivation alter basal extracellular glutamate concentrations and clearance in the mesolimbic system of alcohol-preferring (P) rats. *Addiction Biology, 18*, 297–306.

Domart, M. C., Benyamina, A., Lemoine, A., Bourgain, C., Blecha, L., Debuire, B., et al. (2012). Association between a polymorphism in the promoter of a glutamate receptor subunit gene (GRIN2A) and alcoholism. *Addiction Biology, 17*, 783–785.

Draski, L. J., & Deitrich, R. A. (1996). Initial effects of ethanol on the central nervous system. In R. A. Deitrich & V. G. Erwin (Eds.), *Pharmacological effects of ethanol on the nervous system* (pp. 227–250). Boca Raton, FL: CRC Press.

Drews, E., & Zimmer, A. (1997). Modulation of alcohol and nicotine responses through the endogenous opioid system. *Progress in Neurobiology, 90*, 1–15.

Drummond, D. C., Cooper, T., & Glautier, S. (1990). Conditioned learning in alcohol dependence: Implications for cue exposure treatment. *British Journal of Addiction, 85*, 725–743.

Dumont, Y., Satoh, H., Cadieux, A., et al. (1993). Evaluation of truncated neuropeptide Y analogues with modifications of the tyrosine residue in position 1 on Y1, Y2 and Y3 receptor sub-types. *European Journal of Pharmacology, 238*, 37–45.

Dundon, W., Lynch, K. G., Pettinati, H. M., & Lipkin, C. (2004). Treatment outcomes in type A and B alcohol dependence 6 months after serotonergic pharmacotherapy. *Alcoholism, Clinical and Experimental Research, 28*, 1065–1073.

Dyr, W., & Kostowski, W. (2008). Warsaw high-preferring (WHP) and Warsaw low-preferring (WLP) lines of rats selectively bred for high and low voluntary ethanol intake: Preliminary phenotypic characterization. *Alcohol, 42*, 161–170.

Economidou, D., Hansson, A. C., Weiss, F., Terasmaa, A., Sommer, W. H., Cippitelli, A., et al. (2008). Dysregulation of nociceptin/orphanin FQ activity in the amygdala is linked to excessive alcohol drinking in the rat. *Biological Psychiatry, 64*, 211–218.

Edenberg, H. J., Dick, D. M., Xuei, X., Tian, H., Almasy, L., Bauer, L. O., et al. (2004). Variations in GABRA2, encoding the alpha 2 subunit of the GABA(A) receptor are associated with alcohol dependence and with brain oscillations. *American Journal of Human Genetics, 74*, 705–714.

Ehlers, C. L., Chaplin, R. I., Wall, T. L., Lumeng, L., Li, T.-K., Owens, M. J., et al. (1992). Corticotropin releasing factor (CRF): Studies in alcohol-preferring and -nonpreferring rats. *Psychopharmacology, 106*, 359–364.

Ehlers, C. L., Li, T. K., Lumeng, L., et al. (1998). Neuropeptide Y levels in ethanol-naive alcohol-preferring and nonpreferring rats and in Wistar rats after ethanol exposure. *Alcoholism, Clinical and Experimental Research, 22*, 1778–1782.

Ehlers, C. L., Lind, P. A., & Wilhelmsen, K. C. (2008). Association between single nucleotide polymorphisms in the mu opioid receptor gene (OPRM1) and self-reported responses to alcohol in American Indians. *BMC Medical Genetics, 9*, 35. http://dx.doi.org/10.1186/1471-2350-9-35.

Ehlers, C. L., Walter, N. A. R., Dick, D. M., Buck, K. J., & Crabbe, J. C. (2010). A comparison of selected quantitative trait loci associated with alcohol use phenotypes in humans and mouse models. *Addiction Biology, 15*, 185–199.

Ehrenreich, H., Schuck, J., Stender, N., et al. (1997). Endocrine and hemodynamic effects of stress versus systemic CRF in alcoholics during early and medium term abstinence. *Alcoholism, Clinical and Experimental Research, 21*, 1285–1293.

Eiler, W. A., II, & June, H. L. (2007). Blockade of GABAA receptors within the extended amygdala attenuates D2 regulation of alcohol-motivated behaviors in the ventral tegmental area of alcohol-preferring (P) rats. *Neuropharmacology, 52*, 1570–1579.

Eiler, W. J. A., II, Seyoum, R., Foster, K. L., Mailey, C., & June, H. L. (2003). D1 dopamine receptor regulates alcohol-motivated behaviors in the bed nucleus of the stria terminalis in alcohol-preferring (P) rats. *Synapse, 48*, 45–56.

Elman, I., Borsook, D., & Volkow, N. D. (2013). Pain and suicidality: Insights from reward and addiction neuroscience. *Progress in Neurobiology, 109*, 1–27.

Eng, M. Y., Schuckit, M. A., & Smith, T. L. (2005). The level of response to alcohol in daughters of alcoholics and controls. *Drug and Alcohol Dependence, 79*, 83–93.

Engel, J. A., & Jerlhag, E. (2014). Alcohol: Mechanisms along the mesolimbic dopamine system. *Progress in Brain Research*, *211*, 201–233.

Engleman, E. A., Ding, Z.-M., Oster, S. M., Toalston, J. E., Bell, R. L., Murphy, J. M., et al. (2009). Ethanol is self-administered into the nucleus accumbens shell, but not core: Evidence of genetic sensitivity. *Alcoholism, Clinical and Experimental Research*, *33*, 2162–2171. http://dx.doi.org/10.1111/j.1530-0277.2009.01055.x.

Engleman, E. A., Ingraham, C. M., McBride, W. J., Lumeng, L., & Murphy, J. M. (2006). Extracellular dopamine levels are lower in the medial prefrontal cortex of alcohol-preferring rats compared to Wistar rats. *Alcohol*, *38*, 5–12.

Engleman, E. A., Keen, E. J., Tilford, S. S., Thielen, R. J., & Morzorati, S. L. (2011). Ethanol drinking reduces extracellular dopamine levels in the posterior ventral tegmental area of nondependent alcohol-preferring rats. *Alcohol*, *45*, 549–557.

Engleman, E. A., McBride, W. J., Li, T. K., Lumeng, L., & Murphy, J. M. (2003). Ethanol drinking experience attenuates (-)sulpiride-induced increases in extracellular dopamine levels in the nucleus accumbens of alcohol-preferring (P) rats. *Alcoholism, Clinical and Experimental Research*, *27*, 424–431.

Engleman, E. A., McBride, W. J., Wilber, A. A., Shaikh, S. R., Eha, R. D., Lumeng, L., et al. (2000). Reverse microdialysis of a dopamine uptake inhibitor in the nucleus accumbens of alcohol-preferring rats: Effects on dialysate dopamine levels and ethanol intake. *Alcoholism, Clinical and Experimental Research*, *24*, 795–801.

Engleman, E. A., Rodd, Z. A., Bell, R. L., & Murphy, J. M. (2008). The role of 5-HT3 receptors in drug abuse and as a target for pharmacotherapy. *CNS & Neurological Disorders Drug Targets*, *7*, 454–467.

Enna, S. J. (1997). GABAB receptor agonists and antagonists: Pharmacological properties and therapeutic possibilities. *Expert Opinion on Investigational Drugs*, *6*, 1319–1325.

Enoch, M. A., Schwartz, L., Albaugh, B., Virkkunen, M., & Goldman, D. (2006). Dimensional anxiety mediates linkage of GABRA2 haplotypes with alcoholism. *American Journal of Medical Genetics. Part B, Neuropsychiatric Genetics*, *141B*, 599–607.

Enoch, M. A., Zhou, Z., Kimura, M., Mash, D. C., Yuan, Q., & Goldman, D. (2012). GABAergic gene expression in postmortem hippocampus from alcoholics and cocaine addicts; corresponding findings in alcohol-naive P and NP rats. *PLoS One*, *7*, e29369.

Epstein, E. E., Kahler, C. W., McCrady, B. S., Lewis, K. D., & Lewis, S. (1995). An empirical classification of drinking patterns among alcoholics: Binge, episodic, sporadic, and steady. *Addictive Behaviors*, *20*, 23–41.

Ericson, M., Blomqvist, O., Engel, J. A., & Söderpalm, B. (1998). Voluntary ethanol intake in the rat and the associated accumbal dopamine overflow are blocked by ventral tegmental mecamylamine. *European Journal of Pharmacology*, *358*, 189–196.

Ericson, M., Löf, E., Stomberg, R., & Söderpalm, B. (2009). The smoking cessation medication varenicline attenuates alcohol and nicotine interactions in the rat mesolimbic dopamine system. *Journal of Pharmacology and Experimental Therapeutics*, *329*, 225–230.

Ericson, M., Molander, A., Lof, E., Engel, J. A., & Soderpalm, B. (2003). Ethanol elevates accumbal dopamine levels via indirect activation of ventral tegmental nicotinic acetylcholine receptors. *European Journal of Pharmacology*, *467*, 85–93.

Everitt, B. J., & Robbins, T. W. (2005). Neural systems of reinforcement for drug addiction: From actions to habits to compulsion. *Nature Neuroscience*, *8*, 1481–1489.

Eysenck, H. J., & Eysenck, S. B. G. (1975). *Manual of the Eysenck personality questionnaire*. London: Hodder & Stoughton.

Fadda, F., & Rossetti, Z. L. (1998). Chronic ethanol consumption: From neuroadaptation to neurodegeneration. *Progress in Neurobiology*, *56*, 385–431.

Faria, R. R., Rueda, A. V. L., Sayuri, C., Soares, S. L., Malta, M. B., Carrara-Nascimento, P. F., et al. (2008). Environmental modulation of ethanol-induced locomotor activity: Correlation with neuronal activity in distinct brain regions of adolescent and adult Swiss mice. *Brain Research*, *1239*, 127–140.

Farris, S. P., & Mayfield, R. D. (2014). RNA-seq reveals novel transcriptional reorganization in human alcoholic brain. *International Review of Neurobiology*, *116*, 275–300.

Fetissov, S. O., Kopp, J., & Hokfelt, T. (2004). Distribution of NPY receptors in the hypothalamus. *Neuropeptides*, *38*, 175–188.

Files, F. J., Samson, H. H., Denning, C. E., & Marvin, S. (1998). Comparison of alcohol-preferring and -nonpreferring selectively bred rat lines. II. Operant self-administration in a continuous-access situation. *Alcoholism, Clinical and Experimental Research*, *22*, 2147–2158.

Filip, M., Zaniewska, M., Frankowska, M., Wydra, K., & Fuxe, K. (2012). The importance of adenosine A(2A) receptor-dopamine D(2) receptor interaction in drug addiction. *Current Medicinal Chemistry*, *19*, 317–355.

Finn, P. R., Earleywine, M., & Pihl, R. O. (1992). Sensation seeking, stress reactivity, and alcohol dampening discriminate the density of a family history of alcoholism. *Alcoholism, Clinical and Experimental Research*, *16*, 585–590.

Finn, P. R., & Justus, A. (1997). Physiological responses in sons of alcoholics. *Alcohol Health and Research World*, *21*, 227.

Fitzgerald, G. J., Liu, H., & Morzorati, S. L. (2012). Decreased sensitivity of NMDA receptors on dopaminergic neurons from the posterior ventral tegmental area following chronic nondependent alcohol consumption. *Alcoholism, Clinical and Experimental Research*, *36*, 1710–1719.

Flavin, S. A., & Winder, D. G. (2013). Noradrenergic control of the bed nucleus of the stria terminalis in stress and reward. *Neuropharmacology*, *70*, 324–330.

Flory, K., Lynam, D., Milich, R., Leukefeld, C., & Clayton, R. (2004). Early adolescent through young adult alcohol and marijuana use trajectories: Early predictors, young adult outcomes, and predictive utility. *Development and Psychopathology*, *16*, 193–213.

Floyd, D. W., Jung, K. Y., & McCool, B. A. (2003). Chronic ethanol ingestion facilitates N-methyl-D aspartate receptor function and expression in rat lateral/basolateral amygdala neurons. *Journal of Pharmacology and Experimental Therapeutics*, *307*, 1020–1029.

Foroud, T., Wetherill, L. F., Liang, T., Dick, D. M., Hesselbrock, V., Kramer, J., et al. (2007). Association of alcohol craving with alpha-synuclein (SNCA). *Alcoholism, Clinical and Experimental Research*, *31*, 537–545.

Forray, A., & Sofuoglu, M. (2014). Future pharmacological treatments for substance use disorders. *British Journal of Clinical Pharmacology*, *77*, 382–400.

Foster, K. L., McKay, P. F., Seyoum, R., Milbourne, D., Yin, W., Sarma, P. V. V. S., et al. (2004). $GABA_A$ and opioid receptors of the central nucleus of the amygdala selectively regulates ethanol-maintained behaviors. *Neuropsychopharmacology*, *29*, 269–284.

Frances, F., Guillen, M., Verdu, F., Portoles, O., Castello, A., Sorli, J. V., et al. (2011). The 1258 G>A polymorphism in the neuropeptide Y gene is associated with greater alcohol consumption in a Mediterranean population. *Alcohol*, *45*, 131–136.

Franklin, K. M., Asatryan, L., Jakowec, M. W., Trudell, J. R., Bell, R. L., & Davies, D. L. (2014). P2X4 receptors (P2X4Rs) represent a novel target for the development of drugs to prevent and/or treat alcohol use disorders. *Frontiers in Neuroscience/Neuropharmacology*, *8*, 176. http://dx.doi.org/10.3389/fnins.2014.00176.

Franklin, K. M., Engleman, E. A., Ingraham, C. M., McClaren, J. A., Keith, C. M., McBride, W. J., et al. (2009). A single, moderate ethanol exposure alters extracellular dopamine levels and dopamine receptor function in the nucleus accumbens of Wistar rats. *Alcoholism, Clinical and Experimental Research*, *33*, 1721–1730.

Franklin, K. M., Hauser, S. R., Lasek, A. W., Bell, R. L., & McBride, W. J. (2015). Involvement of purinergic P2X4 receptors in alcohol intake of high alcohol-drinking (HAD) rats. *Alcoholism, Clinical and Experimental Research, 39*, 2022–2031. http://dx.doi.org/10.1111/acer.12836.

Fritschy, J.-M., Panzanelli, P., & Tyagarajan, S. K. (2012). Molecular and functional heterogeneity of GABAergic synapses. *Cellular and Molecular Life Sciences, 69*, 2485–2499.

Froehlich, J. C., Hausauer, B. J., Federoff, D. L., Fischer, S. M., & Rasmussen, D. D. (2013a). Prazosin reduces alcohol drinking throughout prolonged treatment and blocks the initiation of drinking in rats selectively bred for high alcohol intake. *Alcoholism, Clinical and Experimental Research, 37*, 1552–1560.

Froehlich, J. C., Hausauer, B., Fischer, S., Wise, B., & Rasmussen, D. D. (2015). Prazosin reduces alcohol intake in an animal model of alcohol relapse. *Alcoholism, Clinical and Experimental Research, 39*, 1538–1546.

Froehlich, J. C., Hausauer, B. J., & Rasmussen, D. D. (2013b). Combining naltrexone and prazosin in a single oral medication decreases alcohol drinking more effectively than does either drug alone. *Alcoholism, Clinical and Experimental Research, 37*, 1763–1770.

Gaiarsa, J.-L., Kuczewski, N., & Porcher, C. (2011). Contribution of metabotropic GABAB receptors to neuronal network construction. *Pharmacology & Therapeutics, 132*, 170–179.

Gass, J. T., & Olive, M. F. (2008). Glutamatergic substrates of drug addiction and alcoholism. *Biochemical Pharmacology, 75*, 218–265.

Gatto, G. J., Murphy, J. M., Waller, M. B., McBride, W. J., Lumeng, L., & Li, T.-K. (1987). Chronic ethanol tolerance through free-choice drinking in the P line of alcohol-preferring rats. *Pharmacology, Biochemistry, and Behavior, 28*, 111–115.

Gedik, H., Erdal, M. E., Yilmaz, S. G., Sengul, C., Sengul, C. B., & Herken, H. (2015). Association of microRNA biogenesis pathway gene variants and alcohol dependence risk. *DNA and Cell Biology, 34*, 220–226.

Gegelashvili, G., & Schousboe, A. (1997). High affinity glutamate transporters: Regulation of expression and activity. *Molecular Pharmacology, 52*, 6–15.

Geiger, M. J., Neufang, S., Stein, D. J., & Domschke, K. (2014). Arousal and the attentional network in panic disorder. *Human Psychopharmacology, 29*, 599–603.

George, D. T., Gilman, J., Hersh, J., Thorsell, A., Herion, D., Geyer, C., et al. (2008). Neurokinin 1 receptor antagonism as a possible therapy for alcoholism. *Science, 319*, 1536–1539.

Gerra, G., Leonardi, C., Cortese, E., D'Amore, A., Lucchini, A., Strepparola, G., et al. (2007). Human kappa opioid receptor gene (OPRK1) polymorphism is associated with opiate addiction. *American Journal of Medical Genetics. Part B, Neuropsychiatric Genetics, 144B*, 771–775.

Gessa, G. L., Muntoni, F., Collu, M., Vargiu, L., & Mereu, G. (1985). Low doses of ethanol activate dopaminergic neurons in the ventral tegmental area. *Brain Research, 348*, 201–203.

Getachew, B., Hauser, S. R., Taylor, R. E., & Tizabi, Y. (2010). Alcohol-induced depressive-like behavior is associated with cortical norepinephrine reduction. *Pharmacology, Biochemistry, and Behavior, 96*, 395–401.

Gianoulakis, C., Guillaume, P., De Waele, J.-P., & Angelogianni, P. (1995). Effects of stress and alcohol on the proopiomelanocortin/β-endorphin system. In W. A. Hunt & S. Zakhari (Eds.), *Stress, gender, and alcohol-seeking behavior* (pp. 145–165). Bethesda, MD: National Institute on Alcohol Abuse and Alcoholism (Research monograph no. 29).

Gilpin, N. W., Herman, M. A., & Roberto, M. (2015). The central amygdala as an integrative hub for anxiety and alcohol use disorders. *Biological Psychiatry, 77*, 859–869.

Gilpin, N. W., & Koob, G. F. (2010). Effects of β-adrenoceptor antagonists on alcohol drinking by alcohol-dependent rats. *Psychopharmacology, 212*, 431–439.

Gilpin, N. W., Richardson, H. N., & Koob, G. F. (2008). Effects of CRF1-receptor and opioid-receptor antagonists on dependence-induced increases in alcohol drinking by alcohol-preferring (P) rats. *Alcoholism, Clinical and Experimental Research, 32*, 1535–1542.

Gilpin, N. W., Stewart, R. B., & Badia-Elder, N. E. (2008). Neuropeptide Y administration into the amygdala suppresses ethanol drinking in alcohol-preferring (P) rats following multiple deprivations. *Pharmacology, Biochemistry, and Behavior, 90*, 470–474.

Gilpin, N. W., Stewart, R. B., Murphy, J. M., & Badia-Elder, N. E. (2004). Neuropeptide Y in the paraventricular nucleus of the hypothalamus increases ethanol intake in high- and low-alcohol-drinking rats. *Alcoholism, Clinical and Experimental Research, 28*, 1492–1498.

Gilpin, N. W., Stewart, R. B., Murphy, J. M., Li, T. K., & Badia-Elder, N. E. (2003). Neuropeptide Y reduces oral ethanol intake in alcohol-preferring (P) rats following a period of imposed ethanol abstinence. *Alcoholism, Clinical and Experimental Research, 27*, 787–794.

Gobrogge, K. L. (2014). Sex, drugs, and violence: Neuromodulation of attachment and conflict in voles. *Current Topics in Behavioral Neurosciences, 17*, 229–264. http://dx.doi.org/10.1007/7854_2013_264.

Golan, D. E., Tashjian, A. H., Armstrong, E. J., & Armstrong, A. W. (2012). *Principles of pharmacology: The pathophysiologic basis of drug therapy* (3rd ed.). Philadelphia, PA: Lippincott Williams & Wilkins.

Golbidi, S., Frisbee, J. C., & Laher, I. (2015). Chronic stress impacts the cardiovascular system: Animal models and clinical outcomes. *American Journal of Physiology. Heart and Circulatory Physiology, 308*, H1476–H1498. http://dx.doi.org/10.1152/ajpheart.00859.2014.

Gold, P. W. (2015). The organization of the stress system and its dysregulation in depressive illness. *Molecular Psychiatry, 20*, 32–47.

Goldberg, J., Eisen, S. A., True, W. R., & Henderson, W. G. (1990). A twin study of the effects of the Vietnam conflict on alcohol drinking patterns. *American Journal of Public Health, 80*, 570–574.

Goldman, D., Oroszi, G., O'Malley, S., & Anton, R. (2005). COMBINE genetics study: The pharmacogenetics of alcoholism treatment response: Genes and mechanisms. *Journal of Studies on Alcohol*, (15), 56–64.

Gonzalez-Burgos, G. (2010). GABA transporter GAT1: A crucial determinant of GABAB receptor activation in cortical circuits? *Advances in Pharmacology, 58*, 175–204.

Goodman, A. (2008). Neurobiology of addiction. An integrative review. *Biochemical Pharmacology, 75*, 266–322.

Goodwani, S., Rao, P. S., Bell, R. L., & Sari, Y. (2015). Amoxicillin and amoxicillin/clavulanate reduce ethanol intake and increase GLT-1 expression as well as AKT phosphorylation in mesocorticolimbic regions. *Brain Research, 1622*, 397–408.

Gorini, G., Bell, R. L., & Mayfield, R. D. (2011). Molecular targets of alcohol action: Translational research for pharmacotherapy development and screening. In S. Rahman (Ed.), *Progress in molecular biology and translational science: The brain as a drug target: Vol. 98* (pp. 293–347). New York, NY: Elsevier Science.

Gorini, G., Harris, R. A., & Mayfield, R. D. (2014). Proteomic approaches and identification of novel therapeutic targets for alcoholism. *Neuropsychopharmacology, 39*, 104–130.

Gotti, C., Clementi, F., Fornari, A., Gaimarri, A., Guiducci, S., Manfredi, I., et al. (2009). Structural and functional diversity of native brain nicotinic receptors. *Biochemical Pharmacology, 78*, 703–711.

Grace, A. A., Floresco, S. B., Goto, Y., & Lodge, D. J. (2007). Regulation of firing of dopaminergic neurons and control of goal-directed behaviors. *Trends in Neurosciences, 30*, 220–227.

Gray, J. C., & MacKillop, J. (2014). Interrelationships among individual differences in alcohol demand, impulsivity, and alcohol misuse. *Psychology of Addictive Behaviors, 28*, 282–287. http://dx.doi.org/10.1037/a0032766.

Greeley, J. D., Swift, W., Prescott, J., & Heather, N. (1993). Reactivity to alcohol-related cues in heavy and light drinkers. *Journal of Studies on Alcohol, 54*, 359–368.

Griffin, J. F. (2002). A strategic approach to vaccine development: Animal models, monitoring vaccine efficacy, formulation and delivery. *Advanced Drug Delivery Reviews*, *54*, 851–861.
Grzywacz, A., Małecka, I., Suchanecka, A., Bieńkowski, P., & Samochowiec, J. (2013). Family-based and case-control study of glutamate receptor GRIK3 Ser310Ala polymorphism in alcohol dependence. *European Addiction Research*, *19*, 55–59.
Guan, X. M., & McBride, W. J. (1989). Serotonin microinfusion into the ventral tegmental area increases accumbens dopamine release. *Brain Research Bulletin*, *23*, 541–547.
Guillot, C. R., Fanning, J. R., Liang, T., Leventhal, A. M., & Berman, M. E. (2015). An α-synuclein gene (SNCA) polymorphism moderates the association of PTSD symptomatology with hazardous alcohol use, but not with aggression-related measures. *Journal of Anxiety Disorders*, *30*, 41–47.
Guillot, C. R., Pang, R. D., Leventhal, A. M., Liang, T., & Berman, M. E. (2015). The association of SNCA with hazardous alcohol use is mediated by impulsivity. *Psychiatry Research*, *30*(226), 523–524.
Gulley, J. M., & Juraska, J. M. (2013). The effects of abused drugs on adolescent development of corticolimbic circuitry and behavior. *Neuroscience*, *249*, 3–20.
Gupta, T., Syed, Y. M., Revis, A. A., Miller, S. A., Martinez, M., Cohn, K. A., et al. (2008). Acute effects of acamprosate and MPEP on ethanol drinking-in-the-dark in male C57BL/6J mice. *Alcoholism, Clinical and Experimental Research*, *32*, 1992–1998.
Halliday, G. M., & Törk, I. (1989). Serotonin-like immunoreactive cells and fibres in the rat ventromedial mesencephalic tegmentum. *Brain Research Bulletin*, *22*, 725–735.
Hamilton, K. R., Felton, J. W., Risco, C. M., Lejuez, C. W., & MacPherson, L. (2014). Brief report: The interaction of impulsivity with risk-taking is associated with early alcohol use initiation. *Journal of Adolescence*, *37*, 1253–1256. http://dx.doi.org/10.1016/j.adolescence.2014.08.013.
Hanchar, H. J., Dodson, P. D., Olsen, R. W., Otis, T. S., & Wallner, M. (2005). Alcohol-induced motor impairment caused by extrasynaptic GABA(A) receptor activity. *Nature Neuroscience*, *8*, 339–345.
Haughey, H. M., Ray, L. A., Finan, P., Villanueva, R., Niculescu, M., & Hutchison, K. E. (2008). Human gamma-aminobutyric acid A receptor alpha2 gene moderates the acute effects of alcohol and brain mRNA expression. *Genes, Brain, and Behavior*, *7*, 447–454.
Hauser, S. R., Ding, Z. M., Getachew, B., Toalston, J. E., Oster, S. M., McBride, W. J., et al. (2011). The posterior ventral tegmental area mediates alcohol-seeking behavior in alcohol-preferring rats. *Journal of Pharmacology and Experimental Therapeutics*, *336*, 857–865.
Hauser, S. R., Getachew, B., Oster, S. M., Dhaher, R., Ding, Z.-M., Bell, R. L., et al. (2012). Nicotine modulates alcohol-seeking and relapse by alcohol-preferring (P) rats in a time dependent manner. *Alcoholism, Clinical and Experimental Research*, *36*, 43–54.
Hauser, S. R., Hedlund, P. B., Roberts, A. J., Sari, Y., Bell, R. L., & Engleman, E. A. (2014). The 5-HT-7 receptor as a potential target for treating drug and alcohol abuse. *Frontiers in Neuroscience/Neuropharmacology*, *8*, 448. http://dx.doi.org/10.3389/fnins.2014.00448.
Heath, A. C. (1995). Genetic influences on drinking behavior in humans. In H. Begleiter & B. Kissin (Eds.), *The genetics of alcoholism* (pp. 82–121). New York: Oxford University Press.
Heather, N., Tebbutt, J. S., Mattick, R. P., & Zamir, R. (1993). Development of a scale for measuring impaired control over alcoholism: A preliminary report. *Journal of Studies on Alcohol*, *54*, 700–709.
Hegerl, U., & Hensch, T. (2014). The vigilance regulation model of affective disorders and ADHD. *Neuroscience & Biobehavioral Reviews*, *44*, 45–57.
Heilig, M., & Egli, M. (2006). Pharmacological treatment of alcohol dependence: Target symptoms and target mechanisms. *Pharmacology & Therapeutics*, *111*, 855–876.

Heilig, M., Egli, M., Crabbe, J. C., & Becker, H. C. (2010). Acute withdrawal, protracted abstinence and negative affect in alcoholism: Are they linked? *Addiction Biology, 15*, 169–184.

Heilig, M., & Thorsell, A. (2002). Brain neuropeptide Y (NPY) in stress and alcohol dependence. *Reviews in the Neurosciences, 13*, 85–94.

Heilig, M., Thorsell, A., Sommer, W. H., Hansson, A. C., Ramchandani, V. A., George, D. T., et al. (2010). Translating the neuroscience of alcoholism into clinical treatments: From blocking the buzz to curing the blues. *Neuroscience & Biobehavioral Reviews, 35*, 334–344.

Heinz, A. (2002). Dopaminergic dysfunction in alcoholism and schizophrenia—Psychopathological and behavioral correlates. *European Psychiatry: Journal of the Association of Europen Psychiatry, 17*, 9–16.

Heinz, A., Goldman, D., Gallinat, J., Schumann, G., & Puls, I. (2004). Pharmacogenetic insights to monoaminergic dysfunction in alcohol dependence. *Psychopharmacology, 174*, 561–570.

Heit, C., Dong, H., Chen, Y., Thompson, D. C., Deitrich, R. A., & Vasiliou, V. K. (2013). The role of CYP2E1 in alcohol metabolism and sensitivity in the central nervous system. *Sub-Cellular Biochemistry, 67*, 235–247.

Hendrickson, L. M., Guildford, M. J., & Tapper, A. R. (2013). Neuronal nicotinic acetylcholine receptors: Common molecular substrates of nicotine and alcohol dependence. *Frontiers in Psychiatry, 4*, 29.

Hendrickson, L. M., Zhao-Shea, R., & Tapper, A. R. (2009). Modulation of ethanol drinking-in-the-dark by mecamylamine and nicotinic acetylcholine receptor agonists in C57BL/6J mice. *Psychopharmacology, 204*, 563–572.

Henley, J. M., & Wilkinson, K. A. (2013). AMPA receptor trafficking and the mechanisms underlying synaptic plasticity and cognitive aging. *Dialogues in Clinical Neuroscience, 15*, 11–27.

Herman, M. A., Kallupi, M., Luu, G., Oleata, C. S., Heilig, M., Koob, G. F., et al. (2012). Enhanced GABAergic transmission in the central nucleus of the amygdala of genetically selected Marchigian Sardinian rats: Alcohol and CRF effects. *Neuropharmacology, 67C*, 337–348.

Herman, M. A., Varodayan, F. P., Oleata, C. S., Luu, G., Kirson, D., Heilig, M., et al. (2016). Glutamatergic transmission in the central nucleus of the amygdala is selectively altered in Marchigian Sardinian alcohol-preferring rats: Alcohol and CRF effects. *Neuropharmacology, 102*, 21–31.

Hersch, S. M., Gutekunst, C. A., Rees, H. D., Heilman, C. J., & Levey, A. I. (1994). Distribution of m1-m4 muscarinic receptor proteins in the rat striatum: Light and electron microscopic immunocytochemistry using subtype-specific antibodies. *Journal of Neuroscience, 14*, 3351–3363.

Herve, D., Pickel, V. M., Joh, T. H., & Beaudet, A. (1987). Serotonin axon terminals in the ventral tegmental area of the rat: Fine structure and synaptic input to dopaminergic neurons. *Brain Research, 435*, 71–83.

Hesselbrock, M. N. (1995a). Genetic determinants of alcoholic subtypes. In H. Begleiter & B. Kissin (Eds.), *The genetics of alcoholism* (pp. 40–69). New York, NY: Oxford University Press.

Hesselbrock, V. M. (1995b). The genetic epidemiology of alcoholism. In H. Begleiter & B. Kissin (Eds.), *The genetics of alcoholism* (pp. 17–39). New York, NY: Oxford University Press.

Hinckers, A. S., Laucht, M., Schmidt, M. H., Mann, K. F., Schumann, G., Schuckit, M. A., et al. (2006). Low level of response to alcohol as associated with serotonin transporter genotype and high alcohol intake in adolescents. *Biological Psychiatry, 60*, 282–287.

Hrubec, Z., & Neel, J. V. (1978). The National Academy of Sciences—National Research Council Twin Registry: Ten years of operation. *Progress in Clinical and Biological Research, 24*(Pt. B), 153–172.

Hulse, G. K. (2012). Improving clinical outcomes for naltrexone as a management of problem alcohol use. *British Journal of Clinical Pharmacology, 76*(5), 632–641. http://dx.doi.org/10.1111/j.1365-2125.2012.04452.x.

Hunt, W. A. (1993). Neuroscience research: How has it contributed to our understanding of alcohol abuse and alcoholism? A review. *Alcoholism, Clinical and Experimental Research, 17*, 1055–1065.

Hwang, B. H., Lumeng, L., Wu, J.-Y., & Li, T.-K. (1990). Increased number of GABAergic terminals in the nucleus accumbens is associated with alcohol preference in rats. *Alcoholism, Clinical and Experimental Research, 14*, 503–507.

Hwang, B. H., Stewart, R., Zhang, J.-K., Lumeng, L., & Li, T.-K. (2004). Corticotropin-releasing factor gene expression is down-regulated in the central nucleus of the amygdala of alcohol-preferring rats which exhibit high anxiety: A comparison between rat lines selectively bred for high and low alcohol preference. *Brain Research, 1026*, 143–150.

Hwang, B. H., Zhang, J. K., Ehlers, C. L., Lumeng, L., & Li, T. K. (1999). Innate differences of neuropeptide Y (NPY) in hypothalamic nuclei and central nucleus of the amygdala between selectively bred rats with high and low alcohol preference. *Alcoholism, Clinical and Experimental Research, 23*, 1023–1030.

Imperato, A., & Di Chiara, G. (1986). Preferential stimulation of dopamine release in the nucleus accumbens of freely moving rats by ethanol. *Journal of Pharmacology and Experimental Therapeutics, 239*, 219–228.

Irimia, C., Wiskerke, J., Natividad, L. A., Polis, I. Y., de Vries, T. J., Pattij, T., et al. (2013). Increased impulsivity in rats as a result of repeated cycles of alcohol intoxication and abstinence. *Addiction Biology, 20*, 263–274. http://dx.doi.org/10.1111/adb.12119.

Ishii, T., & Mann, G. E. (2014). Redox status in mammalian cells and stem cells during culture in vitro: Critical roles of Nrf2 and cystine transporter activity in the maintenance of redox balance. *Redox Biology, 2*, 786–794.

Iyer, R. N., & Bradberry, C. W. (1996). Serotonin-mediated increase in prefrontal cortex dopamine release: Pharmacological characterization. *Journal of Pharmacology and Experimental Therapeutics, 277*, 40–47.

Jablon, S., Neel, J. V., Gershowitz, H., & Atkinson, G. F. (1967). The NAS-NRC twin panel: Methods of construction of the panel, zygosity diagnosis, and proposed use. *American Journal of Human Genetics, 19*, 133–161.

Jerlhag, E., Egecioglu, E., Dickson, S., & Engel, J. A. (2011). Glutamatergic regulation of ghrelin-induced activation of the mesolimbic dopamine system. *Addiction Biology, 16*, 82–91.

Jerlhag, E., Egecioglu, E., Landgren, S., Salome, N., Heilig, M., Moechars, D., et al. (2009). Requirement of central ghrelin signaling for alcohol reward. *Proceedings of the National Academy of Sciences of the United States of America, 106*, 11318–11323.

Jerlhag, E., Landgren, S., Egecioglu, E., Dickson, S. L., & Engel, J. A. (2011). The alcohol-induced locomotor stimulation and accumbal dopamine release is suppressed in ghrelin knockout mice. *Alcohol, 45*, 341–347.

Johnson, B. A. (2005). Recent advances in the development of treatments for alcohol and cocaine dependence. *CNS Drugs, 19*, 873–896.

Johnson, B. A. (2010). Medication treatment of different types of alcoholism. *The American Journal of Psychiatry, 167*, 630–639.

Johnson, B. A., Ait-Daoud, N., Ma, J. Z., & Wang, Y. (2003). Ondansetron reduces mood disturbance among biologically predisposed, alcohol-dependent individuals. *Alcoholism, Clinical and Experimental Research, 27*, 1773–1779.

Johnston, L. D., O'Malley, P. M., & Bachman, J. G. (1999). *National survey results on drug use from the monitoring the future study, 1975–1997*. Rockville, MD: National Institute on Drug Abuse.

Johnston, L. D., O'Malley, P. M., Bachman, J. G., & Schulenberg, J. E. (2008). National survey results on drug use from the monitoring the future study, 1975–2007. *Secondary school students: Vol. 1*. Bethesda, MD: National Institute on Drug Abuse (NIH Publication No. 08-6418A).

Jonas, D. E., Amick, H. R., Feltner, C., Wines, R., Shanahan, E., Rowe, C. J., et al. (2014). Genetic polymorphisms and response to medications for alcohol use disorders: A systematic review and meta-analysis. *Pharmacogenomics*, *15*, 1687–1700.

June, H. L., Cummings, R., Eiler, W. J. A., II, Foster, K. L., McKay, P. F., Seyoum, R., et al. (2004). Central opioid receptors differentially regulate the nalmefene-induced suppression of ethanol- and saccharin-reinforced behaviors in alcohol-preferring (P) rats. *Neuropsychopharmacology*, *29*, 285–299.

June, H. L., Devaraju, S. L., Eggers, M. W., Williams, J. A., Cason, C. R., Greene, T. L., et al. (1998). Benzodiazepine receptor antagonists modulate the actions of ethanol in alcohol-preferring and -nonpreferring rats. *European Journal of Pharmacology*, *342*, 139–151.

June, H. L., Eggers, M. W., Warren-Reese, C., DeLong, J., Ricks-Cord, A., Durr, L. F., et al. (1998). The effects of the novel benzodiazepine inverse agonist RU 34000 on ethanol-maintained behaviors. *European Journal of Pharmacology*, *350*, 151–158.

June, H. L., Foster, K. L., McKay, P. F., Seyoum, R., Woods, J. E., II, Harvey, S. C., et al. (2003). The reinforcing properties of alcohol are mediated by $GABA_{A1}$ receptors in the ventral pallidum. *Neuropsychopharmacology*, *28*, 2124–2137.

June, H. L., Greene, T. L., Murphy, J. M., Hite, M. L., Williams, J. A., Cason, C. R., et al. (1996). Effects of the benzodiazepine inverse agonist RO19-4603 alone and in combination with the benzodiazepine receptor antagonists flumazenil, ZK 93426 and CGS 8216, on ethanol in take in alcohol-preferring (P) rats. *Brain Research*, *734*, 19–34.

June, H. L., Grey, C., Warren-Reese, C., Durr, L. F., Ricks-Cord, A., Johnson, A., et al. (1998). The opioid receptor antagonist nalmefene reduces responding maintained by ethanol presentation: Preclinical studies in ethanol-preferring and outbred Wistar rats. *Alcoholism, Clinical and Experimental Research*, *22*, 2174–2185.

June, H. L., Lummis, G. H., Colker, R. E., Moore, T. O., & Lewis, M. J. (1991). Ro 15-4513 attenuates the consumption of ethanol in deprived rats. *Alcoholism: Clinical and Experimental Research*, *15*, 406–411.

June, H. L., McCane, S. R., Zink, R. W., Portoghese, P. S., Li, T.-K., & Froehlich, J. C. (1999). The delta-2 opioid receptor antagonist naltriben reduces motivated responding for ethanol. *Psychopharmacology*, *147*, 81–89.

June, H. L., Murphy, J. M., Mellor-Burke, J. J., Lumeng, L., & Li, T.-K. (1994). The benzodiazepine inverse agonist RO19-4603 exerts prolonged and selective suppression of ethanol intake in alcohol-preferring (P) rats. *Psychopharmacology*, *115*, 325–331.

June, H. L., Torres, L., Cason, C. R., Hwang, B. H., Braun, M. R., & Murphy, J. M. (1998). The novel benzodiazepine inverse agonist RO19-4603 antagonizes ethanol motivated behaviors: Neuropharmacological studies. *Brain Research*, *784*, 256–275.

June, H. L., Williams, J. A., Cason, C. R., Devaraju, S., Lin, M., Murphy, J. M., et al. (1995). Low doses of gamma-hydroxybutyric acid (GHB) attenuate ethanol intake in alcohol-preferring (P) rats. *Alcoholism, Clinical and Experimental Research*, *19*, 14A.

June, H. L., Zuccarelli, D., Torres, L., Craig, K. S., DeLong, J., Allen, A., et al. (1998). High-affinity benzodiazepine antagonists reduce responding maintained by ethanol presentation in ethanol-preferring rats. *Journal of Pharmacology and Experimental Therapeutics*, *284*, 1006–1014.

Jupp, B., & Lawrence, A. J. (2010). New horizons for therapeutics in drug and alcohol abuse. *Pharmacology & Therapeutics*, *125*, 138–168.

Kaji, H. (2013). Neuropeptide Y and its receptors: Molecular structure and pathophysiological role in food intake and energy homeostasis. In S. L. Parker (Ed.), *Neuropeptide Y: Molecular structure, role in food intake and direct/indirect effects* (pp. 39–82). New York, NY: Nova Biomedical.

Kampov-Polevoy, A. B., Matthews, D. B., Gause, L., Morrow, A. L., & Overstreet, D. H. (2000). P rats develop physical dependence on alcohol via voluntary drinking: Changes in seizure thresholds, anxiety, and patterns of alcohol drinking. *Alcoholism, Clinical and Experimental Research*, *24*, 278–284.

Kane, C. J., Phelan, K. D., Douglas, J. C., Wagoner, G., Johnson, J. W., Xu, J., et al. (2014). Effects of ethanol on immune response in the brain: Region-specific changes in adolescent versus adult mice. *Alcoholism, Clinical and Experimental Research*, *38*, 384–391.

Kapasova, Z., & Szumlinski, K. K. (2008). Strain differences in alcohol induced neurochemical plasticity: A role for accumbens glutamate in alcohol intake. *Alcoholism, Clinical and Experimental Research*, *32*, 617–631.

Kaplan, R. F., Cooney, N. L., Baker, L. H., Gillespie, R. A., Meyer, R. E., & Pomerleau, O. F. (1985). Reactivity to alcohol-related cues: Physiological and subjective responses in alcoholics and nonproblem drinkers. *Journal of Studies on Alcohol*, *46*, 267–272.

Kaplan, R. F., Meyer, R. E., & Stroebel, C. F. (1983). Alcohol dependence and responsivity to an ethanol stimulus as predictors of alcohol consumption. *British Journal of Addiction*, *78*, 259–267.

Kaprio, J., Rose, R. J., Romanov, K., & Koskenvuo, M. (1991). Genetic and environmental determinants of use and abuse of alcohol: The Finnish twin cohort studies. *Alcohol and Alcoholism*, *1*(Suppl. 1), 131–136.

Katner, S. N., Oster, S. M., Ding, Z. M., Deehan, G. A., Jr., Toalston, J. E., Hauser, S. R., et al. (2011). Alcohol-preferring (P) rats are more sensitive than Wistar rats to the reinforcing effects of cocaine self-administered directly into the nucleus accumbens shell. *Pharmacology Biochemistry Behavior*, *99*, 688–695.

Keating, G. M. (2013). Nalmefene: A review of its use in the treatment of alcohol dependence. *CNS Drugs*, *27*, 761–772.

Keith, L. D., Roberts, A., Wisen, K. M., & Crabbe, J. C. (1995). In W. A. Hunt & S. Zakhari (Eds.), *Stress, gender, and alcohol-seeking behavior* (pp. 181–196). Bethesda, MD: National Institute on Alcohol Abuse and Alcoholism (Research monograph no. 29).

Kelley, S. P., Nannini, M. A., Bratt, A. M., & Hodge, C. W. (2001). Neuropeptide-Y in the paraventricular nucleus increases ethanol self-administration. *Peptides*, *22*, 515–522.

Kiefer, F., Jahn, H., Otte, C., Nakovics, H., & Wiedemann, K. (2006). Effects of treatment with acamprosate on β-endorphin plasma concentration in humans with high alcohol preference. *Neuroscience Letters*, *404*, 103–106.

Kim, S. A., Kim, J. W., Song, J. Y., Park, S., Lee, H. J., & Chung, J. H. (2004). Association of polymorphisms in nicotinic acetylcholine receptor alpha 4 subunit gene (CHRNA4), mu-opioid receptor gene (OPRM1), and ethanol-metabolizing enzyme genes with alcoholism in Korean patients. *Alcohol*, *34*, 115–120.

Kimpel, M. W., Strother, W. N., McClintick, J. N., Carr, L. G., Liang, T., Edenberg, H. J., et al. (2007). Functional gene expression differences between inbred alcohol-preferring and -non-preferring rats in five brain regions. *Alcohol*, *41*, 95–132.

Kirby, L. G., Zeeb, F. D., & Winstanley, C. A. (2011). Contributions of serotonin in addiction vulnerability. *Neuropharmacology*, *61*, 421–432.

Klimek, V., Rajkowska, G., Luker, S. N., Dilley, G., et al. (1999). Brain noradrenergic receptors in major depression. *Neuropsychopharmacology*, *21*, 69–81.

Koob, G. F. (2004). A role for GABA mechanisms in the motivational effects of alcohol. *Biochemical Pharmacology, 68*, 1515–1525.
Koob, G. F. (2009). Dynamics of neuronal circuits in addiction: Reward, antireward, and emotional memory. *Pharmacopsychiatry, 42*(Suppl. 1), S32–S41.
Koob, G. F. (2010). The role of CRF and CRF-related peptides in the dark side of addiction. *Brain Research, 1314*, 3–14.
Koob, G. F. (2013). Theoretical frameworks and mechanistic aspects of alcohol addiction: Alcohol addiction as a reward deficit disorder. *Current Topics in Behavioral Neurosciences, 13*, 3–30.
Koob, G. F., Arends, M. A., & Le Moal, M. (2014). *Drugs, addiction, and the brain*. New York, NY: Elsevier Academic Press.
Koob, G. F., Buck, C. L., Cohen, A., Edwards, S., Park, P. E., Schlosburg, J. E., et al. (2014). Addiction as a stress surfeit disorder. *Neuropharmacology, 76*, 370–382.
Koob, G. F., & Le Moal, M. (2006). *Neurobiology of addiction*. New York, NY: Elsevier Academic Press.
Koob, G. F., & Le Moal, M. (2008). Neurobiological mechanisms for opponent motivational processes in addiction. *Philosophical Transactions of the Royal Society of London. Series B, Biological Sciences, 363*, 3113–3123.
Koob, G. F., & Volkow, N. D. (2010). Neurocircuitry of addiction. *Neuropsychopharmacology, 35*, 217–238.
Korpi, E. R., & Sinkkonen, S. T. (2006). GABA(A) receptor subtypes as targets for neuropsychiatric drug development. *Pharmacology & Therapeutics, 109*, 12–32.
Krishnan, V., & Nestler, E. J. (2008). The molecular neurobiology of depression. *Nature, 455*, 894–902.
Krishnan-Sarin, S., Jing, S.-L., Kurtz, D. L., Zweifel, M., Portoghese, P. S., Li, T.-K., et al. (1995). The delta opioid receptor antagonist naltrindole attenuates both alcohol and saccharin intake in rats selectively bred for alcohol preference. *Psychopharmacology, 120*, 177–185.
Krishnan-Sarin, S., Portoghese, P. S., Li, T.-K., & Froehlich, J. C. (1995). The delta2-opioid receptor antagonist naltriben selectively attenuates alcohol intake in rats bred for alcohol preference. *Pharmacology, Biochemistry, and Behavior, 52*, 153–159.
Kryger, R., & Wilce, P. A. (2010). The effects of alcoholism on the human basolateral amygdala. *Neuroscience, 167*, 361–371.
Kumar, S., Porcu, P., Werner, D. F., Matthews, D. B., Diaz-Granados, J. L., Helfand, R. S., et al. (2009). The role of GABA-A receptors in the acute and chronic effects of ethanol: A decade of progress. *Psychopharmacology, 205*, 529–564.
Kuntsche, E., Rehm, J., & Gmel, G. (2004). Characteristics of binge drinkers in Europe. *Social Science & Medicine, 59*, 113–127.
Kupila, J., Karkkainen, O., Laukkanen, V., Tupala, E., Tiihonen, J., & Storvik, M. (2013). mGluR1/5 receptor densities in the brains of alcoholic subjects: A whole hemisphere autoradiography study. *Psychiatry Research, 212*, 245–250.
Kurtz, D. L., Stewart, R. B., Zweifel, M., Li, T.-K., & Froehlich, J. C. (1996). Genetic differences in tolerance and sensitization to the sedative/hypnotic effects of alcohol. *Pharmacology, Biochemistry, and Behavior, 53*, 585–591.
Lancaster, F. E. (1994). Gender differences in the brain: Implications for the study of human alcoholism. *Alcoholism, Clinical and Experimental Research, 18*, 740–746.
Lankford, M. F., Bjork, A. K., & Myers, R. D. (1996). Differential efficacy of serotonergic drugs FG5974, FG5893, and amperozide in reducing alcohol drinking in P rats. *Alcohol, 13*, 399–404.
Le, A. D., Li, Z., Funk, D., Shram, M., Li, T. K., & Shaham, Y. (2006). Increased vulnerability to nicotine self-administration and relapse in alcohol-naïve offspring of rats selectively bred for high alcohol intake. *Journal of Neuroscience, 26*, 1872–1879.

Lê, A. D., Corrigall, W. A., Harding, J. W., Juzytsch, W., & Li, T.-K. (2000). Involvement of nicotinic receptors in alcohol self-administration. *Alcoholism, Clinical and Experimental Research*, *24*, 155–163.

Lê, A. D., Li, Z., Funk, D., Shram, M., Li, T.-K., & Shaham, Y. (2003). Nicotine increases alcohol self-administration and reinstates alcohol seeking in rats. *Psychopharmacology*, *168*, 216–221.

Lê, A. D., & Mayer, J. M. (1996). Aspects of alcohol tolerance in humans and experimental animals. In R. A. Deitrich & V. G. Erwin (Eds.), *Pharmacological effects of ethanol on the nervous system* (pp. 251–268). Boca Raton, FL: CRC Press.

Leggio, L., & Addolorato, G. (2008). Serotonin transporter (SERT) brain density and neurobiological Cloninger subtypes model: A lesson by human autoradiography studies. *Alcohol and Alcoholism*, *43*, 148–150.

Levy, A. D., Murphy, J. M., McBride, W. J., Lumeng, L., & Li, T.-K. (1991). Microinjection of sulpiride into the nucleus accumbens increases ethanol intake of alcohol-preferring (P) rats. *Alcohol and Alcoholism*, *1*(Suppl. 1), 417–420.

Lesch, O. M., & Walter, H. (1996). Subtypes of alcoholism and their role in therapy. *Alcohol and Alcoholism*, *31*(Suppl. 1), 63–67.

Lester, D., & Freed, E. X. (1973). Criteria for an animal model of alcoholism. *Pharmacology, Biochemistry, and Behavior*, *1*, 103–107.

Lex, B. W., Lukas, S. E., Greenwald, N. E., & Mendelson, H. H. (1988). Alcohol-induced changes in body sway in women at risk for alcoholism: A pilot study. *Journal of Studies on Alcohol*, *49*, 346–356.

Li, T.-K., & Lumeng, L. (1977). Alcohol metabolism of inbred strains of rats with alcohol preference and non-preference. In R. G. Thurman, J. R. Williamson, H. Drott, & B. Chance (Eds.), *Alcohol and aldehyde metabolizing systems: Vol. 3* (pp. 625–633). New York, NY: Academic Press.

Li, T.-K., Lumeng, L., McBride, W. J., & Murphy, J. M. (1987). Rodent lines selected for factors affecting alcohol consumption. *Alcohol and Alcoholism*, *1*, 91–96.

Liang, J. H., Chen, F., Krstew, E., Cowen, M. S., Carroll, F. Y., Crawford, D., et al. (2006). The $GABA_B$ receptor allosteric modulator CGP7930, like baclofen, reduces operant self-administration of ethanol in alcohol-preferring rats. *Neuropharmacology*, *50*, 632–639.

Liang, J., & Olsen, R. W. (2014). Alcohol use disorders and current pharmacological therapies: The role of GABA(A) receptors. *Acta Pharmacologica Sinica*, *35*, 981–993.

Liang, T., Spence, J., Liu, L., Strother, W. N., Chang, H. W., Ellison, J. A., et al. (2003). α-synuclein maps to a quantitative trait locus for alcohol preference and is differentially expressed in alcohol-preferring and -nonpreferring rats. *Proceedings of the National Academy of Sciences of the United States of America*, *100*, 4690–4695.

Litten, R. Z., Egli, M., Heilig, M., Cui, C., Fertig, J. B., Ryna, M. L., et al. (2012). Medications development to treat alcohol dependence: A vision for the next decade. *Addiction Biology*, *17*, 513–527.

Liu, J., Yang, A. R., Kelly, T., Puche, A., Esoga, C., June, H. L., Jr., et al. (2011). Binge alcohol drinking is associated with GABAA α2-regulated toll-like receptor 4 (TLR4) expression in the central amygdala. *Proceedings of the National Academy of Sciences of the United States of America*, *108*, 4465–4470.

Logrip, M. L., Janak, P. H., & Ron, D. (2009). Escalating ethanol intake is associated with altered corticostriatal BDNF expression. *Journal of Neurochemistry*, *109*, 1459–1468.

Lovinger, D. M. (1999). 5-HT3 receptors and the neural actions of alcohols: An increasingly exciting topic. *Neurochemistry International*, *35*, 125–130.

Lovinger, D. M., & Roberto, M. (2013). *Synaptic effects induced by alcohol. Behavioral neurobiology of alcohol addiction.* Berlin: Springer.

Lukas, S. E., Mendelson, J. H., Benedikt, R. A., & Jones, B. (1986). EEG alpha activity increases during transient episodes of ethanol-induced euphoria. *Pharmacology, Biochemistry, and Behavior*, *25*, 889–895.

Lukkes, J. L., Staub, D. R., Dietrich, A., Truitt, W., Neufeld-Cohen, A., Chen, A., et al. (2011). Topographical distribution of corticotropin-releasing factor type 2 receptor-like immunoreactivity in the rat dorsal raphe nucleus: Co-localization with tryptophan hydroxylase. *Neuroscience*, *183*, 47–63.

Lumeng, L., Hawkins, T. D., & Li, T.-K. (1977). New strains of rats with alcohol preference and non-preference. In R. G. Thurman, J. R. Williamson, H. Drott, & B. Chance (Eds.), *Alcohol and aldehyde metabolizing systems: Vol. 3* (pp. 537–544). New York, NY: Academic Press.

Lumeng, L., & Li, T.-K. (1986). The development of metabolic tolerance in the alcohol-preferring P rats: Comparison of forced and free-choice drinking of ethanol. *Pharmacology, Biochemistry, and Behavior*, *25*, 1013–1020.

Lumeng, L., Waller, M. B., McBride, W. J., & Li, T.-K. (1982). Different sensitivities to ethanol in alcohol-preferring and -nonpreferring rats. *Pharmacology, Biochemistry, and Behavior*, *16*, 125–130.

Luo, X., Kranzler, H. R., Zuo, L., Wang, S., Blumberg, H. P., & Gelernter, J. (2005). CHRM2 gene predisposes to alcohol dependence, drug dependence and affective disorders: Results from an extended case-control structured association study. *Human Molecular Genetics*, *14*, 2421–2434.

Lynch, W. J., Bond, C., Breslin, F. J., & Johnson, B. A. (2011). Severity of drinking as a predictor of efficacy of the combination of ondansetron and topiramate in rat models of ethanol consumption and relapse. *Psychopharmacology*, *217*, 3–12.

Lyness, W. H., & Smith, F. (1992). Influence of dopaminergic and serotonergic neurons on intravenous ethanol self-administration in the rat. *Pharmacology, Biochemistry, and Behavior*, *42*, 187–192.

Maccioni, P., & Colombo, G. (2009). Role of the GABA-B receptor in alcohol-seeking and drinking behavior. *Alcohol*, *43*, 555–558.

Maccioni, P., Zaru, A., Loi, B., Lobina, C., Carai, M. A. M., Gessa, G. L., et al. (2012). Comparison of the effect of the $GABA_B$ receptor agonist, baclofen, and the positive allosteric modulator of the $GABA_B$ receptor, GS39783, on alcohol self-administration in three different lines of alcohol-preferring rats. *Alcoholism, Clinical and Experimental Research*, *36*, 1748–1766.

Mackenzie, C. S., El-Gabalaway, R., Chou, K.-L., & Sareen, J. (2014). Prevalence and predictors of persistent versus remitting mood, anxiety, and substance disorders in a national sample of older adults. *American Journal of Geriatric Psychiatry*, *22*(9), 854–865. http://dx.doi.org/10.1016/j.jagp.2013.02.007.

Madsen, K. K., White, H. S., & Schousboe, A. (2010). Neuronal and non-neuronal GABA transporters as targets for antiepileptic drugs. *Pharmacology & Therapeutics*, *125*, 394–401.

Manzardo, A. M., McGuire, A., & Butler, M. G. (2015). Clinically relevant genetic biomarkers from the brain in alcoholism with representation on high resolution chromosome ideograms. *Gene*, *560*, 184–194.

Mardones, J., Segovia, N., & Hederra, A. (1953). Heredity of experimental alcohol preference in rats. II. Coefficient of heredity. *Quarterly Journal of Studies on Alcohol*, *14*, 1–2.

Mardones, J., & Segovia-Riquelme, N. (1983). Thirty-two years of selection of rats by ethanol preference: UChA and UChB strains. *Neurobehavioral Toxicology and Teratology*, *5*, 171–178.

Marini, V., Fucile, C., Zuccoli, M. L., Testino, G., Sumberaz, A., Robbiano, L., et al. (2013). Involvement of the mu-opioid receptor gene polymorphism A118G in the efficacy of detoxification of alcohol dependent patients. *Addictive Behaviors*, *38*, 1669–1671.

Martin, N. G., Oakeshott, J. G., Gibson, J. B., Starmer, G. A., Perl, J., & Wilks, A. V. (1985). A twin study of psychomotor and physiological responses to an acute dose of alcohol. *Behavior Genetics*, *15*, 305–347.

Mason, G. A., Rezvani, A. H., Grady, D. R., & Garbutt, J. C. (1994). The subchronic effects of the TRH analog TA-0910 and bromocriptine on alcohol preference in alcohol-preferring rats: Development of tolerance and cross-tolerance. *Alcoholism: Clinical and Experimental Research*, *18*, 1196–1201.

Mason, G. A., Rezvani, A. H., Overstreet, D. H., Hamedi, M., Walker, C. H., Yang, Y., et al. (1997). Involvement of dopamine D2 receptors in the suppressive effect of the thyrotropin-releasing hormone analog TA-0910 on alcohol intake in alcohol-preferring rats. *Alcoholism, Clinical and Experimental Research*, *21*, 1623–1629.

McBride, W. J., Bell, R. L., Rodd, Z. A., Strother, W. N., & Murphy, J. M. (2005). Adolescent alcohol drinking and its long-range consequences: Studies with animal models. In M. Galanter (Ed.), *Alcohol problems in adolescents and young adults: Epidemiology, neurobiology, prevention, and treatment* (pp. 123–142). New York, NY: Springer.

McBride, W. J., Chernet, E., Dyr, W., Lumeng, L., & Li, T.-K. (1993a). Densities of dopamine D2 receptors are reduced in CNS regions of alcohol-preferring P rats. *Alcohol*, *10*, 387–390.

McBride, W. J., Chernet, E., Rabold, J. A., Lumeng, L., & Li, T.-K. (1993b). Serotonin-2 receptors in the CNS of alcohol-preferring and -nonpreferring rats. *Pharmacology, Biochemistry, and Behavior*, *46*, 631–636.

McBride, W. J., Chernet, E., Russell, R. N., Wong, D. T., Guan, X.-M., Lumeng, L., et al. (1997). Regional CNS densities of monoamine receptors in alcohol-naive alcohol-preferring P and -nonpreferring NP rats. *Alcohol*, *14*, 141–148.

McBride, W. J., Guan, X.-M., Chernet, E., Lumeng, L., & Li, T.-K. (1994). Regional serotonin$_{1A}$ receptors in the CNS of alcohol-preferring and -nonpreferring rats. *Pharmacology, Biochemistry, and Behavior*, *49*, 7–12.

McBride, W. J., Kimpel, M. W., McClintick, J. N., Ding, Z.-M., Edenberg, H. J., Liang, T., et al. (2014). Changes in gene expression within the extended amygdala following binge-like alcohol drinking by adolescent alcohol-preferring (P) rats. *Pharmacology, Biochemistry, and Behavior*, *117*, 52–60.

McBride, W. J., Kimpel, M. W., McClintick, J. N., Ding, Z.-M., Hauser, S. R., Edenberg, H. J., et al. (2013a). Changes in gene expression within the ventral tegmental area following repeated excessive binge-like alcohol drinking by alcohol-preferring (P) rats. *Alcohol*, *47*, 367–380. http://dx.doi.org/10.1016/j.alcohol.2013.04.002.

McBride, W. J., Kimpel, M. W., McClintick, J. N., Ding, Z.-M., Hyytia, P., Colombo, G., et al. (2012). Gene expression in the ventral tegmental area of 5 pairs of rat lines selectively bred for high or low ethanol consumption. *Pharmacology, Biochemistry & Behavior*, *102*, 275–285. http://dx.doi.org/10.1016/j.pbb.2012.04.016.

McBride, W. J., Kimpel, M. W., McClintick, J. N., Ding, Z.-M., Hyytia, P., Colombo, G., et al. (2013b). Gene expression within the extended amygdala of 5 pairs of rat lines selectively bred for high or low ethanol consumption. *Alcohol*, *47*, 517–529. http://dx.doi.org/10.1016/j.alcohol.2013.08.004.

McBride, W. J., Kimpel, M. W., Schultz, J. A., McClintick, J. N., Edenberg, H. J., & Bell, R. L. (2010). Changes in gene expression in regions of the extended amygdala of alcohol-preferring rats after binge-like alcohol drinking. *Alcohol*, *44*, 171–183.

McBride, W. J., & Li, T. K. (1998). Animal models of alcoholism: Neurobiology of high alcohol-drinking behavior in rodents. *Critical Reviews in Neurobiology*, *12*, 339–369.

McBride, W. J., Murphy, J. M., Lumeng, L., & Li, T.-K. (1988). Effects of Ro 15-4513, fluoxetine and desipramine on the intake of ethanol, water and food by the alcohol-preferring (P) and -nonpreferring (NP) lines of rats. *Pharmacology, Biochemistry, and Behavior*, *30*, 1045–1050.

McBride, W. J., Murphy, J. M., Lumeng, L., & Li, T.-K. (1990). Serotonin, dopamine and GABA involvement in alcohol drinking of selectively bred rats. *Alcohol*, *7*, 199–205.

McBride, W. J., Rodd, Z. A., Bell, R. L., Lumeng, L., & Li, T.-K. (2014). The alcohol-preferring (P) and high-alcohol-drinking (HAD) rats—Animal models of alcoholism. *Alcohol*, *48*, 209–215. http://dx.doi.org/10.1016/j.alcohol.2013.09.044.

McBride, W. J., Schultz, J. A., Kimpel, M. W., McClintick, J. N., Wang, M., You, J., et al. (2009). Differential effects of ethanol in the nucleus accumbens shell of alcohol-preferring (P), alcohol non-preferring (NP) and Wistar rats: A proteomic study. *Pharmacology, Biochemistry, and Behavior*, *92*, 304–313.

McCairn, K. W., & Isoda, M. (2013). Pharmacological animal models of tic disorders. *International Review of Neurobiology*, *112*, 179–203. http://dx.doi.org/10.1016/B978-0-12-411546-0.00007-X.

McCarson, K. E., & Enna, S. J. (2014). GABA pharmacology: The search for analgesics. *Neurochemical Research*, *39*, 1948–1963.

McClintick, J. N., McBride, W. J., Bell, R. L., Ding, Z.-M., Liu, Y., Xuei, X., et al. (2015). Gene expression changes in serotonin, GABA-A receptors, neuropeptides and ion channels in the dorsal raphe nucleus of adolescent alcohol-preferring (P) rats following binge-like alcohol drinking. *Pharmacology, Biochemistry, and Behavior*, *129*, 87–96.

McGehee, D. S., & Role, L. W. (1995). Physiological diversity of nicotinic acetylcholine receptors expressed by vertebrate neurons. *Annual Review of Physiology*, *57*, 521–546.

McGonigle, P., & Ruggeri, B. (2014). Animal models of human disease: Challenges in enabling translation. *Biochemical Pharmacology*, *87*(1), 162.171. http://dx.doi.org/10.1016/j.bcp.2013.08.006.

McHugh, R. K., Hofmann, S. G., Asnaani, A., Sawyer, A. T., & Otto, M. W. (2010). The serotonin transporter gene and risk for alcohol dependence: A meta-analytic review. *Drug and Alcohol Dependence*, *108*, 1–6.

McKinney, W. T. (2001). Overview of the past contributions of animal models and their changing place in psychiatry. *Seminars in Clinical Psychiatry*, *6*, 68–78.

McLarnon, J. G. (2014). Correlated inflammatory responses and neurodegeneration in peptide-injected animal models of Alzheimer's disease. *BioMed Research International*, *2014*, 923670. http://dx.doi.org/10.1155/2014/923670.

Meinhardt, M. W., Hansson, A. C., Perreau-Lenz, S., Bauder-Wenz, C., Stahlin, O., Heilig, M., et al. (2013). Rescue of infralimbic mGluR2 deficit restores control over drug-seeking behavior in alcohol Dependence. *Journal of Neuroscience*, *33*, 2794–2806.

Melendez, R. I., Rodd-Henricks, Z. A., Engleman, E. A., Li, T.-K., McBride, W. J., & Murphy, J. M. (2002). Microdialysis of dopamine in the nucleus accumbens of alcohol-preferring (P) rats during anticipation and operant self-administration of ethanol. *Alcoholism, Clinical and Experimental Research*, *26*, 318–325.

Melendez, R. I., Rodd, Z. A., McBride, W. J., & Murphy, J. M. (2005). Dopamine receptor regulation of ethanol intake and extracellular dopamine levels in the ventral pallidum of alcohol preferring (P) rats. *Drug and Alcohol Dependence*, 77, 293–301.

Melis, M., Camarini, R., Ungless, M. A., & Bonci, A. (2002). Long-lasting potentiation of GABAergic synapses in dopamine neurons after a single in vivo ethanol exposure. *Journal of Neuroscience*, *22*, 2074–2082.

Menkes, D. B., Baraban, J. M., & Aghajanian, G. K. (1981). Prazosin selectively antagonizes neuronal responses mediated by alpha1-adrenoceptors in brain. *Naunyn-Schmiedeberg's Archives of Pharmacology*, *317*, 273–275.

Milner, L. C., & Buck, K. J. (2010). Identifying quantitative trait loci (QTL) and genes (QTGs) for alcohol-related phenotypes in mice. *International Review of Neurobiology*, *91*, 173–204.

Mitani, A., & Tanaka, K. (2003). Functional changes of glial glutamate transporter GLT-1 during ischemia: An in vivo study in the hippocampal CA1 of normal mice and mutant mice lacking GLT-1. *Journal of Neuroscience*, *23*, 7176–7182.

Miyata, I., Shiota, C., Chaki, S., Okuyama, S., & Inagami, T. (2001). Localization and characterization of a short isoform of the corticotropin-releasing factor receptor type 2 alpha (CRF(2)alpha-tr) in the rat brain. *Biochemical and Biophysical Research Communications, 280*, 553–557.

Mokdad, A., Marks, J., Stroup, D., & Gerberding, J. (2004). Actual causes of death in the United States, 2000. *Journal of American Medical Association, 291*, 1238–1245.

Moonat, S., Starkman, B. G., Sakharkar, A., & Pandey, S. C. (2010). Neuroscience of alcoholism: Molecular and cellular mechanisms. *Cellular and Molecular Life Sciences, 67*, 73–88.

Morean, M. E., & Corbin, W. R. (2010). Subjective response to alcohol: A critical review of the literature. *Alcoholism, Clinical and Experimental Research, 34*, 385–395. http://dx.doi.org/10.1111/j.1530-0277.2009.01103.x.

Moreira, F. A., Jupp, B., Belin, D., & Dalley, J. W. (2015). Endocannabinoids and striatal function: Implications for addiction-related behaviors. *Behavioural Pharmacology, 26*, 59–72.

Moret, C., & Briley, M. (2011). The importance of norepinephrine in depression. *Neuropsychiatric Disease and Treatment*, 7(Suppl. 1), 9–13.

Morganstern, I., Liang, S., Ye, Z., Karatayev, O., & Leibowitz, S. F. (2012). Disturbances in behavior and cortical encephalin gene expression during the anticipation of ethanol in rats characterized as high drinkers. *Alcohol, 46*, 559–568.

Morozova, T. V., Mackay, T. F. C., & Anholt, R. R. H. (2014). Genetics and genomics of alcohol sensitivity. *Molecular Genetics and Genomics, 289*, 253–269. http://dx.doi.org/10.1007/s00438-013-0808-y.

Morris, R. G. M. (2013). NMDA receptors and memory encoding. *Neuropharmacology, 74*, 32–40.

Morzorati, S. L. (1998). VTA dopamine neuron activity distinguishes alcohol-preferring (P) rats from Wistar rats. *Alcoholism, Clinical and Experimental Research, 22*, 854–857.

Morzorati, S. L., & Marunde, R. L. (2006). Comparison of VTA dopamine neuron activity in lines of rats selectively bred to prefer or avoid alcohol. *Alcoholism, Clinical and Experimental Research, 30*, 991–997.

Morzorati, S. L., Marunde, R. L., & Downey, D. (2010). Limited access to ethanol increases the number of spontaneously active dopamine neurons in the posterior ventral tegmental area of nondependent P rats. *Alcohol, 44*, 257–264.

Morzorati, S. L., Ramchandani, V. A., Flury, L., Li, T.-K., & O'Connor, S. (2002). Self-reported subjective perception of intoxication reflects family history of alcoholism when breath alcohol levels are constant. *Alcoholism, Clinical and Experimental Research, 26*, 1299–1306.

Moss, H. B., Chen, C. M., & Yi, H.-Y. (2007). Subtypes of alcohol dependence in a nationally representative sample. *Drug and Alcohol Dependence, 91*, 149–158.

Murphy, J. M., Gatto, G. J., McBride, W. J., Lumeng, L., & Li, T.-K. (1989). Operant responding for oral ethanol in the alcohol-preferring P and alcohol-nonpreferring NP lines of rats. *Alcohol, 6*, 127–131.

Murphy, J. M., Gatto, G. J., Waller, M. B., McBride, W. J., Lumeng, L., & Li, T.-K. (1986). Effects of scheduled access on ethanol intake by the alcohol-preferring (P) line of rats. *Alcohol, 3*, 331–336.

Murphy, J. M., McBride, W. J., Lumeng, L., & Li, T.-K. (1982). Regional brain levels of monoamines in alcohol-preferring and -nonpreferring rats. *Pharmacology, Biochemistry, and Behavior, 16*, 145–149.

Murphy, J. M., McBride, W. J., Lumeng, L., & Li, T.-K. (1987). Contents of monoamines in forebrain regions of alcohol-preferring (P) and -nonpreferring (NP) lines of rats. *Pharmacology, Biochemistry, and Behavior, 26*, 389–392.

Murphy, J. M., Stewart, R. B., Bell, R. L., Badia-Elder, N. E., Carr, L. G., McBride, W. J., et al. (2002). Phenotypic and genotypic characterization of the Indiana University rat lines selectively bred for high and low alcohol preference. *Behavior Genetics*, *32*, 363–388.

Murphy, J. M., Waller, M. B., Gatto, G. J., McBride, W. J., Lumeng, L., & Li, T.-K. (1985). Monoamine uptake inhibitors attenuate ethanol intake in alcohol-preferring (P) rats. *Alcohol*, *2*, 349–352.

Murphy, J. M., Waller, M. B., Gatto, G. J., McBride, W. J., Lumeng, L., & Li, T.-K. (1988). Effects of fluoxetine on the intragastric self-administration of ethanol in the alcohol preferring P line of rats. *Alcohol*, *5*, 283–286.

Nam, H. W., Bruner, R. C., & Choi, D. S. (2013). Adenosine signaling in striatal circuits and alcohol use disorders. *Molecules and Cells*, *36*, 195–202.

Nestler, E. J., & Hyman, S. E. (2010). Animal models of neuropsychiatric disorders. *Nature Neuroscience*, *13*, 1161–1169.

Nevo, I., & Hamon, M. (1995). Neurotransmitter and neuromodulatory mechanisms involved in alcohol abuse and alcoholism. *Neurochemistry International*, *26*, 305–342.

Newman, E. L., Gupta, K., Climer, J. R., Monaghan, C. K., & Hasselmo, M. E. (2012). Cholinergic modulation of cognitive processing: Insights drawn from computational models. *Frontiers in Behavioral Neuroscience*, *6*, 24.

NIAAA. (2012). *Underage drinking. NIAAA Fact Sheet.* Available at: http://pubs.niaaa.nih.gov/publications/UnderageDrinking/Underage_Fact.pdf. Published March 2012.

NIAAA National Advisory Council (2004). NIAAA Council approves definition of binge drinking. *NIAAA Newsletter*, *3*, 5.

Niciu, M. J., Kelmendi, B., & Sanacora, G. (2012). Overview of glutamatergic neurotransmission in the nervous system. *Pharmacology, Biochemistry, and Behavior*, *100*, 656–664.

Nie, Z., Madamba, S. G., & Siggins, G. R. (1994). Ethanol inhibits glutamatergic neurotransmission in nucleus accumbens neurons by multiple mechanisms. *Journal of Pharmacology and Experimental Therapeutics*, *271*, 1566–1573.

Nie, Z., Yuan, X., Madamba, S. G., & Siggins, G. R. (1993). Ethanol decreases glutamatergic synaptic transmission in rat nucleus accumbens in vitro: Naloxone reversal. *Journal of Pharmacology and Experimental Therapeutics*, *266*, 1705–1712.

Nisell, M., Nomikos, G. G., & Svensson, T. H. (1994). Systemic nicotine-induced dopamine release in the rat nucleus accumbens is regulated by nicotinic receptors in the ventral tegmental area. *Synapse*, *16*, 36–44.

Noble, E. P., Zhnag, X., Ritchie, T., Lawford, B. R., Grosser, S. C., Young, R. M., et al. (1998). D2 dopamine receptor and GABA(A) receptor beta 3 subunit genes and alcoholism. *Psychiatry Research*, *81*, 133–147.

Nogueira, L., Kalivas, P. W., & Lavin, A. (2006). Long-term neuroadaptations produced by withdrawal from repeated cocaine treatment: Role of dopaminergic receptors in modulating cortical excitability. *Journal of Neuroscience*, *26*, 12308–12313.

Nordberg, A., Larsson, C., Perdahl, E., & Winblad, B. (1983). Changes in cholinergic activity in human hippocampus following chronic alcohol abuse. *Pharmacology, Biochemistry, and Behavior*, *18*, 379–400.

Noronha, A. B. C., Cui, C., Harris, R. A., & Crabbe, J. C. (2014). *Neurobiology of alcohol dependence.* New York, NY: Elsevier Academic Press.

Nowak, K. L., McBride, W. J., Lumeng, L., Li, T. K., & Murphy, J. M. (1998). Blocking GABA(A) receptors in the anterior ventral tegmental area attenuates ethanol intake of the alcohol-preferring P rat. *Psychopharmacology*, *139*, 108–116.

Nowak, K. L., McBride, W. J., Lumeng, L., Li, T. K., & Murphy, J. M. (2000). Involvement of dopamine D2 autoreceptors in the ventral tegmental area on alcohol and saccharin intake of the alcohol preferring P rat. *Alcoholism, Clinical and Experimental Research*, *24*, 476–483.

Nunez, Y. O., Truitt, J. M., Gorini, G., Ponomareva, O. N., Bldenov, Y. A., Harris, R. A., et al. (2013). Positively correlated miRNA-mRNA regulatory networks in mouse frontalcortex during early stages of alcohol dependence. *BMC Genomics*, *14*, 725.
Nutt, D. J., Lingfor-Hughes, A., Erritzoe, D., & Stokes, P. R. A. (2015). The dopamine theory of addiction: 40 years of highs and lows. *Nature Reviews. Neuroscience*, *16*, 305–312.
Obara, I., Bell, R. L., Goulding, S. P., Reyes, C. M., Larson, L. A., Ary, A. W., et al. (2009). Differential effects of chronic ethanol consumption and withdrawal on Homer/glutamate receptor expression in subregions of the accumbens and amygdala of P rats. *Alcoholism, Clinical and Experimental Research*, *33*, 1924–1934. http://dx.doi.org/10.1111/j.1530-0277.2009.01030.x.
O'Brien, C. P., Childress, A. R., McLellan, A. T., & Ehrman, R. (1992). A learning model of addiction. In C. P. O'Brien & J. H. Jaffe (Eds.), *Addictive States* (pp. 157–177). New York: Raven Press.
Okahisa, Y., Ujike, H., Kotaka, T., Morita, Y., Kodama, M., Iwata, N., et al. (2009). Association between neuropeptide Y gene and its receptor Y1 gene and methamphetamine dependence. *Psychiatry and Clinical Neurosciences*, *63*, 417–422.
Olney, J. J., Navarro, M., & Thiele, T. E. (2014). Targeting central melanocortin receptors: A promising novel approach for treating alcohol abuse disorders. *Frontiers in Neuroscience/Neuropharmacology*, *8*, 128. http://dx.doi.org/10.3389/fnins.2014.00128.
Oswald, L. M., & Wand, G. S. (2004). Opioids and alcoholism. *Physiology & Behavior*, *81*, 339–358.
Othman, T., Sinclair, C. J., Haughey, N., Geiger, J. D., & Parkinson, F. E. (2002). Ethanol alters glutamate but not adenosine uptake in rat astrocytes: Evidence for protein kinase C involvement. *Neurochemical Research*, *27*, 289–296.
Overstreet, D. H., McArthur, R. A., Rezvani, A. H., & Post, C. (1997). Selective inhibition of alcohol intake in diverse alcohol-preferring rats strains by 5-HT2A antagonists amperozide and FG 5974. *Alcoholism, Clinical and Experimental Research*, *21*, 1448–1454.
Overstreet, D. H., Rezvani, A. H., Djouma, E., Parsian, A., & Lawrence, A. J. (2007). Depressive-like behavior and high alcohol drinking co-occur in the FH/WJD rat but appear to be under independent genetic control. *Neuroscience & Biobehavioral Reviews*, *31*, 103–114.
Palmer, A. A., Low, M. J., Grandy, D. K., & Phillips, T. J. (2003). Effects of a Drd2 deletion mutation on ethanol-induced locomotor stimulation and sensitization suggest a role for epistasis. *Behavior Genetics*, *33*, 311–324.
Pandey, S. C. (2003). Anxiety and alcohol abuse disorders: A common role for CREB and its target, the neuropeptide Y gene. *Trends in Pharmacological Sciences*, *24*, 455–460.
Pandey, S. C., Carr, L. G., Heilig, M., Ilveskoski, E., & Thiele, E. (2003). Neuropeptide Y and alcoholism: Genetic, molecular, and pharmacological evidence. *Alcoholism, Clinical and Experimental Research*, *27*, 149–154.
Pandey, S. C., Chartoff, E. H., Carlezon, W. A., Jr., et al. (2005). CREB gene transcription factors: Role in molecular mechanisms of alcohol and drug addiction. *Alcoholism, Clinical and Experimental Research*, *29*, 176–184.
Pandey, S. C., Lumeng, L., & Li, T.-K. (1996). Serotonin$_{2C}$ receptors and serotonin$_{2C}$ receptor-mediated phosphoinositide hydrolysis in the brain of alcohol-preferring and alcohol-nonpreferring rats. *Alcoholism, Clinical and Experimental Research*, *20*, 1038–1042.
Panula, P., & Nuutinen, S. (2011). Histamine and H_3 receptor in alcohol-related behaviors. *Journal of Pharmacology and Experimental Therapeutics*, *336*, 9–16.
Parent, A., Descarries, L., & Beaudet, A. (1981). Organization of ascending serotonin systems in the adult rat brain. A radioautographic study after intraventricular administration of [3H] 5 hydroxytryptamine. *Neuroscience*, *6*, 115–138.
Parker, S. L. (2013). *Neuropeptide Y: Molecular structure, role in food intake and direct/indirect effects*. New York, NY: Nova Biomedical.

Parks, M. H., Dawant, B. M., Riddle, W. R., Hartmann, S. L., Dietrich, M. S., Nickel, M. K., et al. (2002). Longitudinal brain metabolic characterization of chronic alcoholics with proton magnetic resonance spectroscopy. *Alcoholism, Clinical and Experimental Research, 26*, 1368–1380.

Partanen, J., Bruun, K., & Markkanen, T. (1966). Inheritance of drinking behavior: A study on intelligence, personality, and use of alcohol of adult twins. *Alcohol research in the Northern Countries. The Finnish foundation for alcohol studies: Vol. 14*. Stockholm: Amquist & Wiksell.

Pautassi, R. M., Camarini, R., Quadros, I. M., Miczek, K. A., & Israel, Y. (2010). Genetic and environmental influences on ethanol consumption: Perspectives from preclinical research. *Alcoholism, Clinical and Experimental Research, 34*, 976–987.

Pessia, M., Jiang, Z. G., North, R. A., & Johnson, S. W. (1994). Actions of 5-hydroxytryptamine on ventral tegmental area neurons of the rat in vitro. *Brain Research, 654*, 324–330.

Peterson, J. B., Pihl, R. O., Gianoulakis, C., Conrod, P., Finn, P. R., Stewart, S. H., et al. (1996). Ethanol-induced change in cardiac and endogenous opiate function and risk for alcoholism. *Alcoholism, Clinical and Experimental Research, 20*, 1542–1552.

Peterson, J. B., Pihl, R. O., Seguin, J. R., Finn, P. R., & Stewart, S. H. (1993). Heart-rate reactivity and alcohol consumption among sons of male alcoholics and sons of non-alcoholics. *Journal of Psychiatry & Neuroscience, 18*, 190–198.

Pierce, R. C., & Kenny, P. J. (2013). *Addiction*. New York, NY: Cold Spring Harbor Laboratory Press.

Piercy, K. T., Bjork, A. K., & Myers, R. D. (1996). The mixed 5HT1A/2A receptor drug FG5938 suppresses alcohol drinking while enhancing feeding in P rats. *Alcohol, 13*, 521–527.

Pivac, N., Mück-Seler, D., Mustapić, M., Nenadić-Sviglin, K., & Kozarić-Kovacić, D. (2004). Platelet serotonin concentration in alcoholic subjects. *Life Sciences, 76*, 521–531.

Plemenitas, A., Kastelic, M., Porcelli, S., Serretti, A., Dolžan, V., & Kores, P. B. (2015). Alcohol dependence and genetic variability in the serotonin pathway among currently and formerly alcohol-dependent males. *Neuropsychobiology, 72*, 57–64.

Ponomarev, I., Wang, S., Zhang, L., Harris, R. A., & Mayfield, R. D. (2012). Gene coexpression networks in human brain identify epigenetic modifications in alcohol dependence. *Journal of Neuroscience, 32*, 1884–1897.

Prelipceanu, D., & Mihailescu, R. (2005). Non A non B variants in the binary typology of alcoholics. *Addictive Disorders and Their Treatment, 4*, 149–156.

Presley, C. A., Meilman, P. W., & Lyerla, R. (1994). Development of the core alcohol and drug survey: Initial findings and future directions. *Journal of American College Health, 42*, 248–255.

Preuss, U. W., Watzke, S., & Wurst, F. M. (2014). Dimensionality and stages of severity of DSM-5 criteria in an international sample of alcohol-consuming individuals. *Psychological Medicine, 44*, 3303–3314. http://dx.doi.org/10.1017/S0033291714000889.

Preuss, U. W., Wurst, F. M., Ridinger, M., Rujescu, D., Fehr, C., Koller, G., et al. (2013). Association of functional DBH genetic variants with alcohol dependence risk and related depression and suicide attempt phenotypes: Results from a large multicenter association study. *Drug and Alcohol Dependence, 133*, 459–467.

Rahman, S. (2013). Nicotinic receptors as therapeutic targets for drug addictive disorders. *CNS & Neurological Disorders Drug Targets, 12*, 633–640.

Rahman, S., Engleman, E. A., & Bell, R. L. (2014). Nicotinic receptor modulation to treat alcohol and drug dependence. *Frontiers in Neuroscience/Neuropharmacology, 8*, 426. http://dx.doi.org/10.3389/fnins.2014.00426.

Rahman, S., Engleman, E. A., & Bell, R. L. (2015). Nicotinic receptor modulation to treat alcohol and drug dependence. *Frontiers in Neuroscience/Neuropharmacology, 8*, 426. http://dx.doi.org/10.3389/fnins.2014.00426.

Rahman, S., Engleman, E. A., & Bell, R. L. (2016). Recent advances in nicotinic receptor signaling in alcohol abuse and alcoholism. *Progress in Molecular Biology and Translational Science*, *137*, 183–201. http://dx.doi.org/10.1016/bs.pmbts.2015.10.004.

Rahman, S., & McBride, W. J. (2002). Involvement of GABA and cholinergic receptors in the nucleus accumbens on feedback control of somatodendritic dopamine release in the ventral tegmental area. *Journal of Neurochemistry*, *80*, 646–654.

Rahman, S., & Prendergast, M. A. (2012). Brain cholinergic receptor system as a target for treating alcohol dependence. *Recent Patents on CNS Drug Discovery*, 7, 145–150.

Rajan, I., Naga Venkatesha Murthy, P. J., Ramakrishnan, A. G., & Gangadhar, B. N. (1998). Heart rate variability as an index of cue reactivity in alcoholics. *Biological Psychiatry*, *43*, 544–546.

Rao, P. S., Bell, R. L., Engleman, E. A., & Sari, Y. (2015). Targeting glutamate uptake to treat alcohol use disorders. *Frontiers in Neuroscience/Neuropharmacology*, *9*, 144.

Rao, P. S., Goodwani, S., Bell, R. L., Wei, Y., Boddu, S. H., & Sari, Y. (2015). Effects of ampicillin, cefazolin and cefoperazone treatments on GLT-1 expressions in the mesocorticolimbic system and ethanol intake in alcohol-preferring rats. *Neuroscience*, *295*, 164–174.

Rao, P., & Sari, Y. (2014a). Effectiveness of Ceftriaxone treatment in preventing relapse-like drinking behavior following long-term ethanol dependence in P rats. *Journal of Addiction Research & Therapy*, *5*, 183. http://dx.doi.org/10.4172/2155-6105.1000183.

Rao, P. S., & Sari, Y. (2014b). Effects of ceftriaxone on chronic ethanol consumption: A potential role for xCT and GLT1 modulation of glutamate levels in male P rats. *Journal of Molecular Neuroscience*, *54*, 71–77.

Rao, P. S., Saternos, H., Goodwani, S., & Sari, Y. (2015). Effects of ceftriaxone on GLT1 isoforms, xCT and associated signaling pathways in P rats exposed to ethanol. *Psychopharmacology*, *232*, 2333–2342.

Rasmussen, D. D., Alexander, L., Malone, J., Federoff, D., & Froehlich, J. C. (2014). The α2 adrenergic receptor agonist, clonidine, reduces alcohol drinking in alcohol-preferring (P) rats. *Alcohol*, *48*, 543–549.

Rasmussen, D. D., Alexander, L. L., Raskind, M. A., & Froehlich, J. C. (2009). The alpha1-adrenergic receptor antagonist, prazosin, reduces alcohol drinking in alcohol-preferring (P) rats. *Alcoholism, Clinical and Experimental Research*, *33*, 264–272.

Rasmussen, D. D., Beckwith, L. E., Kincaid, C. L., & Froehlich, J. C. (2014). Combining the α1-adrenergic receptor antagonist, prazosin, with the β-adrenergic receptor antagonist, propranolol, reduces alcohol drinking more effectively than either drug alone. *Alcoholism, Clinical and Experimental Research*, *38*, 1532–1539.

Rasmussen, D. D., Boldt, B. M., Wilkinson, C. W., & Mitton, D. R. (2002). Chronic daily ethanol and withdrawal: 3. Forebrain pro-opiomelanocortin gene expression and implications for dependence, relapse, and deprivation effect. *Alcoholism, Clinical and Experimental Research*, *26*, 534–546.

Rasmussen, D. D., Kincaid, C. L., & Froehlich, J. C. (2015). Prazosin + naltrexone decreases alcohol drinking more effectively than does either drug alone in P rats with a protracted history of extensive voluntary alcohol drinking, dependence, and multiple withdrawals. *Alcoholism, Clinical and Experimental Research*, *39*, 1832–1841.

Reed, T., Page, W. F., Viken, R. J., & Christian, J. C. (1996). Genetic predisposition to organ-specific endpoints of alcoholism. *Alcoholism, Clinical and Experimental Research*, *20*, 1528–1533.

Rehm, J. (2011). The risks associated with alcohol use and alcoholism. *Alcohol Health and Research World*, *34*, 135–143.

Rehm, J., Room, R., Graham, K., Monteiro, M., Gmel, G., & Sempos, C. T. (2003). The relationship of average volume of alcohol consumption and patterns of drinking to burden of disease—An overview. *Addiction*, *98*, 1209–1228.

Renthal, W., & Nestler, E. J. (2009). Histone acetylation in drug addiction. *Seminars in Cellular and Developmental Biology*, *20*, 387–394.
Research Society on Alcoholism. (2009). *Impact of alcoholism and alcohol induced disease on America*. Austin, TX: Research Society on Alcoholism.
Research Society on Alcoholism. (2015). *Impact of alcoholism and alcohol induced disease on America*. Austin, TX: Research Society on Alcoholism.
Reser, J. E. (2014). Solitary mammals provide an animal model for autism spectrum disorders. *Journal of Comparative Psychology*, *128*(1), 99–113. http://dx.doi.org/10.1037/a0034519.
Reul, J. M., & Holsboer, F. (2002). Corticotropin-releasing factor receptors 1 and 2 in anxiety and depression. *Current Opinion in Pharmacology*, *2*, 23–33.
Reyes, B. A., Bangasser, D. A., Valentino, R. J., & Van Bockstaele, E. J. (2014). Using high resolution imaging to determine trafficking of corticotropin-releasing factor receptors in noradrenergic neurons of the rat locus coeruleus. *Life Sciences*, *112*, 2–9.
Rezvani, A. H., Overstreet, D. H., & Lee, Y.-W. (1995). Attenuation of alcohol intake by ibogaine in three strains of alcohol-preferring rats. *Pharmacology Biochemistry and Behavior*, *52*, 615–620.
Rezvani, A. H., Overstreet, D. H., Mason, G. A., Janowsky, D. S., Hamedi, M., Clark, E., et al. (2000). Combination pharmacotherapy: A mixture of small doses of naltrexone, fluoxetine, and a thyrotropin-releasing hormone analogue reduces alcohol intake in three strains of alcohol-preferring rats. *Alcohol and Alcoholism*, *35*, 76–83.
Rezvani, A. H., Slade, S., Wells, C., Petro, A., Lumeng, L., Li, T.-K., et al. (2010). Effects of sazetidine-A, a selective alpha-4-beta-2 nicotinic acetylcholine receptor desensitizing agent, on alcohol and nicotine self-administration in selectively bred alcohol-preferring (P) rats. *Psychopharmacology*, *211*, 161–174.
Richardson, H. N., Lee, S. Y., O'Dell, L. E., Koob, G. F., & Rivier, C. L. (2008). Alcohol self-administration acutely stimulates the hypothalamic-pituitary-adrenal axis, but alcohol dependence leads to a dampened neuroendocrine state. *European Journal of Neuroscience*, *28*, 1641–1653.
Richter, C. P., & Campbell, K. H. (1940). Alcohol taste thresholds and concentrations of solutions preferred by rats. *Science*, *9*, 507–508.
Riihioja, P., Jaatinen, P., Oksanen, H., et al. (1997). Dexmedetomidine alleviates ethanol withdrawal symptoms in the rat. *Alcohol*, *14*, 537–544.
Robbins, T. W., Everitt, B. J., & Nutt, D. J. (2010). *The neurobiology of addiction: New vistas*. New York, NY: Oxford University Press.
Roberto, M., Gilpin, N. W., & Siggins, G. R. (2012). The central amygdala and alcohol: Role of gamma-aminobutyric acid, glutamate, and neuropeptides. *Cold Spring Harbor Perspectives in Medicine*, *2*, a012195.
Roberts, A. J., Heyser, C. J., Cole, M., Griffin, P., & Koob, G. F. (2000). Excessive ethanol drinking following a history of dependence: Animal model of allostasis. *Neuropsychopharmacology*, *22*, 581–594.
Robinson, T. E., & Berridge, K. C. (2008). The incentive sensitization theory of addiction: Some current issues. *Philosophical Transactions of the Royal Society of London. Series B, Biological Sciences*, *363*, 3137–3146.
Robinson, G., Most, D., Ferguson, L. B., Mayfield, J., Harris, R. A., & Blednov, Y. A. (2014). Neuroimmune pathways in alcohol consumption: Evidence from behavioral and genetic studies in rodents and humans. *International Review of Neurobiology*, *118*, 13–39.
Rodd, Z. A., Bell, R. L., Kuc, K. A., Murphy, J. M., Lumeng, L., Li, T.-K., et al. (2003). Effects of repeated alcohol deprivations on operant ethanol self-administration by alcohol-preferring (P) rats. *Neuropsychopharmacology*, *28*, 1614–1621.
Rodd, Z. A., Bell, R. L., McKinzie, D. L., Webster, A. A., Murphy, J. M., Lumeng, L., et al. (2004a). Low-dose stimulatory effects of ethanol during adolescence in rat lines selectively-bred for high alcohol intake. *Alcoholism, Clinical and Experimental Research*, *28*, 535–543.

Rodd, Z. A., Bell, R. L., Oster, S. M., Toalston, J. E., Pommer, T. J., McBride, W. J., et al. (2010). Serotonin-3 receptors in the posterior ventral tegmental area regulate ethanol self-administration of alcohol-preferring (P) rats. *Alcohol, 44*, 245–255.

Rodd, Z. A., Bell, R. L., Sable, H. J., Murphy, J. M., & McBride, W. J. (2004b). Recent advances in animal models of alcohol craving and relapse. *Pharmacology, Biochemistry, and Behavior, 79*, 439–450.

Rodd, Z. A., Kimpel, M. W., Edenberg, H. J., Bell, R. L., Strother, W. N., McClintick, J. N., et al. (2008). Differential gene expression in the nucleus accumbens with ethanol self-administration in inbred alcohol-preferring rats. *Pharmacology, Biochemistry, and Behavior, 89*, 481–498.

Rodd, Z. A., McKinzie, D. L., Bell, R. L., McQueen, V. K., Murphy, J. M., Schoepp, D. D., et al. (2006). The metabotropic glutamate 2/3 receptor agonist LY404039 reduces alcohol-seeking but not alcohol self-administration in alcohol-preferring (P) rats. *Behavioural Brain Research, 171*, 207–215.

Rodd-Henricks, Z. A., Bell, R. L., Kuc, K. A., Murphy, J. M., McBride, W. J., Lumeng, L., et al. (2002a). Effects of ethanol exposure on subsequent acquisition and extinction of ethanol self-administration and expression of alcohol-seeking behavior in adult alcohol-preferring (P) rats: I. Periadolescent exposure. *Alcoholism, Clinical and Experimental Research, 26*, 1632–1641.

Rodd-Henricks, Z. A., Bell, R. L., Kuc, K. A., Murphy, J. M., McBride, W. J., Lumeng, L., et al. (2002b). Effects of ethanol exposure on subsequent acquisition and extinction of ethanol self-administration and expression of alcohol-seeking behavior in adult alcohol-preferring (P) rats: II. Adult exposure. *Alcoholism, Clinical and Experimental Research, 26*, 1642–1652.

Rodd-Henricks, Z. A., Bell, R. L., Murphy, J. M., McBride, W. J., Lumeng, L., & Li, T.-K. (2001). Effects of concurrent access to multiple ethanol concentrations and repeated deprivations on alcohol intake of alcohol-preferring (P) rats. *Alcoholism, Clinical and Experimental Research, 24*, 747–753.

Rodd-Henricks, Z. A., McKinzie, D. L., Edmundson, V. E., Dagon, C. L., Murphy, J. M., McBride, W. J., et al. (2000a). Effects of 5-HT3 receptor antagonists on daily alcohol intake under acquisition, maintenance, and relapse conditions in alcohol-preferring (P) rats. *Alcohol, 21*, 73–85.

Rodd-Henricks, Z. A., McKinzie, D. L., Shaikh, S. R., Murphy, J. M., McBride, W. J., Lumeng, L., et al. (2000b). The alcohol deprivation effect is prolonged in the alcohol preferring (P) rat following repeated deprivations. *Alcoholism, Clinical and Experimental Research, 24*, 8–16.

Rothstein, J. D., Van Kammen, M., Levey, A. I., Martin, L. J., & Kuncl, R. W. (1995). Selective loss of glial glutamate transporter GLT-1 in amyotrophic lateral sclerosis. *Annals of Neurology, 38*, 73–84.

Russell, R. N., McBride, W. J., Lumeng, L., Li, T.-K., & Murphy, J. M. (1996). Apomorphine and 7-OH DPAT reduce ethanol intake of P and HAD rats. *Alcohol, 13*, 515–519.

Saba, L. M., Flink, S. L., Vanderlinden, L. A., Israel, Y., Tampier, L., Colombo, G., et al. (2015). The sequenced rat brain transcriptome, its use in identifying networks predisposing alcohol consumption. *The FEBS Journal, 282*(18), 3556–3578. http://dx.doi.org/10.1111/febs.13358.

Sable, H. J. K., Bell, R. L., Rodd, Z. A., & McBride, W. J. (2006). Effects of naltrexone on the acquisition of alcohol intake in male and female periadolescent and adult alcohol-preferring (P) rats. *International Journal of Adolescent Medicine and Health, 18*, 139–149.

Saccone, S. F., Hinrichs, A. L., Saccone, N. L., Chase, G. A., Konvicka, K., Madden, P. A., et al. (2007). Cholinergic nicotinic receptor genes implicated in a nicotine dependence association study targeting 348 candidate genes with 3713 SNPs. *Human Molecular Genetics, 16*, 36–49.

Sahr, A. E., Thielen, R. J., Lumeng, L., Li, T.-K., & McBride, W. J. (2004). Long-lasting alterations of the mesolimbic dopamine system after periadolescent ethanol drinking by alcohol-preferring rats. *Alcoholism: Clinical and Experimental Research*, *28*, 702–711.

Sajdyk, T. J., Shekhar, A., & Gehlert, D. R. (2004). Interactions between NPY and CRF in the amygdala to regulate emotionality. *Neuropeptides*, *38*, 225–234.

Sajja, R. K., Dwivedi, C., & Rahman, S. (2010). Nicotinic ligands modulate ethanol-induced dopamine functions in mice. *Pharmacology*, *86*, 168–173.

Sajja, R. K., & Rahman, S. (2011). Lobeline and cytisine reduce voluntary ethanol drinking behavior in male C57BL/6J mice. *Progress in Neuropsychopharmacology & Biological Psychiatry*, *35*, 257–264.

Sajja, R. K., & Rahman, S. (2012). Neuronal nicotinic receptor ligands modulate chronic nicotine-induced ethanol consumption in C57BL/6J mice. *Pharmacology, Biochemistry, and Behavior*, *102*, 36–43.

Sajja, R. K., & Rahman, S. (2013). Nicotinic receptor partial agonists modulate alcohol deprivation effect in C57BL/6J mice. *Pharmacology, Biochemistry, and Behavior*, *110*, 161–167.

Samson, H. H., & Czachowski, C. L. (2003). Behavioral measures of alcohol self-administration and intake control: Rodent models. *International Reviews in Neurobiology*, *54*, 107–143.

Samson, H. H., Files, F. J., Denning, C., & Marvin, S. (1998). Comparison of alcohol-preferring and -nonpreferring selectively bred rat lines. I. Ethanol initiation and limited access operant self-administration. *Alcoholism, Clinical and Experimental Research*, *22*, 2133–2146.

Sari, Y. (2014). Role of glutamate transporter 1 in the attenuation of alcohol intake. *Frontiers in Neuroscience*, *8*, 200.

Sari, Y., Bell, R. L., & Zhou, F. C. (2006). Effects of chronic alcohol and repeated deprivations on dopamine D1 and D2 receptor levels in the extended amygdala of inbred alcohol-preferring rats. *Alcoholism: Clinical and Experimental Research*, *30*, 46–56.

Sari, Y., Sakai, M., Weedman, J. M., Rebec, G. V., & Bell, R. L. (2011). Ceftriaxone, a beta-lactam antibiotic, reduces ethanol consumption in alcohol-preferring rats. *Alcohol and Alcoholism*, *46*, 239–246. http://dx.doi.org/10.1093/alcalc/agr023.

Sari, Y., & Sreemantula, S. N. (2012). Neuroimmunophilin GPI-1046 reduces ethanol consumption in part through activation of GLT1 in alcohol-preferring rats. *Neuroscience*, *227*, 327–335.

Sari, Y., Sreemantula, S. N., Lee, M. R., & Choi, D. S. (2013). Ceftriaxone treatment affects the levels of GLT1 and ENT1 as well as ethanol intake in alcohol-preferring rats. *Journal of Molecular Neuroscience*, *51*, 779–787.

Sartor, C. E., Kranzler, H. R., & Gelernter, J. (2014). Rate of progression from first use to dependence on cocaine or opioids: A cross-substance examination of associated demographic, psychiatric, and childhood risk factors. *Addictive Behaviors*, *39*, 473–479. http://dx.doi.org/10.1016/j.addbeh.2013.10.021.

Sato, N., Kageyama, S., Chen, R., Suzuki, M., Mori, H., Tanioka, F., et al. (2010). Association between neuropeptide Y receptor 2 polymorphism and the smoking behavior of elderly Japanese. *Journal of Human Genetics*, *55*, 755–760.

Sauvage, M., & Steckler, T. (2001). Detection of corticotropin-releasing hormone receptor 1 immunoreactivity in cholinergic, dopaminergic and noradrenergic neurons of the murine basal forebrain and brainstem nuclei—Potential implication for arousal and attention. *Neuroscience*, *104*, 643–652.

Schlaepfer, I. R., Hoft, N. R., Collins, A. C., Corley, R. P., Hewitt, J. K., Hopfer, C. J., et al. (2008). The CHRNA5/A3/B4 gene cluster variability as an important determinant of early alcohol and tobacco initiation in young adults. *Biological Psychiatry*, *63*, 1039–1046.

Schmidt, L. G., Dufeu, P., Heinz, A., Kuhn, S., & Rommelspacher, H. (1997). Serotonergic dysfunction in addiction: Effects of alcohol, cigarette smoking and heroin on platelet 5-HT content. *Psychiatry Research, 72*, 177–185.

Schroeder, J. P., Overstreet, D. H., & Hodge, C. W. (2005a). The mGluR5 antagonist MPEP decreases operant ethanol self-administration during maintenance and after repeated alcohol deprivations in alcohol-preferring (P) rats. *Psychopharmacology, 179*, 262–270.

Schroeder, J. P., Overstreet, D. H., & Hodge, C. W. (2005b). The neuropeptide-Y Y5 receptor antagonist L152,804 decreases alcohol self-administration in inbred alcohol-preferring (iP) rats. *Alcohol, 36*, 179–186.

Schuckit, M. A. (1985). Ethanol-induced changes in body sway in men of high alcoholism risk. *Archives of General Psychiatry, 42*, 3375–3379.

Schuckit, M. A. (1986). Genetic aspects of alcoholism. *Annals of Emergency Medicine, 15*, 991–996.

Schuckit, M. A. (1994). Low level of response to alcohol as a predictor of future alcoholism. *American Journal of Psychiatry, 151*, 184–189.

Schuckit, M. A. (2009). An overview of genetic influences in alcoholism. *Journal of Substance Abuse Treatment, 36*, S5–S14.

Schuckit, M. A. (2014). A brief history of research on the genetics of alcohol and other drug use disorders. *Journal of Studies on Alcohol and Drugs. Supplement, 17*, 59–67.

Schuckit, M. A., & Gold, E. O. (1988). A simultaneous evaluation of multiple markers of ethanol/placebo challenges in sons of alcoholics and controls. *Archives of General Psychiatry, 45*, 211–216.

Seal, R. P., & Amara, S. G. (1999). Excitatory amino acid transporters: A family in flux. *Annual Review of Pharmacology and Toxicology, 39*, 431–456.

Self, D. W., & Staley, J. K. (2010). *Behavioral neuroscience of drug addiction*. New York, NY: Springer.

Shield, K. D., Rehm, J., Gmel, G., Rehm, M. X., & Allamani, A. (2013). Alcohol consumption, alcohol dependence, and related mortality in Italy in 2004: Effects of treatment-based interventions on alcohol dependence. *Substance Abuse Treatment, Prevention, and Policy, 8*, 21. http://dx.doi.org/10.1186/1747-597X-8-21.

Shin, J. H., Adrover, M. F., Wess, J., & Alvarez, V. A. (2015). Muscarinic regulation of dopamine and glutamate transmission in the nucleus accumbens. *Proceedings of the National Academy of Sciences of the United States of America, 112*, 8124–8129.

Silveri, M. M. (2012). Adolescent brain development and underage drinking in the United States: Identifying risks of alcohol use in college populations. *Harvard Review of Psychiatry, 20*, 189–200. http://dx.doi.org/10.3109/10673229.2012.714642.

Silveri, M. M. (2014). GABAergic contributions to alcohol responsivity during adolescence: Insights from preclinical and clinical studies. *Pharmacology & Therapeutics, 143*, 197–216. http://dx.doi.org/10.1016/j.pharmthera.2014.03.001.

Simpson, T. L., Saxon, A. J., Meredith, C. W., et al. (2009). A pilot trial of the alpha-1 adrenergic antagonist, prazosin, for alcohol dependence. *Alcoholism, Clinical and Experimental Research, 33*, 255–263.

Smialowska, M., Wieronska, J. M., & Wedzony, K. (2002). A search for colocalization of mglu1a receptors with CRF or NPY in the rat brain amygdala. *Folia Histochemica et Cytobiologica, 40*, 153–154.

Smith, T. L. (1997). Regulation of glutamate uptake in astrocytes continuously exposed to ethanol. *Life Sciences, 61*, 2499–2505.

Smith, J. E., Co, C., McIntosh, S., & Cunningham, C. C. (2008). Chronic binge-like moderate ethanol drinking in rats results in widespread decreases in serotonin, dopamine, and norepinephrine turnover rates reversed by ethanol intake. *Journal of Neurochemistry, 105*, 2134–2155.

Smith, A. D., & Weiss, F. (1999). Ethanol exposure differentially alters central monoamine neurotransmission in alcohol-preferring versus -nonpreferring rats. *Journal of Pharmacology and Experimental Therapeutics*, *288*, 1223–1228.

Smith, T. L., & Zsigo, A. (1996). Increased Na(+)-dependent high affinity uptake of glutamate in astrocytes chronically exposed to ethanol. *Neuroscience Letters*, *218*, 142–144.

Soderpalm, B., & Ericson, M. (2013). Neurocircuitry involved in the development of alcohol addiction: The dopamine system and its access points. *Current Topics in Behavioral Neurosciences*, *13*, 127–161.

Soderpalm, B., Ericson, M., Olausson, P., Blomqvist, O., & Engel, J. A. (2000). Nicotinic mechanisms involved in the dopamine activating and reinforcing properties of ethanol. *Behavioural Brain Research*, *113*, 85–96.

Sofuoglu, M., Rosenheck, R., & Petrakis, I. (2014). Pharmacological treatment of comorbid PTSD and substance use disorder: Recent progress. *Addictive Behaviors*, *39*, 428–433.

Sommer, W. H., & Spanagel, R. (2013). *Behavioral neurobiology of alcohol addiction*. New York, NY: Springer.

Sorbel, J., Morzorati, S., O'Connor, S., Li, T.-K., & Christian, J. C. (1996). Alcohol effects on the heritability of EEG spectral power. *Alcoholism, Clinical and Experimental Research*, *20*, 1523–1527.

Spanagel, R. (2009). Alcoholism: A systems approach from molecular physiology to addictive behavior. *Physiological Reviews*, *89*, 649–705.

Spanagel, R., Durstewitz, D., Hansson, A., Heinz, A., Kiefer, F., Kohr, G., et al. (2013). A systems medicine research approach for studying alcohol addiction. *Addiction Biology*, *18*, 883–896.

Spear, L. P. (2010). *The behavioral neuroscience of adolescence*. New York, NY: WW Norton.

Spear, L. P. (2014). Adolescents and alcohol: Acute sensitivities, enhanced intake, and later consequences. *Neurotoxicology and Teratology*, *41*, 51–59. http://dx.doi.org/10.1016/j.ntt.2013.11.006.

Spence, J. P., Liang, T., Habegger, K., & Carr, L. G. (2005). Effect of polymorphism on expression of the neuropeptide Y gene in inbred alcohol-preferring and -nonpreferring rats. *Neuroscience*, *131*, 871–876.

Spence, J. P., Liang, T., Liu, L., Johnson, P. L., Foroud, T., Carr, L. G., et al. (2009). From QTL to candidate gene: A genetic approach to alcoholism research. *Current Drug Abuse Reviews*, *2*, 127–134.

Stanley, S. A., Murphy, K. G., Bewick, G. A., Kong, W. M., Opacka-Juffry, J., Gardiner, J. V., et al. (2004). Regulation of rat pituitary cocaine- and amphetamine-regulated transcript (CART) by CRH and glucocorticoids. *American Journal of Physiology. Endocrinology and Metabolism*, *287*, E583–E590.

Steensland, P., Simms, J. A., Holgate, J., Richards, J. K., & Bartlett, S. E. (2007). Varenicline, an alpha4beta2 nicotinic acetylcholine receptor partial agonist, selectively decreases ethanol consumption and seeking. *Proceedings of the National Academy of Sciences of the United States of America*, *104*, 12518–12523.

Stewart, R. B., Gatto, G. J., Lumeng, L., Li, T.-K., & Murphy, J. M. (1993). Comparison of alcohol-preferring (P) and nonpreferring (NP) rats on tests of anxiety and for the anxiolytic effects of ethanol. *Alcohol*, *10*, 1–10.

Stewart, R. B., Kurtz, D. L., Zweifel, M., Li, T.-K., & Froehlich, J. C. (1992). Differences in the hypothermic response to ethanol in rats selectively bred for oral ethanol preference and nonpreference. *Psychopharmacology*, *106*, 169–174.

Stewart, R. B., McBride, W. J., Lumeng, L., Li, T.-K., & Murphy, J. M. (1991). Chronic alcohol consumption in alcohol-preferring P rats attenuates subsequent conditioned taste aversion produced by ethanol injections. *Psychopharmacology*, *105*, 530–534.

Stewart, R. B., Murphy, J. M., McBride, W. J., Lumeng, L., & Li, T.-K. (1996). Place conditioning with alcohol in alcohol-preferring and -nonpreferring rats. *Pharmacology Biochemistry Behavior, 53*, 487–491.

Stormark, K. M., Laberg, J. C., Nordby, H., & Hugdahl, K. (1998). Heart rate responses indicate locked-in attention in alcoholics immediately prior to drinking. *Addictive Behaviors, 23*, 251–255.

Strobel, C., Hunt, S., Sullivan, R., Sun, J., & Sah, P. (2014). Emotion regulation of pain: The role of noradrenaline in the amygdala. *Science China. Life Sciences, 57*, 384–390.

Strother, W. N., Chernet, E. J., Lumeng, L., Li, T.-K., & McBride, W. J. (2001). Regional central nervous system densities of delta-opioid receptors in alcohol-preferring P, alcohol-nonpreferring NP, and unselected Wistar rats. *Alcohol, 25*, 31–38.

Strother, W. N., Lumeng, L., Li, T.-K., & McBride, W. J. (2003). Regional CNS densities of serotonin1A and dopamine D2 receptors in peri-adolescent alcohol-preferring P and alcohol-nonpreferring NP rat pups. *Pharmacology, Biochemistry, and Behavior, 74*, 335–342.

Strother, W. N., Lumeng, L., Li, T.-K., & McBride, W. J. (2005). Dopamine and serotonin content in select brain regions of weanling and adult alcohol drinking rat lines. *Pharmacology, Biochemistry, and Behavior, 80*, 229–237.

Szumlinski, K. K., Diab, M. E., Friedman, R., Henze, L. M., Lominac, K. D., & Bowers, M. S. (2007). Accumbens neurochemical adaptations produced by binge-like alcohol consumption. *Psychopharmacology, 190*, 415–431.

Tabakoff, B., & Hoffman, P. L. (2013). The neurobiology of alcohol consumption and alcoholism: An integrative history. *Pharmacology, Biochemistry, and Behavior, 113*, 20–37.

Tabakoff, B., Saba, L., Printz, M., Flodman, P., Hodgkinson, C., Goldman, D., et al. (2009). Genetical genomic determinants of alcohol consumption in rats and humans. *BMC Biology*, 7, 70. http://dx.doi.org/10.1186/1741-7007-7-70.

Tanchuck, M. A., Yoneyama, N., Ford, M. M., Fretwell, A. M., & Finn, D. A. (2011). Assessment of GABA-B, metabotropic glutamate, and opioid receptor involvement in an animal model of binge drinking. *Alcohol, 45*, 33–44.

Terunuma, M., Revilla-Sanchez, R., Quadros, I. M., Deng, Q., Deeb, T. Z., Lumb, M., et al. (2014). Postsynaptic GABAB receptor activity regulates excitatory neuronal architecture and spatial memory. *Journal of Neuroscience, 34*, 804–816.

Thanos, P. K., Katana, J. K., Ashby, C. R., Jr., Michaelides, M., Gardner, E. L., Heidbreder, C. A., et al. (2005). The selective dopamine D3 receptor antagonist SB-277011-A attenuates ethanol consumption in ethanol preferring (P) and non-preferring (NP) rats. *Pharmacology Biochemistry and Behavior, 81*, 190–197.

Thiele, E., & Badia-Elder, N. E. (2003). A role for neuropeptide Y in alcohol intake control: Evidence from human and animal research. *Physiology & Behavior, 79*, 95–101.

Thielen, R. J., Engleman, E. A., Rodd, Z. A., Murphy, J. M., Lumeng, L., Li, T.-K., et al. (2004). Ethanol drinking and deprivation alter dopaminergic and serotonergic function in the nucleus accumbens of alcohol-preferring rats. *Journal of Pharmacology and Experimental Therapeutics, 309*, 216–225.

Thielen, R. J., McBride, W. J., Chernet, E., Lumeng, L., & Li, T.-K. (1997). Regional densities of benzodiazepine sites in the CNS of alcohol-naive P and NP rats. *Pharmacology, Biochemistry, and Behavior, 57*, 875–882.

Thoma, R., Mullins, P., Ruhl, D., Monnig, M., Yeo, R. A., Caprihan, A., et al. (2011). Perturbation of the glutamate-glutamine system in alcohol dependence and remission. *Neuropsychopharmacology, 36*, 1359–1365.

Thorsell, A. (2010). Brain neuropeptide Y and corticotropin-releasing hormone in mediating stress and anxiety. *Experimental Biology and Medicine, 235*, 1163–1167.

Tizabi, Y., Bai, L., Copeland, R. L., Jr., & Taylor, R. E. (2007). Combined effects of systemic alcohol and nicotine on dopamine release in the nucleus accumbens shell. *Alcohol and Alcoholism, 42*, 413–416.

Tizabi, Y., Getachew, B., Davila-Garcia, M., & Taylor, R. E. (2001). Alcohol preference: Association with reduced striatal nicotinic receptors. *Alcohol and Alcoholism, 36*, 318–322.

Toalston, J. E., Deehan, J. A., Jr., Hauser, S. R., Engleman, E. A., Bell, R. L., Murphy, J. M., et al. (2014). The reinforcing properties and neurochemical response of ethanol within the posterior ventral tegmental area are enhanced in adulthood by peri-adolescent ethanol consumption. *The Journal of Pharmacology and Experimental Therapeutics, 351*, 317–326. http://dx.doi.org/10.1124/jpet.114.218172. [PMCID: PMC4201272].

Toalston, J. E., Oster, S. M., Kuc, K. A., Ding, Z.-M., Pommer, T. J., Murphy, J. M., et al. (2008). Effects of alcohol and saccharin deprivations on concurrent ethanol and saccharin operant self-administration by alcohol-preferring (P) rats. *Alcohol, 42*, 277–284.

Traynelis, S. F., Wollmuth, L. P., McBain, C. J., Menniti, F. S., Vance, K. M., Ogden, K. K., et al. (2010). Glutamate receptor ion channels: Structure, regulation, and function. *Pharmacological Reviews, 62*, 405–496.

Trim, R. S., Simmons, A. N., Tolentino, N. J., Hall, S. A., Matthews, S. C., Robinson, S. K., et al. (2010). Acute ethanol effects on brain activation in low- and high-level responders to alcohol. *Alcoholism: Clinical and Experimental Research, 34*, 1162–1170.

Truitt, W. A., Hauser, S. R., Deehan, G. A., Jr., Toalston, J. E., Wilden, J. A., Bell, R. L., et al. (2015). Ethanol and nicotine interaction within the posterior ventral tegmental area in male and female alcohol-preferring rats: Evidence of synergy and differential gene activation in the nucleus accumbens shell. *Psychopharmacology, 232*, 639–649.

Van Bockstaele, E. J., Cestari, D. M., & Pickel, V. M. (1994). Synaptic structure and connectivity of serotonin terminals in the ventral tegmental area: Potential sites for modulation of mesolimbic dopamine neurons. *Brain Research, 647*, 307–322.

Van Pett, K., Viau, V., Bittencourt, J. C., Chan, R. K., Li, H. Y., Arias, C., et al. (2000). Distribution of mRNAs encoding CRF receptors in brain and pituitary of rat and mouse. *Journal of Comparative Neurology, 428*, 191–212.

Van Rizen, M. J., & Dishion, T. J. (2014). Adolescent deviant peer clustering as an amplifying mechanism underlying the progression from early substance use to late adolescent dependence. *Journal of Child Psychology & Psychiatry, 55*(10), 1153–1161. http://dx.doi.org/10.1111/jcpp.12211.

Varlinskaya, E. I., & Spear, L. P. (2009). Ethanol-induced social facilitation in adolescent rats: Role of endogenous activity at mu opioid receptors. *Alcoholism, Clinical and Experimental Research, 33*, 991–1000.

Varlinskaya, E. I., & Spear, L. P. (2010). Sensitization to social anxiolytic effects of ethanol in adolescent and adult Sprague-Dawley rats after repeated ethanol exposure. *Alcohol, 44*, 99–110.

Vengeliene, V., Bilbao, A., Molander, A., & Spanagel, R. (2008). Neuropharmacology of alcohol addiction. *British Journal of Pharmacology, 154*, 299–315.

Vengeliene, V., Bilbao, A., & Spanagel, R. (2014). The alcohol deprivation effect model for studying relapse behavior: A comparison between rats and mice. *Alcohol, 48*, 313–320.

Verplaetse, T. L., & Czachowski, C. L. (2015). Low-dose prazosin alone and in combination with propranolol or naltrexone: Effects on ethanol and sucrose seeking and self-administration in the P rat. *Psychopharmacology, 232*, 2647–2657.

Verplaetse, T. L., Rasmussen, D. D., Froehlich, J. C., & Czachowski, C. L. (2012). Effects of prazosin, an α1-adrenergic receptor antagonist, on the seeking and intake of alcohol and sucrose in alcohol-preferring (P) rats. *Alcoholism, Clinical and Experimental Research, 36*, 881–886.

Vetreno, R. P., Broadwater, M., Liu, W., Spear, L. P., & Crews, F. T. (2014). Adolescent, but not adult, binge ethanol exposure leads to persistent global reductions of choline acetyltransferase expressing neurons in brain. *PLoS One*, *9*, e113421.

Viken, R. J., Rose, R. J., Morzorati, S. L., Christian, J. C., & Li, T.-K. (2003). Subjective intoxication in response to alcohol challenge: Heritability and covariation with personality, breath alcohol level, and drinking history. *Alcoholism, Clinical and Experimental Research*, *27*, 795–803.

Vilaró, M. T., Palacios, J. M., & Mengod, G. (1990). Localization of m5 muscarinic receptor mRNA in rat brain examined by in situ hybridization histochemistry. *Neuroscience Letters*, *114*, 154–159.

Volkow, N. D., & Li, T.-K. (2005). Drugs and alcohol: Treating and preventing abuse, addiction and their medical consequences. *Pharmacology & Therapeutics*, *108*, 3–17.

Volkow, N. D., & Morales, M. (2015). The brain on drugs: From reward to addiction. *Cell*, *162*, 712–725. http://dx.doi.org/10.1016/j.cell.2015.07.046.

Walker, B. M., Rasmussen, D. D., Raskind, M. A., & Koob, G. F. (2008). Alpha1-noradrenergic receptor antagonism blocks dependence-induced increases in responding for ethanol. *Alcohol*, *42*, 91–97.

Waller, M. B., McBride, W. J., Gatto, G. J., Lumeng, L., & Li, T.-K. (1984). Intragastric self-administration of ethanol by ethanol-preferring and -nonpreferring lines of rats. *Science*, *225*, 78–80.

Waller, M. B., McBride, W. J., Lumeng, L., & Li, T.-K. (1982). Induction of dependence on ethanol by free-choice drinking in alcohol-preferring rats. *Pharmacology, Biochemistry, and Behavior*, *16*, 501–507.

Waller, M. B., Murphy, J. M., McBride, W. J., Lumeng, L., & Li, T.-K. (1986). Effect of low dose ethanol on spontaneous motor activity in alcohol-preferring and -nonpreferring lines of rats. *Pharmacology, Biochemistry, and Behavior*, *24*, 617–623.

Wand, G. (2005). The anxious amygdala: CREB signaling and predisposition to anxiety and alcoholism. *Journal of Clinical Investigation*, *115*, 2697–2699.

Wang, J. C., Grucza, R., Cruchaga, C., Hinrichs, A. L., Bertelsen, S., Budde, J. P., et al. (2009). Genetic variation in the CHRNA5 gene affects mRNA levels and is associated with risk for alcohol dependence. *Molecular Psychiatry*, *14*, 501–510.

Wang, J. C., Hinrichs, A. L., Stock, H., Budde, J., Allen, R., Bertelsen, S., et al. (2004). Evidence of common and specific genetic effects: Association of the muscarinic acetylcholine receptor M2 (CHRM2) gene with alcohol dependence and major depressive syndrome. *Human Molecular Genetics*, *13*, 1903–1911.

Wang, J. C., Kapoor, M., & Goate, A. M. (2012). The genetics of substance dependence. *Annual Review of Genomics and Human Genetics*, *13*, 241–261.

Wang, Y., & Qin, Z. H. (2010). Molecular and cellular mechanisms of excitotoxic neuronal death. *Apoptosis*, *15*, 1382–1402. http://dx.doi.org/10.1007/s10495-010-0481-0.

Wang, Z., et al. (2006). Dopaminergic control of corticostriatal long-term synaptic depression in medium spiny neurons is mediated by cholinergic interneurons. *Neuron*, *50*, 443–452.

Warburton, E. C., Barker, G. R. I., & Brown, M. W. (2013). Investigations into the involvement of NMDA mechanisms in recognition memory. *Neuropharmacology*, *74*, 41–47.

Ward, R. J., Colivicchi, M. A., Allen, R., Schol, F., Lallemand, F., de Witte, P., et al. (2009). Neuro-inflammation induced in the hippocampus of 'binge drinking' rats may be mediated by elevated extracellular glutamate content. *Journal of Neurochemistry*, *111*, 1119–1128.

Ward, R. J., Lallemand, F., & de Witte, P. (2014). Influence of adolescent heavy session drinking on the systemic and brain innate immune system. *Alcohol and Alcoholism*, *49*, 193–197.

Wechsler, H., Lee, J. E., Kuo, M., & Lee, H. (2000). College binge drinking in the 1990s: A continuing problem. Results of the Harvard School of Public Health 1999 College Alcohol Study. *Journal of American College Health*, *48*, 199–210.

Weiner, D. M., Levey, A. I., & Brann, M. R. (1990). Expression of muscarinic acetylcholine and dopamine receptor mRNAs in rat basal ganglia. *Proceedings of the National Academy of Sciences of the United States of America*, *87*, 7050–7054.

Weiss, F., Mitchiner, M., Bloom, F. E., & Koob, G. F. (1990). Free-choice ethanol responding for ethanol versus water in alcohol-preferring (P) and unselected Wistar rats is differentially modified by naloxone, bromocriptine, and methysergide. *Psychopharmacology*, *101*, 178–186.

Weitlauf, C., & Woodward, J. J. (2008). Ethanol selectively attenuates NMDAR-mediated synaptic transmission in the prefrontal cortex. *Alcoholism, Clinical and Experimental Research*, *32*, 690–698.

Wess, J. (2003). Novel insights into muscarinic acetylcholine receptor function using gene targeting technology. *Trends in Pharmacological Sciences*, *24*, 414–420.

Wess, J., Duttaroy, A., Zhang, W., Gomeza, J., Cui, Y., Miyakawa, T., et al. (2003). M1-M5 muscarinic receptor knockout mice as novel tools to study the physiological roles of the muscarinic cholinergic system. *Receptors & Channels*, *9*, 279–290.

Wetherill, L., Schuckit, M. A., Hesselbrock, V., Xuei, X., Liang, T., Dick, D. M., et al. (2008). Neuropeptide Y receptor genes are associated with alcohol dependence, alcohol withdrawal phenotypes, and cocaine dependence. *Alcoholism, Clinical and Experimental Research*, *32*, 2031–2040.

White, A. M., Kraus, C. L., & Swartzwelder, H. (2006). Many college freshmen drink at levels far beyond the binge threshold. *Alcoholism, Clinical and Experimental Research*, *30*, 1006–1010.

Whiteside, G. T., Pomonis, J. D., & Kennedy, J. D. (2013). An industry perspective on the role and utility of animal models of pain in drug discovery. *Neuroscience Letters*, *557*(Pt. A), 65–72. http://dx.doi.org/10.1016/j.neulet.2013.08.033.

Williams, R. J., Berry, L. J., & Beerstecher, E., Jr. (1949). Individual metabolic patterns, alcoholism, genotrophic diseases. *Proceedings of the National Academy of Sciences of the United States of America*, *35*, 265–271.

Williams, T. J., LaForge, K. S., Gordon, D., Bart, G., Kellog, S., Ott, J., et al. (2007). Prodynorphin gene promoter repeat associated with cocaine/alcohol codependence. *Addiction Biology*, *12*, 496–502.

Wimalawansa, S. J. (2014). Mechanisms of developing post-traumatic stress disorder: New targets for drug development and other potential interventions. *CNS & Neurological Disorders Drug Targets*, *13*, 807–816.

Windle, M., & Scheidt, D. M. (2004). Alcoholic subtypes: Are two sufficient? *Addiction*, *99*, 1508–1519.

Wise, R. A., & Bozarth, M. A. (1987). A psychomotor stimulant theory of addiction. *Psychological Review*, *94*, 469–492.

Witkin, J. M., Statnick, M. A., Rorick-Kehn, L. M., Pintar, J. E., Ansonoff, M., Chen, Y., et al. (2014). The biology of nociceptin/orphanin FQ (N/OFQ) related to obesity, stress, anxiety, mood, and drug dependence. *Pharmacology & Therapeutics*, *141*, 283–299.

Witt, E. D. (1994). Mechanisms of alcohol abuse and alcoholism in adolescents: A case for developing animal models. *Behavioral and Neural Biology*, *62*, 168–177.

Witt, E. D. (2006). Neurobiology. In M. Galanter (Ed.), *Alcohol problems in adolescents and young adults* (pp. 119–122). New York, NY: Springer.

Witt, E. D. (2010). Research on alcohol and adolescent brain development: Opportunities and future directions. *Alcohol*, *44*, 119–124.

Wolak, M. L., DeJoseph, M. R., Cator, A. D., Mokashi, A. S., Brownfield, M. S., & Urban, J. H. (2003). Comparative distribution of neuropeptide Y Y1 and Y5 receptors in the rat brain by using immunohistochemistry. *Journal of Comparative Neurology*, *464*, 285–311.

Wrase, J., Reimold, M., Puls, I., Kienast, T., & Heinz, A. (2006). Serotonergic dysfunction: Brain imaging and behavioral correlates. *Cognitive, Affective, & Behavioral Neuroscience*, *6*, 53–61.

Xuei, X., Dick, D., Flury-Wetherill, L., Tian, H. J., Agrawal, A., Bierut, L., et al. (2006). Association of the kappa-opioid system with alcohol dependence. *Molecular Psychiatry*, *11*, 1016–1024.

Yamada, M., Basile, A. S., Fedorova, I., Zhang, W., Duttaroy, A., Cui, Y., et al. (2003). Novel insights into M5 muscarinic acetylcholine receptor function by the use of gene targeting technology. *Life Sciences*, *74*, 345–353.

Yan, Z., & Surmeier, D. J. (1996). Muscarinic (m2/m4) receptors reduce N- and P-type Ca2 + currents in rat neostriatal cholinergic interneurons through a fast, membrane- delimited, G-protein pathway. *Journal of Neuroscience*, *16*, 2592–2604.

Yang, A. R. S. T., Yi, H. S., Warnock, K. T., Mamczarz, J., June, H. L., Jr., Mallick, N., et al. (2012). Effects of the triple monoamine uptake inhibitor DOV 102,677 on alcohol-motivated responding and antidepressant activity in alcohol-preferring (P) rats. *Alcoholism, Clinical and Experimental Research*, *36*, 863–873.

Yong, W., Spence, J. P., Eskay, R., Fitz, S. D., Damadzic, R., Jr., Lai, D., et al. (2014). Alcohol-preferring rats show decreased corticotropin-releasing hormone-2 receptor expression and differences in HPA activation compared to alcohol-nonpreferring rats. *Alcoholism: Clinical and Experimental Research*, *38*, 1275–1283.

Yoshimoto, K., & McBride, W. J. (1992). Regulation of nucleus accumbens dopamine release by the dorsal raphe nucleus in the rat. *Neurochemical Research*, *17*, 401–407.

Zeitzer, J. M. (2013). Control of sleep and wakefulness in health and disease. *Progress in Molecular Biology and Translational Science*, *119*, 137–154.

Zhang, W., et al. (2002). Characterization of central inhibitory muscarinic autoreceptors by the use of muscarinic acetylcholine receptor knock-out mice. *Journal of Neuroscience*, *22*, 1709–1717.

Zhou, F. C., Bledsoe, S., Lumeng, L., & Li, T.-K. (1991a). Immunostained serotonergic fibers are decreased in selected brain regions of alcohol-preferring rats. *Alcohol*, *8*, 425–431.

Zhou, F. C., Bledsoe, S., Lumeng, L., & Li, T.-K. (1991b). Serotonergic immunostained terminal fibers are lower in selected forebrain regions of alcohol-preferring rats. *Alcohol*, *8*, 1–7.

Zhou, Z., Karlsson, C., Liang, T., Xiong, W., Kimura, M., Tapocik, J. D., et al. (2013). Loss of metabotropic glutamate receptor 2 escalates alcohol consumption. *Proceedings of the National Academy of Sciences of the United States of America*, *110*, 16963–16968.

Zhou, F. C., McKinzie, D. L., Patel, T. D., Lumeng, L., & Li, T.-K. (1998). Additive reduction of alcohol drinking by 5-HT1A antagonist WAY 100635 and serotonin uptake blocker fluoxetine in alcohol-preferring P rats. *Alcoholism, Clinical and Experimental Research*, *22*, 266–269.

Zhou, F. C., Zhang, J. K., Lumeng, L., & Li, T.-K. (1995). Mesolimbic dopamine system in alcohol-preferring rats. *Alcohol*, *12*, 403–412.

Zlojutro, M., Manz, N., Rangaswamy, M., Xuei, X., Flury-Wetherill, L., Koller, D., et al. (2010). Genome-wide association study of theta band event-related oscillations identifies serotonin receptor gene HTR7 influencing risk of alcohol dependence. *American Journal of Medical Genetics. Part B, Neuropsychiatric Genetics*, *156*, 44–58.

Zucker, R. A. (1987). Alcohol and addictive behaviors. In P. C. Rivers (Ed.), *Nebraska symposium on motivation: Vol. 34* (pp. 27–84). Lincoln, NE: University of Nebraska Press.

Zucker, R. A. (1995). Pathways to alcohol problems and alcoholism: A developmental account of the evidence for multiple alcoholisms and for contextual contributions to risk. In R. A. Zucker, G. M. Boyd, & J. Howard (Eds.), *The development of alcohol problems: Exploring the biopsychosocial matrix of risk* (pp. 255–289). Rockville, MD: U.S. Dept. of Health and Human Services, National Institute on Alcohol Abuse and Alcoholism (NIAAA research monograph 26).

Zuo, L., Lu, L., Tan, Y., Pan, X., Cai, Y., Wang, X., et al. (2014). Genome-wide association discoveries of alcohol dependence. *American Journal of Addictions*, *23*, 526–539.

Zuo, L., Zhang, X. Y., Wang, F., Li, C. S., Lu, L., Ye, L., et al. (2013). Genome-wide significant association signals in IPO11-HTR1A region specific for alcohol and nicotine codependence. *Alcoholism, Clinical and Experimental Research*, *37*, 730–739.

CHAPTER EIGHT

Animal Models and the Development of Vaccines to Treat Substance Use Disorders

O. Ohia-Nwoko, T.A. Kosten, C.N. Haile[1]
University of Houston, Houston, TX, United States
Texas Institute for Measurement, Evaluation and Statistics (TIMES), University of Houston, Houston, TX, United States
[1]Corresponding author: e-mail address: cnhaile@central.uh.edu

Contents

Abstract

The development of pharmacotherapies for substance use disorders (SUDs) is a high priority in addiction research. At present, there are no approved pharmacotherapies for cocaine and methamphetamine use disorders, while treatments for nicotine and opioid use are moderately effective. Indeed, many of these treatments can cause adverse drug side effects and have poor medication compliance, which often results in increased drug relapse rates. An alternative to these traditional pharmacological interventions is immunotherapy or vaccines that can target substances associated with SUDs. In this chapter, we discuss the current knowledge on the efficacy of preclinical vaccines, particularly immunogens that target methamphetamine, cocaine, nicotine, or opioids to attenuate drug-induced behaviors in animal models of SUDs. We also review vaccines (and antibodies) against cocaine, nicotine, and methamphetamine that

International Review of Neurobiology, Volume 126
ISSN 0074-7742
http://dx.doi.org/10.1016/bs.irn.2016.02.009

have been assessed in human clinical trials. While preclinical studies indicate that several vaccines show promise, these findings have not necessarily translated to the clinical population. Thus, continued effort to design more effective vaccine immunogens using SUD animal models is necessary in order to support the use of immunotherapy as a viable option for individuals with SUDs.

1. INTRODUCTION

For the past several decades, the ongoing quest to develop safe and effective pharmacotherapeutics for substance use disorders (SUDs) has remained a high priority in addiction research. This is not surprising, as an estimated 22.7 million Americans aged 12 or older sought treatment for a SUD in 2013 (Center for Behavioral Health Statistics and Quality, 2015). While several FDA-approved pharmacotherapies for certain SUDs (eg, opiates, nicotine, and alcohol) have some therapeutic success, these treatments are often accompanied by several limitations including adverse drug side effects and poor medication compliance that is associated with high relapse rates (Kahan, Srivastava, & Conway, 2011; Kreek, Borg, Ducat, & Ray, 2010; Ling, Rawson, Shoptaw, & Ling, 2006; Veilleux, Colvin, Anderson, York, & Heinz, 2010). An alternative to traditional pharmacological intervention is immunotherapy or vaccines that target the particular substance. Vaccines have the potential to sustain long-term drug abstinence while possessing fewer of the aforementioned limitations of traditional medications (Haney & Kosten, 2004; Kosten & Domingo, 2013; Orson et al., 2007). Here, we review current knowledge on vaccines for SUDs, with the primary focus on the integral use of animal models for the development of vaccines for methamphetamine, cocaine, tobacco, and opiate use disorders. We also discuss evidence from clinical trials assessing immunotherapy as potential treatments for cocaine, methamphetamine, and tobacco use disorders (TUDs).

1.1 Neural Substrates of Drug Reinforcement

Substances associated with SUDs are generally small lipophilic molecules that rapidly cross the blood–brain barrier (BBB) and activate mesocorticolimbic circuitry that plays an important role in drug reinforcement (Fowler, Arends, & Kenny, 2008; Volkow, Wang, Fowler, Tomasi, & Telang, 2011). Methamphetamine (METH) and cocaine are both potent psychostimulants (Gonzalez Castro, Barrington, Walton, & Rawson, 2000; Haile, Kosten, & Kosten, 2007) that exert their actions at monoaminergic synapses (Rothman et al., 2001). METH elevates synaptic dopamine

(DA) and norepinephrine (NE) levels through a number of mechanisms including inducing DA release from the cytosol, and reversing the DA and vesicular monoamine transporter. In contrast, cocaine elevates synaptic neurotransmitter levels by blocking presynaptic reuptake through DA, NE, and serotonin (5-HT) transporters (Hall et al., 2004). Unlike METH and cocaine, nicotine elevates mesocorticolimbic DA levels through activation of nicotinic acetylcholine receptors on DAergic neurons, while opiates (eg, heroin and morphine) stimulate opioid receptors on GABAergic interneurons, disinhibiting DAergic cell bodies in the VTA (Koob & Le Moal, 2005, chap. 4). The rapidity with which these drugs enter the brain and activate their receptor targets correlates with their positive subjective effects in humans (Fowler et al., 2008; Volkow et al., 2000).

1.2 Innovative Therapeutic Strategy: Vaccines for SUDs

As shown in Fig. 1, the primary strategy for developing a therapeutic vaccine for SUDs is to induce high levels of drug-specific antibodies sequestering the

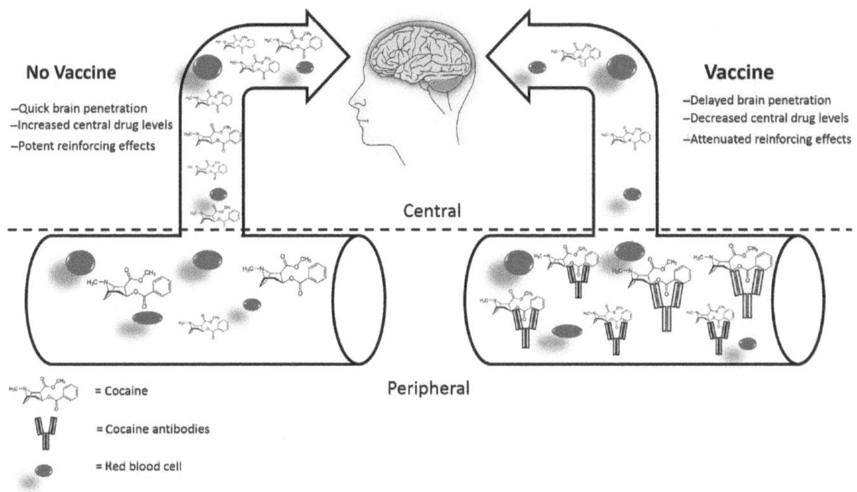

Fig. 1 Mechanism of action for a therapeutic vaccine. Hypothetical representation depicting mechanism of action using a cocaine vaccine as an example. Without vaccination, cocaine molecules present in the peripheral bloodstream rapidly penetrate the BBB, entering the central nervous system. This leads to increased levels of cocaine in the brain and subsequent activation of limbic circuitry responsible for the subjective and reinforcing effects of the drug. In contrast, anticocaine antibodies generated following vaccination delay drug penetration and reduce cocaine brain levels by sequestering the drug in the periphery. In animal models, these vaccination effects are generally observed as a reduction in locomotor hyperactivity, antinociception, conditioned place preference, and cocaine self-administration.

drug target in the periphery, thereby slowing penetration into the CNS. By reducing the levels of targeted drug that enters the CNS and activates mesocorticolimbic circuits, the drug's reinforcing capacity is diminished. However, the major challenge in stimulating the production of drug-specific antibodies is that these molecules are not inherently immunogenic. So the development of effective vaccines requires conjugating chemical derivatives of target drugs (drug-like haptens) to an immunogenic carrier protein via a chemical linker. The majority of preclinical vaccines developed for SUDs are conjugated with the carrier protein keyhole limpet hemocyanin (KLH). KLH is highly immunogenic, and most preclinical studies indicate that its incorporation induces efficient production of specific antibodies to cocaine (eg, Brimijoin et al., 2013; Ettinger, Ettinger, & Harless, 1997; Johnson & Ettinger, 2000; Kosten, Shen, Kinsey, Kosten, & Orson, 2014), METH (Miller et al., 2015, 2013; Ruedi-Bettschen et al., 2013; Shen et al., 2013), nicotine (de Villiers et al., 2002; Lindblom et al., 2002), and opioids (eg, Kosten, Shen, et al., 2013; Li et al., 2011; Stowe et al., 2011). Other drug-associated carrier proteins include tetanus toxoid (TT), diphtheria toxoid, and cholera toxin B-subunit. These carrier proteins have been integrated into vaccines against heroin (Anton & Leff, 2006; Matyas et al., 2013, 2014), nicotine (Cerny et al., 2002; McCluskie et al., 2013), and cocaine (Haney, Gunderson, Jiang, Collins, & Foltin, 2010; Martell et al., 2009).

In addition to hapten–carrier conjugation, vaccine immunogenicity is further enhanced with the addition of an adjuvant. A majority of clinically available vaccines are formulated with aluminum salt adjuvants (Baylor, Egan, & Richman, 2002), primarily because of their safety and efficacy. Accordingly, aluminum salts are widely used as adjuvants for SUD vaccine development and preclinical testing (Alving, Matyas, Torres, Jalah, & Beck, 2014). Oil emulsion adjuvants (water-in-oil and oil-in-water) have also been incorporated into a multitude of preclinical vaccines for cocaine, METH, nicotine, and opiates (Alving et al., 2014).

Unlike traditional SUD medications (eg, naloxone, methadone, and varenicline), vaccines do not prevent drug binding to specific neuronal receptors to antagonize drug effects. In contrast to this pharmacodynamic approach, antidrug antibodies are pharmacokinetic antagonists that reduce the concentration of drug available to multiple organ systems, most importantly, the brain. Additionally, since antidrug antibodies do not cross the BBB this limits off-target effects of the vaccine, likely decreasing potential harmful side effects (eg, Kosten, Shen, et al., 2014; Martell et al., 2009).

Vaccine development and production are also relatively less expensive than traditional pharmacotherapies (Andre, 2002). Because of these advantages, testing of candidate vaccines in animal models is a fundamental step toward developing safe and effective vaccinations for SUD patients.

2. ANIMAL MODELS AND PRECLINICAL DEVELOPMENT OF SUD VACCINES

Typically, preclinical testing of potential SUD pharmacotherapeutics involves rodent animal models. Rats or mice are either administered the target drug passively—eg, intraperitoneally (IP), subcutaneously (SC), or intravenously (IV)—or allowed to self-administer the drug via a surgically implanted intrajugular catheter. Preclinical SUD vaccines have been tested in assessments of drug-induced locomotor activation and conditioned place preference (CPP), as well as several aspects of operant self-administration: acquisition, maintenance, extinction, and reinstatement. Here, we review the preclinical studies that have investigated the effectiveness of vaccines against cocaine, METH, nicotine, and opioids.

2.1 Cocaine

There are no approved pharmacotherapies for cocaine use disorder (CUD). Fortunately, there has been significant progress toward developing an effective CUD vaccine; however, further research is needed (see below). Early animal studies investigating potential vaccines examined the efficacy of treatment on cocaine-induced behaviors (see Table 1). One of the first cocaine–KLH conjugate vaccines developed reduced cocaine-induced antinociception, but did not completely block these effects in rats (Bagasra et al., 1992). This was probably due to very low anticocaine antibody levels, as only half of the animals displayed concentrations above 10 μg/mL. Subsequent attempts were made to boost vaccine effectiveness by adding the cross-linker, *N*-hydroxysuccinimide-4-azidobenzoate (HSAB; Ettinger et al., 1997; Johnson & Ettinger, 2000) or by incorporating the cocaine-like hapten, GNC (Carrera et al., 1995). In rats, cocaine–HSAB–KLH successfully blocked cocaine antinociception (Ettinger et al., 1997) and cocaine discrimination (Johnson & Ettinger, 2000), underscoring the importance of immunogen design and preparation in vaccine development.

Although cocaine-conjugated vaccines displayed some effectiveness, concerns were raised about the stability and efficacy of the cocaine molecule as a hapten (Gallacher, 1994), since cocaine degrades spontaneously

Table 1 Vaccines for Cocaine Use Disorder

Vaccine	Animal Model (Sex)	Cocaine Administration	Results	References
Cocaine–KLH	Fisher rats (M)	25 mg/kg (IP)	↓ Hot plate analgesia	Bagasra, Forman, Howeedy, and Whittle (1992)
Cocaine–HSAB–KLH	Long Evans rats (F)	10 mg/kg (IP) 25 mg/kg (IP) 5 mg/kg (IP)	CPP ↓ Hot plate analgesia	Ettinger et al. (1997)
	Hooded rats (F)		↓ Drug discrimination	Johnson and Ettinger (2000)
GNC–KLH	Wistar rats (M)	15 mg/kg (IP)	↓ Locomotor activity	Carrera et al. (1995)
	Wistar rats (M)	0.25 mg/kg/infusion (IV)	↓ IVSA reinstatement	Carrera et al. (2000)
GND–KLH	Wistar rats (M)	15 mg/kg (IP)	↓ Locomotor activity	Carrera, Ashley, Wirsching, Koob, and Janda (2001)
GNE–KLH	Swiss Webster mice	5, 10 mg/kg (IP)	↓ Locomotor activity	Cai et al. (2013)
SNC–rCTB (TA-CD)	Wistar rats (M)	1 mg/kg/infusion (IV)	↓ IVSA	Kantak et al. (2000) and Kantak, Collins, Bond, and Fox (2001)
SNC–KLH	BALB/c mice (M, F)	20 mg/kg (IP)	↓ Locomotor activity (greater effect in F)	Kosten, Shen, et al. (2014)
dAd5GNC	BALB/c mice (F)	25, 50 μg/kg (IV)	↓ Locomotor activity	Hicks et al. (2011)
dAd5GNE	BALB/c mice (M)	15 mg/kg (IP)	↓ Locomotor activity	Koob et al. (2011)
dAd5GNE	Wistar rats (M)	15 mg/kg (IP)	↓ Locomotor activity	Wee et al. (2012)
		0.5 mg/kg/infusion	↓ IVSA reinstatement	

CPP, conditioned place preference; *dAd5−*, disrupted serotype 5 $E1^{-}E3^{-}$ adenovirus gene transfer vector; *GNC, GND,* and *GNE,* cocaine-like haptens; *HSAB*, *N*-hydroxysuccinimide-4-azidobenzoate; *IVSA,* intravenous self-administration; *KLH,* keyhole limpet hemocyanin; *SNC,* succinylnorcocaine.

(in vitro and in vivo: Garrett & Seyda, 1983; Stewart, Inaba, Tang, & Kalow, 1977) or by nonspecific esterases (in vivo: Matsubara, Kagawa, & Fukui, 1984) into its nonpsychoactive metabolite benzoylecgonine. This may have contributed to the low antibody titers reported in earlier studies. To address this problem, another vaccine incorporated GNC (a cocaine derivative) conjugated to KLH, which provided a structurally analogous, but more stable hapten. GNC–KLH reduced cocaine-induced hyperactivity and stereotyped behavior in rats compared to immunized KLH-only controls (Carrera et al., 1995). Additionally, appreciable levels of anticocaine antibody serum titers were observed as well as a 60–80% decrease in cocaine concentrations in the brain (Carrera et al., 1995). GNC–KLH was also effective at blocking drug-induced reinstatement of cocaine self-administration (Carrera et al., 2000). Other cocaine-like hapten-conjugated vaccines have been developed and tested that also blocked cocaine-induced locomotor activity in rodents: GND–KLH (Carrera et al., 2001) and GNE–KLH (Cai et al., 2013).

Another cocaine-like hapten that has shown promise is succinylnorcocaine (SNC). Using this hapten, a cocaine-conjugate vaccine composed of SNC and recombinant cholera toxin B [SNC–rCTB: TA-CD (Heading, 2002; Kantak et al., 2001, 2000)] has displayed considerable efficacy. Immunization with TA-CD resulted in significant levels of serum anticocaine antibodies, while also attenuating cocaine self-administration in rats (Kantak et al., 2001, 2000). Recently, our group demonstrated that a similar immunoconjugate, SNC–KLH, was also effective. We observed that SNC–KLH blocked cocaine-induced locomotor activity in female mice (vs female controls), while the vaccine effect was marginal in male mice (Kosten, Shen, et al., 2014). Our work was the first to demonstrate the impact of sex on cocaine vaccine effectiveness. While many preclinical SUD vaccination investigations are single-sex studies, our findings indicate the utility of testing both females and males. Indeed, there are sex differences in the course of cocaine use (Kosten, Gawin, Kosten, & Rounsaville, 1993) and females show greater craving responses to cocaine cues than males in a laboratory setting (Robbins, Ehrman, Childress, & O'Brien, 1999). These differences could impact responses to potential pharmacological treatments. In fact, sex differences have been observed in adaptive and innate immune responses (Klein, Jedlicka, & Pekosz, 2010), which could impact vaccine immunogenicity. Evidence of this comes from the observation that gonadal hormone levels alter immunogenic responses to herpes simplex virus vaccination in mice (Gillgrass, Tang, Towarnicki, Rosenthal, & Kaushic, 2005). In light of this evidence, incorporating both sexes should be a key consideration in future studies on SUD vaccine development.

In addition to the traditional SUD vaccine design (drug hapten–[linker]–protein carrier), adenovirus (Ad) gene transfer vectors have been incorporated into vaccines to improve immunogenic response (Chirmule et al., 1999; Harvey, Worgall, Ely, Leopold, & Crystal, 1999). For example, by developing a disrupted serotype 5 $E1^-E3^-$ Ad gene transfer vector (dAd5) conjugated to GNC (dAd5GNC), vaccination with dAd5GNC reduced brain cocaine levels in mice challenged with IV cocaine (Hicks et al., 2011). Additionally, vaccination with dAd5GNC, and a similar compound dAd5GNE, blocked cocaine-induced hyperactivity (Hicks et al., 2011; Koob et al., 2011; Wee et al., 2012) and cocaine self-administration in rats (Koob et al., 2011; Wee et al., 2012).

2.2 Methamphetamine

Like CUD, there are no FDA-approved medications for METH use disorder (MUD; Vocci & Appel, 2007). However, preclinical work incorporating animal models of MUD indicates that a METH vaccine is a promising therapeutic strategy (see Table 2). Thus far, the most efficacious MUD vaccines have incorporated METH and METH-like haptens conjugated to the carrier protein KLH. These vaccines produce significant METH antibody levels that display high affinities for METH (Byrnes-Blake, Carroll, Abraham, & Owens, 2001; Hambuchen et al., 2015; Miller et al., 2013; Moreno, Mayorov, & Janda, 2011; Ruedi-Bettschen et al., 2013). Additionally, KLH-conjugated METH vaccines have shown therapeutic efficacy by reducing the behavioral effects of METH in rodents. Administration of the MH6 (METH derivative)–KLH vaccine (Moreno et al., 2011) attenuated METH-induced hypothermia and locomotor activity in rats, an effect most likely resulting from pharmacokinetic antagonism (Miller et al., 2013). Studies from our lab corroborate these findings, as vaccination with succinyl METH (SMA)–KLH reduced METH-induced locomotor activity and CPP in mice (Shen et al., 2013). Additional studies with the same hapten but a different carrier protein (SMA-TT) also attenuated acquisition and reinstatement of METH-induced CPP and decreased brain levels of METH (Haile et al., 2015).

Duryee and colleagues (2009) developed several unique METH vaccines that incorporated a METH-like hapten, T-cell epitope from tetanus toxoid ($TT_{593-599}$), and peptide-based molecular adjuvant (EP54). Two of these vaccines achieved significant levels of anti-METH antibodies in rats: a METH monovalent vaccine (with one METH hapten) and a METH

Table 2 Vaccines for Methamphetamine Use Disorder

Vaccine	Animal Model (Sex)	Methamphetamine Administration	Results	References
METH–EP54: monovalent and divalent	Wistar rats (M)	0.05 mg/kg/ infusion (IV)	↑ IVSA acquisition	Duryee et al. (2009)
MH6–KLH	Sprague Dawley rats (M)	0.5, 5.6 mg/kg (SC)	↓ Locomotor activity	Miller et al. (2013)
		0.1, 0.05 mg/kg/ infusion (IV)	↓ IVSA acquisition	Miller et al. (2015)
IC_{KLH}-SMO9	Sprague Dawley rats (M)	3.0 mg/kg (SC)	↓ Impaired food responding	Ruedi-Bettschen et al. (2013)
SMA–KLH	BALB/c mice (F)	3.0 mg/kg (SC)	↓ Locomotor activity, CPP	Shen et al. (2013)
SMA–TT	BALB/c mice (F)	0.5, 2 mg/kg (SC)	↓ CPP acquisition and reinstatement	Haile et al. (2015)

CPP, conditioned place preference; *EP54,* YSFKPMPMLaR molecular adjuvant; *IVSA,* intravenous self-administration; *KLH,* keyhole limpet hemocyanin; *MH6,* methamphetamine-like hapten; *SMA,* succinyl methamphetamine; *TT,* tetanus toxoid.

divalent vaccine (Duryee et al., 2009). The monovalent vaccine was shown to have a greater stimulation index in a T-cell proliferation assay and was subsequently used to test its impact on IV self-administration of METH. Surprisingly, this vaccine increased acquisition of operant self-administration of METH in rats, which the authors concluded was a result of pharmacokinetic antagonism (Duryee et al., 2009). In some instances, rodents initially increase self-administration in order to compensate for the lack of drug effects due to the sequestration of METH in the peripheral circulation. This phenomenon has been observed with pharmacodynamic antagonists (Barrett, Miller, Dohrmann, & Caine, 2004; Weissenborn, Deroche, Koob, & Weiss, 1996) and also supports the notion that the primary mechanism of action for SUD vaccines is pharmacokinetic antagonism.

Another vaccine (IC_{KLH}-SMO9) prevented impairments in food responding resulting from a high dose of METH administration (3.0 mg/kg, SC) in rats (Ruedi-Bettschen et al., 2013). In a more recent study by

Miller and colleagues (2015), MH6–KLH attenuated acquisition of operant self-administration of METH in rats. Interestingly, results from this study prompted the authors to speculate the potential benefits of vaccines as prophylactic treatments (Miller et al., 2015), a topic that has garnered some skepticism from others (Bevins, Wilkinson, & Sanderson, 2008; Vocci & Chiang, 2001).

2.3 Nicotine

Pharmacotherapy for TUD is well established, with several first- and second-line medications available as moderately effective smoking cessation aids (see Cahill, Stevens, Perera, & Lancaster, 2013). First-line therapies, such as nicotine replacement therapy, bupropion, and varenicline, attenuate nicotine withdrawal while maintaining abstinence from smoking (Gonzalez et al., 2002; Gross & Stitzer, 1989; Hurt et al., 1997; Jorenby et al., 2006, 1999; Molander, Lunell, & Fagerstrom, 2000; Shiffman, 2008; Tonstad et al., 2006). Nicotine vaccines could offer an additional treatment option for TUD, as they have a limited side effect profile, unlike bupropion (Hughes, Stead, Hartmann-Boyce, Cahill, & Lancaster, 2014) and varenicline (Savage, Zekarias, & Caduff-Janosa, 2015). A summary of nicotine vaccines that have been developed and tested in SUD animal models is shown in Table 3.

Paul Pentel's group implemented several preliminary in vivo studies on nicotine vaccination in rodents. They developed two similar nicotine derivatives, 6-(carboxymethylureido)-(±)-nicotine (CMUNic) and

Table 3 Vaccines for Tobacco Use Disorder

Vaccine	Animal Model (Sex)	Nicotine Administration	Results	References
CMUNic-rEPA	Holtzman rats (M)	0.01 mg/kg/ infusion (IV)	↓ IVSA maintenance	LeSage et al. (2006)
AMNic–rEPA + Nic311	Holtzman rats (M)	0.3 mg/kg (SC)	↓ Locomotor activity	Roiko et al. (2008)
IP18-KLH	Wistar rats (M)	0.03 mg/kg/ infusion (IV)	↓ IVSA reinstatement	Lindblom et al. (2002)
trans-3′-Aminomethyl-nicotine	Sprague Dawley rats (M)	2.0 mg/kg (IP)	↓ Nicotine-induced seizures	Tuncok et al. (2001)

AMNic, 3-aminomethylnicotine; *CMUNic,* 6-(carboxymethylureido)-(±)-nicotine; *Nic311,* nicotine-specific antibody; *KLH,* keyhole limpet hemocyanin; *rEPA, Pseudomonas aeruginosa* exoprotein A.

3-aminomethylnicotine (AmNic), which have been utilized as vaccine haptens (Hieda et al., 1997; Pentel et al., 2000). In one study, following immunization with CMUNic–KLH, rats displayed decreased brain/plasma ratios of nicotine compared to KLH-treated controls (Hieda et al., 1997, 1999; Keyler, Hieda, St Peter, & Pentel, 1999). Later studies indicated that vaccination with a similar vaccine conjugated to *Pseudomonas aeruginosa* exoprotein A, CMUNic-rEPA, was sufficient to significantly reduce maintenance, but not acquisition of nicotine self-administration in rats (LeSage et al., 2006). Also, when rabbits were immunized with AmNic conjugated to rEPA (AmNic–rEPA), antiserum containing nicotine-IgG was collected, purified, and administered to rats. Passive immunization with nicotine-IgG blocked the effects of nicotine, resulting in a dose-dependent decrease in blood pressure and locomotor hyperactivity (Pentel et al., 2000). Combining AMNic–rEPA with a nicotine-specific antibody (Nic311) resulted in greater circulating antibodies and was more effective in reducing nicotine-induced locomotor sensitization in rats when administered together compared to either treatment administered alone (Roiko et al., 2008). Results suggest supplementing active immunization with concurrent passive immunization may be an effective strategy to increase nicotine-specific antibodies and improve potential therapeutic efficacy.

Torgny Svensson's group developed the IP18-KLH immunoconjugate, which consisted of the nicotine derivative, IP18 (de Villiers et al., 2002, 2004; Lindblom et al., 2002, 2005). Immunization with IP18–KLH prevented nicotine-induced reinstatement (Lindblom et al., 2002) of IV self-administration and attenuated withdrawal symptoms in rats (Lindblom et al., 2005). Immunization also resulted in significant plasma levels of antinicotine antibodies and decreased brain nicotine levels (Lindblom et al., 2002, 2005) an indication that treatment reduced central bioavailability of nicotine.

More recent work by Pentel and colleagues investigated the use of divalent or trivalent active vaccination, which involved concurrent administration of two or three different nicotine vaccines (Cornish, de Villiers, Pravetoni, & Pentel, 2013; de Villiers, Cornish, Troska, Pravetoni, & Pentel, 2013; Keyler, Roiko, Earley, Murtaugh, & Pentel, 2008). Rats that were vaccinated with CMUNic–KLH and AMNic–rEPA achieved serum antinicotine antibody levels that were roughly twice that of administration of either vaccine separately (Keyler et al., 2008). Similarly, when three vaccines were administered—CMUNic–KLH, AMNic–rEPA, and 1′-N-(2-mercaptoethyl) pentanamide-(−)-nicotine (SNic)–KLH (Pravetoni, Keyler, et al., 2012)—high levels of antinicotine antibodies were also achieved (de Villiers et al., 2013). Since each vaccine activates several unique B-cell populations, this alternative strategy provides a way to produce a polyclonal antibody response that cannot be achieved with traditional monovalent vaccination procedures. Perhaps this strategy could be employed in future clinical vaccines, which have been largely unsuccessful, partly due to low vaccine immunogenicity in humans (Cornuz et al., 2008; Hoogsteder, Kotz, van Spiegel, Viechtbauer, & van Schayck, 2014; Tonstad, Heggen, et al., 2013; Tonstad, Job, et al., 2013). Conceivably, this approach could also be used to target other drugs associated with SUDs.

2.4 Opioids

Pharmaceutical agents that have been approved for opioid use disorder (OUD) include substitution therapies (eg, methadone and buprenorphine) and opioid antagonists (eg, naloxone and naltrexone). Substitution therapies are effective at attenuating craving for heroin and illicit opioid use (Fareed et al., 2011; Lobmaier, Kunoe, Gossop, Katevoll, & Waal, 2010), while opioid antagonists can be used for detoxification and relapse prevention (Lobmaier et al., 2010). Both types of treatments are useful for maintaining abstinence; however, there are several disadvantages that may outweigh their benefits. For example, methadone and buprenorphine both have abuse potential (Fareed et al., 2011; Miller & Gold, 2007) and possess inherent overdose risk (Bell, Butler, Lawrance, Batey, & Salmelainen, 2009). Furthermore, concerns have been raised about the side effects of long-term naltrexone administration (Crowley, Wagner, Zerbe, & Macdonald, 1985; Malcolm, O'Neil, Von, & Dickerson, 1987), including dysregulation of endogenous opioids and hormones (Bronstein, Gutstein, & Akil, 1993; Kosten, Kreek, Ragunath, & Kleber, 1986). Given these disadvantages, immunotherapy for OUD could provide a way to reduce long-term opioid use and overdose with a minimal risk of side effects. Hence, there has been a wealth of preclinical research investigating OUD vaccines (see Table 4).

The concept of active immunization for OUD is not new. Over 40 years ago, Berkowitz and Spector (1972) reported the first evidence of active immunization to morphine in rodents. The morphine immunogen, 3-carboxymethylmorphine-BSA, selectively reduced morphine-induced analgesia in mice (Berkowitz & Spector, 1972). Around the same time Bonese and colleagues (1974) reported that immunization with morphine-6-hemisuccinyl-BSA blocked reinstatement of heroin vs cocaine self-administration in a Rhesus monkey (Bonese et al., 1974). Surprisingly,

Table 4 Vaccines for Opioid Use Disorders

Vaccine	Animal Model (Sex)	Opioid Administration	Results	References
3-Carboxymethylmorphine-BSA	Mice (M)	0.75 mg/kg (IP) morphine	↓ Hot plate analgesia	Berkowitz and Spector (1972)
Morphine-6-hemisuccinyl-BSA	Rhesus monkey (M)	6 μg/kg/infusion (IV) heroin	↓ IVSA	Bonese, Wainer, Fitch, Rothberg, and Schuster (1974)
Morphine-TT	Rats (M)	0.06 mg/kg/infusion (IV) heroin	↓ IVSA reinstatement	Anton and Leff (2006)
6-Glutaryl-morphine-KLH	Sprague Dawley rats (M)	10 mg/kg (SC) morphine	↓ Locomotor activity	Li et al. (2011)
		0.05 mg/kg/infusion (IV) heroin	↓ IVSA reinstatement	
Morphine-KLH	Holtzman rats (M)	1 mg/kg (SC) heroin, 2.25 mg/kg (SC) methadone, 2.25 mg/kg (SC) oxycodone	↓ Thermal antinociception	Raleigh, Pravetoni, Harris, Birnbaum, and Pentel (2013)
		0.25 mg/kg (SC) heroin	↓ Locomotor activity	
Morphine-KLH	Holtzman rats (M)	0.06 mg/kg/infusion (IV) heroin	↓ IVSA acquisition and reinstatement	Raleigh, Pentel, and LeSage (2014)
Her-KLH	Wistar rats (M)	1 mg/kg (SC) heroin	↓ Thermal and mechanical antinociception	Stowe et al. (2011)
		1 mg/kg/infusion (IV) heroin	↓ IVSA acquisition	

Continued

Table 4 Vaccines for Opioid Use Disorders—cont'd

Vaccine	Animal Model (Sex)	Opioid Administration	Results	References
Her-KLH	Wistar rats (M)	0.4, 1 mg/kg (SC) heroin	↓ Thermal and mechanical antinociception	Schlosburg et al. (2013)
		0.06 mg/kg/infusion (IV) heroin	↓ IVSA acquisition and reinstatement	
KLH-6-SM	Sprague Dawley rats (M)	1, 2 mg/kg (SC) morphine	↓ CPP	Kosten, Shen, et al. (2013)
			↓ Thermal antinociception	
DiAmHap-TT	BALB/c mice (F)	0.75 mg/kg (SC) heroin	↓ Thermal antinociception	Li et al. (2014) and Matyas et al. (2014)
6-Glutaryl-morphine-TFCS-KLH	Sprague Dawley rats (M)	10 mg/kg (SC) morphine	↓ Locomotor activity	Li et al. (2015)
		0.05 mg/kg/infusion (IV) heroin	↓ IVSA reinstatement	
OXY(Gly)4-KLH	Holtzman rats (M)	2.5 mg/kg (SC) oxycodone	↓ Thermal antinociception	Pravetoni, Keyler, et al. (2012)
OXY(Gly)4-KLH	Holtzman rats (M)	2.25 mg/kg (SC) oxycodone	↓ Thermal antinociception	Pravetoni, Raleigh, et al. (2012)
OXY(Gly)4-KLH	Holtzman rats (M)	0.06 mg/kg/infusion (IV) oxycodone	↓ IVSA acquisition	Pravetoni et al. (2014)

CPP, conditioned place preference; *IVSA,* intravenous self- administration; *KLH,* keyhole limpet hemocyanin.

despite promising preliminary evidence, no further progress was made on opioid vaccine development until about 10 years ago (see Anton & Leff, 2006).

Immunogen design for opioid vaccines has largely focused on haptens that recognize heroin, morphine, and their psychoactive metabolites. Within seconds of IV administration, heroin undergoes rapid enzymatic degradation into 6-acetylmorphine (6-AM) and subsequently morphine. Although these compounds are structurally analogous, they vary in their ability to cross the BBB. Heroin and 6-AM rapidly cross the BBB, while the less lipophilic morphine crosses more slowly (Oldendorf, Hyman, Braun, & Oldendorf, 1972; Umans & Inturrisi, 1981). Stowe and colleagues initially developed two heroin-derived vaccines, Her-KLH and Mor-KLH, and were able to show that Her-KLH was superior at blocking heroin-induced thermal antinociception in rats (Stowe et al., 2011). Furthermore, Her-KLH reduced the likelihood of heroin IV self-administration acquisition, while Mor-KLH did not (Stowe et al., 2011). Subsequent confirmatory studies showed that vaccination with Her-KLH blocked heroin-induced CPP and drug-induced reinstatement of heroin-seeking behavior in an operant procedure (Schlosburg et al., 2013). These effects were most likely due to higher heroin and 6-AM affinities observed with Her-KLH ($K_d \approx 4.19$ μM and $K_d \approx 0.035$ μM, respectively) vs Mor-KLH administration ($K_d \approx 14.18$ μM and $K_d > 100$ μM, respectively). Indeed, these observations also support the notion that 6-AM is an important mediator of heroin's central effects (Pravetoni, Raleigh, et al., 2012; Raleigh et al., 2014, 2013), as Her-KLH administration produced antibodies with exceptional affinity for 6-AM (Stowe et al., 2011) and selectively increased serum concentrations of heroin and 6-AM but not morphine (Schlosburg et al., 2013).

Others have also reported efficacious vaccines that incorporate morphine-like haptens. Anton and Leff (2006) were the first to demonstrate that a morphine-TT vaccine could suppress drug-induced reinstatement of heroin-seeking behavior in rats (Anton & Leff, 2006). Another morphine-like immunoconjugate, morphine-KLH, blunted opioid (heroin, methadone, and oxycodone) antinociception and heroin self-administration acquisition in rats (Raleigh et al., 2014, 2013). Furthermore, recent work from our laboratory demonstrated that vaccination of rats with KLH-6-succinyl morphine (KLH-6-SM) could produce high titers of antibodies that display exceptional affinities to morphine and 6-AM (Kosten, Shen, et al., 2013). These authors reported that morphine-induced antinociception (measured by tail flick and hot plate tests) and morphine CPP were also reduced. Presumably, this vaccine could also block the effects of heroin given its ability to produce high-affinity antibodies to 6-AM, thus supporting its use as a potential treatment for opioid dependence.

The newest category of opioid vaccines targets the prescription opioids, oxycodone and hydrocodone. Over the past 10 years, there has been a dramatic increase in illicit use of prescription opioids that has surpassed heroin (Center for Behavioral Health Statistics and Quality, 2015). Therefore, there has been some interest in developing targeted vaccines that can reduce the effects of prescription opioids. Much of this work has been implemented by Pravetoni and colleagues who developed OXY$(Gly)_4$-KLH and HYDROC$(Gly)_4$-KLH vaccines, which target oxycodone and hydrocodone, respectively. Evidence from their investigations is hopeful, as they demonstrated that OXY$(Gly)_4$-KLH can reduce brain/serum ratios of oxycodone (Pravetoni, Le Naour, et al., 2012; Pravetoni, Raleigh, et al., 2012) and hydrocodone (Pravetoni, Raleigh, et al., 2012), blunt oxycodone-induced analgesia (Pravetoni, Keyler, et al., 2012; Pravetoni, Raleigh, et al., 2012), and prevent acquisition of oxycodone self-administration in rats (Pravetoni et al., 2014). These preliminary data suggest that vaccination can reduce the reinforcing effects of oxycodone, supporting its possible role in treating OUDs.

3. TRANSLATION OF PRECLINICAL FINDINGS TO HUMANS

Evidence from preclinical studies clearly indicates that vaccines can successfully block the behavioral effects of various substances associated with SUDs. However, data from human clinical trials assessing vaccines as potential therapies for SUDs suggest marginal efficacy at best. In this section we review vaccines (and antibodies) that have been evaluated in human clinical trials.

3.1 Cocaine

One of the first clinical trials testing a therapeutic vaccine for a SUD assessed the safety and immunogenicity of SNC conjugated to rCTB (TA-CD) in a randomized, double-blind, placebo-controlled, Phase I clinical trial (Kosten et al., 2002). Three cohorts of participants received one of three doses of the vaccine (13, 82, and 709 μg) at baseline and at 2 months. Immunization resulted in elevated serum levels of anticocaine antibodies that showed specificity for cocaine over benzoylecgonine and lidocaine. Additionally, serum antibody levels were positively correlated with vaccine dose. The vaccine was well tolerated, with no serious adverse events noted. A subsequent open-label, Phase IIa, 14-week, dose-escalation follow-up study tested 400 μg ($N=10$) and 2000 μg ($N=8$) of TA-CD (Martell, Mitchell, Poling, Gonsai, & Kosten, 2005). The vaccine appeared safe, and individuals who received the high dose of vaccine were more likely to produce cocaine-free urines compared to participants who received the lower vaccine dose. Anecdotal reports from participants noted attenuation of cocaine's subjective effects upon relapse (Martell et al., 2005).

This same group assessed the TA-CD vaccine in a 24-week, Phase IIb, randomized, double-blind, placebo-controlled trial in individuals with cocaine and OUDs on methadone maintenance (Martell et al., 2009). Generation of IgG cocaine-specific antibodies following vaccination was highly variable, with only 38% of the participants producing antibodies above a target concentration (43 μg/mL). Between weeks 9 and 16, individuals with high antibody levels were more likely to produce cocaine-free urines compared to participants who generated low antibody levels and placebo treatment groups. As antibody levels declined (between 16 and 24 weeks), there was no significant difference between the groups regarding cocaine-free urines. Interestingly, a reanalysis of these data indicated that a positive vaccine response may be influenced by genotype (Kosten, Domingo, Hamon, & Nielsen, 2013) and the presence of anticocaine IgM antibodies prior to vaccination (Orson et al., 2013).

Another group assessed the TA-CD vaccine in a human laboratory study that was conducted to determine conclusively that anticocaine antibody levels could attenuate the subjective effects of self-administered cocaine (Haney et al., 2010). The impact of two doses (82 μg, $N=4$; 360 μg, $N=6$) of the vaccine administered over a 13-week period was determined on the subjective effects of smoked cocaine (25 and 50 mg; 2 days/week). Results indicated that peak (at week 13) plasma anticocaine antibody levels varied greatly between participants. Additionally, cocaine antibody levels predicted cocaine's subjective effects ratings (eg, "Good Drug Effect," "Cocaine Quality"). That is, higher cocaine antibody levels were associated with lower subjective ratings: individuals with high antibody levels rated the subjective effects of cocaine 55–81% lower compared to baseline (Haney et al., 2010).

In another 24-week, Phase III, randomized double-blind, placebo-controlled outpatient clinical trial, Kosten and colleagues investigated the potential efficacy of the TA-CD vaccine to reduce cocaine use in nontreatment-seeking individuals with CUD (Kosten, Domingo, et al., 2014). Groups received either TA-CD vaccine (400 μg, at weeks 1, 3, 5, 9, and 13, $N=152$) or placebo ($N=148$), and urine screens were assessed from 8 to 16 weeks. Consistent with previous studies, antibody levels significantly varied between participants. No significant differences in cocaine-free urines were observed between the vaccine and placebo groups. However, continuous abstinence from cocaine for at least 2 weeks occurred more often in the vaccine group compared to placebo (Kosten, Domingo, et al., 2014).

3.2 Methamphetamine

To date, there have been no clinical studies investigating the use of vaccines for MUD. However, Stevens, Henry, Owens, Schutz, and Gentry (2014) were the first to assess the safety and pharmacokinetics of an anti-METH monoclonal antibody (ch-mAb7f9) in healthy volunteers (Stevens et al., 2014). This Phase I, double-blind, randomized, placebo-controlled, ascending IV single-dose (0.2–20 mg/kg) study employed 42 healthy, non-METH users followed for 147 days after dosing. Results revealed no serious adverse reactions. Assessment of pharmacokinetic parameters indicated that the antibody half-life of the three highest doses (2, 6, and 20 mg/kg) was approximately 17–19 days. A volume of distribution of 5–6 L was observed, indicative of vascular sequestration of the antibody. Indeed, ch-mAb7f9 antibody concentrations remained above 50 μg/mL following 6 mg/kg dosing for 1–2 weeks and for approximately 5 weeks after the 20 mg/kg dose (Stevens et al., 2014). Somewhat concerning is that 12.5% of the participants developed human antichimeric antibodies which may potentially affect the efficacy of the antibodies upon multiple dosing. Nonetheless, this potential therapeutic METH antibody was found to be safe and tolerable for further research in individuals with MUD.

3.3 Nicotine

Tobacco use continues to be the leading cause of preventable death with quit rates despairingly low despite approved pharmacotherapies (Cahill et al., 2013). Accordingly, nicotine vaccines offer a novel method of potentially increasing the probability of smoking cessation. Similar to preclinical studies for cocaine and METH, early animal studies assessing a wide variety of nicotine vaccines appeared promising (de Villiers et al., 2002; Pentel et al., 2000), as did initial human clinical trials.

Two Phase I clinical trials have been conducted assessing the safety and immunogenicity of 3′-AmNic–rEPA (NicVax®) (Hatsukami et al., 2005; Wagena, de Vos, Horwith, & van Schayck, 2008). Overall, appreciable antinicotine antibody levels were achieved in both studies following vaccination, with no serious adverse events. Hatsukami and colleagues (2005) noted that 30-day abstinence rate was significantly greater in individuals vaccinated with the high dose of the vaccine (200 μg; Hatsukami et al., 2005). A randomized ($N=301$ smokers), double-blinded, placebo-controlled, multicenter, Phase II clinical trial assessed the efficacy of NicVax® (200 and 400 μg) to decrease smoking (Hatsukami et al., 2011). The highest vaccine dose produced greater antinicotine levels and was associated with significantly higher abstinence rates compared to placebo. These results supported the assessment of NicVax® in a Phase III clinical trial, which unfortunately did not demonstrate efficacy to decrease smoking (Fahim, Kessler, & Kalnik, 2013). Since the vaccine had showed potential in Phase II studies, a follow-up study was conducted to determine whether NicVax® (400 μg) combined with an approved medication for smoking cessation, varenicline (Jorenby et al., 2006), would increase probability of smoking cessation (Hoogsteder et al., 2014). In short, this randomized, placebo-controlled (NicVax®, $N=278$; placebo, $N=280$) study showed that coadministration of the vaccine in combination with varenicline did not significantly improve smoking cessation (Hoogsteder et al., 2014). The authors speculated that insufficient antibody response to the vaccine may have contributed to the lack of efficacy.

As noted, the underlying mechanism by which therapeutic vaccines act is by preventing the target drug from penetrating the CNS. To demonstrate proof of concept, a small displacement SPECT (single-photon emission tomography) study was conducted in smokers ($N=11$) vaccinated with NicVax® (400 μg, X4) (Esterlis et al., 2013). Results revealed a 12.5% reduction in nicotine binding following an IV nicotine infusion, which related to approximately 25% reduction in brain nicotine levels. In contrast, no effects of immunization or nicotine challenge were observed in a more recent study using the same vaccine and vaccination schedule (Havermans, Vuurman, van den Hurk, Hoogsteder, & van Schayck, 2014). Contrasting results between these studies may be due to the imaging method employed (SPECT vs regional blood oxygenated level-dependent [BOLD] responses).

In addition to NicVax®, two other nicotine vaccines (Niccine®, Nicotine-Qβ) have been assessed in Phase II clinical trials (Cornuz et al., 2008; Tonstad, Heggen, et al., 2013; Tonstad, Job, et al., 2013).

Overall, the vaccines appeared to be well tolerated, but did not significantly increase continuous abstinence rates. Again, lack of sufficiently high antibody levels likely played a significant role in these negative outcomes.

At present, evidence that nicotine vaccines enhance long-term smoking cessation is lacking (Hartmann-Boyce, Cahill, Hatsukami, & Cornuz, 2012). It appears that the primary factors that adversely affect vaccine therapeutic efficacy include: (1) ineffective generation of peripheral antibody levels and (2) broad variability in antibody production among participants. Thus, further development might focus on a vaccine that can engender persistent high levels of antibodies against the target drug with reduced interindividual variability.

4. CONCLUSIONS

It is clear that preclinical work toward vaccine development for SUDs has proven successful at designing agents that can mitigate drug-induced behaviors in animal models. Many of the vaccines outlined here reduced drug-induced locomotor activation, CPP, and IV self-administration, which are all recognized as modeling various facets of SUDs. While many of these vaccines demonstrate great efficacy preclinically, it is unfortunate that these findings have not translated well to the clinical population. Thus, efforts to improve vaccine efficacy in humans—particularly vaccine immunogenicity—are imperative. When designing and testing potential vaccines, investigators should consider the influence of sex on vaccine immunogenicity as indicated by our work on a potential cocaine vaccine (Kosten, Shen, et al., 2014). Moreover, pharmacogenetics may also influence vaccine effects. For example, individuals with a variant of the dopamine β-hydroxylase (DBH) gene (which lowers DBH levels) displayed better treatment outcomes to a cocaine vaccine than those with normal DBH levels (Kosten, Domingo, et al., 2013). Additionally, cocaine users that had high levels of IgM anticocaine antibodies before cocaine vaccination had reduced peak level IgG anticocaine antibodies generated after vaccination (Orson et al., 2013). Taken together, these studies reveal how sex, pharmacogenetics, and prior immune responses to a SUD-associated substance could influence vaccine immunogenicity, stressing the need for rigorous testing of these factors when designing future SUD vaccines. With this in mind, immunotherapy is poised to be a viable option for individuals seeking treatment for a SUD.

ACKNOWLEDGMENTS

This work was supported in part by National Institutes of Health (NIH) Grants U01-AA013476 and DP1DA033502. The views expressed in this chapter do not represent those of NIH and solely those of the authors.

The authors declare no conflicts of interest.

REFERENCES

Alving, C. R., Matyas, G. R., Torres, O., Jalah, R., & Beck, Z. (2014). Adjuvants for vaccines to drugs of abuse and addiction. *Vaccine*, *32*(42), 5382–5389. http://dx.doi.org/10.1016/j.vaccine.2014.07.085.

Andre, F. E. (2002). How the research-based industry approaches vaccine development and establishes priorities. *Developmental Biology (Basel)*, *110*, 25–29.

Anton, B., & Leff, P. (2006). A novel bivalent morphine/heroin vaccine that prevents relapse to heroin addiction in rodents. *Vaccine*, *24*(16), 3232–3240. http://dx.doi.org/10.1016/j.vaccine.2006.01.047.

Bagasra, O., Forman, L. J., Howeedy, A., & Whittle, P. (1992). A potential vaccine for cocaine abuse prophylaxis. *Immunopharmacology*, *23*(3), 173–179.

Barrett, A. C., Miller, J. R., Dohrmann, J. M., & Caine, S. B. (2004). Effects of dopamine indirect agonists and selective D1-like and D2-like agonists and antagonists on cocaine self-administration and food maintained responding in rats. *Neuropharmacology*, *47*(Suppl. 1), 256–273. http://dx.doi.org/10.1016/j.neuropharm.2004.07.007.

Baylor, N. W., Egan, W., & Richman, P. (2002). Aluminum salts in vaccines—US perspective. *Vaccine*, *20*(Suppl. 3), S18–S23.

Bell, J. R., Butler, B., Lawrance, A., Batey, R., & Salmelainen, P. (2009). Comparing overdose mortality associated with methadone and buprenorphine treatment. *Drug and Alcohol Dependence*, *104*(1–2), 73–77. http://dx.doi.org/10.1016/j.drugalcdep.2009.03.020.

Berkowitz, B., & Spector, S. (1972). Evidence for active immunity to morphine in mice. *Science*, *178*(4067), 1290–1292.

Bevins, R. A., Wilkinson, J. L., & Sanderson, S. D. (2008). Vaccines to combat smoking. *Expert Opinion on Biological Therapy*, *8*(4), 379–383. http://dx.doi.org/10.1517/14712598.8.4.379.

Bonese, K. F., Wainer, B. H., Fitch, F. W., Rothberg, R. M., & Schuster, C. R. (1974). Changes in heroin self-administration by a rhesus monkey after morphine immunisation. *Nature*, *252*(5485), 708–710.

Brimijoin, S., Orson, F., Kosten, T. R., Kinsey, B., Shen, X. Y., White, S. J., & Gao, Y. (2013). Anti-cocaine antibody and butyrylcholinesterase-derived cocaine hydrolase exert cooperative effects on cocaine pharmacokinetics and cocaine-induced locomotor activity in mice. *Chemico-Biological Interactions*, *203*(1), 212–216. http://dx.doi.org/10.1016/j.cbi.2012.08.015.

Bronstein, D. M., Gutstein, H. B., & Akil, H. (1993). Effects of chronic morphine treatment on beta-endorphin-related peptides in the caudal medulla and spinal cord. *Journal of Neurochemistry*, *60*(6), 2304–2307.

Byrnes-Blake, K. A., Carroll, F. I., Abraham, P., & Owens, S. M. (2001). Generation of anti-(+)methamphetamine antibodies is not impeded by (+)methamphetamine administration during active immunization of rats. *International Immunopharmacology*, *1*(2), 329–338.

Cahill, K., Stevens, S., Perera, R., & Lancaster, T. (2013). Pharmacological interventions for smoking cessation: An overview and network meta-analysis. *Cochrane Database of Systematic Reviews*, *5*, CD009329. http://dx.doi.org/10.1002/14651858.CD009329.pub2.

Cai, X., Whitfield, T., Moreno, A. Y., Grant, Y., Hixon, M. S., Koob, G. F., & Janda, K. D. (2013). Probing the effects of hapten stability on cocaine vaccine immunogenicity. *Molecular Pharmaceutics*, *10*(11), 4176–4184. http://dx.doi.org/10.1021/mp400214w.

Carrera, M. R., Ashley, J. A., Parsons, L. H., Wirsching, P., Koob, G. F., & Janda, K. D. (1995). Suppression of psychoactive effects of cocaine by active immunization. *Nature*, *378*(6558), 727–730.

Carrera, M. R., Ashley, J. A., Wirsching, P., Koob, G. F., & Janda, K. D. (2001). A second-generation vaccine protects against the psychoactive effects of cocaine. *Proceedings of the National Academy of Sciences of the United States of America*, *98*(4), 1988–1992.

Carrera, M. R., Ashley, J. A., Zhou, B., Wirsching, P., Koob, G. F., & Janda, K. D. (2000). Cocaine vaccines: Antibody protection against relapse in a rat model. *Proceedings of the National Academy of Sciences of the United States of America*, *97*(11), 6202–6206.

Center for Behavioral Health Statistics and Quality (2015). *Behavioral health trends in the United States: Results from the 2014 National Survey on Drug use and Health*. (HHS Publication No. SMA 15-4927, NSDUH series H-50). Retrieved from http://www.samhsa.gov/data/.

Cerny, E. H., Levy, R., Mauel, J., Mpandi, M., Mutter, M., Henzelin-Nkubana, C., … Cerny, T. (2002). Preclinical development of a vaccine 'against smoking'. *Onkologie*, *25*(5), 406–411.

Chirmule, N., Propert, K., Magosin, S., Qian, Y., Qian, R., & Wilson, J. (1999). Immune responses to adenovirus and adeno-associated virus in humans. *Gene Therapy*, *6*(9), 1574–1583.

Cornish, K. E., de Villiers, S. H., Pravetoni, M., & Pentel, P. R. (2013). Immunogenicity of individual vaccine components in a bivalent nicotine vaccine differ according to vaccine formulation and administration conditions. *PLoS One*, *8*(12), e82557. http://dx.doi.org/10.1371/journal.pone.0082557.

Cornuz, J., Zwahlen, S., Jungi, W. F., Osterwalder, J., Klingler, K., van Melle, G., … Cerny, T. (2008). A vaccine against nicotine for smoking cessation: A randomized controlled trial. *PLoS One*, *3*(6), e2547. http://dx.doi.org/10.1371/journal.pone.0002547.

Crowley, T. J., Wagner, J. E., Zerbe, G., & Macdonald, M. (1985). Naltrexone-induced dysphoria in former opioid addicts. *The American Journal of Psychiatry*, *142*(9), 1081–1084.

de Villiers, S. H., Cornish, K. E., Troska, A. J., Pravetoni, M., & Pentel, P. R. (2013). Increased efficacy of a trivalent nicotine vaccine compared to a dose-matched monovalent vaccine when formulated with alum. *Vaccine*, *31*(52), 6185–6193. http://dx.doi.org/10.1016/j.vaccine.2013.10.051.

de Villiers, S. H., Lindblom, N., Kalayanov, G., Gordon, S., Johansson, A. M., & Svensson, T. H. (2004). Active immunization against nicotine alters the distribution of nicotine but not the metabolism to cotinine in the rat. *Naunyn-Schmiedeberg's Archives of Pharmacology*, *370*(4), 299–304. http://dx.doi.org/10.1007/s00210-004-0960-3.

de Villiers, S. H., Lindblom, N., Kalayanov, G., Gordon, S., Malmerfelt, A., Johansson, A. M., & Svensson, T. H. (2002). Active immunization against nicotine suppresses nicotine-induced dopamine release in the rat nucleus accumbens shell. *Respiration*, *69*(3), 247–253.

Duryee, M. J., Bevins, R. A., Reichel, C. M., Murray, J. E., Dong, Y., Thiele, G. M., & Sanderson, S. D. (2009). Immune responses to methamphetamine by active immunization with peptide-based, molecular adjuvant-containing vaccines. *Vaccine*, *27*(22), 2981–2988. http://dx.doi.org/10.1016/j.vaccine.2009.02.105.

Esterlis, I., Hannestad, J. O., Perkins, E., Bois, F., D'Souza, D. C., Tyndale, R. F., … O'Malley, S. S. (2013). Effect of a nicotine vaccine on nicotine binding to beta2*-nicotinic acetylcholine receptors in vivo in human tobacco smokers. *The American Journal of Psychiatry*, *170*(4), 399–407. http://dx.doi.org/10.1176/appi.ajp.2012.12060793.

Ettinger, R. H., Ettinger, W. F., & Harless, W. E. (1997). Active immunization with cocaine-protein conjugate attenuates cocaine effects. *Pharmacology, Biochemistry, and Behavior*, *58*(1), 215–220.

Fahim, R. E., Kessler, P. D., & Kalnik, M. W. (2013). Therapeutic vaccines against tobacco addiction. *Expert Review of Vaccines*, *12*(3), 333–342. http://dx.doi.org/10.1586/erv.13.13.

Fareed, A., Vayalapalli, S., Stout, S., Casarella, J., Drexler, K., & Bailey, S. P. (2011). Effect of methadone maintenance treatment on heroin craving, a literature review. *Journal of Addictive Diseases*, *30*(1), 27–38. http://dx.doi.org/10.1080/10550887.2010.531672.

Fowler, C. D., Arends, M. A., & Kenny, P. J. (2008). Subtypes of nicotinic acetylcholine receptors in nicotine reward, dependence, and withdrawal: Evidence from genetically modified mice. *Behavioural Pharmacology*, *19*(5–6), 461–484. http://dx.doi.org/10.1097/FBP.0b013e32830c360e.

Gallacher, G. (1994). A potential vaccine for cocaine abuse prophylaxis? *Immunopharmacology*, *27*(1), 79–84.

Garrett, E. R., & Seyda, K. (1983). Prediction of stability in pharmaceutical preparations XX: Stability evaluation and bioanalysis of cocaine and benzoylecgonine by high-performance liquid chromatography. *Journal of Pharmaceutical Sciences*, *72*(3), 258–271.

Gillgrass, A. E., Tang, V. A., Towarnicki, K. M., Rosenthal, K. L., & Kaushic, C. (2005). Protection against genital herpes infection in mice immunized under different hormonal conditions correlates with induction of vagina-associated lymphoid tissue. *Journal of Virology*, *79*(5), 3117–3126. http://dx.doi.org/10.1128/JVI.79.5.3117-3126.2005.

Gonzalez Castro, F., Barrington, E. H., Walton, M. A., & Rawson, R. A. (2000). Cocaine and methamphetamine: Differential addiction rates. *Psychology of Addictive Behaviors*, *14*(4), 390–396.

Gonzalez, S., Cascio, M. G., Fernandez-Ruiz, J., Fezza, F., Di Marzo, V., & Ramos, J. A. (2002). Changes in endocannabinoid contents in the brain of rats chronically exposed to nicotine, ethanol or cocaine. *Brain Research*, *954*(1), 73–81.

Gross, J., & Stitzer, M. L. (1989). Nicotine replacement: Ten-week effects on tobacco withdrawal symptoms. *Psychopharmacology*, *98*(3), 334–341.

Haile, C. N., Kosten, T. R., & Kosten, T. A. (2007). Genetics of dopamine and its contribution to cocaine addiction. *Behavior Genetics*, *37*(1), 119–145. http://dx.doi.org/10.1007/s10519-006-9115-2.

Haile, C. N., Kosten, T. A., Shen, X. Y., O'Malley, P. W., Winoske, K. J., Kinsey, B. M., … Orson, F. M. (2015). Altered methamphetamine place conditionng in mice vaccinated with a succinyl-methamphetamine-tetanus-toxoid vaccine. *The American Journal on Addictions*, *24*(8), 748–755. http://dx.doi.org/10.1111/ajad.12307.

Hall, F. S., Sora, I., Drgonova, J., Li, X. F., Goeb, M., & Uhl, G. R. (2004). Molecular mechanisms underlying the rewarding effects of cocaine. *Annals of the New York Academy of Sciences*, *1025*, 47–56. http://dx.doi.org/10.1196/annals.1316.006.

Hambuchen, M. D., Carroll, F. I., Ruedi-Bettschen, D., Hendrickson, H. P., Hennings, L. J., Blough, B. E., … Owens, S. M. (2015). Combining active immunization with monoclonal antibody therapy to facilitate early initiation of a long-acting anti-methamphetamine antibody response. *Journal of Medicinal Chemistry*, *58*(11), 4665–4677. http://dx.doi.org/10.1021/acs.jmedchem.5b00220.

Haney, M., Gunderson, E. W., Jiang, H., Collins, E. D., & Foltin, R. W. (2010). Cocaine-specific antibodies blunt the subjective effects of smoked cocaine in humans. *Biological Psychiatry*, *67*(1), 59–65. http://dx.doi.org/10.1016/j.biopsych.2009.08.031.

Haney, M., & Kosten, T. R. (2004). Therapeutic vaccines for substance dependence. *Expert Review of Vaccines*, *3*(1), 11–18. http://dx.doi.org/10.1586/14760584.3.1.11.

Hartmann-Boyce, J., Cahill, K., Hatsukami, D., & Cornuz, J. (2012). Nicotine vaccines for smoking cessation. *Cochrane Database of Systematic Reviews*, *8*, CD007072. http://dx.doi.org/10.1002/14651858.CD007072.pub2.

Harvey, B. G., Worgall, S., Ely, S., Leopold, P. L., & Crystal, R. G. (1999). Cellular immune responses of healthy individuals to intradermal administration of an $E1^-E3^-$ adenovirus gene transfer vector. *Human Gene Therapy*, *10*(17), 2823–2837.

Hatsukami, D. K., Jorenby, D. E., Gonzales, D., Rigotti, N. A., Glover, E. D., Oncken, C. A., … Rennard, S. I. (2011). Immunogenicity and smoking-cessation outcomes for a novel nicotine immunotherapeutic. *Clinical Pharmacology and Therapeutics*, *89*(3), 392–399. http://dx.doi.org/10.1038/clpt.2010.317.

Hatsukami, D. K., Rennard, S., Jorenby, D., Fiore, M., Koopmeiners, J., de Vos, A., … Pentel, P. R. (2005). Safety and immunogenicity of a nicotine conjugate vaccine in current smokers. *Clinical Pharmacology and Therapeutics*, *78*(5), 456–467. http://dx.doi.org/10.1016/j.clpt.2005.08.007.

Havermans, A., Vuurman, E. F., van den Hurk, J., Hoogsteder, P., & van Schayck, O. C. (2014). Treatment with a nicotine vaccine does not lead to changes in brain activity during smoking cue exposure or a working memory task. *Addiction*, *109*(8), 1260–1267. http://dx.doi.org/10.1111/add.12577.

Heading, C. E. (2002). TA-CD. Xenova. *IDrugs*, *5*(11), 1070–1074.

Hicks, M. J., De, B. P., Rosenberg, J. B., Davidson, J. T., Moreno, A. Y., Janda, K. D., … Crystal, R. G. (2011). Cocaine analog coupled to disrupted adenovirus: A vaccine strategy to evoke high-titer immunity against addictive drugs. *Molecular Therapy*, *19*(3), 612–619. http://dx.doi.org/10.1038/mt.2010.280.

Hieda, Y., Keyler, D. E., Vandevoort, J. T., Kane, J. K., Ross, C. A., Raphael, D. E., … Pentel, P. R. (1997). Active immunization alters the plasma nicotine concentration in rats. *The Journal of Pharmacology and Experimental Therapeutics*, *283*(3), 1076–1081.

Hieda, Y., Keyler, D. E., VanDeVoort, J. T., Niedbala, R. S., Raphael, D. E., Ross, C. A., & Pentel, P. R. (1999). Immunization of rats reduces nicotine distribution to brain. *Psychopharmacology*, *143*(2), 150–157.

Hoogsteder, P. H., Kotz, D., van Spiegel, P. I., Viechtbauer, W., & van Schayck, O. C. (2014). Efficacy of the nicotine vaccine 3′-AmNic-rEPA (NicVAX) co-administered with varenicline and counselling for smoking cessation: A randomized placebo-controlled trial. *Addiction*, *109*(8), 1252–1259. http://dx.doi.org/10.1111/add.12573.

Hughes, J. R., Stead, L. F., Hartmann-Boyce, J., Cahill, K., & Lancaster, T. (2014). Antidepressants for smoking cessation. *Cochrane Database of Systematic Reviews*, *1*, CD000031. http://dx.doi.org/10.1002/14651858.CD000031.pub4.

Hurt, R. D., Sachs, D. P., Glover, E. D., Offord, K. P., Johnston, J. A., Dale, L. C., … Sullivan, P. M. (1997). A comparison of sustained-release bupropion and placebo for smoking cessation. *The New England Journal of Medicine*, *337*(17), 1195–1202.

Johnson, M. W., & Ettinger, R. H. (2000). Active cocaine immunization attenuates the discriminative properties of cocaine. *Experimental and Clinical Psychopharmacology*, *8*(2), 163–167.

Jorenby, D. E., Hays, J. T., Rigotti, N. A., Azoulay, S., Watsky, E. J., Williams, K. E., … Varenicline Phase 3 Study Group. (2006). Efficacy of varenicline, an alpha4beta2 nicotinic acetylcholine receptor partial agonist, vs placebo or sustained-release bupropion for smoking cessation: A randomized controlled trial. *JAMA*, *296*(1), 56–63. http://dx.doi.org/10.1001/jama.296.1.56.

Jorenby, D. E., Leischow, S. J., Nides, M. A., Rennard, S. I., Johnston, J. A., Hughes, A. R., … Baker, T. B. (1999). A controlled trial of sustained-release bupropion, a nicotine patch, or both for smoking cessation. *The New England Journal of Medicine*, *340*(9), 685–691.

Kahan, M., Srivastava, A., & Conway, B. (2011). Is there a need for heroin substitution treatment in Vancouver's Downtown Eastside? *Canadian Journal of Public Health*, *102*(2), 84–86.

Kantak, K. M., Collins, S. L., Bond, J., & Fox, B. S. (2001). Time course of changes in cocaine self-administration behavior in rats during immunization with the cocaine vaccine IPC-1010. *Psychopharmacology*, *153*(3), 334–340.

Kantak, K. M., Collins, S. L., Lipman, E. G., Bond, J., Giovanoni, K., & Fox, B. S. (2000). Evaluation of anti-cocaine antibodies and a cocaine vaccine in a rat self-administration model. *Psychopharmacology*, *148*(3), 251–262.

Keyler, D. E., Hieda, Y., St Peter, J., & Pentel, P. R. (1999). Altered disposition of repeated nicotine doses in rats immunized against nicotine. *Nicotine & Tobacco Research*, *1*(3), 241–249.

Keyler, D. E., Roiko, S. A., Earley, C. A., Murtaugh, M. P., & Pentel, P. R. (2008). Enhanced immunogenicity of a bivalent nicotine vaccine. *International Immunopharmacology*, *8*(11), 1589–1594. http://dx.doi.org/10.1016/j.intimp.2008.07.001.

Klein, S. L., Jedlicka, A., & Pekosz, A. (2010). The Xs and Y of immune responses to viral vaccines. *The Lancet Infectious Diseases*, *10*(5), 338–349. http://dx.doi.org/10.1016/S1473-3099(10)70049-9.

Koob, G., Hicks, M. J., Wee, S., Rosenberg, J. B., De, B. P., Kaminsky, S. M., … Crystal, R. G. (2011). Anti-cocaine vaccine based on coupling a cocaine analog to a disrupted adenovirus. *CNS & Neurological Disorders Drug Targets*, *10*(8), 899–904. http://dx.doi.org/10.2174/187152711799219334.

Koob, G. F., & Le Moal, M. (2005). *Neurobiology of addiction* (1st ed.). London: Academic Press.

Kosten, T. R., & Domingo, C. B. (2013). Can you vaccinate against substance abuse? *Expert Opinion on Biological Therapy*, *13*(8), 1093–1097. http://dx.doi.org/10.1517/14712598.2013.791278.

Kosten, T. R., Domingo, C. B., Hamon, S. C., & Nielsen, D. A. (2013). DBH gene as predictor of response in a cocaine vaccine clinical trial. *Neuroscience Letters*, *541*, 29–33. http://dx.doi.org/10.1016/j.neulet.2013.02.037.

Kosten, T. R., Domingo, C. B., Shorter, D., Orson, F., Green, C., Somoza, E., … Kampman, K. (2014). Vaccine for cocaine dependence: A randomized double-blind placebo-controlled efficacy trial. *Drug and Alcohol Dependence*, *140*, 42–47. http://dx.doi.org/10.1016/j.drugalcdep.2014.04.003.

Kosten, T. A., Gawin, F. H., Kosten, T. R., & Rounsaville, B. J. (1993). Gender differences in cocaine use and treatment response. *Journal of Substance Abuse Treatment*, *10*(1), 63–66.

Kosten, T. R., Kreek, M. J., Ragunath, J., & Kleber, H. D. (1986). A preliminary study of beta endorphin during chronic naltrexone maintenance treatment in ex-opiate addicts. *Life Sciences*, *39*(1), 55–59.

Kosten, T. R., Rosen, M., Bond, J., Settles, M., Roberts, J. S., Shields, J., … Fox, B. (2002). Human therapeutic cocaine vaccine: Safety and immunogenicity. *Vaccine*, *20*(7–8), 1196–1204.

Kosten, T. A., Shen, X. Y., Kinsey, B. M., Kosten, T. R., & Orson, F. M. (2014). Attenuation of cocaine-induced locomotor activity in male and female mice by active immunization. *The American Journal on Addictions*, *23*(6), 604–607. http://dx.doi.org/10.1111/j.1521-0391.2014.12152.x.

Kosten, T. A., Shen, X. Y., O'Malley, P. W., Kinsey, B. M., Lykissa, E. D., Orson, F. M., & Kosten, T. R. (2013). A morphine conjugate vaccine attenuates the behavioral effects of morphine in rats. *Progress in Neuro-Psychopharmacology & Biological Psychiatry*, *45*, 223–229. http://dx.doi.org/10.1016/j.pnpbp.2013.05.012.

Kreek, M. J., Borg, L., Ducat, E., & Ray, B. (2010). Pharmacotherapy in the treatment of addiction: Methadone. *Journal of Addictive Diseases*, *29*(2), 200–216. http://dx.doi.org/10.1080/10550881003684798.

LeSage, M. G., Keyler, D. E., Hieda, Y., Collins, G., Burroughs, D., Le, C., & Pentel, P. R. (2006). Effects of a nicotine conjugate vaccine on the acquisition and maintenance of nicotine self-administration in rats. *Psychopharmacology*, *184*(3–4), 409–416. http://dx.doi.org/10.1007/s00213-005-0027-2.

Li, F., Cheng, K., Antoline, J. F., Iyer, M. R., Matyas, G. R., Torres, O. B., ... Rice, K. C. (2014). Synthesis and immunological effects of heroin vaccines. *Organic and Biomolecular Chemistry*, *12*(37), 7211–7232. http://dx.doi.org/10.1039/c4ob01053a.

Li, Q. Q., Luo, Y. X., Sun, C. Y., Xue, Y. X., Zhu, W. L., Shi, H. S., ... Lu, L. (2011). A morphine/heroin vaccine with new hapten design attenuates behavioral effects in rats. *Journal of Neurochemistry*, *119*(6), 1271–1281. http://dx.doi.org/10.1111/j.1471-4159.2011.07502.x.

Li, Q. Q., Sun, C. Y., Luo, Y. X., Xue, Y. X., Meng, S. Q., Xu, L. Z., ... Sun, H. Q. (2015). A conjugate vaccine attenuates morphine- and heroin-induced behavior in rats. *The International Journal of Neuropsychopharmacology*, *18*(5). http://dx.doi.org/10.1093/ijnp/pyu093.

Lindblom, N., de Villiers, S. H., Kalayanov, G., Gordon, S., Johansson, A. M., & Svensson, T. H. (2002). Active immunization against nicotine prevents reinstatement of nicotine-seeking behavior in rats. *Respiration*, *69*(3), 254–260.

Lindblom, N., de Villiers, S. H., Semenova, S., Kalayanov, G., Gordon, S., Schilstrom, B., ... Svensson, T. H. (2005). Active immunisation against nicotine blocks the reward facilitating effects of nicotine and partially prevents nicotine withdrawal in the rat as measured by dopamine output in the nucleus accumbens, brain reward thresholds and somatic signs. *Naunyn-Schmiedeberg's Archives of Pharmacology*, *372*(3), 182–194. http://dx.doi.org/10.1007/s00210-005-0019-0.

Ling, W., Rawson, R., Shoptaw, S., & Ling, W. (2006). Management of methamphetamine abuse and dependence. *Current Psychiatry Reports*, *8*(5), 345–354. http://dx.doi.org/10.1007/s11920-006-0035-x.

Lobmaier, P. P., Kunoe, N., Gossop, M., Katevoll, T., & Waal, H. (2010). Naltrexone implants compared to methadone: Outcomes six months after prison release. *European Addiction Research*, *16*(3), 139–145. http://dx.doi.org/10.1159/000313336.

Malcolm, R., O'Neil, P. M., Von, J. M., & Dickerson, P. C. (1987). Naltrexone and dysphoria: A double-blind placebo controlled trial. *Biological Psychiatry*, *22*(6), 710–716.

Martell, B. A., Mitchell, E., Poling, J., Gonsai, K., & Kosten, T. R. (2005). Vaccine pharmacotherapy for the treatment of cocaine dependence. *Biological Psychiatry*, *58*(2), 158–164. http://dx.doi.org/10.1016/j.biopsych.2005.04.032.

Martell, B. A., Orson, F. M., Poling, J., Mitchell, E., Rossen, R. D., Gardner, T., & Kosten, T. R. (2009). Cocaine vaccine for the treatment of cocaine dependence in methadone-maintained patients: A randomized, double-blind, placebo-controlled efficacy trial. *Archives of General Psychiatry*, *66*(10), 1116–1123. http://dx.doi.org/10.1001/archgenpsychiatry.2009.128.

Matsubara, K., Kagawa, M., & Fukui, Y. (1984). In vivo and in vitro studies on cocaine metabolism: Ecgonine methyl ester as a major metabolite of cocaine. *Forensic Science International*, *26*(3), 169–180.

Matyas, G. R., Mayorov, A. V., Rice, K. C., Jacobson, A. E., Cheng, K., Iyer, M. R., ... Alving, C. R. (2013). Liposomes containing monophosphoryl lipid A: A potent adjuvant system for inducing antibodies to heroin hapten analogs. *Vaccine*, *31*(26), 2804–2810. http://dx.doi.org/10.1016/j.vaccine.2013.04.027.

Matyas, G. R., Rice, K. C., Cheng, K., Li, F., Antoline, J. F., Iyer, M. R., ... Alving, C. R. (2014). Facial recognition of heroin vaccine opiates: Type 1 cross-reactivities of antibodies induced by hydrolytically stable haptenic surrogates of heroin,

6-acetylmorphine, and morphine. *Vaccine*, *32*(13), 1473–1479. http://dx.doi.org/10.1016/j.vaccine.2014.01.028.

McCluskie, M. J., Pryde, D. C., Gervais, D. P., Stead, D. R., Zhang, N., Benoit, M., … Davis, H. L. (2013). Enhancing immunogenicity of a 3′aminomethylnicotine-DT-conjugate anti-nicotine vaccine with CpG adjuvant in mice and non-human primates. *International Immunopharmacology*, *16*(1), 50–56. http://dx.doi.org/10.1016/j.intimp.2013.03.021.

Miller, M. L., Aarde, S. M., Moreno, A. Y., Creehan, K. M., Janda, K. D., & Taffe, M. A. (2015). Effects of active anti-methamphetamine vaccination on intravenous self-administration in rats. *Drug and Alcohol Dependence*, *153*, 29–36. http://dx.doi.org/10.1016/j.drugalcdep.2015.06.014.

Miller, N. S., & Gold, M. S. (2007). Opiate prescription medication dependence and pain perceptions. *Journal of Addictive Diseases*, *26*(Suppl. 1), 65–71. http://dx.doi.org/10.1300/J069v26S01_07.

Miller, M. L., Moreno, A. Y., Aarde, S. M., Creehan, K. M., Vandewater, S. A., Vaillancourt, B. D., … Taffe, M. A. (2013). A methamphetamine vaccine attenuates methamphetamine-induced disruptions in thermoregulation and activity in rats. *Biological Psychiatry*, *73*(8), 721–728. http://dx.doi.org/10.1016/j.biopsych.2012.09.010.

Molander, L., Lunell, E., & Fagerstrom, K. O. (2000). Reduction of tobacco withdrawal symptoms with a sublingual nicotine tablet: A placebo controlled study. *Nicotine & Tobacco Research*, *2*(2), 187–191.

Moreno, A. Y., Mayorov, A. V., & Janda, K. D. (2011). Impact of distinct chemical structures for the development of a methamphetamine vaccine. *Journal of the American Chemical Society*, *133*(17), 6587–6595. http://dx.doi.org/10.1021/ja108807j.

Oldendorf, W. H., Hyman, S., Braun, L., & Oldendorf, S. Z. (1972). Blood–brain barrier: Penetration of morphine, codeine, heroin, and methadone after carotid injection. *Science*, *178*(4064), 984–986.

Orson, F. M., Kinsey, B. M., Singh, R. A., Wu, Y., Gardner, T., & Kosten, T. R. (2007). The future of vaccines in the management of addictive disorders. *Current Psychiatry Reports*, *9*(5), 381–387. http://dx.doi.org/10.1007/s11920-007-0049-z.

Orson, F. M., Rossen, R. D., Shen, X., Lopez, A. Y., Wu, Y., & Kosten, T. R. (2013). Spontaneous development of IgM anti-cocaine antibodies in habitual cocaine users: Effect on IgG antibody responses to a cocaine cholera toxin B conjugate vaccine. *The American Journal on Addictions*, *22*(2), 169–174. http://dx.doi.org/10.1111/j.1521-0391.2013.00314.x.

Pentel, P. R., Malin, D. H., Ennifar, S., Hieda, Y., Keyler, D. E., Lake, J. R., … Fattom, A. (2000). A nicotine conjugate vaccine reduces nicotine distribution to brain and attenuates its behavioral and cardiovascular effects in rats. *Pharmacology, Biochemistry, and Behavior*, *65*(1), 191–198.

Pravetoni, M., Keyler, D. E., Pidaparthi, R. R., Carroll, F. I., Runyon, S. P., Murtaugh, M. P., … Pentel, P. R. (2012). Structurally distinct nicotine immunogens elicit antibodies with non-overlapping specificities. *Biochemical Pharmacology*, *83*(4), 543–550. http://dx.doi.org/10.1016/j.bcp.2011.11.004.

Pravetoni, M., Le Naour, M., Harmon, T. M., Tucker, A. M., Portoghese, P. S., & Pentel, P. R. (2012). An oxycodone conjugate vaccine elicits drug-specific antibodies that reduce oxycodone distribution to brain and hot-plate analgesia. *The Journal of Pharmacology and Experimental Therapeutics*, *341*(1), 225–232. http://dx.doi.org/10.1124/jpet.111.189506.

Pravetoni, M., Pentel, P. R., Potter, D. N., Chartoff, E. H., Tally, L., & LeSage, M. G. (2014). Effects of an oxycodone conjugate vaccine on oxycodone self-administration and oxycodone-induced brain gene expression in rats. *PLoS One*, *9*(7), e101807. http://dx.doi.org/10.1371/journal.pone.0101807.

Pravetoni, M., Raleigh, M. D., Le Naour, M., Tucker, A. M., Harmon, T. M., Jones, J. M., ... Pentel, P. R. (2012). Co-administration of morphine and oxycodone vaccines reduces the distribution of 6-monoacetylmorphine and oxycodone to brain in rats. *Vaccine*, *30*(31), 4617–4624. http://dx.doi.org/10.1016/j.vaccine.2012.04.

Raleigh, M. D., Pentel, P. R., & LeSage, M. G. (2014). Pharmacokinetic correlates of the effects of a heroin vaccine on heroin self-administration in rats. *PLoS One*, *9*(12), e115696. http://dx.doi.org/10.1371/journal.pone.0115696.

Raleigh, M. D., Pravetoni, M., Harris, A. C., Birnbaum, A. K., & Pentel, P. R. (2013). Selective effects of a morphine conjugate vaccine on heroin and metabolite distribution and heroin-induced behaviors in rats. *The Journal of Pharmacology and Experimental Therapeutics*, *344*(2), 397–406. http://dx.doi.org/10.1124/jpet.112.201194.

Robbins, S. J., Ehrman, R. N., Childress, A. R., & O'Brien, C. P. (1999). Comparing levels of cocaine cue reactivity in male and female outpatients. *Drug and Alcohol Dependence*, *53*(3), 223–230.

Roiko, S. A., Harris, A. C., Keyler, D. E., Lesage, M. G., Zhang, Y., & Pentel, P. R. (2008). Combined active and passive immunization enhances the efficacy of immunotherapy against nicotine in rats. *The Journal of Pharmacology and Experimental Therapeutics*, *325*(3), 985–993. http://dx.doi.org/10.1124/jpet.107.135111.

Rothman, R. B., Baumann, M. H., Dersch, C. M., Romero, D. V., Rice, K. C., Carroll, F. I., & Partilla, J. S. (2001). Amphetamine-type central nervous system stimulants release norepinephrine more potently than they release dopamine and serotonin. *Synapse*, *39*(1), 32–41.

Ruedi-Bettschen, D., Wood, S. L., Gunnell, M. G., West, C. M., Pidaparthi, R. R., Carroll, F. I., ... Owens, S. M. (2013). Vaccination protects rats from methamphetamine-induced impairment of behavioral responding for food. *Vaccine*, *31*(41), 4596–4602. http://dx.doi.org/10.1016/j.vaccine.2013.07.038.

Savage, R. L., Zekarias, A., & Caduff-Janosa, P. (2015). Varenicline and abnormal sleep related events. *Sleep*, *38*(5), 833–837. http://dx.doi.org/10.5665/sleep.4686.

Schlosburg, J. E., Vendruscolo, L. F., Bremer, P. T., Lockner, J. W., Wade, C. L., Nunes, A. A., ... Koob, G. F. (2013). Dynamic vaccine blocks relapse to compulsive intake of heroin. *Proceedings of the National Academy of Sciences of the United States of America*, *110*(22), 9036–9041. http://dx.doi.org/10.1073/pnas.1219159110.

Shen, X. Y., Kosten, T. A., Lopez, A. Y., Kinsey, B. M., Kosten, T. R., & Orson, F. M. (2013). A vaccine against methamphetamine attenuates its behavioral effects in mice. *Drug and Alcohol Dependence*, *129*(1–2), 41–48. http://dx.doi.org/10.1016/j.drugalcdep.2012.09.007.

Shiffman, S. (2008). Effect of nicotine lozenges on affective smoking withdrawal symptoms: Secondary analysis of a randomized, double-blind, placebo-controlled clinical trial. *Clinical Therapeutics*, *30*(8), 1461–1475. http://dx.doi.org/10.1016/j.clinthera.2008.07.019.

Stevens, M. W., Henry, R. L., Owens, S. M., Schutz, R., & Gentry, W. B. (2014). First human study of a chimeric anti-methamphetamine monoclonal antibody in healthy volunteers. *MAbs*, *6*(6), 1649–1656. http://dx.doi.org/10.4161/19420862.2014.976431.

Stewart, D. J., Inaba, T., Tang, B. K., & Kalow, W. (1977). Hydrolysis of cocaine in human plasma by cholinesterase. *Life Sciences*, *20*(9), 1557–1563.

Stowe, G. N., Vendruscolo, L. F., Edwards, S., Schlosburg, J. E., Misra, K. K., Schulteis, G., ... Janda, K. D. (2011). A vaccine strategy that induces protective immunity against heroin. *Journal of Medicinal Chemistry*, *54*(14), 5195–5204. http://dx.doi.org/10.1021/jm200461m.

Tonstad, S., Heggen, E., Giljam, H., Lagerback, P. A., Tonnesen, P., Wikingsson, L. D., ... Fagerstrom, K. O. (2013). Niccine(R), a nicotine vaccine, for relapse prevention: A phase II, randomized, placebo-controlled, multicenter clinical trial. *Nicotine & Tobacco Research*, *15*(9), 1492–1501. http://dx.doi.org/10.1093/ntr/ntt003.

Tonstad, S., Job, J. S., Batech, M., Yel, D., Kheam, T., & Singh, P. N. (2013). Adult tobacco cessation in Cambodia: I. Determinants of quitting tobacco use. *Asia-Pacific Journal of Public Health*, *25*(5 Suppl.), 10S–19S. http://dx.doi.org/10.1177/1010539512451853.

Tonstad, S., Tonnesen, P., Hajek, P., Williams, K. E., Billing, C. B., Reeves, K. R., & Varenicline Phase 3 Study, G. (2006). Effect of maintenance therapy with varenicline on smoking cessation: A randomized controlled trial. *JAMA*, *296*(1), 64–71. http://dx.doi.org/10.1001/jama.296.1.64.

Tuncok, Y., Hieda, Y., Keyler, D. E., Brown, S., Ennifar, S., Fattom, A., & Pentel, P. R. (2001). Inhibition of nicotine-induced seizures in rats by combining vaccination against nicotine with chronic nicotine infusion. *Experimental and Clinical Psychopharmacology*, *9*(2), 228–234.

Umans, J. G., & Inturrisi, C. E. (1981). Pharmacodynamics of subcutaneously administered diacetylmorphine, 6-acetylmorphine and morphine in mice. *The Journal of Pharmacology and Experimental Therapeutics*, *218*(2), 409–415.

Veilleux, J. C., Colvin, P. J., Anderson, J., York, C., & Heinz, A. J. (2010). A review of opioid dependence treatment: Pharmacological and psychosocial interventions to treat opioid addiction. *Clinical Psychology Review*, *30*(2), 155–166. http://dx.doi.org/10.1016/j.cpr.2009.10.006.

Vocci, F. J., & Appel, N. M. (2007). Approaches to the development of medications for the treatment of methamphetamine dependence. *Addiction*, *102*(Suppl. 1), 96–106. http://dx.doi.org/10.1111/j.1360-0443.2007.01772.x.

Vocci, F. J., & Chiang, C. N. (2001). Vaccines against nicotine: How effective are they likely to be in preventing smoking? *CNS Drugs*, *15*(7), 505–514.

Volkow, N. D., Wang, G. J., Fischman, M. W., Foltin, R., Fowler, J. S., Franceschi, D., … Pappas, N. (2000). Effects of route of administration on cocaine induced dopamine transporter blockade in the human brain. *Life Sciences*, *67*(12), 1507–1515.

Volkow, N. D., Wang, G. J., Fowler, J. S., Tomasi, D., & Telang, F. (2011). Addiction: Beyond dopamine reward circuitry. *Proceedings of the National Academy of Sciences of the United States of America*, *108*(37), 15037–15042. http://dx.doi.org/10.1073/pnas.1010654108.

Wagena, E. J., de Vos, A., Horwith, G., & van Schayck, C. P. (2008). The immunogenicity and safety of a nicotine vaccine in smokers and nonsmokers: Results of a randomized, placebo-controlled phase 1/2 trial. *Nicotine & Tobacco Research*, *10*(1), 213–218. http://dx.doi.org/10.1080/14622200701704921.

Wee, S., Hicks, M. J., De, B. P., Rosenberg, J. B., Moreno, A. Y., Kaminsky, S. M., … Koob, G. F. (2012). Novel cocaine vaccine linked to a disrupted adenovirus gene transfer vector blocks cocaine psychostimulant and reinforcing effects. *Neuropsychopharmacology*, *37*(5), 1083–1091. http://dx.doi.org/10.1038/npp.2011.200.

Weissenborn, R., Deroche, V., Koob, G. F., & Weiss, F. (1996). Effects of dopamine agonists and antagonists on cocaine-induced operant responding for a cocaine-associated stimulus. *Psychopharmacology*, *126*(4), 311–322.

CHAPTER NINE

Genes and Alcohol Consumption: Studies with Mutant Mice

J. Mayfield*, M.A. Arends†, R.A. Harris*[,1], Y.A. Blednov*

*Waggoner Center for Alcohol and Addiction Research, The University of Texas at Austin, Austin, TX, United States

†Committee on the Neurobiology of Addictive Disorders, The Scripps Research Institute, La Jolla, CA, United States

[1]Corresponding author: e-mail address: harris@austin.utexas.edu

Contents

Abstract

In this chapter, we review the effects of global null mutant and overexpressing transgenic mouse lines on voluntary self-administration of alcohol. We examine

ISSN 0074-7742
http://dx.doi.org/10.1016/bs.irn.2016.02.014

approximately 200 publications pertaining to the effects of 155 mouse genes on alcohol consumption in different drinking models. The targeted genes vary in function and include neurotransmitter, ion channel, neuroimmune, and neuropeptide signaling systems. The alcohol self-administration models include operant conditioning, two- and four-bottle choice continuous and intermittent access, drinking in the dark limited access, chronic intermittent ethanol, and scheduled high alcohol consumption tests. Comparisons of different drinking models using the same mutant mice are potentially the most informative, and we will highlight those examples. More mutants have been tested for continuous two-bottle choice consumption than any other test; of the 137 mouse genes examined using this model, 97 (72%) altered drinking in at least one sex. Overall, the effects of genetic manipulations on alcohol drinking often depend on the sex of the mice, alcohol concentration and time of access, genetic background, as well as the drinking test.

1. INTRODUCTION

Alcohol use disorder (AUD) is a multifactorial disease, and its risk factors are determined by the interplay of genetic and environmental factors, combined with neuroadaptations following acute and repeated alcohol exposure. Alcohol targets ion channels and signaling cascades, producing intoxication, anxiolysis, and a sense of reward. After prolonged, repeated exposure, alcohol-induced changes in gene expression and synaptic function are thought to contribute to the development tolerance, sensitization, and compulsive consumption and drug seeking.

More than 100 genes have been shown to affect alcohol consumption and other alcohol-related behaviors in mouse models (Crabbe, Phillips, Harris, Arends, & Koob, 2006). Excessive alcohol consumption is a common model of addictive behavior, and animal models of voluntary self-administration are valuable for profiling genetic determinants of AUD (Green & Grahame, 2008). The preference to drink alcohol is a reliable measure that depends upon mouse genotype and has been consistent across laboratories despite variations of the drinking protocol used. Preclinical models, in conjunction with human genetic studies, may expose overlapping target genes and identify the most relevant drinking models and biological systems associated with AUD.

In this review, we focus on a single phenotype, voluntary alcohol self-administration, and summarize the global genetic manipulations in mice published to date on this behavior. Most of the studies used two-bottle choice (2BC) tests, where mice had a choice between water and ethanol and access was usually measured in continuous 24-h periods. In some cases,

2BC access to ethanol was intermittent (eg, every other day), which typically (Hwa et al., 2011; Melendez, 2011; Rosenwasser, Fixaris, Crabbe, Brooks, & Ascheid, 2013), but not always (Crabbe, Harkness, Spence, Huang, & Metten, 2012), results in higher ethanol intake compared to continuous access. A few studies used four-bottle choice (4BC) access, where mice have simultaneous access to water and three different concentrations of ethanol. Because rodents distribute their drinking across the circadian cycle and because limited access to ethanol tends to increase intake, restricted access during the dark cycle is often used to study periods of high consumption and to model binge-like drinking in humans (Thiele & Navarro, 2014). In the classic mouse drinking in the dark (DID) test, drinking session times begin a few hours after the start of the dark cycle and usually last 2–4 h over a few days. High levels of ethanol drinking and pharmacologically relevant blood ethanol concentrations (BECs) are achieved using this model (Thiele & Navarro, 2014). The scheduled high access consumption (SHAC) test uses fluid restriction to promote drinking of a low ethanol concentration (Finn et al., 2005). This chronic drinking model can also produce high BECs. Fluid access is first restricted and then gradually relaxed until the effects of fluid limitation are minimized. In operant self-administration tests, mice are trained to self-administer quantities of ethanol that produce moderate to high BECs. Removal of access to alcohol followed by restored access transiently increases consumption in dependent mice, and the effects of mutant genes on this alcohol deprivation effect (ADE; Rodd, Bell, Sable, Murphy, & McBride, 2004; Vengeliene, Bilbao, & Spanagel, 2014), a model of relapse drinking, are also presented. As shown in the tables throughout the chapter, the effects of some mutants can depend on the drinking test used as well as the ethanol concentration, time of access, genetic background, and sex of the mice (Vanderlinden, Saba, Bennett, Hoffman, & Tabakoff, 2015).

There are important considerations regarding genetic engineering methods, including the potential alteration of genes other than the mutated gene and the influence of the background strain carrying the genotype. C57BL/6J (B6) mice are a high alcohol-drinking strain and, as the tables in this chapter demonstrate, occupy a central role in voluntary drinking studies. This review focuses on global homozygous knockouts, although a few studies used hypofunctional or overexpressing transgenic lines. Strategies to reduce confounding and compensatory effects of null mutations include the use of knockin mice and brain regional or cell-specific knockouts, and a few of these studies are noted. In this review, we do not debate the genetic engineering methods used; instead, our aim is to provide a summary of homozygous null or overexpressing mutants and their role (or lack thereof) in alcohol consumption in mice. Current mouse gene and protein names from Uniprot (http://www.uniprot.org) are listed in the tables, which in some cases differ from the nomenclature used in the published studies. We first review the effects of mutant neurotransmitter receptor subunits on alcohol drinking in mice and then examine mutations in other ion channel receptors, cannabinoid and opioid receptors, neuropeptides, kinases/enzymes, and immune-related genes.

2. NEUROTRANSMITTER SYSTEMS

2.1 γ-Aminobutyric Acid

Alcohol potentiates γ-aminobutyric acid type A ($GABA_A$) receptor-mediated responses and enhances inhibitory neurotransmission, and some of the top candidate genes implicated in alcohol consumption code for specific $GABA_A$ receptor subunits (Trudell, Messing, Mayfield, & Harris, 2014). Deletion of the α1 subunit decreased ethanol consumption in operant and 2BC tests (Blednov et al., 2003; June et al., 2007), and knockdown of α5 reduced drinking in male (Boehm, Ponomarev, et al., 2004) but not female mice (Stephens, Pistovcakova, Worthing, Atack, & Dawson, 2005; Table 1). However, loss of α2 or β2 subunits had no effect on drinking (Blednov et al., 2003; Boehm, Ponomarev, et al., 2004). Decreased drinking was also observed in mice lacking δ or ρ1 subunits of $GABA_A$ receptors (Blednov et al., 2014; Mihalek et al., 2001). In rats, local knockdown of the δ or α4 subunit in select regions of the nucleus accumbens reduced ethanol intake (Nie, Rewal, Gill, Ron, & Janak, 2011; Rewal et al., 2009). Deletion of glutamic acid decarboxylase, the principal enzyme for synthesis of GABA, had no effect on ethanol consumption if the mutation was expressed on a B6 background; however, consumption increased in *Gad2* knockout mice on a B6 × 129/SvJ (N2) background (Blednov, Walker, Iyer, Homanics, & Harris, 2010). Knockin mice with ethanol-insensitive α1 subunits did not differ from control in ethanol intake (Werner et al., 2006), whereas α2 subunits were linked with alcohol consumption in some drinking models using a knockin strategy (Blednov, Borghese, et al., 2011). Furthermore, human genetic association studies have nominated the $GABA_A$ receptor α2 subunit as a top candidate in alcohol-dependent individuals, reviewed in Enoch (2008). A recent meta-analysis also identified a strong association

Table 1 Alcohol Consumption in GABA Mutant Mice

Gene Knockout/Knockin[#]	Background	Operant	2BC	DID	SHAC	References
γ-Aminobutyric acid type A receptor ($GABA_AR$) subunit α-1 (*Gabra1*)	Not specified	↓ (30–60 min), males/females		↓ (2 h) 1B, postoperant		June et al. (2007)
	B6 × 129SvEv		↓ females			Blednov et al. (2003)
Gabra1[#]	B6 × 129SvJ		—			Werner et al. (2006)
$GABA_AR$ subunit α-2 (*Gabra2*)	B6 × 129SvEv		— males — females			Boehm, Ponomarev, et al. (2004)
$GABA_AR$ subunit α-5 (*Gabra5*)	B6 × 129SvEv		↓ males — females			Boehm, Ponomarev, et al. (2004)
	B6 × 129SvEv	— (1 h) females	— females			Stephens et al. (2005)
$GABA_AR$ subunit β-2 (*Gabrb2*)	B6 × 129SvEv		— females			Blednov et al. (2003)
$GABA_AR$ subunit δ (*Gabrd*)	B6 × 129/Sv/SvJ		↓ males/females			Mihalek et al. (2001)
$GABA_AR$ subunit ρ-1 (*Gabrr1*)	B6 × 129S4		↓ males — females — intermittent, males/females	— (2, 4 h), males/females		Blednov et al. (2014)

Continued

Table 1 Alcohol Consumption in GABA Mutant Mice—cont'd

Gene Knockout/Knockin	Background	Operant	2BC	DID	SHAC	References
Sodium- and chloride-dependent GABA transporter 1 (*Slc6a1*)	B6		—			Cai et al. (2006)
Glutamic acid decarboxylase 2 (*Gad2*)	B6		—			Blednov et al. (2010)
	B6 × 129/SvJ N1		—			
	B6 × 129/SvJ N2		↑	— (3 h) 2BC — (2, 4 h) 1B	— (30 min)	

[#]indicates gene knockin.

—, ↓, ↑: no significant difference, decreased, or increased ethanol intake and/or preference, respectively, in mutant vs wild-type mice. Male mice were tested unless indicated otherwise. Ethanol intake in the two-bottle choice (2BC) tests was measured in 24-h sessions. Drinking session times for the other tests are indicated in *parenthesis*. *DID*, drinking in the dark; *1B*, one bottle; *SHAC*, scheduled high alcohol concentration. Recommended mouse protein and gene (in *italics*) names are from Uniprot. B6 refers to C57BL/6J mice.

of α2 with alcohol dependence along with evidence for α6 and γ2 subunits (Li et al., 2014).

2.2 Glutamate

Ionotropic and some metabotropic glutamate receptors are known targets for alcohol action (D'Souza, 2015). Mice lacking *Gria1* (glutamate receptor ionotropic AMPA 1), ie, the GluR1 subunit, did not differ from wild type in voluntary ethanol consumption, stress-induced drinking, or in the expression of ADE (Cowen, Schroff, Gass, Sprengel, & Spanagel, 2003; Table 2). Mice lacking *Gria3* (encoding the GluR3 subunit) did not differ from wild type in the number of reinforced lever presses during operant ethanol self-administration, although cue-induced reinstatement was reduced (Sanchis-Segura et al., 2006). There were no genotype differences in voluntary ethanol consumption or preference at any concentration tested (2–16%) in a 2BC test. Both genotypes showed increased ethanol consumption after a deprivation period following 14 weeks of 16% ethanol exposure, but the increase was initially blunted in the knockout mice. Although self-administration was not different between wild type and mice lacking GluR1 or GluR3, ethanol-seeking behavior after operant administration and resumption of self-administration after alcohol deprivation were reduced in GluR3 knockout mice, suggesting a role for the GluR3 subunit in alcohol seeking and relapse.

Mice lacking *Grm2*, which encodes the metabotropic glutamate receptor 2, increased consumption and preference for high ethanol concentrations in a 2BC model using an escalation procedure in which the ethanol concentration was increased from 3% to 17% over 80 days (Zhou et al., 2013; Table 2). However, *Grm4* knockout mice did not differ from wild type in ethanol intake or preference in a 2BC test using 3–9% ethanol (Blednov, Walker, Osterndorf-Kahanek, & Harris, 2004). *Grm5* knockout mice differed from wild type in some drinking models (Blednov & Harris, 2008). There were no differences in ethanol (3–12%) intake and preference in male or female mice in a 2BC test. Because the effects of deleting *Grm5* were somewhat greater in females, only females were tested in 4BC, 1-bottle and 2BC DID, and SHAC tests. Female *Grm5* knockout mice showed lowered ethanol intake and preference in the 4BC and 2BC DID tests, but there were no genotype differences in the other tests (Blednov & Harris, 2008). Decreased ethanol intake and preference were also reported

Table 2 Alcohol Consumption in Glutamate Mutant Mice

Gene Knockout/ Overexpression*	Background	Operant	2BC	4BC	DID	SHAC	References
Glutamate receptor 1 (*Gria1*)	B6N		— — ADE — poststress				Cowen, Schroff, et al. (2003)
Glutamate receptor 3 (*Gria3*)	B6N	—	— ↓ ADE, day 1				Sanchis-Segura et al. (2006)
Glutamate receptor ionotropic, NMDA 2A (*Grin2a*)	B6		—				Boyce-Rustay and Holmes (2006)
Metabotropic glutamate receptor 2, mGluR2 (*Grm2*)	CD1		↑ intake				Zhou et al. (2013)
mGluR4 (*Grm4*)	CD1 × 129/SvJ		— males/females				Blednov, Walker, Osterndorf-Kahanek, et al. (2004)
mGluR5 (*Grm5*)	B6 × 129/SvJ		— males/females	↓ females	— (2 h) 1B, females ↓ (3 h) 2BC, females	— (30 min), females	Blednov and Harris (2008)
	B6		↓				Bird et al. (2008)

Excitatory amino acid transporter 1, GLAST, EAAT1 (*Slc1a3*)	B6		↓ males/females		Karlsson et al. (2012)
Homer protein homolog 2 (*Homer2*)	Not specified		↓		Szumlinski et al. (2003)
	B6 × 129Xi/SvJ		↓ 12%, males/females		Szumlinski et al. (2005)
	B6 × 129Xi/SvJ			— (2 h)	Lum et al. (2014)
*Homer2b**	B6	↑ (21 min), NAc			Szumlinski et al. (2008)
Neuronal pentraxin-2, NARP (*Nptx2*)	129Sv × B6		↓ intake, no escalation		Ary et al. (2012)

*indicates gene overexpression.

—, ↓, ↑: no significant difference, decreased, or increased ethanol intake and/or preference, respectively, in mutant vs wild-type mice. Males were tested unless otherwise indicated. Ethanol intake in the two- and four-bottle choice (2BC, 4BC) tests was measured in continuous 24-h sessions. Drinking session times for the other tests are indicated in parenthesis. *DID*, drinking in the dark; *SHAC*, scheduled high alcohol consumption; *1B*, one bottle; *ADE*, alcohol deprivation effect; *NAc*, nucleus accumbens. Recommended mouse protein and gene (in *italics*) names are from Uniprot. B6 refers to C57BL/6J mice.

in male *Grm5* knockout mice consuming either 5% or 10% ethanol in a 2BC model (Bird, Kirchhoff, Djouma, & Lawrence, 2008).

The Homer family of postsynaptic scaffolding proteins modulates metabotropic glutamate receptors and is involved in addiction-related neuroplasticity. *Homer2* knockout mice drank less ethanol than wild type (Szumlinski, Toda, Middaugh, Worley, & Kalivas, 2003) and showed aversion to 12% ethanol, with total ethanol intake less than half of wild-type mice in a 2BC test (Szumlinski et al., 2005; Table 2). Infusion of Homer 2 into the nucleus accumbens reversed the effects on ethanol preference and intake in knockout mice. Furthermore, Homer 2b overexpression in the nucleus accumbens increased operant self-administration of 6% and 12% alcohol in B6 mice (Szumlinski, Ary, Lominac, Klugmann, & Kippin, 2008). In contrast to continuous access, there was no difference in intake between *Homer2* knockout and wild-type mice in a limited-access DID model using 20% ethanol (Lum, Campbell, Rostock, & Szumlinski, 2014). These authors reported that infusion of a metabotropic glutamate receptor 1 negative allosteric modulator into the nucleus accumbens reduced drinking in the DID test in control but not *Homer2* knockout mice, indicating a role for metabotropic glutamate receptors scaffolded by Homer 2 proteins in alcohol consumption.

A family of neuronal pentraxins regulates AMPA receptor aggregation and synaptogenesis at the postsynaptic density. Neuronal activity-regulated pentraxin (*Narp*), also known as *Nptx2*, is an immediate-early gene that is induced by synaptic activity and produces an AMPA receptor binding protein that enhances AMPA receptor clustering and function. *Narp* knockout mice did not escalate their daily ethanol intake after repeated exposure, and total ethanol intake in these mice was significantly lower compared to wild type after 10 days of testing (Ary et al., 2012; Table 2). Further study showed that *Narp* deletion shifted intake and preference away from the high alcohol concentration (12%) with repeated alcohol exposure. *Narp* induction may thus be important for escalating alcohol consumption under free-choice conditions.

Chronic alcohol exposure and withdrawal produce a hyperexcitable or hyperglutamatergic state, and restoring the balance of glutamate responses could be relevant for treating dependence (Rao, Bell, Engleman, & Sari, 2015). Acamprosate (FDA approved to treat AUD) may work by inhibiting a hyperglutamatergic state in the brain and helping to balance inhibitory and excitatory transmission. In mouse models, acamprosate was effective in preventing the development of withdrawal (Blednov & Harris, 2008) and reducing free-choice ethanol intake (Brager, Prosser, & Glass, 2011a, 2011b; Spanagel et al., 2005). The anticonvulsant gabapentin is a calcium channel blocker that can also modulate glutamatergic (and GABAergic) activity and showed some promise in treating AUD in a clinical trial (Mason et al., 2014).

2.3 Dopamine

The mesolimbic dopamine pathway and its role in the rewarding properties of drugs of abuse have been a long-term focus for addiction research (van Huijstee & Mansvelder, 2014). Studies consistently show that ethanol self-administration in operant and 2BC tests is decreased in D_1- and D_2-receptor-deficient mice (Delis et al., 2013; El-Ghundi et al., 1998; Palmer, Low, Grandy, & Phillips, 2003; Phillips et al., 1998; Risinger, Freeman, Rubinstein, Low, & Grandy, 2000; Thanos, Rivera, et al., 2005); however, deletion of the D_2 long receptor increased DID (Bulwa et al., 2011; Table 3). Overexpression of D_2 transiently increased intake in D_2 receptor knockout mice and decreased intake in wild-type mice (Thanos, Rivera, et al., 2005). Effects of D_3 and D_4 dopamine receptor deletion are less well studied, but there is a report of substantially decreased alcohol consumption in 2BC and DID tests in D_3 knockout mice (Leggio et al., 2014). An example of the complex role of sex and genetic background is illustrated by the DAT knockout, which decreased drinking in females but not males on a B6 background and increased drinking in males but not females on a mixed genetic background (Table 3).

2.4 Serotonin

The serotonergic system has important roles in mood and impulse control and is implicated in the development and maintenance of alcohol and other drug abuse disorders (Bauer, Graham, Soares, & Nielsen, 2015). Ethanol intake and preference increased in a knockin mouse line (Tph2KI) expressing a hypofunctional variant of the rate-limiting serotonin (5-HT) synthesis enzyme, tryptophan hydroxylase 2 (Sachs, Salahi, & Caron, 2014; Table 4). Increased ethanol self-administration has also been reported in mice lacking 5-HT_{1B} receptors (Bouwknecht et al., 2000; Crabbe et al., 1996; Risinger, Doan, & Vickrey, 1999). Knockout of 5-HT_6 receptors had no effect on 2BC drinking (Bonasera, Chu, Brennan, & Tecott, 2006). However, decreased intake was observed in B6 mice overexpressing 5-HT_{3A} receptors (Metz, Chynoweth, & Allan, 2006). Mice overexpressing 5-HT_3 receptors in the forebrain also drank less alcohol than nontransgenic mice in a 2BC test

Table 3 Alcohol Consumption in Dopamine Mutant Mice

Gene Knockout/ Overexpression*	Background	Operant	2BC	DID	SIP	References
D_{1A} dopamine receptor (*Drd1*)	B6		↓ limited/continuous access			El-Ghundi et al. (1998)
D_2 dopamine receptor (*Drd2*)	B6	↓ (23 h)				Risinger et al. (2000)
	B6		↓ males/females — ethanol exposed			Palmer et al. (2003)
	B6		↓			Thanos, Rivera, et al. (2005)
	B6		↓ males/females			Phillips et al. (1998)
	B6		↓ ↑ chronic stress			Delis et al. (2013)
D_2 long receptor	B6			↑ (4 h), males/females		Bulwa et al. (2011)
*Drd2**	B6		↓ WT ↑ transient KO			Thanos, Rivera, et al. (2005)
D_3 dopamine receptor (*Drd3*)	B6	— (23 h)	—			Boyce-Rustay and Risinger (2003)
	B6		↓	↓ (4 h)		Leggio et al. (2014)

D_4 dopamine receptor (*Drd4*)	B6129/Ola × B6	—		Falzone et al. (2002)
Sodium-dependent dopamine transporter, DAT (*Slc6a3*)	B6	— males ↓ females		Savelieva, Caudle, Findlay, Caron, and Miller (2002)
	B6		↓ (40 min), males/females	Mittleman et al. (2011)
	B6 × 129Sv	↑ 24%, males — females		Hall, Sora, and Uhl (2003)
Dopamine β-hydroxylase (*Dbh*)	B6 × 129/SvEv	↓		Weinshenker, Rust, Miller, and Palmiter (2000)

*indicates gene overexpression.

—, ↓, ↑: no significant difference, decreased, or increased ethanol intake and/or preference, respectively, in knockout (KO) or mice overexpressing D_2 vs wild-type (WT)/control mice. Male mice were tested unless indicated otherwise. Deletion of the long form of D_2 produces overexpression of the short relative to the long form. Ethanol intake in the two-bottle choice (2BC) tests was measured in 24-h sessions, unless otherwise indicated. Drinking session times for the other tests are indicated in *parenthesis*. *DID*, drinking in the dark; *SIP*, scheduled-induced polydipsia. Recommended mouse protein and gene (in *italics*) names are from Uniprot. B6 refers to C57BL/6J mice.

Table 4 Alcohol Consumption in Serotonin Mutant Mice

Gene Knockout/Knockin#/ Overexpression*	Background	Operant	2BC	References
Tryptophan 5-hydroxylase 2 (*Tph2*)#	B6 × 129S6/ SvEvTac		↑	Sachs et al. (2014)
5-Hydroxytryptamine receptor (5-HTR), 5-HTR_{1B} (*Htr1b*)	Not specified	↑ (23 h) 10%, phase 1, males/females		Risinger et al. (1999)
	129/Sv-Ter		↑ males/females	Crabbe et al. (1996)
	129/Sv		— males	Bouwknecht et al. (2000)
5-HTR_{3A} (*Htr3a*)*	B6 DBA/2J		↓ —	Metz et al. (2006)
5-HTR_{6} (*Htr6*)	B6		—	Bonasera et al. (2006)
Sodium-dependent serotonin transporter (*Slc6a4*)	B6	↓ (90 min)		Lamb and Daws (2013)
	B6		↓	Kelai et al. (2003)
	B6		↓ 7%, females	Boyce-Rustay et al. (2006)

#indicates gene knockin.
*indicates gene overexpression.
—, ↓, ↑: no significant difference, decreased, or increased ethanol intake and/or preference, respectively, in mutant vs wild-type mice. Male mice were tested unless indicated otherwise. Ethanol intake in the two-bottle choice (2BC) tests was measured in 24-h sessions. Drinking session times for the operant tests are indicated in *parenthesis*. Recommended mouse protein and gene (in *italics*) names are from Uniprot. B6 refers to C57BL/6J mice.

(Engel, Lyons, & Allan, 1998). The 5-HT transporter is the main means for regulating 5-HT levels in brain, and deletion of the transporter decreased ethanol intake in B6 mice (Boyce-Rustay et al., 2006; Kelai et al., 2003; Lamb & Daws, 2013).

Although selective 5-HT reuptake inhibitors and the 5-HT_3 antagonist ondansetron were not promising as first-line treatments for AUD, they may provide some benefit in treating select populations of alcoholics. A review of pharmacotherapies for AUD in preclinical and clinical models can be found in Zindel and Kranzler (2014).

2.5 Adenosine

Adenosine signaling is implicated in drug addiction and ethanol-related behaviors in preclinical models (Nam, Bruner, & Choi, 2013). Deletion of the adenosine A_1 receptor did not alter operant responding for ethanol in B6 × 129 mice (Houchi, Persyn, Legastelois, & Naassila, 2013; Table 5). Although male and female adenosine A_{2A} receptor (A2AR) knockout mice generated on a CD1 background drank more ethanol compared to wild type in 2BC tests (Houchi et al., 2008; Naassila, Ledent, & Daoust, 2002), no differences were found between A2AR knockout and wild-type mice on a B6 background (Houchi et al., 2008).

Table 5 Alcohol Consumption in Adenosine Knockout Mice

Gene Knockout	Background	Operant	2BC	References
Adenosine receptor A_1 (*Adora1*)	B6 × 129	— (2 h)		Houchi et al. (2013)
Adenosine receptor A_{2A} (*Adora2a*)	CD1		↑ males/females	Naassila et al. (2002)
	CD1 B6		↑ males — males/females	Houchi et al. (2008)
Equilibrative nucleoside transporter-1, ENT1 (*Slc29a1*)	B6 × 129X1/SvJ	↑		Chen et al. (2010)
	B6 × 129X1/SvJ		↑	Nam, Hinton, et al. (2013); Nam et al. (2011)
	B6 × 129X1/SvJ		↑	Choi et al. (2004)
	B6 × 129X1/SvJ		↑ sex not specified	Lee et al. (2013); Ruby et al. (2014)
	B6 × 129X1/SvJ		↑ females	Ruby et al. (2011)

—, ↓, ↑: no significant difference, decreased, or increased ethanol intake and/or preference, respectively, in knockout vs wild-type mice. Male mice were tested unless indicated otherwise. Ethanol intake in the operant and two-bottle choice (2BC) tests was measured in 30-min or 24-h sessions, respectively, unless otherwise indicated. Recommended mouse protein and gene (in *italics*) names are from Uniprot. B6 refers to C57BL/6J mice.

Type 1 equilibrative nucleoside transporter (ENT1) is an ethanol-sensitive adenosine transporter and a main regulator of adenosine levels in the brain. Table 5 shows that ENT1 knockout mice generated on a B6 × 129 background consume more ethanol compared to wild-type littermates in operant and 2BC drinking tests (Chen et al., 2010; Choi et al., 2004; Lee et al., 2013; Nam, Hinton, et al., 2013; Nam et al., 2011; Ruby et al., 2014; Ruby, Walker, An, Kim, & Choi, 2011). Thus, studies from multiple labs support a role for adenosine in controlling alcohol-seeking behavior.

3. CANNABINIOIDS AND OPIOIDS

The endocannabinoid system is involved in brain reward signaling and drug-seeking behavior (Panagis, Mackey, & Vlachou, 2014). In several studies of different genetic backgrounds, male and female cannabinoid 1 receptor (CB1R) knockout mice showed reduced ethanol intake and/or preference for ethanol than wild-type mice (Hungund, Szakall, Adam, Basavarajappa, & Vadasz, 2003; Lallemand & de Witte, 2005; Naassila, Pierrefiche, Ledent, & Daoust, 2004; Poncelet, Maruani, Calassi, & Soubrie, 2003; Racz et al., 2003; Thanos, Dimitrakakis, Rice, Gifford, & Volkow, 2005; Vinod, Yalamanchili, et al., 2008; Table 6). In contrast, CB2R knockout mice showed increased ethanol consumption in two different drinking models (Ortega-Alvaro et al., 2015). Fatty acid amide hydrolase (FAAH) metabolizes endocannabinoids, and FAAH inhibition increases anandamide levels in brain. Increased ethanol intake and preference were reported in FAAH knockout mice on B6 and B6 × 129/SvJ backgrounds (Basavarajappa, Yalamanchili, Cravatt, Cooper, & Hungund, 2006; Blednov, Cravatt, Boehm, Walker, & Harris, 2007; Vinod, Sanguino, Yalamanchili, Manzanares, & Hungund, 2008).

Endogenous opioids are expressed throughout the reinforcement circuitry in the brain, and cannabinoid and opioid systems may interact to regulate the effects of drugs of abuse (Charbogne, Kieffer, & Befort, 2014). Table 6 shows the effects of opioid receptor mutations on voluntary alcohol drinking. Mice lacking the μ-opioid receptor (MOR) did not self-administer ethanol (Roberts et al., 2000). This was demonstrated in two different operant procedures (nose-poke and lever press) and in 2BC drinking with or without prior ethanol experience. Other studies using MOR knockout mice reported decreased drinking compared to wild-type mice (Becker et al., 2002; Contet et al., 2014; Hall, Sora, & Uhl, 2001). Decreased

Table 6 Alcohol Consumption in Cannabinoid and Opioid Knockout Mice

Gene Knockout	Background	Operant	2BC	DID	References
Cannabinoid receptor 1 (*Cnr1*)	B6 × 129/Ola		↓ (6 h)		Poncelet et al. (2003)
			↓ preference		Lallemand and de Witte (2005)
	B6		↓ males/females		Hungund et al. (2003)
	CD1		↓ males/females		Naassila et al. (2004)
	CD1			↓ (8 h)	Thanos, Dimitrakakis, et al. (2005)
	B6 and DBA/2J		↓		Vinod, Yalamanchili, et al. (2008)
	B6		— after 1 week		Racz et al. (2003)
Cannabinoid receptor 2 (*Cnr2*)	CD1 and B6	↑ (1 h)	↑		Ortega-Alvaro et al. (2015)
Fatty acid amide hydrolase (*Faah*)	B6 × 129/SvJ		↑ males/females		Blednov et al. (2007)
	B6		↑ females — males		Basavarajappa et al. (2006)
	B6		↑		Vinod, Sanguino, et al. (2008)

Continued

Table 6 Alcohol Consumption in Cannabinoid and Opioid Knockout Mice—cont'd

Gene Knockout	Background	Operant	2BC	DID	References
μ-type opioid receptor (*Oprm1*)	B6 × 129/Sv	↓ nose-poke ↓ lever operant	↓ postoperant ↓ postoperant ↓ postforced ethanol		Roberts et al. (2000)
	B6 × 129/Sv		↓ females — males		Hall et al. (2001)
	Not specified		↓ after several weeks		Becker et al. (2002)
	B6			↓ (2 h) — (2 h), ADE	Contet et al. (2014)
	B6			— (4 h)	van Rijn and Whistler (2009)
δ-type opioid receptor (*Oprd1*)	B6			↑ (4 h)	van Rijn and Whistler (2009)
	B6Orl × 129/Sv	↑ post-2BC	— naive ↑ postoperant		Roberts et al. (2001)
κ-type opioid receptor (*Oprk1*)	B6			↓ (4 h)	van Rijn and Whistler (2009)
	B6Orl		↓ males/females		Kovacs et al. (2005)
β-Endorphin (*Pomc*)	B6		↑ 7%, males/females — 10%, males/females		Grisel et al. (1999)

	B6	— ↑ 2 days, ADE ↑ (2 h) days 1–10 ↑ day 2, ADE	Grahame, Mosemiller, Low, and Froehlich (2000)
	B6	↓ females — preference, males	Racz et al. (2008)
Prodynorphin (*Pdyn*)	B6	↑ males/females	Racz, Markert, Mauer, Stoffel-Wagner, and Zimmer (2013)
	B6	—	Sperling, Gomes, Sypek, Carey, and McLaughlin (2010)
	B6	↑	Femenia and Manzanares (2012)
	B6 × 129/SvEvTac	↓ females — males	Blednov, Walker, Martinez, and Harris (2006)
Preproenkephalin (*Penk1*)	E6	—	Koenig and Olive (2002)
	B6	— females ↑ 1–3 weeks, males	Racz et al. (2008)

—, ↓, ↑: no significant difference, decreased, or increased ethanol intake and/or preference, respectively, in knockout vs wild-type mice. Male mice were tested unless otherwise indicated. Ethanol intake in the operant and two-bottle choice (2BC) tests was measured in 30-min and 24-h sessions, respectively, unless otherwise indicated. Wild-type, but not knockout, mice showed a stress-induced increase in ethanol intake/preference (Racz et al., 2003, 2013, 2008; Sperling et al., 2010). Drinking in the dark (DID) session times are noted in *parenthesis*. *ADE*, alcohol deprivation effect. Recommended mouse protein and gene (in *italics*) names are from Uniprot. B6 refers to C57BL/6J mice.

consumption was also observed in κ-opioid receptor (KOR) knockout mice (Kovacs et al., 2005; van Rijn & Whistler, 2009). In contrast, mice lacking δ-type opioid receptors (DORs) showed increased alcohol-drinking behavior (Roberts et al., 2001; van Rijn & Whistler, 2009). The effects of deleting β-endorphin, prodynorphin, or preproenkaphalin varied depending on the ethanol concentration, time of exposure, sex, and genotype (Table 6).

Consistent with the preclinical research showing that knockout of select opioid receptors reduces drinking in mice, human genetic studies further support a role for polymorphisms in genes coding for MOR1, DOR1, KOR1, and other opioid receptors in clinical populations of alcoholics (Levran, Yuferov, & Kreek, 2012). In addition, opioids are used in the treatment of AUD. The pan-opioid antagonist naltrexone is already FDA approved to treat AUD and may reduce craving and relapse to heavy drinking in humans. Nalmefene, another pan-opioid receptor antagonist structurally similar to naltrexone, is currently used to treat opioid overdose in the United States and is approved in Europe for as-needed use to reduce heavy drinking. It has several advantages over naltrexone, including longer duration of action, lack of dose-dependent liver toxicity, and higher affinity of binding to MORs and KORs. Nalmefene reduced the total amount of alcohol consumed and the number of heavy drinking days in alcohol-dependent patients (van den Brink et al., 2014).

4. IMMUNE-RELATED GENES

The interplay between brain, behavior, and immune responses in the etiology and progression of drug abuse is a current area of interest in addiction research (http://www.arcr.niaaa.nih.gov/arcr/arcr372/toc37_2.htm). The neuroimmune system, encompassing innate immune signaling within the peripheral and central nervous systems, is important in the pathophysiology and potential treatment of alcohol abuse and dependence (Crews & Vetreno, 2015; Mayfield, Ferguson, & Harris, 2013; Robinson et al., 2014). Alcohol may increase neuroimmune gene and protein expression through both peripheral–central signaling molecules and direct actions in the brain (Crews & Vetreno, 2015).

Deletion of many different immune-related genes lowered ethanol intake and preference in male and female mice on a B6 background (Table 7). In some cases, the effect depended on the drinking test used.

Table 7 Alcohol Consumption in Immune-Related Receptor Knockout Mice

Gene Knockout	Background	2BC	DID	References
Cathepsin F (*Ctsf*)	B6 × 129/SvJ	↓ males — females		Blednov et al. (2012)
Cathepsin S (*Ctss*)	B6	↓ males/females	— 2BC, males ↓ 1B, males ↓ 1B, 2BC, females	Blednov et al. (2012)
C–C motif chemokine 2 (*Cc12*)	B6	— males ↓ females		Blednov et al. (2005)
C–C motif chemokine 3 (*Cc13*)	B6	↓ males/females		Blednov et al. (2005)
C–C chemokine receptor type 2 (*Ccr2*)	B6	↓ males/females		Blednov et al. (2005)
C–C chemokine receptor type 5 (*Ccr5*)	B6	— males/females		Blednov et al. (2005)
Apolipoprotein E (*Apoe*)	B6	— females		Bechtholt, Smith, Raber, and Cunningham (2004)
Monocyte differentiation antigen CD14 (*Cd14*)	B6	↓ males/females	— 1B, 2BC, males — 1B, females ↓ 2BC, females	Blednov et al. (2012)
Interleukin-1 receptor antagonist (*Il1rn*)	B6 × 129/SvJ	↓ males/females	↓ 1B, 2BC, males/females	Blednov et al. (2012)
Interleukin-6 (*Il6*)	B6	↓ males/females	— 1B, 2BC, males/females	Blednov et al. (2012)
β-2-Microglobulin (*B2m*)	B6	↓ males/females	— 1B, 2BC, males/females	Blednov et al. (2012)
B2 bradykinin receptor (*Bdkrb2*)	B6	—		Maul et al. (2005)
Toll-like receptor 4 (*Tlr4*)	B6	—		Pascual et al. (2011)

—, ↓, ↑: no significant difference, decreased ethanol intake and/or preference, or increased ethanol intake and/or preference, respectively, in knockout vs wild-type mice. Male mice were tested unless indicated otherwise. Ethanol intake in the two-bottle choice (2BC) tests was measured in 24-h sessions. Drinking in the dark (DID) session times were 2 and 4 h; *1B*, one bottle. Recommended mouse protein and gene (in *italics*) names are from Uniprot. B6 refers to C57BL/6J mice.

Deletion of certain chemokine and chemokine receptor genes reduced ethanol preference and intake (Blednov et al., 2005). Knockout mice were also used to study other immune-associated genes previously linked to alcohol consumption in a gene expression analysis of mouse brain (Mulligan et al., 2006). The following null mutations were examined based on this transcriptome meta-analysis: *B2m* (β-2 microglobulin), *Cd14* (cluster of differentiation 14), *Il1rn* (interleukin 1 receptor antagonist), *Il6* (interleukin-6), *Ctss* (cathepsin S), and *Ctsf* (cathepsin F; Blednov et al., 2012). These candidate targets were subsequently validated in behavioral tests because knockout of these individual genes reduced ethanol consumption and preference in 2BC tests (Blednov et al., 2012). In contrast, a transgenic mouse line overexpressing *Il6* showed increased alcohol preference in females but not males (Harris & Blednov, 2013). C3H/HeJ mice are naturally TLR4-deficient and showed decreased operant self-administration of ethanol compared to the control strain (Harris & Blednov, 2013); however, ethanol consumption was not changed in TLR4 knockout mice in a 2BC test (Pascual, Balino, Alfonso-Loeches, Aragon, & Guerri, 2011).

Pharmacological inhibition of immune signaling reduces alcohol drinking in mice, whereas activation of the innate immune system using lipopolysaccharide produces prolonged increases in drinking (Blednov, Benavidez, et al., 2011). Immune molecules, acting peripherally and/or centrally, appear to be involved in preference for alcohol and the behavioral evidence agrees with expression profiling of mouse brain (Robinson et al., 2014). Unique gene expression patterns in human postmortem brain provide additional support for immune gene and related network effects in alcohol-dependent individuals (Liu et al., 2006; Ponomarev, Wang, Zhang, Harris, & Mayfield, 2012). Furthermore, genetic association studies in humans link immune-associated genes with alcohol dependence (Blednov, Benavidez, Black, Ferguson, et al., 2015). The use of agents that regulate immune responses as potential therapeutics for AUD is discussed in Section 11.

5. ION CHANNELS

Some of the rapid-onset actions of alcohol are likely mediated by direct action on ion channels (Howard, Trudell, & Harris, 2014; Trudell et al., 2014). In addition to the prominent and well-studied neurotransmitter systems in alcohol dependence discussed previously (eg, GABA, glutamate), other ion channels that have been implicated in alcohol intake and preference are described below.

5.1 Transient Receptor Potential Cation Channels

Transient receptor potential (TRP) channels are a group of nonselective ion channels that mediate sensations, tastes, temperature, pain, and other chemical and physical stimuli. There is a positive correlation between ethanol intake and consumption of sweet solutions in rodents and humans (Kampov-Polevoy et al., 2014). Deletion of TRP cation channel subfamily M, member 5, substantially decreased ethanol intake and preference (Blednov et al., 2008; Table 8). Altered taste perception may explain the reduced preference for both saccharin and ethanol. Mice lacking TRP cation channel subfamily V, member 1 (TRPV1), also known as the vanilloid or capsaicin receptor, showed increased ethanol preference but no change in saccharin or quinine preference (Blednov & Harris, 2009). The endocannabinoid, anandamide, is an endogenous activator of this channel, and ethanol also directly activates TRPV1 at high concentrations. The increased ethanol consumption in *Trpv1* knockout mice may simply be due to a reduction in the ethanol-induced burning sensation, although some central actions, including endocannabinoid signaling, may be involved.

5.2 N-Type Calcium Channels

Voltage-dependent calcium channels regulate neuronal excitability, neurotransmitter release, and gene expression. Several types exist with unique pharmacological and electrophysiological characteristics. L-, T-, and N-type voltage-gated calcium channels have all been implicated in the behavioral effects of ethanol. Reduced ethanol intake and preference were reported in mice lacking functional α_{1B} subunits of N-type calcium channels (Newton et al., 2004; Table 8), suggesting these subunits as a potential target to control drinking. The calcium channel blocker gabapentin targets channels with $\alpha_2\delta$ subunits (Geisler, Schopf, & Obermair, 2015) and, as previously mentioned, offers some promise as a therapeutic for AUD (Mason et al., 2014).

5.3 GIRK Channels

G protein-coupled inwardly rectifying K^+ (GIRK) channels are widely expressed in brain and are activated by ligand-stimulated G protein-coupled receptors, resulting in an outward K^+ current that hyperpolarizes neuronal membranes and decreases neuronal excitability. Genetic evidence from mouse models has provided extensive insight into the significance of GIRK channels in drug addiction, and their role in responses to ethanol and other drugs of abuse is reviewed in Bodhinathan and Slesinger (2014) and Mayfield, Blednov, and Harris (2015).

As shown in Table 8, ethanol consumption and preference did not differ in wild-type and *Girk2* knockout mice in the standard 2BC test, where the bottle positions were alternated daily to control for position preferences. However, when the ethanol bottles were always available in the preferred location, *Girk2* knockout mice consumed more ethanol compared to wild type (Blednov, Stoffel, Chang, & Harris, 2001). Moreover, single-nucleotide polymorphisms of the *KCJN6* gene (GIRK2) have been associated with alcohol dependence in humans (Clarke et al., 2011; Kang et al., 2012). Deletion of *Girk3* increased limited- but not continuous-access voluntary drinking, and overexpression of *Girk3* in the VTA reversed the binge-drinking phenotype and reduced drinking in wild-type mice (Herman et al., 2015). These results point to a role for GIRK subunits in the rewarding properties of ethanol and as a potential target for regulating binge-like drinking.

5.4 BK Channels

BK channels are calcium-activated potassium channels that are characterized by their large conductance of potassium ions. The β subunits influence BK channel responses to acute and chronic ethanol, but effects on one system cannot easily be extrapolated to another. In 2BC continuous- and intermittent-access tests, nondependent mice lacking BK β1 or β4 subunits did not alter their drinking compared to wild type (Kreifeldt, Le, Treistman, Koob, & Contet, 2013; Table 8). Weeks of voluntary 2BC drinking alternated with weeks of chronic intermittent ethanol (CIE) vapor escalated drinking in dependent wild-type mice to a greater extent than BK β4 knockout mice. In contrast, β1 knockout mice drank more after fewer CIE cycles than did wild-type mice. Thus loss of β4 decreased, while loss of β1 increased, the escalation of drinking during withdrawal from CIE (Kreifeldt et al., 2013).

5.5 Nicotinic Acetylcholine Receptors

Neuronal nicotinic acetylcholine receptors (nAChRs) are involved in several alcohol-mediated behaviors (Rahman, Engleman, & Bell, 2014). Varenicline (which has partial agonist activity at nAChRs and is used in smoking cessation programs) reduced drinking in both preclinical and clinical studies, and human genetic studies showed a strong association of polymorphisms in *CHRNA5* (which encodes the α5 subunit) with risk of developing alcohol dependence, reviewed in Rahman et al. (2014). Mice lacking *Chrna5* did not differ from wild type in ethanol consumption (Santos, Chatterjee, Henry, Holgate, & Bartlett, 2013), but transgenic mice overexpressing α5α3β4 showed reduced intake (Gallego et al., 2012; Table 8). Knockout of α6 nACh subunits had no effect on 2BC drinking (Kamens, Hoft, Cox, Miyamoto, & Ehringer, 2012). However, deletion of α4 decreased DID of a high ethanol concentration (Hendrickson, Zhao-Shea, Pang, Gardner, & Tapper, 2010), and loss of α7 decreased 2BC drinking in female mice (Kamens, Andersen, & Picciotto, 2010). Knockout of β2 (Dawson, Miles, & Damaj, 2013; Kamens et al., 2010) or β3 (Kamens et al., 2012) nAChR subunits had no effect in 2BC tests in male or female mice.

5.6 Glycine Receptors

Supraspinal glycine receptors are found in reward pathways and play a key role in alcohol inhibitory responses in some neurons (Perkins, Trudell, Crawford, Alkana, & Davies, 2010). The glycine receptor α1 subunit has proven difficult to engineer and studies are complicated by the reduced glycinergic function and lethality of homozygous mutants. Deletion of glycine receptor subunits α2 or α3 had no effect on limited-access DID, while 2BC continuous-access drinking decreased in α2- but not α3-null mutants (Blednov, Benavidez, Black, Leiter, et al., 2015; Table 8). However, intermittent-access drinking increased in α3- but not α2-null mutants. A linkage analysis in humans also supports a role for the α3 subunit in alcohol dependence (Han, Gelernter, Kranzler, & Yang, 2013).

Table 8 Alcohol Consumption in Ion Channel Mutant Mice

Gene Knockout/Overexpression*	Background	2BC	CIE	DID	References
Transient receptor potential cation channel, subfamily M member 5 (*Trpm5*)	B6	↓			Blednov et al. (2008)
Transient receptor potential cation channel subfamily V member 1 (*Trpv1*)	B6 × 129X1/SvJ	↑			Blednov and Harris (2009)
G protein-activated inward rectifier potassium channel 2, GIRK2 (*Kcnj6*)	B6 × 129Sv	— with bottle alternation, males/females ↑ without bottle alternation, females			Blednov et al. (2001)
G protein-activated inward rectifier potassium channel 3, GIRK3 (*Kcnj9*)	B6	↑ (2 h) —			Herman et al. (2015)
Voltage-dependent N-type calcium channel subunit α-1B (*Cacna1b*)	B6 × 129SvJae	↓			Newton et al. (2004)
Calcium-activated potassium channel subunit β-1, BK β1 (*Kcnmb1*)	B6	— continuous, intermittent	↑[a]	— (2 h)	Kreifeldt et al. (2013)
Calcium-activated potassium channel subunit β-4, BK β4 (*Kcnmb4*)	B6	— continuous, intermittent	↓[a]	— (2 h)	Kreifeldt et al. (2013)
	B6			↑ (2 h) males/females	Martin et al. (2008)
P2X purinoceptor 4 (*P2rx4*)	B6	↑ intake		↑ (4 h) 5%, 20% — (4 h) 10%	Wyatt et al. (2014)
Glycine receptor subunit α-2 (*Glra2*)	B6	↓ — intermittent		— (2, 4 h) 1B	Blednov, Benavidez, Black, Leiter, et al. (2015)

Glycine receptor subunit α-3 (*Glra3*)	B6	— ↑ intermittent	— (2, 4 h) 1B	Blednov, Benavidez, Black, Leiter, et al. (2015)
Nicotinic acetylcholine receptor (nAChR) α4 subunit (*Chrna4*)	B6		↓ (2 h) 20% — (2 h) 2%	Hendrickson et al. (2010)
nAChR α5 subunit (*Chrna5*)	B6		— (4 h)	Santos et al. (2013)
nAChR α6 subunit (*Chrna6*)	B6	— males/females		Kamens et al. (2012)
nAChR α7 subunit (*Chrna7*)	B6	↓ females — males		Kamens et al. (2010)
nAChR β2 subunit (*Chrnb2*)	B6	— males/females		Kamens et al. (2010)
	B6	— intermittent		Dawson et al. (2013)
nAChR β3 subunit (*Chrnb3*)	B6	— males/females		Kamens et al. (2012)
nAChR α5α3β4*	B6SJLF1/J	↓ intake		Gallego et al. (2012)

*indicates gene overexpression.

[a]These tests involved weekly limited access (2 h/day) 2BC drinking alternated with weekly CIE vapor (16 h/day) to create alcohol dependence/withdrawal. Deletion of BK β4 attenuated, while deletion of BK β1 accelerated, the escalation of ethanol drinking in dependent mice during withdrawal from CIE.

—, ↓, ↑: no difference, decreased, or increased ethanol intake and/or preference, respectively, in mutant vs wild-type mice. Male mice were tested unless otherwise indicated. Ethanol intake in the two-bottle (2BC) tests was measured in 24-h sessions unless otherwise indicated. Drinking session times for the other tests are indicated in *parenthesis*. *CIE*, chronic intermittent ethanol; *DID*, drinking in the dark; *1B*, one bottle. Recommended mouse protein and gene (in *italics*) names are from Uniprot. B6 refers to C57BL/6J mice.

6. PROTEIN KINASES

In addition to direct effects on ion channels, ethanol indirectly modulates channel function via phosphorylation and other posttranslational processing mechanisms (Trudell et al., 2014). Mice lacking protein kinase C type ε (PKCε) drank less ethanol than wild type, and this effect has been observed in different drinking tests across multiple labs (Besheer, Lepoutre, Mole, & Hodge, 2006; Choi, Wang, Dadgar, Chang, & Messing, 2002; Hodge et al., 1999; Olive et al., 2005; Olive, Mehmert, Messing, & Hodge, 2000; Wallace et al., 2007; Table 9). Selective knockdown of PKCε in the amygdala also reduced 2-h ethanol consumption in a DID model (Lesscher et al., 2009). PKCε is known to reduce the positive allosteric effects of benzodiazepines and ethanol at $GABA_A$ receptors, which may in turn regulate reward signaling. The effects of deleting other PKC types on alcohol intake and preference have not been studied as extensively, although drinking in 2BC tests increased in mice lacking PKCγ (Bowers & Wehner, 2001) or ζ (Lee et al., 2014).

7. ENZYMES

Table 10 shows the effects of other assorted enzymes on ethanol drinking in mice. In particular, aldehyde dehydrogenase (ALDH) is one of the few known genes to affect risk of developing AUD in humans (Chen, Ferreira, Gross, & Mochly-Rosen, 2014). ALDH2 plays a major role in the detoxification of ethanol-derived acetaldehyde, and inhibition of ALDH is the mechanism of action of disulfiram, an FDA-approved drug for AUD. A mutation in *ALDH2* produces an enzyme incapable of metabolizing ethanol, resulting in severe adverse reactions to even small amounts of ingested alcohol in individuals who are homozygous for the mutation. UChA and UChB rat lines show similar *Aldh2* polymorphisms that affect alcohol consumption (Quintanilla, Israel, Sapag, & Tampier, 2006). Knockout of the *Aldh2* gene in mice was also shown to limit drinking in 2BC tests (Fernandez et al., 2006; Isse et al., 2002). Thus, genetic polymorphisms that alter ethanol metabolism and susceptibility to its effects may decrease ethanol-induced metabolic tolerance and risk for developing AUD.

8. NEUROPEPTIDES/HORMONES

Table 11 summarizes the effects of deletion or overexpression of classical or putative neuropeptides and hormones and their receptors on

Table 9 Alcohol Consumption in Kinase Mutant Mice

Gene Knockout/Overexpression*	Background	Operant	2BC	DID	References
cAMP-dependent protein kinase type II-β regulatory subunit (*Prkar2b*)	B6 × 129/SvJ		↑ males/females		Thiele, Willis, et al. (2000)
	B6	Genotype/sex interactions	↑ males/females		Ferraro, Sparta, Knapp, Breese, and Thiele (2006)
	B6		↑ males/females		Fee et al. (2004)
	B6 × 129/SvEv		↑ males/females		Fee et al. (2004)
Protein kinase C (PKC) ε type (*Prkce*)	B6 × 129/SvJae	↓ ↓ ADE			Olive et al. (2000)
	B6 × 129/SvJae ×FVB/N		↓		Choi et al. (2002)
	B6 × 129/SvJae		↓ (23 h)		Hodge et al. (1999)
	B6 × 129/SvJae		↓ (16 h)		Olive et al. (2005)
	B6 × 129S4		↓ ↓ posttolerance		Wallace et al. (2007)
	B6 × 129/SvJae		↓		Besheer et al. (2006)
	B6 × 129S4/SvJae			↓ (2 h) amygdala	Lesscher et al. (2009)
PKC γ type (*Prkcg*)	B6 × 129/SvEvTac		↑ males/females		Bowers and Wehner (2001)
PKC ζ type (*Prkcz*)	B6 × 129S6/SvEvTac		— continuous ↑ (4, 24 h) intermittent	— (4 h)	Lee et al. (2014)

Continued

Table 9 Alcohol Consumption in Kinase Mutant Mice—cont'd

Gene Knockout/Overexpression	Background	Operant	2BC	DID	References
Tyrosine-protein kinase Fyn (*Fyn*)	B6 × 129SF2/J		↓		Boehm, Peden, Chang, Harris, and Blednov (2003)
	129/SvJ		— (23 h)		Yaka, Tang, Camarini, Janak, and Ron (2003)
	B6 × CBA		— — ADE — poststress		Cowen, Schumann, Yagi, and Spanagel (2003)
*Fyn**	B6 × 129/Sv		↓ males/females		Boehm, Peden, et al. (2004)
	B6		↓		Stork, Kojima, Stork, Kume, and Obata (2002)
cGMP-dependent protein kinase 2 (*Prkg2*)	B6N and 129/SvN		↑		Werner et al. (2004)
Serine/threonine protein kinase, TAO2 (*Taok2*)	B6			↑ (4 h)	Kapfhamer et al. (2013)
ALK tyrosine kinase receptor (*Alk*)	B6			↑ (4 h), intermittent	Lasek et al. (2011)

*indicates gene overexpression.

—, ↓, ↑: no significant difference, decreased, or increased ethanol intake and/or preference, respectively, in mutant vs wild-type mice. Male mice were tested unless indicated otherwise. Ethanol intake in the two-bottle choice (2BC) tests was measured in 24-h sessions, unless otherwise indicated. Session times for operant tests were 16 h, and drinking in the dark (DID) session times are indicated in *parenthesis*. *ADE*, alcohol deprivation effect. Recommended mouse protein and gene (in *italics*) names are from Uniprot. B6 refers to C57BL/6J mice.

Table 10 Alcohol Consumption in Enzyme Knockout Mice

Gene Knockout	Background	Operant	2BC	DID	References
Adenylate cyclase (AC) type 1 (*Adcy1*)	B6		—		Maas et al. (2005)
AC type 5 (*Adcy5*)	B6		↑ males/ females		Kim, Kim, Baek, Lee, and Han (2011)
AC type 8 (*Adcy8*)	B6		↓		Maas et al. (2005)
Adcy1/Adcy8 double knockout	B6		↓		Maas et al. (2005)
Pituitary adenylate cyclase-activating polypeptide (*Adcyap1*)	Crlj:CD1		↑		Tanaka et al. (2010)
Aldehyde dehydrogenase, mitochondrial (*Aldh2*)	B6		↓		Isse et al. (2002)
	B6 × 129Sv/lex		↓ sex not specified		Fernandez et al. (2006)
Glutamyl aminopeptidase (Aminopeptidase A) (*Enpep*)	B6 × 129Sv		— — poststress[a]		Faber et al. (2006)
Catechol-*O*-methyltransferase (*Comt*)	B6		↑ males — females		Tammimaki, Forsberg, Karayiorgou, Gogos, and Mannisto (2008)
Amine oxidase [flavin-containing] A (*Maoa*)	Tg8 and C3H/HeJ		— (2, 24 h)		Popova, Vishnivetskaya, Ivanova, Skrinskaya, and Seif (2000)

Continued

Table 10 Alcohol Consumption in Enzyme Knockout Mice—cont'd

Gene Knockout	Background	Operant	2BC	DID	References
Neprilysin, NEP (*Mme*)	B6N		— ↑ poststress		Maul et al. (2012)
	Not specified		↑		Siems et al. (2000)
Nitric oxide synthase, brain (*Nos1*)	B6 × 129X1/SvJ		↑ 8–16%, sex not specified		Spanagel et al. (2002)
Cytochrome P450 2E1 (*Cyp2e1*)	129S1/SV-Ter		↓ preference 4–8%, females		Correa et al. (2009)
Protein phosphatase 1 regulatory subunit 1B, DARPP-32 (*Ppp1r1b*)	B6	↓ (23 h)			Risinger, Freeman, Greengard, and Fienberg (2001)
Histidine decarboxylase (*Hdc*)	129/Sv and B6		—	— (4 h), males/females	Vanhanen et al. (2013)
Tyrosine-protein phosphatase nonreceptor type 5, STEP (*Ptpn5*)	B6		↑		Legastelois, Darcq, Wegner, Lombroso, and Ron (2015)
Ubiquitin carboxyl-terminal hydrolase 46 (*Usp46*)	B6		↓		Imai, Kano, Nonoyama, and Ebihara (2013)

[a]Social stress reduced alcohol consumption in both knockout and wild-type mice, but there was no genotype difference. Ethanol intake in the two-bottle choice (2BC) tests was measured in 24-h sessions, unless indicated otherwise.

—, ↓, ↑: no significant difference, decreased, or increased ethanol intake and/or preference, respectively, in knockout vs wild-type mice. Male mice were tested unless otherwise indicated. Drinking session times for the other tests are indicated in *parenthesis*. *DID*, drinking in the dark; *1B*, one bottle. Recommended mouse protein and gene (in *italics*) names are from Uniprot. B6 refers to C57BL/6J mice.

Table 11 Alcohol Consumption in Neuropeptide/Hormone Mutant Mice

Gene Knockout/Overexpression*	Background	Operant	2BC	DID	References
Neuropeptide Y, NPY (*Npy*)	129/SvEv		— 3–10%; ↑ 20%		Thiele, Miura, Marsh, Bernstein, and Palmiter (2000)
	B6 × 129Sv		↑		
*Npy**	B6 × 129Sv		↓		Thiele, Marsh, Ste Marie, Bernstein, and Palmiter (1998)
NPY receptor type 1 (*Npy1r*)	B6		↑ 3–10%, males ↑ 10%, females		Thiele, Koh, and Pedrazzini (2002)
NPY receptor type 2 (*Npy2r*)	129/SvJ × Balb/cJ		↓		Thiele, Naveilhan, and Ernfors (2004)
	Balb/cJ		—		Thiele et al. (2004)
NPY receptor Y5 (*Npy5r*)	129/SvEv		—		Thiele, Miura, et al. (2000)
Agouti-related protein (*Agrp*)	B6	↓ (2 h), males/females	↓ (2 h), females	— (2 h), males/females ↓ (4 h), males/females	Navarro, Cubero, Ko, and Thiele (2009)
Angiotensinogen (*Agt*)	Not specified		↓		Maul et al. (2001)
*Agt**	Not specified		↑		Maul et al. (2001)
Type-1A angiotensin II receptor, AT1A (*Agtr1a*)	Not specified		↓ sex not specified		Maul et al. (2005)

Continued

Table 11 Alcohol Consumption in Neuropeptide/Hormone Mutant Mice—cont'd

Gene Knockout/Overexpression	Background	Operant	2BC	DID	References
*Agtr1a**	Not specified		—		Moore, Krstew, Kirchhoff, Davisson, and Lawrence (2007)
Type-2 angiotensin II receptor, AT2 (*Agtr2*)	Not specified		— sex not specified		Maul et al. (2005)
Cholecystokinin receptor type A, CCK-A (*Cckar*)	B6		— preference		Miyasaka et al. (2005)
Gastrin/cholecystokinin type B receptor, CCK-B (*Cckbr*)	B6		—		Miyasaka et al. (2005)
	B6		— males — preference, females		Abramov et al. (2006)
Cocaine- and amphetamine-regulated transcript protein (*Cartpt*)	B6		↓ males/females		Salinas, Nguyen, Ahmadi-Tehrani, and Morrisett (2014)
Corticotropin-releasing factor (*Crh*)	B6 × 129S		↑ (2, 23 h)		Olive et al. (2003)
	B6			↓ (2, 4 h), males/females	Kaur, Li, Stenzel-Poore, and Ryabinin (2012)
Corticotropin-releasing factor (CRF) receptor 1 (*Crhr1*)	B6			↓ (2, 4 h), males/females	Kaur et al. (2012)

	B6 × 129SvJ	— (1 h) ↓ (1 h) ADE			Chu, Koob, Cole, Zorrilla, and Roberts (2007)
	129/Ola × CD1		— ↑ poststress		Sillaber et al. (2002)
	B6, males/females		— 3–10%, ↓ 20% — (21 h) 10%, pre- and poststress ↓ (21 h) 10%, chronic stress		Pastor et al. (2011)
	B6			↓ intake (2, 4 h), males/females	Giardino and Ryabinin (2013)
	129 × 1/SvJ ×CD1		— ↑ poststress ↑ escalation, postdependence — ADE		Molander et al. (2012)
Crhr1^{NestinCre}	129S2/Sv × B6 × (B6 × SJL)		— ↓ poststress ↓ escalation, postdependence — ADE		Molander et al. (2012)
CRF receptor 2 (*Crhr2*)	B6			↓ (2 h) day 1, males/females	Kaur et al. (2012)

Continued

Table 11 Alcohol Consumption in Neuropeptide/Hormone Mutant Mice—cont'd

Gene Knockout/Overexpression	Background	Operant	2BC	DID	References
	B6		— females	↑ (30 min), males/females — (2 h), males/females	Sharpe et al. (2005)
Crhr1/Crhr2 double knockout	B6, males/females		— (21 h) 10%, pre- and poststress ↓ (21 h) 10%, chronic stress		Pastor et al. (2011)
	B6			— (2, 4 h), males/females	Kaur et al. (2012)
Galanin peptides (*Gal*)	129Ola/Hsd		↓ 15%, females — males		Karatayev et al. (2010)
*Gal**	B6		↑ 15%, pre- and postfood deprivation; males — pre- and postfood deprivation; females		Karatayev, Baylan, and Leibowitz (2009)
Neurokinin-1 receptor, Substance P receptor (*Tacr1*)	B6		↓		Thorsell, Schank, Singley, Hunt, and Heilig (2010)
Nociceptin receptor (*Oprl1*)	B6		↓ 20%, females		Sakoori and Murphy (2008)
Relaxin-3 receptor 1 (*Rxfp3*)	B6		— ↓ poststress		Walker et al. (2015)
Melanin-concentrating hormone receptor 1 (*Mchr1*)	B6		↑ males — females	— (1 h), males/females	Duncan et al. (2007)

Neurotensin receptor type 1 (*Ntsr1*)	B6 × 129X1/SvJ	↑		Lee et al. (2010)
Neurotensin receptor type 2 (*Ntsr2*)	B6 × 129X1/SvJ	↑		Lee, Hinton, Unal, Richelson, and Choi (2011)
Urocortin (*Ucn*)	B6		— (2, 4 h), males/females	Kaur et al. (2012)
	B6	↓ preference		Giardino, Cocking, Kaur, Cunningham, and Ryabinin (2011)
Vasopressin V1a receptor (*Avpr1a*)	B6Cr Slc × 129/Sv and B6Cr Slc	↑ males > females		Sanbe et al. (2008)
	B6 × 129/Sv-CP, females	— homozygous mutant — poststress		Caldwell et al. (2006)
Vasopressin V1b receptor (*Avpr1b*)	B6 × 129/SvJ, females	— — poststress		Caldwell et al. (2006)
Adiponectin receptor protein 2 (*Adipor2*)	B6		— (2 h) ↓ (2 h) CIE	Repunte-Canonigo et al. (2010)
Atrial natriuretic peptide receptor 1 (*Npr1*)	B6 × 129/SvJ	— ↑ poststress		Mutschler et al. (2010)
Ghrelin (*Ghrl*)	B6		— (90 min)	Jerlhag, Landgren, Egecioglu, Dickson, and Engel (2011)
	Not specified	↓		Bahi et al. (2013)

Continued

Table 11 Alcohol Consumption in Neuropeptide/Hormone Mutant Mice—cont'd

Gene Knockout/Overexpression	Background	Operant	2BC	DID	References
Leptin (*Lep*)	B6-Lep^{ob}		↓ males/females		Blednov, Walker, and Harris (2004)
Leptin receptor (*Lepr*)	B6-m Lepr^{db}/J		↓ males/females		Blednov, Walker, and Harris, (2004)
Melanocortin receptor 3 (*Mc3r*)	B6		— males/females		Navarro et al. (2005)
	B6			↓ (1 h) — (4 h), males/females	Olney, Sprow, Navarro, and Thiele (2014)
Melanocortin receptor 4 (*Mc4r*)	B6		— (6, 24 h), males/females		Navarro et al. (2011)

*indicates gene overexpression.

—, ↓, ↑: no significant difference, decreased, or increased intake and/or preference, respectively, in mutant vs wild-type/control mice. Male mice were tested unless indicated otherwise. Ethanol intake in the two-bottle choice (2BC) tests was measured in 24-h sessions unless indicated otherwise. Drinking session times for operant and drinking in the dark (DID) tests are indicated in *parenthesis*. Stress and alcohol dependence tend to increase ethanol intake in both control and knockout mice, and the *arrows* indicate if the drinking in knockout mice was lower or higher compared to control under these conditions (Molander et al., 2012); however, the knockout mice in Pastor et al. (2011) showed reduced intake following chronic stress, whereas wild-type mice showed increased intake. *ADE*, alcohol deprivation effect; *CIE*, chronic intermittent ethanol vapor. Recommended mouse protein and gene (in *italics*) names are from Uniprot. B6 refers to C57BL/6J mice. Classical neuropeptides are listed followed by a putative neuropeptide and hormones and hormone receptors.

voluntary ethanol administration in mice. Of the genes represented here, the corticotropin-releasing factor/urocortin family and other stress-related neuromodulators are promising for future studies of genetic determinants of AUD (Schank, Ryabinin, Giardino, Ciccocioppo, & Heilig, 2012).

9. OTHER GENE TARGETS

Table 12 represents an assortment of genes that did not specifically fit into the previous categories. These genes are associated with synaptic function, development, circadian regulation, and other cellular regulatory functions but are not discussed individually herein. One of the most pronounced phenotypes among mutant mice is the almost complete blockade of alcohol consumption in mice lacking any one of the three taste genes (*Gnat3*, *Tas1r3*, and *Trpm5*) (Blednov et al., 2008).

10. CONCLUDING REMARKS

Mouse models of voluntary ethanol administration have been instrumental for profiling putative behavioral and genetic determinants in human alcoholics, who exhibit excessive consumption as a hallmark of the disease. The impact of more than 150 genes on alcohol consumption has been evaluated by construction of mutant mice. The global knockout strategy has been used extensively in addiction research to link proteins with behavior, and most studies presented in this chapter used this approach. Collectively speaking, the studies suggest the potential involvement of a large number of different mouse genes in voluntary alcohol drinking. However, many of the effects are modest and global knockout or overexpression of a specific gene does not determine if the effect on drinking is directly linked to effects of decreased or increased expression of the respective protein. Other technologies such as knockin animal models and conditional and cell-specific knockouts will provide more discriminating tools for linking ethanol-sensitive sites on proteins to behaviors (Blednov, Borghese, et al., 2011). The goal of the knockin approach is to remove ethanol action on a specific protein without otherwise altering its function, but this strategy (like the knockout model) can still be complicated by compensatory actions from other proteins. Selectively targeting an ethanol site could also affect other modulatory sites of the protein despite the appearance of normal function (Harris, Osterndorff-Kahanek, Ponomarev, Homanics, & Blednov, 2011).

Table 12 Alcohol Consumption in Other Mutant Mice

Gene Knockout/Overexpression*	Background	Operant	2BC	DID	References
α-Synuclein (*Snca*)	B6		↑		Lopez-Jimenez et al. (2013)
β-Arrestin-1 (*Arrb1*)	B6		—		Bjork et al. (2008)
β-Arrestin-2 (*Arrb2*)	B6		↓ 9–15%		Bjork et al. (2008)
	B6		↑		Li, Tao, Ma, Liu, and Ma (2013)
Gap junction delta-2 protein, Connexin-36 (*Gjd2*)	B6			↓ (2 h) 20%	Steffensen et al. (2011)
G1/S-specific cyclin-D2 (*Ccnd2*)	129X1/SvJ × B6 × BALB/cAnNCrl		— 2%, 4% ↑ 8%, 16%		Jaholkowski et al. (2011)
α-Gustducin (*Gnat3*)	B6		↓		Blednov et al. (2008)
Histamine H3 receptor (*Hrh3*)	B6		↓	↓ (2, 4 h)	Nuutinen et al. (2011)
LIM domain only protein 3 (*Lmo3*)	B6		— males/females	↑ (2, 4 h), males/females	Savarese, Zou, Kharazia, Maiya, and Lasek (2014)
Metallothionein-1/2 (*Mt1*, *Mt2* double knockout)	129S7/SvEvBrd and 129S1/SvImJ		↑ males/females		Loney, Uddin, and Singh (2006)
Period circadian protein homolog 1 (*Per1*)	*Per1*^{Brdm1} B6-Tyr$^{c\text{-}Brd}$ × 129S7		— drink-o-meter ↑ stress		Dong et al. (2011)
	B6-Tyr$^{c\text{-}Brd}$ × 129SvEv^{Brd}	—	— — ADE		Zghoul et al. (2007)
Period circadian protein homolog 2 (*Per2*)	*Per2*^{Brdm1} B6-Tyr$^{c\text{-}Brd}$ × 129SvEv^{Brd}	↑	↑ 8–16% ↓ after acamprosate		Spanagel et al. (2005)
	B6		↑ drink-o-meter		Brager et al. (2011a)
	B6		↑		Brager et al. (2011b)

Disks large homolog 4, PSD-95 (*Dlg4*)	B6	↓ males/females ↑ day 1, ADE	Camp et al. (2011)
Protein fosB (*Fosb*)	129Sv × BALB/c	— males/females	Korkosz et al. (2004)
Protransforming growth factor α (*Tgfa*)*	CD1	↑	Hilakivi-Clarke and Goldberg (1995)
Epidermal growth factor receptor kinase substrate 8 (*Eps8*)	B6	↑ males/females	Offenhauser et al. (2006)
Ras-related protein Rab-3A (*Rab3a*)	B6	— (23 h)	Kapfhamer et al. (2008)
Ras-specific guanine nucleotide-releasing factor 2 (*Rasgrf2*)	B6	↓	Stacey et al. (2012)
Regulator of G-protein signaling 6 (*Rgs6*)	B6 × 129/Sv	↓ males/females	Stewart et al. (2015)
Taste receptor type 1 member 3 (*Tas1r3*)	B6	↓	Blednov et al. (2008)
	B6	↓ males/females	Brasser, Norman, and Lemon (2010)
Trace amine-associated receptor 1 (*Taar1*)	B6 × 129S1/Sv	↑ females	Lynch et al. (2013)
Protein unc-79 homolog (*Unc79*)	B6 and B6 × DBA/2J	↑	Speca et al. (2010)

*indicates gene overexpression.

—, ↓, ↑: no significant difference, decreased, or increased ethanol intake and/or preference, respectively, in mutant vs wild-type mice. Male mice were tested unless indicated otherwise. Ethanol intake in the operant and two-bottle choice (2BC) tests was measured in 30-min and 24-h sessions, respectively, unless otherwise indicated. Drinking in the dark (DID) session times are indicated in *parenthesis*. *ADE*, alcohol deprivation effect. Recommended mouse protein and gene (in *italics*) names are from Uniprot. B6 refers to C57BL/6J mice.

When focusing on genes that have more profound effects on consumption and have consistently been implicated across multiple drinking tests, the number of targets diminishes. For example, null mutant mice for specific $GABA_A$ (Table 1), dopamine (Table 3), cannabinoid and opioid receptors (Table 6), immune-related genes (Table 7), and PKCε (Table 9) have demonstrated decreased drinking and preference in several drinking tests. These systems either have established roles in addiction research or represent emerging targets, as in the case of immune pathways. Further consideration should be given to the impact of the drinking paradigms on peripheral and central genes. For example, even when similar amounts of total ethanol were consumed in the continuous- and intermittent-access tests by B6 mice, brain and liver transcriptomes were differentially affected (Osterndorff-Kahanek, Ponomarev, Blednov, & Harris, 2013). The distinct genomic effects induced by the alcohol exposure protocol may explain the differing effects observed in some drinking tests. The peripheral effects of alcohol may be particularly important for neuroimmune modulation of drinking because production of cytokines outside the brain can influence brain function. This peripheral–central neuroimmune communication may also be important in alcohol craving and dependence in humans (Leclercq, De Saeger, Delzenne, de Timary, & Starkel, 2014).

11. MEDICATION DEVELOPMENT

FDA-approved drugs for AUD have provided only modest benefit and are not routinely prescribed, and so the search continues for more effective drugs. Identifying existing drugs that could be repurposed to treat AUD is a current goal for researchers and, if successful, would fast track therapeutic options for the disease. A strategy for prioritizing relevant genes from a large list of potential targets is to examine the preclinical evidence in combination with genetic association studies in human alcoholics. Overlapping evidence from animal and human studies will help pinpoint genes and their biological networks that could be most relevant in the treatment of AUD. Mouse drinking models could then be used to test viable targets using approved drugs known to modulate the target network.

Proinflammatory neuroimmune signaling pathways have been increasingly implicated in the etiology and progression of AUD and offer unexplored avenues in its treatment. Currently, this is an emphasis area of the Integrative Neuroscience Initiative on Alcoholism (INIA) sponsored by the National Institutes of Health, National Institute on Alcohol Abuse and Alcoholism. Several FDA-approved drugs with antiinflammatory and immune inhibitory effects regulate alcohol responses in animal models, lending support that some of these could potentially be repurposed to treat AUD. For example, the peroxisome proliferator-activated receptor (PPAR) agonists, fenofibrate and tesaglitazar, reduced drinking in a PPARα-dependent manner (Blednov, Black, Benavidez, Stamatakis, & Harris, 2016), altered expression of immune genes in the liver, and altered neuronal gene expression in mouse brain (Ferguson, Most, Blednov, & Harris, 2014). There is overlapping evidence in mice and humans for specific PPAR genes in alcohol consumption and dependence (Blednov, Benavidez, Black, Ferguson, et al., 2015). In addition, minocycline reduced DID in adult mice, and a bioinformatics pathway analysis revealed an overexpression of neuroimmune-related pathways in these mice (Agrawal et al., 2014). Mouse models will continue to be essential for identifying candidate genes and screening medications to modulate the affected pathways.

12. FUTURE DIRECTIONS

While the study of individual genes is informative, combining gene network and systems biology approaches to identify interrelated networks and pathways is critical in the future treatment of AUD. Because complex trait diseases involve coordinated expression changes in multiple gene families, examining gene clusters is an important research direction, as supported by the INIA studies showing that coexpression patterns can distinguish gene modules related to alcohol consumption in animal models (Iancu et al., 2013; Nunez et al., 2013; Saba et al., 2011, 2015). Another INIA study reported that RNA-Seq profiling of postmortem human prefrontal cortex also revealed disrupted gene networks in alcohol-dependent individuals compared to matched controls (Farris, Arasappan, Hunicke-Smith, Harris, & Mayfield, 2015). The subnetworks related to lifetime alcohol consumption in humans contained known alcohol targets, from the preclinical literature, and were overrepresented for genes involved in synaptic function. These genes were also enriched in a transcriptome meta-analysis of mouse drinking behavior (Mulligan et al., 2006), suggesting an overlapping set of alcohol-related genes across species. Comparison of human and animal model networks, combined with single gene-based approaches, will be important in understanding the neurobiology of AUD (Farris, Pietrzykowski, et al., 2015).

Genomic and proteomic techniques are capable of analyzing complex data sets into related biological processes (Gorini, Bell, & Mayfield, 2011;

Gorini, Harris, & Mayfield, 2014; Gorini, Roberts, & Mayfield, 2013; Sikela et al., 2006). Ideally, integration of genetics and transcriptomics with convergent biological processes and phenotypic behaviors will reveal prospective therapeutic targets for the disease (Gorini et al., 2011, 2014). Drug addiction is mediated by both genetic and environmental determinants, and neuroadaptations in reward circuitry are related to disease progression. Therefore, in addition to individual candidate genes, future research must consider the gene and protein networks and associated biological system changes observed at different stages of the disease. These network-centric approaches can summarize complex gene lists into interrelated components and provide a systems-level picture of the disease. This biological systems framework will be key in determining causal factors of AUD and providing a more integrated, functional approach to its treatment.

ACKNOWLEDGMENTS

The authors acknowledge funding from NIAAA grants AA006399, AA013520, AA020926, and AA012404. The authors have no conflicts of interest with this material. The views expressed herein are solely those of the authors and do not necessarily represent those of the funding agencies.

REFERENCES

Abramov, U., Raud, S., Innos, J., Koks, S., Matsui, T., & Vasar, E. (2006). Gender specific effects of ethanol in mice, lacking CCK2 receptors. *Behavioural Brain Research*, *175*(1), 149–156. http://dx.doi.org/10.1016/j.bbr.2006.08.015.

Agrawal, R. G., Owen, J. A., Levin, P. S., Hewetson, A., Berman, A. E., Franklin, S. R., … Bergeson, S. E. (2014). Bioinformatics analyses reveal age-specific neuroimmune modulation as a target for treatment of high ethanol drinking. *Alcoholism, Clinical and Experimental Research*, *38*(2), 428–437. http://dx.doi.org/10.1111/acer.12288.

Ary, A. W., Cozzoli, D. K., Finn, D. A., Crabbe, J. C., Dehoff, M. H., Worley, P. F., & Szumlinski, K. K. (2012). Ethanol up-regulates nucleus accumbens neuronal activity dependent pentraxin (Narp): Implications for alcohol-induced behavioral plasticity. *Alcohol*, *46*(4), 377–387. http://dx.doi.org/10.1016/j.alcohol. 2011.10.003.

Bahi, A., Tolle, V., Fehrentz, J. A., Brunel, L., Martinez, J., Tomasetto, C. L., & Karam, S. M. (2013). Ghrelin knockout mice show decreased voluntary alcohol consumption and reduced ethanol-induced conditioned place preference. *Peptides*, *43*, 48–55. http://dx.doi.org/10.1016/j.peptides.2013.02.008.

Basavarajappa, B. S., Yalamanchili, R., Cravatt, B. F., Cooper, T. B., & Hungund, B. L. (2006). Increased ethanol consumption and preference and decreased ethanol sensitivity in female FAAH knockout mice. *Neuropharmacology*, *50*(7), 834–844. http://dx.doi.org/10.1016/j.neuropharm.2005.12.005.

Bauer, I. E., Graham, D. P., Soares, J. C., & Nielsen, D. A. (2015). Serotonergic gene variation in substance use pharmacotherapy: A systematic review. *Pharmacogenomics*, *16*(11), 1307–1314. http://dx.doi.org/10.2217/pgs.15.72.

Bechtholt, A. J., Smith, R., Raber, J., & Cunningham, C. L. (2004). Enhanced ethanol-, but not cocaine-induced, conditioned place preference in Apoe(-/-) mice. *Pharmacology, Biochemistry, and Behavior, 77*(4), 783–792. http://dx.doi.org/10.1016/j.pbb.2004.02.002.

Becker, A., Grecksch, G., Kraus, J., Loh, H. H., Schroeder, H., & Hollt, V. (2002). Rewarding effects of ethanol and cocaine in mu opioid receptor-deficient mice. *Naunyn-Schmiedeberg's Archives of Pharmacology, 365*(4), 296–302. http://dx.doi.org/10.1007/s00210-002-0533-2.

Besheer, J., Lepoutre, V., Mole, B., & Hodge, C. W. (2006). GABAA receptor regulation of voluntary ethanol drinking requires PKCepsilon. *Synapse, 60*(6), 411–419. http://dx.doi.org/10.1002/syn.20314.

Bird, M. K., Kirchhoff, J., Djouma, E., & Lawrence, A. J. (2008). Metabotropic glutamate 5 receptors regulate sensitivity to ethanol in mice. *The International Journal of Neuropsychopharmacology, 11*(6), 765–774. http://dx.doi.org/10.1017/S1461145708008572.

Bjork, K., Rimondini, R., Hansson, A. C., Terasmaa, A., Hyytia, P., Heilig, M., & Sommer, W. H. (2008). Modulation of voluntary ethanol consumption by beta-arrestin 2. *The FASEB Journal, 22*(7), 2552–2560. http://dx.doi.org/10.1096/fj.07-102442.

Blednov, Y. A., Benavidez, J. M., Black, M., Ferguson, L. B., Schoenhard, G. L., Goate, A. M., … Harris, R. A. (2015). Peroxisome proliferator-activated receptors alpha and gamma are linked with alcohol consumption in mice and withdrawal and dependence in humans. *Alcoholism, Clinical and Experimental Research, 39*(1), 136–145. http://dx.doi.org/10.1111/acer.12610.

Blednov, Y. A., Benavidez, J. M., Black, M., Leiter, C. R., Osterndorff-Kahanek, E., & Harris, R. A. (2015). Glycine receptors containing alpha2 or alpha3 subunits regulate specific ethanol-mediated behaviors. *The Journal of Pharmacology and Experimental Therapeutics, 353*(1), 181–191. http://dx.doi.org/10.1124/jpet.114.221895.

Blednov, Y. A., Benavidez, J. M., Black, M., Leiter, C. R., Osterndorff-Kahanek, E., Johnson, D., … Harris, R. A. (2014). GABAA receptors containing rho1 subunits contribute to in vivo effects of ethanol in mice. *PLoS One, 9*(1), e85525. http://dx.doi.org/10.1371/journal.pone.0085525.

Blednov, Y. A., Benavidez, J. M., Geil, C., Perra, S., Morikawa, H., & Harris, R. A. (2011). Activation of inflammatory signaling by lipopolysaccharide produces a prolonged increase of voluntary alcohol intake in mice. *Brain, Behavior, and Immunity, 25*(Suppl. 1), S92–S105. http://dx.doi.org/10.1016/j.bbi.2011.01.008.

Blednov, Y. A., Bergeson, S. E., Walker, D., Ferreira, V. M., Kuziel, W. A., & Harris, R. A. (2005). Perturbation of chemokine networks by gene deletion alters the reinforcing actions of ethanol. *Behavioural Brain Research, 165*(1), 110–125. http://dx.doi.org/10.1016/j.bbr.2005.06.026.

Blednov, Y. A., Black, M., Benavidez, J. M., Stamatakis, E. E., & Harris, R. A. (2016). PPAR agonists: I. Role of receptor subunits in alcohol consumption in male and female mice. *Alcoholism, Clinical and Experimental Research, 40*(3), 553–562. http://dx.doi.org/10.1111/acer.12976.

Blednov, Y. A., Borghese, C. M., McCracken, M. L., Benavidez, J. M., Geil, C. R., Osterndorff-Kahanek, E., … Harris, R. A. (2011). Loss of ethanol conditioned taste aversion and motor stimulation in knockin mice with ethanol-insensitive alpha2-containing GABA(A) receptors. *The Journal of Pharmacology and Experimental Therapeutics, 336*(1), 145–154. http://dx.doi.org/10.1124/jpet.110.171645.

Blednov, Y. A., Cravatt, B. F., Boehm, S. L., 2nd, Walker, D., & Harris, R. A. (2007). Role of endocannabinoids in alcohol consumption and intoxication: Studies of mice lacking fatty acid amide hydrolase. *Neuropsychopharmacology, 32*(7), 1570–1582. http://dx.doi.org/10.1038/sj.npp.1301274.

Blednov, Y. A., & Harris, R. A. (2008). Metabotropic glutamate receptor 5 (mGluR5) regulation of ethanol sedation, dependence and consumption: Relationship to acamprosate actions. *The International Journal of Neuropsychopharmacology, 11*(6), 775–793. http://dx.doi.org/10.1017/S1461145708008584.

Blednov, Y. A., & Harris, R. A. (2009). Deletion of vanilloid receptor (TRPV1) in mice alters behavioral effects of ethanol. *Neuropharmacology, 56*(4), 814–820. Retrieved from, http://www.ncbi.nlm.nih.gov/pubmed/19705551.

Blednov, Y. A., Ponomarev, I., Geil, C., Bergeson, S., Koob, G. F., & Harris, R. A. (2012). Neuroimmune regulation of alcohol consumption: Behavioral validation of genes obtained from genomic studies. *Addiction Biology, 17*(1), 108–120. http://dx.doi.org/10.1111/j.1369-1600.2010.00284.x.

Blednov, Y. A., Stoffel, M., Chang, S. R., & Harris, R. A. (2001). Potassium channels as targets for ethanol: Studies of G-protein-coupled inwardly rectifying potassium channel 2 (GIRK2) null mutant mice. *The Journal of Pharmacology and Experimental Therapeutics, 298*(2), 521–530. Retrieved from, http://www.ncbi.nlm.nih.gov/pubmed/11454913.

Blednov, Y. A., Walker, D., Alva, H., Creech, K., Findlay, G., & Harris, R. A. (2003). GABAA receptor alpha 1 and beta 2 subunit null mutant mice: Behavioral responses to ethanol. *The Journal of Pharmacology and Experimental Therapeutics, 305*(3), 854–863. http://dx.doi.org/10.1124/jpet.103.049478.

Blednov, Y. A., Walker, D., & Harris, R. A. (2004). Blockade of the leptin-sensitive pathway markedly reduces alcohol consumption in mice. *Alcoholism, Clinical and Experimental Research, 28*(11), 1683–1692. Retrieved from, http://www.ncbi.nlm.nih.gov/pubmed/15547455.

Blednov, Y. A., Walker, D. L., Iyer, S. V., Homanics, G., & Harris, A. R. (2010). Mice lacking Gad2 show altered behavioral effects of ethanol, flurazepam and gabaxadol. *Addiction Biology, 15*(1), 45–61. http://dx.doi.org/10.1111/j.1369-1600.2009.00186.x.

Blednov, Y. A., Walker, D., Martinez, M., & Harris, R. A. (2006). Reduced alcohol consumption in mice lacking preprodynorphin. *Alcohol, 40*(2), 73–86. http://dx.doi.org/10.1016/j.alcohol.2006.12.002.

Blednov, Y. A., Walker, D., Martinez, M., Levine, M., Damak, S., & Margolskee, R. F. (2008). Perception of sweet taste is important for voluntary alcohol consumption in mice. *Genes, Brain, and Behavior,* 7(1), 1–13. http://dx.doi.org/10.1111/j.1601-183X.2007.00309.x.

Blednov, Y. A., Walker, D., Osterndorf-Kahanek, E., & Harris, R. A. (2004). Mice lacking metabotropic glutamate receptor 4 do not show the motor stimulatory effect of ethanol. *Alcohol, 34*(2–3), 251–259. Retrieved from, http://www.ncbi.nlm.nih.gov/pubmed/15902920.

Bodhinathan, K., & Slesinger, P. A. (2014). Alcohol modulation of G-protein-gated inwardly rectifying potassium channels: From binding to therapeutics. *Frontiers in Physiology, 5*, 76. http://dx.doi.org/10.3389/fphys.2014.00076.

Boehm, S. L., 2nd., Peden, L., Chang, R., Harris, R. A., & Blednov, Y. A. (2003). Deletion of the fyn-kinase gene alters behavioral sensitivity to ethanol. *Alcoholism, Clinical and Experimental Research,* 27(7), 1033–1040. http://dx.doi.org/10.1097/01.ALC.0000075822.80583.71.

Boehm, S. L., 2nd., Peden, L., Jennings, A. W., Kojima, N., Harris, R. A., & Blednov, Y. A. (2004). Over-expression of the fyn-kinase gene reduces hypnotic sensitivity to ethanol in mice. *Neuroscience Letters, 372*(1–2), 6–11. http://dx.doi.org/10.1016/j.neulet.2004.08.028.

Boehm, S. L., 2nd., Ponomarev, I., Jennings, A. W., Whiting, P. J., Rosahl, T. W., Garrett, E. M., … Harris, R. A. (2004). gamma-Aminobutyric acid A receptor subunit mutant mice: New perspectives on alcohol actions. *Biochemical Pharmacology, 68*(8), 1581–1602. http://dx.doi.org/10.1016/j.bcp.2004.07.023.

Bonasera, S. J., Chu, H. M., Brennan, T. J., & Tecott, L. H. (2006). A null mutation of the serotonin 6 receptor alters acute responses to ethanol. *Neuropsychopharmacology*, *31*(8), 1801–1813. http://dx.doi.org/10.1038/sj.npp.1301030.

Bouwknecht, J. A., Hijzen, T. H., van der Gugten, J., Maes, R. A., Hen, R., & Olivier, B. (2000). Ethanol intake is not elevated in male 5-HT(1B) receptor knockout mice. *European Journal of Pharmacology*, *403*(1–2), 95–98. Retrieved from, http://www.ncbi.nlm.nih.gov/pubmed/10969149.

Bowers, B. J., & Wehner, J. M. (2001). Ethanol consumption and behavioral impulsivity are increased in protein kinase Cgamma null mutant mice. *The Journal of Neuroscience*, *21*(21), RC180. Retrieved from, http://www.ncbi.nlm.nih.gov/pubmed/11606660.

Boyce-Rustay, J. M., & Holmes, A. (2006). Ethanol-related behaviors in mice lacking the NMDA receptor NR2A subunit. *Psychopharmacology*, *187*(4), 455–466. http://dx.doi.org/10.1007/s00213-006-0448-6.

Boyce-Rustay, J. M., & Risinger, F. O. (2003). Dopamine D3 receptor knockout mice and the motivational effects of ethanol. *Pharmacology, Biochemistry, and Behavior*, *75*(2), 373–379. Retrieved from, http://www.ncbi.nlm.nih.gov/pubmed/12873629.

Boyce-Rustay, J. M., Wiedholz, L. M., Millstein, R. A., Carroll, J., Murphy, D. L., Daws, L. C., & Holmes, A. (2006). Ethanol-related behaviors in serotonin transporter knockout mice. *Alcoholism, Clinical and Experimental Research*, *30*(12), 1957–1965. http://dx.doi.org/10.1111/j.1530-0277.2006.00241.x.

Brager, A. J., Prosser, R. A., & Glass, J. D. (2011a). Circadian and acamprosate modulation of elevated ethanol drinking in mPer2 clock gene mutant mice. *Chronobiology International*, *28*(8), 664–672. http://dx.doi.org/10.3109/07420528.2011.601968.

Brager, A., Prosser, R. A., & Glass, J. D. (2011b). Acamprosate-responsive brain sites for suppression of ethanol intake and preference. *American Journal of Physiology. Regulatory, Integrative and Comparative Physiology*, *301*(4), R1032–R1043. http://dx.doi.org/10.1152/ajpregu.00179.2011.

Brasser, S. M., Norman, M. B., & Lemon, C. H. (2010). T1r3 taste receptor involvement in gustatory neural responses to ethanol and oral ethanol preference. *Physiological Genomics*, *41*(3), 232–243. http://dx.doi.org/10.1152/physiolgenomics.00113.2009.

Bulwa, Z. B., Sharlin, J. A., Clark, P. J., Bhattacharya, T. K., Kilby, C. N., Wang, Y., & Rhodes, J. S. (2011). Increased consumption of ethanol and sugar water in mice lacking the dopamine D2 long receptor. *Alcohol*, *45*(7), 631–639. http://dx.doi.org/10.1016/j.alcohol.2011.06.004.

Cai, Y. Q., Cai, G. Q., Liu, G. X., Cai, Q., Shi, J. H., Shi, J., … Fei, J. (2006). Mice with genetically altered GABA transporter subtype I (GAT1) expression show altered behavioral responses to ethanol. *Journal of Neuroscience Research*, *84*(2), 255–267. http://dx.doi.org/10.1002/jnr.20884.

Caldwell, H. K., Stewart, J., Wiedholz, L. M., Millstein, R. A., Iacangelo, A., Holmes, A., … Wersinger, S. R. (2006). The acute intoxicating effects of ethanol are not dependent on the vasopressin 1a or 1b receptors. *Neuropeptides*, *40*(5), 325–337. http://dx.doi.org/10.1016/j.npep.2006.08.001.

Camp, M. C., Feyder, M., Ihne, J., Palachick, B., Hurd, B., Karlsson, R. M., … Holmes, A. (2011). A novel role for PSD-95 in mediating ethanol intoxication, drinking and place preference. *Addiction Biology*, *16*(3), 428–439. http://dx.doi.org/10.1111/j.1369-1600.2010.00282.x.

Charbogne, P., Kieffer, B. L., & Befort, K. (2014). 15 years of genetic approaches in vivo for addiction research: Opioid receptor and peptide gene knockout in mouse models of drug abuse. *Neuropharmacology*, *76*(Pt. B), 204–217. http://dx.doi.org/10.1016/j.neuropharm.2013.08.028.

Chen, C. H., Ferreira, J. C., Gross, E. R., & Mochly-Rosen, D. (2014). Targeting aldehyde dehydrogenase 2: New therapeutic opportunities. *Physiological Reviews*, *94*(1), 1–34. http://dx.doi.org/10.1152/physrev.00017.2013.

Chen, J., Nam, H. W., Lee, M. R., Hinton, D. J., Choi, S., Kim, T., … Choi, D. S. (2010). Altered glutamatergic neurotransmission in the striatum regulates ethanol sensitivity and intake in mice lacking ENT1. *Behavioural Brain Research*, *208*(2), 636–642. http://dx.doi.org/10.1016/j.bbr.2010.01.011.

Choi, D. S., Cascini, M. G., Mailliard, W., Young, H., Paredes, P., McMahon, T., … Messing, R. O. (2004). The type 1 equilibrative nucleoside transporter regulates ethanol intoxication and preference. *Nature Neuroscience*, 7(8), 855–861. http://dx.doi.org/10.1038/nn1288.

Choi, D. S., Wang, D., Dadgar, J., Chang, W. S., & Messing, R. O. (2002). Conditional rescue of protein kinase C epsilon regulates ethanol preference and hypnotic sensitivity in adult mice. *The Journal of Neuroscience*, *22*(22), 9905–9911. Retrieved from, http://www.ncbi.nlm.nih.gov/pubmed/12427847.

Chu, K., Koob, G. F., Cole, M., Zorrilla, E. P., & Roberts, A. J. (2007). Dependence-induced increases in ethanol self-administration in mice are blocked by the CRF1 receptor antagonist antalarmin and by CRF1 receptor knockout. *Pharmacology, Biochemistry, and Behavior*, *86*(4), 813–821. http://dx.doi.org/10.1016/j.pbb.2007.03.009.

Clarke, T. K., Laucht, M., Ridinger, M., Wodarz, N., Rietschel, M., Maier, W., … Schumann, G. (2011). KCNJ6 is associated with adult alcohol dependence and involved in gene x early life stress interactions in adolescent alcohol drinking. *Neuropsychopharmacology*, *36*(6), 1142–1148. http://dx.doi.org/10.1038/npp.2010.247.

Contet, C., Kim, A., Le, D., Iyengar, S. K., Kotzebue, R. W., Yuan, C. J., … Mandyam, C. D. (2014). mu-Opioid receptors mediate the effects of chronic ethanol binge drinking on the hippocampal neurogenic niche. *Addiction Biology*, *19*(5), 770–780. http://dx.doi.org/10.1111/adb.12040.

Correa, M., Viaggi, C., Escrig, M. A., Pascual, M., Guerri, C., Vaglini, F., … Corsini, G. U. (2009). Ethanol intake and ethanol-induced locomotion and locomotor sensitization in Cyp2e1 knockout mice. *Pharmacogenetics and Genomics*, *19*(3), 217–225. http://dx.doi.org/10.1097/FPC.0b013e328324e726.

Cowen, M. S., Schroff, K. C., Gass, P., Sprengel, R., & Spanagel, R. (2003). Neurobehavioral effects of alcohol in AMPA receptor subunit (GluR1) deficient mice. *Neuropharmacology*, *45*(3), 325–333. Retrieved from, http://www.ncbi.nlm.nih.gov/pubmed/12871650.

Cowen, M. S., Schumann, G., Yagi, T., & Spanagel, R. (2003). Role of Fyn tyrosine kinase in ethanol consumption by mice. *Alcoholism, Clinical and Experimental Research*, *27*(8), 1213–1219. http://dx.doi.org/10.1097/01.ALC.0000081630.14159.02.

Crabbe, J. C., Harkness, J. H., Spence, S. E., Huang, L. C., & Metten, P. (2012). Intermittent availability of ethanol does not always lead to elevated drinking in mice. *Alcohol and Alcoholism*, *47*(5), 509–517. http://dx.doi.org/10.1093/alcalc/ags067.

Crabbe, J. C., Phillips, T. J., Feller, D. J., Hen, R., Wenger, C. D., Lessov, C. N., & Schafer, G. L. (1996). Elevated alcohol consumption in null mutant mice lacking 5-HT1B serotonin receptors. *Nature Genetics*, *14*(1), 98–101. http://dx.doi.org/10.1038/ng0996-98.

Crabbe, J. C., Phillips, T. J., Harris, R. A., Arends, M. A., & Koob, G. F. (2006). Alcohol-related genes: Contributions from studies with genetically engineered mice. *Addiction Biology*, *11*(3–4), 195–269. http://dx.doi.org/10.1111/j.1369-1600.2006.00038.x.

Crews, F. T., & Vetreno, R. P. (2015). Mechanisms of neuroimmune gene induction in alcoholism. *Psychopharmacology*. http://dx.doi.org/10.1007/s00213-015-3906-1[Epub ahead of print].

Dawson, A., Miles, M. F., & Damaj, M. I. (2013). The beta2 nicotinic acetylcholine receptor subunit differentially influences ethanol behavioral effects in the mouse. *Alcohol*, *47*(2), 85–94. http://dx.doi.org/10.1016/j.alcohol.2012.12.004.

Delis, F., Thanos, P. K., Rombola, C., Rosko, L., Grandy, D., Wang, G. J., & Volkow, N. D. (2013). Chronic mild stress increases alcohol intake in mice with low dopamine D2 receptor levels. *Behavioral Neuroscience*, *127*(1), 95–105. http://dx.doi.org/10.1037/a0030750.

Dong, L., Bilbao, A., Laucht, M., Henriksson, R., Yakovleva, T., Ridinger, M., ... Schumann, G. (2011). Effects of the circadian rhythm gene period 1 (per1) on psychosocial stress-induced alcohol drinking. *The American Journal of Psychiatry*, *168*(10), 1090–1098. http://dx.doi.org/10.1176/appi.ajp.2011.10111579.

D'Souza, M. S. (2015). Glutamatergic transmission in drug reward: Implications for drug addiction. *Frontiers in Neuroscience*, *9*, 404. http://dx.doi.org/10.3389/fnins.2015.00404.

Duncan, E. A., Sorrell, J. E., Adamantidis, A., Rider, T., Jandacek, R. J., Seeley, R. J., ... Woods, S. C. (2007). Alcohol drinking in MCH receptor-1-deficient mice. *Alcoholism, Clinical and Experimental Research*, *31*(8), 1325–1337. http://dx.doi.org/10.1111/j.1530-0277.2007.00427.x.

El-Ghundi, M., George, S. R., Drago, J., Fletcher, P. J., Fan, T., Nguyen, T., ... O'Dowd, B. F. (1998). Disruption of dopamine D1 receptor gene expression attenuates alcohol-seeking behavior. *European Journal of Pharmacology*, *353*(2–3), 149–158. Retrieved from, http://www.ncbi.nlm.nih.gov/pubmed/9726645.

Engel, S. R., Lyons, C. R., & Allan, A. M. (1998). 5-HT3 receptor over-expression decreases ethanol self administration in transgenic mice. *Psychopharmacology*, *140*(2), 243–248. Retrieved from, http://www.ncbi.nlm.nih.gov/pubmed/9860116.

Enoch, M. A. (2008). The role of GABA(A) receptors in the development of alcoholism. *Pharmacology, Biochemistry, and Behavior*, *90*(1), 95–104. http://dx.doi.org/10.1016/j.pbb.2008.03.007.

Faber, F., Gembardt, F., Sun, X., Mizutani, S., Siems, W. E., & Walther, T. (2006). Lack of angiotensin II conversion to angiotensin III increases water but not alcohol consumption in aminopeptidase A-deficient mice. *Regulatory Peptides*, *136*(1–3), 130–137. http://dx.doi.org/10.1016/j.regpep.2006.06.001.

Falzone, T. L., Gelman, D. M., Young, J. I., Grandy, D. K., Low, M. J., & Rubinstein, M. (2002). Absence of dopamine D4 receptors results in enhanced reactivity to unconditioned, but not conditioned, fear. *The European Journal of Neuroscience*, *15*(1), 158–164. Retrieved from, http://www.ncbi.nlm.nih.gov/pubmed/11860516.

Farris, S. P., Arasappan, D., Hunicke-Smith, S., Harris, R. A., & Mayfield, R. D. (2015). Transcriptome organization for chronic alcohol abuse in human brain. *Molecular Psychiatry*, *20*(11), 1438–1447. http://dx.doi.org/10.1038/mp.2014.159.

Farris, S. P., Pietrzykowski, A. Z., Miles, M. F., O'Brien, M. A., Sanna, P. P., Zakhari, S., ... Harris, R. A. (2015). Applying the new genomics to alcohol dependence. *Alcohol*, *49*(8), 825–836. http://dx.doi.org/10.1016/j.alcohol.2015.03.001.

Fee, J. R., Sparta, D. R., Knapp, D. J., Breese, G. R., Picker, M. J., & Thiele, T. E. (2004). Predictors of high ethanol consumption in RIIbeta knock-out mice: Assessment of anxiety and ethanol-induced sedation. *Alcoholism, Clinical and Experimental Research*, *28*(10), 1459–1468. Retrieved from, http://www.ncbi.nlm.nih.gov/pubmed/15597077.

Femenia, T., & Manzanares, J. (2012). Increased ethanol intake in prodynorphin knockout mice is associated to changes in opioid receptor function and dopamine transmission. *Addiction Biology*, *17*(2), 322–337. http://dx.doi.org/10.1111/j.1369-1600.2011. 00378.x.

Ferguson, L. B., Most, D., Blednov, Y. A., & Harris, R. A. (2014). PPAR agonists regulate brain gene expression: Relationship to their effects on ethanol consumption. *Neuropharmacology*, *86*, 397–407. http://dx.doi.org/10.1016/j.neuropharm.2014.06.024.

Fernandez, E., Koek, W., Ran, Q., Gerhardt, G. A., France, C. P., & Strong, R. (2006). Monoamine metabolism and behavioral responses to ethanol in mitochondrial aldehyde dehydrogenase knockout mice. *Alcoholism, Clinical and Experimental Research*, *30*(10), 1650–1658. http://dx.doi.org/10.1111/j.1530-0277.2006.00200.x.

Ferraro, F. M., 3rd, Sparta, D. R., Knapp, D. J., Breese, G. R., & Thiele, T. E. (2006). Increased consumption but not operant self-administration of ethanol in mice lacking the RIIbeta subunit of protein kinase A. *Alcoholism, Clinical and Experimental Research*, *30*(5), 825–835. http://dx.doi.org/10.1111/j.1530-0277.2006.00096.x.

Finn, D. A., Belknap, J. K., Cronise, K., Yoneyama, N., Murillo, A., & Crabbe, J. C. (2005). A procedure to produce high alcohol intake in mice. *Psychopharmacology*, *178*(4), 471–480. http://dx.doi.org/10.1007/s00213-004-2039-8.

Gallego, X., Ruiz-Medina, J., Valverde, O., Molas, S., Robles, N., Sabria, J., … Dierssen, M. (2012). Transgenic over expression of nicotinic receptor alpha 5, alpha 3, and beta 4 subunit genes reduces ethanol intake in mice. *Alcohol*, *46*(3), 205–215. http://dx.doi.org/10.1016/j.alcohol.2011.11.005.

Geisler, S., Schopf, C. L., & Obermair, G. J. (2015). Emerging evidence for specific neuronal functions of auxiliary calcium channel alpha(2)delta subunits. *General Physiology and Biophysics*, *34*(2), 105–118. http://dx.doi.org/10.4149/gpb_2014037.

Giardino, W. J., Cocking, D. L., Kaur, S., Cunningham, C. L., & Ryabinin, A. E. (2011). Urocortin-1 within the centrally-projecting Edinger-Westphal nucleus is critical for ethanol preference. *PLoS One*, *6*(10), e26997. http://dx.doi.org/10.1371/journal.pone.0026997.

Giardino, W. J., & Ryabinin, A. E. (2013). CRF1 receptor signaling regulates food and fluid intake in the drinking-in-the-dark model of binge alcohol consumption. *Alcoholism, Clinical and Experimental Research*, *37*(7), 1161–1170. http://dx.doi.org/10.1111/acer.12076.

Gorini, G., Bell, R. L., & Mayfield, R. D. (2011). Molecular targets of alcohol action: Translational research for pharmacotherapy development and screening. *Progress in Molecular Biology and Translational Science*, *98*, 293–347. http://dx.doi.org/10.1016/B978-0-12-385506-0.00007-7.

Gorini, G., Harris, R. A., & Mayfield, R. D. (2014). Proteomic approaches and identification of novel therapeutic targets for alcoholism. *Neuropsychopharmacology*, *39*(1), 104–130. http://dx.doi.org/10.1038/npp.2013.182.

Gorini, G., Roberts, A. J., & Mayfield, R. D. (2013). Neurobiological signatures of alcohol dependence revealed by protein profiling. *PLoS One*, *8*(12), e82656. http://dx.doi.org/10.1371/journal.pone.0082656.

Grahame, N. J., Mosemiller, A. K., Low, M. J., & Froehlich, J. C. (2000). Naltrexone and alcohol drinking in mice lacking beta-endorphin by site-directed mutagenesis. *Pharmacology, Biochemistry, and Behavior*, *67*(4), 759–766. Retrieved from, http://www.ncbi.nlm.nih.gov/pubmed/11166066.

Green, A. S., & Grahame, N. J. (2008). Ethanol drinking in rodents: Is free-choice drinking related to the reinforcing effects of ethanol? *Alcohol*, *42*(1), 1–11. http://dx.doi.org/10.1016/j.alcohol.2007.10.005.

Grisel, J. E., Mogil, J. S., Grahame, N. J., Rubinstein, M., Belknap, J. K., Crabbe, J. C., & Low, M. J. (1999). Ethanol oral self-administration is increased in mutant mice with decreased beta-endorphin expression. *Brain Research*, *835*(1), 62–67. Retrieved from, http://www.ncbi.nlm.nih.gov/pubmed/10448196.

Hall, F. S., Sora, I., & Uhl, G. R. (2001). Ethanol consumption and reward are decreased in mu-opiate receptor knockout mice. *Psychopharmacology*, *154*(1), 43–49. Retrieved from, http://www.ncbi.nlm.nih.gov/pubmed/11292005.

Hall, F. S., Sora, I., & Uhl, G. R. (2003). Sex-dependent modulation of ethanol consumption in vesicular monoamine transporter 2 (VMAT2) and dopamine transporter (DAT) knockout mice. *Neuropsychopharmacology*, *28*(4), 620–628. http://dx.doi.org/10.1038/sj.npp.1300070.

Han, S., Gelernter, J., Kranzler, H. R., & Yang, B. Z. (2013). Ordered subset linkage analysis based on admixture proportion identifies new linkage evidence for alcohol dependence

in African-Americans. *Human Genetics*, *132*(4), 397–403. http://dx.doi.org/10.1007/s00439-012-1255-2.
Harris, R. A., & Blednov, Y. A. (2013). Neuroimmune genes and alcohol drinking behavior. In C. Cui, L. Grandison, & A. Noronha (Eds.), *Neural-immune interactions in brain function and alcohol related disorders* (pp. 425–440). New York: Springer.
Harris, R. A., Osterndorff-Kahanek, E., Ponomarev, I., Homanics, G. E., & Blednov, Y. A. (2011). Testing the silence of mutations: Transcriptomic and behavioral studies of GABA(A) receptor alpha1 and alpha2 subunit knock-in mice. *Neuroscience Letters*, *488*(1), 31–35. http://dx.doi.org/10.1016/j.neulet.2010.10.075.
Hendrickson, L. M., Zhao-Shea, R., Pang, X., Gardner, P. D., & Tapper, A. R. (2010). Activation of alpha4* nAChRs is necessary and sufficient for varenicline-induced reduction of alcohol consumption. *The Journal of Neuroscience*, *30*(30), 10169–10176. http://dx.doi.org/10.1523/JNEUROSCI.2601-10.2010.
Herman, M. A., Sidhu, H., Stouffer, D. G., Kreifeldt, M., Le, D., Cates-Gatto, C., ... Contet, C. (2015). GIRK3 gates activation of the mesolimbic dopaminergic pathway by ethanol. *Proceedings of the National Academy of Sciences of the United States of America*, *112*(22), 7091–7096. http://dx.doi.org/10.1073/pnas.1416146112.
Hilakivi-Clarke, L., & Goldberg, R. (1995). Gonadal hormones and aggression-maintaining effect of alcohol in male transgenic transforming growth factor-alpha mice. *Alcoholism, Clinical and Experimental Research*, *19*(3), 708–713. Retrieved from, http://www.ncbi.nlm.nih.gov/pubmed/7573797.
Hodge, C. W., Mehmert, K. K., Kelley, S. P., McMahon, T., Haywood, A., Olive, M. F., ... Messing, R. O. (1999). Supersensitivity to allosteric GABA(A) receptor modulators and alcohol in mice lacking PKCepsilon. *Nature Neuroscience*, *2*(11), 997–1002. http://dx.doi.org/10.1038/14795.
Houchi, H., Persyn, W., Legastelois, R., & Naassila, M. (2013). The adenosine A2A receptor agonist CGS 21680 decreases ethanol self-administration in both non-dependent and dependent animals. *Addiction Biology*, *18*(5), 812–825. http://dx.doi.org/10.1111/adb.12032.
Houchi, H., Warnault, V., Barbier, E., Dubois, C., Pierrefiche, O., Ledent, C., ... Naassila, M. (2008). Involvement of A2A receptors in anxiolytic, locomotor and motivational properties of ethanol in mice. *Genes, Brain, and Behavior*, *7*(8), 887–898. Retrieved from, http://www.ncbi.nlm.nih.gov/pubmed/19097273.
Howard, R. J., Trudell, J. R., & Harris, R. A. (2014). Seeking structural specificity: Direct modulation of pentameric ligand-gated ion channels by alcohols and general anesthetics. *Pharmacological Reviews*, *66*(2), 396–412. http://dx.doi.org/10.1124/pr.113.007468.
Hungund, B. L., Szakall, I., Adam, A., Basavarajappa, B. S., & Vadasz, C. (2003). Cannabinoid CB1 receptor knockout mice exhibit markedly reduced voluntary alcohol consumption and lack alcohol-induced dopamine release in the nucleus accumbens. *Journal of Neurochemistry*, *84*(4), 698–704. Retrieved from, http://www.ncbi.nlm.nih.gov/pubmed/12562514.
Hwa, L. S., Chu, A., Levinson, S. A., Kayyali, T. M., DeBold, J. F., & Miczek, K. A. (2011). Persistent escalation of alcohol drinking in C57BL/6J mice with intermittent access to 20% ethanol. *Alcoholism, Clinical and Experimental Research*, *35*(11), 1938–1947. http://dx.doi.org/10.1111/j.1530-0277.2011.01545.x.
Iancu, O. D., Oberbeck, D., Darakjian, P., Metten, P., McWeeney, S., Crabbe, J. C., & Hitzemann, R. (2013). Selection for drinking in the dark alters brain gene coexpression networks. *Alcoholism, Clinical and Experimental Research*, *37*(8), 1295–1303. http://dx.doi.org/10.1111/acer.12100.
Imai, S., Kano, M., Nonoyama, K., & Ebihara, S. (2013). Behavioral characteristics of ubiquitin-specific peptidase 46-deficient mice. *PLoS One*, *8*(3), e58566. http://dx.doi.org/10.1371/journal.pone.0058566.

Isse, T., Oyama, T., Kitagawa, K., Matsuno, K., Matsumoto, A., Yoshida, A., …Kawamoto, T. (2002). Diminished alcohol preference in transgenic mice lacking aldehyde dehydrogenase activity. *Pharmacogenetics*, *12*(8), 621–626. Retrieved from, http://www.ncbi.nlm.nih.gov/pubmed/12439222.

Jaholkowski, P., Mierzejewski, P., Zatorski, P., Scinska, A., Sienkiewicz-Jarosz, H., Kaczmarek, L., …Bienkowski, P. (2011). Increased ethanol intake and preference in cyclin D2 knockout mice. *Genes, Brain, and Behavior*, *10*(5), 551–556. http://dx.doi.org/10.1111/j.1601-183X.2011.00692.x.

Jerlhag, E., Landgren, S., Egecioglu, E., Dickson, S. L., & Engel, J. A. (2011). The alcohol-induced locomotor stimulation and accumbal dopamine release is suppressed in ghrelin knockout mice. *Alcohol*, *45*(4), 341–347. http://dx.doi.org/10.1016/j.alcohol.2010.10.002.

June, H. L., Sr., Foster, K. L., Eiler, W. J., 2nd., Goergen, J., Cook, J. B., Johnson, N., …Homanics, G. E. (2007). Dopamine and benzodiazepine-dependent mechanisms regulate the EtOH-enhanced locomotor stimulation in the GABAA alpha1 subunit null mutant mice. *Neuropsychopharmacology*, *32*(1), 137–152. http://dx.doi.org/10.1038/sj.npp.1301097.

Kamens, H. M., Andersen, J., & Picciotto, M. R. (2010). Modulation of ethanol consumption by genetic and pharmacological manipulation of nicotinic acetylcholine receptors in mice. *Psychopharmacology*, *208*(4), 613–626. http://dx.doi.org/10.1007/s00213-009-1759-1.

Kamens, H. M., Hoft, N. R., Cox, R. J., Miyamoto, J. H., & Ehringer, M. A. (2012). The alpha6 nicotinic acetylcholine receptor subunit influences ethanol-induced sedation. *Alcohol*, *46*(5), 463–471. http://dx.doi.org/10.1016/j.alcohol.2012.03.001.

Kampov-Polevoy, A., Lange, L., Bobashev, G., Eggleston, B., Root, T., & Garbutt, J. C. (2014). Sweet-liking is associated with transformation of heavy drinking into alcohol-related problems in young adults with high novelty seeking. *Alcoholism, Clinical and Experimental Research*, *38*(7), 2119–2126. http://dx.doi.org/10.1111/acer.12458.

Kang, S. J., Rangaswamy, M., Manz, N., Wang, J. C., Wetherill, L., Hinrichs, T., …Porjesz, B. (2012). Family-based genome-wide association study of frontal theta oscillations identifies potassium channel gene KCNJ6. *Genes, Brain, and Behavior*, *11*(6), 712–719. http://dx.doi.org/10.1111/j.1601-183X.2012.00803.x.

Kapfhamer, D., Bettinger, J. C., Davies, A. G., Eastman, C. L., Smail, E. A., Heberlein, U., & McIntire, S. L. (2008). Loss of RAB-3/A in *Caenorhabditis elegans* and the mouse affects behavioral response to ethanol. *Genes, Brain, and Behavior*, 7(6), 669–676. http://dx.doi.org/10.1111/j.1601-183X.2008.00404.x.

Kapfhamer, D., Taylor, S., Zou, M. E., Lim, J. P., Kharazia, V., & Heberlein, U. (2013). Taok2 controls behavioral response to ethanol in mice. *Genes, Brain, and Behavior*, *12*(1), 87–97. http://dx.doi.org/10.1111/j.1601-183X.2012.00834.x.

Karatayev, O., Baylan, J., & Leibowitz, S. F. (2009). Increased intake of ethanol and dietary fat in galanin overexpressing mice. *Alcohol*, *43*(8), 571–580. http://dx.doi.org/10.1016/j.alcohol.2009.09.025.

Karatayev, O., Baylan, J., Weed, V., Chang, S., Wynick, D., & Leibowitz, S. F. (2010). Galanin knockout mice show disturbances in ethanol consumption and expression of hypothalamic peptides that stimulate ethanol intake. *Alcoholism, Clinical and Experimental Research*, *34*(1), 72–80. http://dx.doi.org/10.1111/j.1530-0277.2009.01068.x.

Karlsson, R. M., Adermark, L., Molander, A., Perreau-Lenz, S., Singley, E., Solomon, M., …Heilig, M. (2012). Reduced alcohol intake and reward associated with impaired endocannabinoid signaling in mice with a deletion of the glutamate transporter GLAST. *Neuropharmacology*, *63*(2), 181–189. http://dx.doi.org/10.1016/j.neuropharm.2012.01.027.

Kaur, S., Li, J., Stenzel-Poore, M. P., & Ryabinin, A. E. (2012). Corticotropin-releasing factor acting on corticotropin-releasing factor receptor type 1 is critical for binge alcohol drinking in mice. *Alcoholism, Clinical and Experimental Research, 36*(2), 369–376. http://dx.doi.org/10.1111/j.1530-0277.2011.01610.x.

Kelai, S., Aissi, F., Lesch, K. P., Cohen-Salmon, C., Hamon, M., & Lanfumey, L. (2003). Alcohol intake after serotonin transporter inactivation in mice. *Alcohol and Alcoholism, 38*(4), 386–389. Retrieved from, http://www.ncbi.nlm.nih.gov/pubmed/12814910.

Kim, K. S., Kim, H., Baek, I. S., Lee, K. W., & Han, P. L. (2011). Mice lacking adenylyl cyclase type 5 (AC5) show increased ethanol consumption and reduced ethanol sensitivity. *Psychopharmacology, 215*(2), 391–398. http://dx.doi.org/10.1007/s00213-010-2143-x.

Koenig, H. N., & Olive, M. F. (2002). Ethanol consumption patterns and conditioned place preference in mice lacking preproenkephalin. *Neuroscience Letters, 325*(2), 75–78. Retrieved from, http://www.ncbi.nlm.nih.gov/pubmed/12044625.

Korkosz, A., Kolomanska, P., Kowalska, K., Rogowski, A., Radwanska, K., Kaczmarek, L., ... Bienkowski, P. (2004). Dissociation of ethanol and saccharin preference in fosB knockout mice. *Physiology & Behavior, 82*(2–3), 391–395. http://dx.doi.org/10.1016/j.physbeh.2004.04.054.

Kovacs, K. M., Szakall, I., O'Brien, D., Wang, R., Vinod, K. Y., Saito, M., ... Vadasz, C. (2005). Decreased oral self-administration of alcohol in kappa-opioid receptor knock-out mice. *Alcoholism, Clinical and Experimental Research, 29*(5), 730–738. Retrieved from, http://www.ncbi.nlm.nih.gov/pubmed/15897716.

Kreifeldt, M., Le, D., Treistman, S. N., Koob, G. F., & Contet, C. (2013). BK channel beta1 and beta4 auxiliary subunits exert opposite influences on escalated ethanol drinking in dependent mice. *Frontiers in Integrative Neuroscience, 7*, 105. http://dx.doi.org/10.3389/fnint.2013.00105.

Lallemand, F., & de Witte, P. (2005). Ethanol induces higher BEC in CB1 cannabinoid receptor knockout mice while decreasing ethanol preference. *Alcohol and Alcoholism, 40*(1), 54–62. http://dx.doi.org/10.1093/alcalc/agh115.

Lamb, R. J., & Daws, L. C. (2013). Ethanol self-administration in serotonin transporter knockout mice: Unconstrained demand and elasticity. *Genes, Brain, and Behavior, 12*(7), 741–747. http://dx.doi.org/10.1111/gbb.12068.

Lasek, A. W., Lim, J., Kliethermes, C. L., Berger, K. H., Joslyn, G., Brush, G., ... Heberlein, U. (2011). An evolutionary conserved role for anaplastic lymphoma kinase in behavioral responses to ethanol. *PLoS One, 6*(7), e22636. http://dx.doi.org/10.1371/journal.pone.0022636.

Leclercq, S., De Saeger, C., Delzenne, N., de Timary, P., & Starkel, P. (2014). Role of inflammatory pathways, blood mononuclear cells, and gut-derived bacterial products in alcohol dependence. *Biological Psychiatry, 76*(9), 725–733. http://dx.doi.org/10.1016/j.biopsych.2014.02.003.

Lee, M. R., Hinton, D. J., Song, J. Y., Lee, K. W., Choo, C., Johng, H., ... Choi, D. S. (2010). Neurotensin receptor type 1 regulates ethanol intoxication and consumption in mice. *Pharmacology, Biochemistry, and Behavior, 95*(2), 235–241. http://dx.doi.org/10.1016/j.pbb.2010.01.012.

Lee, M. R., Hinton, D. J., Unal, S. S., Richelson, E., & Choi, D. S. (2011). Increased ethanol consumption and preference in mice lacking neurotensin receptor type 2. *Alcoholism, Clinical and Experimental Research, 35*(1), 99–107. http://dx.doi.org/10.1111/j.1530-0277.2010.01326.x.

Lee, M. R., Ruby, C. L., Hinton, D. J., Choi, S., Adams, C. A., Young Kang, N., & Choi, D. S. (2013). Striatal adenosine signaling regulates EAAT2 and astrocytic AQP4 expression and alcohol drinking in mice. *Neuropsychopharmacology, 38*(3), 437–445. http://dx.doi.org/10.1038/npp.2012.198.

Lee, A. M., Zou, M. E., Lim, J. P., Stecher, J., McMahon, T., & Messing, R. O. (2014). Deletion of Prkcz increases intermittent ethanol consumption in mice. *Alcoholism, Clinical and Experimental Research, 38*(1), 170–178. http://dx.doi.org/10.1111/acer.12211.

Legastelois, R., Darcq, E., Wegner, S. A., Lombroso, P. J., & Ron, D. (2015). Striatal-enriched protein tyrosine phosphatase controls responses to aversive stimuli: Implication for ethanol drinking. *PLoS One*, *10*(5), e0127408. http://dx.doi.org/10.1371/journal.pone.0127408.

Leggio, G. M., Camillieri, G., Platania, C. B., Castorina, A., Marrazzo, G., Torrisi, S. A., … Salomone, S. (2014). Dopamine D3 receptor is necessary for ethanol consumption: An approach with buspirone. *Neuropsychopharmacology*, *39*(8), 2017–2028. http://dx.doi.org/10.1038/npp.2014.51.

Lesscher, H. M., Wallace, M. J., Zeng, L., Wang, V., Deitchman, J. K., McMahon, T., … Newton, P. M. (2009). Amygdala protein kinase C epsilon controls alcohol consumption. *Genes, Brain, and Behavior*, *8*(5), 493–499. http://dx.doi.org/10.1111/j.1601-183X.2009.00485.x.

Levran, O., Yuferov, V., & Kreek, M. J. (2012). The genetics of the opioid system and specific drug addictions. *Human Genetics*, *131*(6), 823–842. http://dx.doi.org/10.1007/s00439-012-1172-4.

Li, D., Sulovari, A., Cheng, C., Zhao, H., Kranzler, H. R., & Gelernter, J. (2014). Association of gamma-aminobutyric acid A receptor alpha2 gene (GABRA2) with alcohol use disorder. *Neuropsychopharmacology*, *39*(4), 907–918. http://dx.doi.org/10.1038/npp.2013.291.

Li, H., Tao, Y., Ma, L., Liu, X., & Ma, L. (2013). beta-Arrestin-2 inhibits preference for alcohol in mice and suppresses Akt signaling in the dorsal striatum. *Neuroscience Bulletin*, *29*(5), 531–540. http://dx.doi.org/10.1007/s12264-013-1350-y.

Liu, J., Lewohl, J. M., Harris, R. A., Iyer, V. R., Dodd, P. R., Randall, P. K., & Mayfield, R. D. (2006). Patterns of gene expression in the frontal cortex discriminate alcoholic from nonalcoholic individuals. *Neuropsychopharmacology*, *31*(7), 1574–1582. http://dx.doi.org/10.1038/sj.npp.1300947.

Loney, K. D., Uddin, R. K., & Singh, S. M. (2006). Analysis of metallothionein brain gene expression in relation to ethanol preference in mice using cosegregation and gene knockouts. *Alcoholism, Clinical and Experimental Research*, *30*(1), 15–25. http://dx.doi.org/10.1111/j.1530-0277.2006.00015.x.

Lopez-Jimenez, A., Walter, N. A., Gine, E., Santos, A., Echeverry-Alzate, V., Buhler, K. M., … Lopez-Moreno, J. A. (2013). A spontaneous deletion of alpha-synuclein is associated with an increase in CB1 mRNA transcript and receptor expression in the hippocampus and amygdala: Effects on alcohol consumption. *Synapse*, *67*(6), 280–289. http://dx.doi.org/10.1002/syn.21639.

Lum, E. N., Campbell, R. R., Rostock, C., & Szumlinski, K. K. (2014). mGluR1 within the nucleus accumbens regulates alcohol intake in mice under limited-access conditions. *Neuropharmacology*, *79*, 679–687. http://dx.doi.org/10.1016/j.neuropharm.2014.01.024.

Lynch, L. J., Sullivan, K. A., Vallender, E. J., Rowlett, J. K., Platt, D. M., & Miller, G. M. (2013). Trace amine associated receptor 1 modulates behavioral effects of ethanol. *Substance Abuse*, 7, 117–126. http://dx.doi.org/10.4137/SART.S12110.

Maas, J. W., Jr., Vogt, S. K., Chan, G. C., Pineda, V. V., Storm, D. R., & Muglia, L. J. (2005). Calcium-stimulated adenylyl cyclases are critical modulators of neuronal ethanol sensitivity. *The Journal of Neuroscience*, *25*(16), 4118–4126. http://dx.doi.org/10.1523/JNEUROSCI.4273-04.2005.

Martin, G. E., Hendrickson, L. M., Penta, K. L., Friesen, R. M., Pietrzykowski, A. Z., Tapper, A. R., & Treistman, S. N. (2008). Identification of a BK channel auxiliary protein controlling molecular and behavioral tolerance to alcohol. *Proceedings of the National Academy of Sciences of the United States of America*, *105*(45), 17543–17548. http://dx.doi.org/10.1073/pnas.0801068105.

Mason, B. J., Quello, S., Goodell, V., Shadan, F., Kyle, M., & Begovic, A. (2014). Gabapentin treatment for alcohol dependence: A randomized clinical trial. *JAMA Internal Medicine, 174*(1), 70–77. http://dx.doi.org/10.1001/jamainternmed.2013.11950.

Maul, B., Becker, M., Gembardt, F., Becker, A., Schultheiss, H. P., Siems, W. E., & Walther, T. (2012). Genetic deficiency in neprilysin or its pharmacological inhibition initiate excessive stress-induced alcohol consumption in mice. *PLoS One*, 7(11), e50187. http://dx.doi.org/10.1371/journal.pone.0050187.

Maul, B., Krause, W., Pankow, K., Becker, M., Gembardt, F., Alenina, N., … Siems, W. E. (2005). Central angiotensin II controls alcohol consumption via its AT1 receptor. *The FASEB Journal, 19*(11), 1474–1481. http://dx.doi.org/10.1096/fj.05-3742com.

Maul, B., Siems, W. E., Hoehe, M. R., Grecksch, G., Bader, M., & Walther, T. (2001). Alcohol consumption is controlled by angiotensin II. *The FASEB Journal, 15*(9), 1640–1642. Retrieved from, http://www.ncbi.nlm.nih.gov/pubmed/11427512.

Mayfield, J., Blednov, Y. A., & Harris, R. A. (2015). Behavioral and genetic evidence for GIRK channels in the CNS: Role in physiology, pathophysiology, and drug addiction. *International Review of Neurobiology, 123*, 279–313. http://dx.doi.org/10.1016/bs.irn.2015.05.016.

Mayfield, J., Ferguson, L., & Harris, R. A. (2013). Neuroimmune signaling: A key component of alcohol abuse. *Current Opinion in Neurobiology, 23*(4), 513–520. http://dx.doi.org/10.1016/j.conb.2013.01.024.

Melendez, R. I. (2011). Intermittent (every-other-day) drinking induces rapid escalation of ethanol intake and preference in adolescent and adult C57BL/6J mice. *Alcoholism, Clinical and Experimental Research, 35*(4), 652–658. http://dx.doi.org/10.1111/j.1530-0277.2010.01383.x.

Metz, A. V., Chynoweth, J., & Allan, A. M. (2006). Influence of genetic background on alcohol drinking and behavioral phenotypes of 5-HT3 receptor over-expressing mice. *Pharmacology, Biochemistry, and Behavior, 84*(1), 120–127. http://dx.doi.org/10.1016/j.pbb.2006.04.016.

Mihalek, R. M., Bowers, B. J., Wehner, J. M., Kralic, J. E., VanDoren, M. J., Morrow, A. L., & Homanics, G. E. (2001). GABA(A)-receptor delta subunit knockout mice have multiple defects in behavioral responses to ethanol. *Alcoholism, Clinical and Experimental Research, 25*(12), 1708–1718. Retrieved from, http://www.ncbi.nlm.nih.gov/pubmed/11781502.

Mittleman, G., Call, S. B., Cockroft, J. L., Goldowitz, D., Matthews, D. B., & Blaha, C. D. (2011). Dopamine dynamics associated with, and resulting from, schedule-induced alcohol self-administration: Analyses in dopamine transporter knockout mice. *Alcohol, 45*(4), 325–339. http://dx.doi.org/10.1016/j.alcohol.2010.12.006.

Miyasaka, K., Hosoya, H., Takano, S., Ohta, M., Sekime, A., Kanai, S., … Funakoshi, A. (2005). Differences in ethanol ingestion between cholecystokinin-A receptor deficient and -B receptor deficient mice. *Alcohol and Alcoholism, 40*(3), 176–180. http://dx.doi.org/10.1093/alcalc/agh143.

Molander, A., Vengeliene, V., Heilig, M., Wurst, W., Deussing, J. M., & Spanagel, R. (2012). Brain-specific inactivation of the Crhr1 gene inhibits post dependent and stress-induced alcohol intake, but does not affect relapse-like drinking. *Neuropsychopharmacology, 37*(4), 1047–1056. http://dx.doi.org/10.1038/npp.2011.297.

Moore, R., Krstew, E. V., Kirchhoff, J., Davisson, R. L., & Lawrence, A. J. (2007). Central overexpression of angiotensin AT(1A) receptors prevents dopamine D(2) receptor regulation of alcohol consumption in mice. *Alcoholism, Clinical and Experimental Research, 31*(7), 1128–1137. http://dx.doi.org/10.1111/j.1530-0277.2007.00399.x.

Mulligan, M. K., Ponomarev, I., Hitzemann, R. J., Belknap, J. K., Tabakoff, B., Harris, R. A., … Bergeson, S. E. (2006). Toward understanding the genetics of alcohol drinking through transcriptome meta-analysis. *Proceedings of the National Academy of Sciences of the United States of America, 103*(16), 6368–6373. http://dx.doi.org/10.1073/pnas.0510188103.

Mutschler, J., Bilbao, A., von der Goltz, C., Demiralay, C., Jahn, H., Wiedemann, K., ... Kiefer, F. (2010). Augmented stress-induced alcohol drinking and withdrawal in mice lacking functional natriuretic peptide-A receptors. *Alcohol and Alcoholism, 45*(1), 13–16. http://dx.doi.org/10.1093/alcalc/agp065.

Naassila, M., Ledent, C., & Daoust, M. (2002). Low ethanol sensitivity and increased ethanol consumption in mice lacking adenosine A2A receptors. *The Journal of Neuroscience, 22*(23), 10487–10493. Retrieved from, http://www.ncbi.nlm.nih.gov/pubmed/12451148.

Naassila, M., Pierrefiche, O., Ledent, C., & Daoust, M. (2004). Decreased alcohol self-administration and increased alcohol sensitivity and withdrawal in CB1 receptor knockout mice. *Neuropharmacology, 46*(2), 243–253. Retrieved from, http://www.ncbi.nlm.nih.gov/pubmed/14680762.

Nam, H. W., Bruner, R. C., & Choi, D. S. (2013). Adenosine signaling in striatal circuits and alcohol use disorders. *Molecules and Cells, 36*(3), 195–202. http://dx.doi.org/10.1007/s10059-013-0192-9.

Nam, H. W., Hinton, D. J., Kang, N. Y., Kim, T., Lee, M. R., Oliveros, A., ... Choi, D. S. (2013). Adenosine transporter ENT1 regulates the acquisition of goal-directed behavior and ethanol drinking through A2A receptor in the dorsomedial striatum. *The Journal of Neuroscience, 33*(10), 4329–4338. http://dx.doi.org/10.1523/JNEUROSCI.3094-12.2013.

Nam, H. W., Lee, M. R., Zhu, Y., Wu, J., Hinton, D. J., Choi, S., ... Choi, D. S. (2011). Type 1 equilibrative nucleoside transporter regulates ethanol drinking through accumbal N-methyl-D-aspartate receptor signaling. *Biological Psychiatry, 69*(11), 1043–1051. http://dx.doi.org/10.1016/j.biopsych.2011.02.013.

Navarro, M., Cubero, I., Chen, A. S., Chen, H. Y., Knapp, D. J., Breese, G. R., ... Thiele, T. E. (2005). Effects of melanocortin receptor activation and blockade on ethanol intake: A possible role for the melanocortin-4 receptor. *Alcoholism, Clinical and Experimental Research, 29*(6), 949–957. Retrieved from, http://www.ncbi.nlm.nih.gov/pubmed/15976520.

Navarro, M., Cubero, I., Ko, L., & Thiele, T. E. (2009). Deletion of agouti-related protein blunts ethanol self-administration and binge-like drinking in mice. *Genes, Brain, and Behavior, 8*(4), 450–458. http://dx.doi.org/10.1111/j.1601-183X.2009.00493.x.

Navarro, M., Lerma-Cabrera, J. M., Carvajal, F., Lowery, E. G., Cubero, I., & Thiele, T. E. (2011). Assessment of voluntary ethanol consumption and the effects of a melanocortin (MC) receptor agonist on ethanol intake in mutant C57BL/6J mice lacking the MC-4 receptor. *Alcoholism, Clinical and Experimental Research, 35*(6), 1058–1066. http://dx.doi.org/10.1111/j.1530-0277.2011.01438.x.

Newton, P. M., Orr, C. J., Wallace, M. J., Kim, C., Shin, H. S., & Messing, R. O. (2004). Deletion of N-type calcium channels alters ethanol reward and reduces ethanol consumption in mice. *The Journal of Neuroscience, 24*(44), 9862–9869. http://dx.doi.org/10.1523/JNEUROSCI.3446-04.2004.

Nie, H., Rewal, M., Gill, T. M., Ron, D., & Janak, P. H. (2011). Extrasynaptic delta-containing GABAA receptors in the nucleus accumbens dorsomedial shell contribute to alcohol intake. *Proceedings of the National Academy of Sciences of the United States of America, 108*(11), 4459–4464. http://dx.doi.org/10.1073/pnas.1016156108.

Nunez, Y. O., Truitt, J. M., Gorini, G., Ponomareva, O. N., Blednov, Y. A., Harris, R. A., & Mayfield, R. D. (2013). Positively correlated miRNA-mRNA regulatory networks in mouse frontal cortex during early stages of alcohol dependence. *BMC Genomics, 14*, 725. http://dx.doi.org/10.1186/1471-2164-14-725.

Nuutinen, S., Lintunen, M., Vanhanen, J., Ojala, T., Rozov, S., & Panula, P. (2011). Evidence for the role of histamine H3 receptor in alcohol consumption and alcohol reward in mice. *Neuropsychopharmacology*, *36*(10), 2030–2040. http://dx.doi.org/10.1038/npp.2011.90.

Offenhauser, N., Castelletti, D., Mapelli, L., Soppo, B. E., Regondi, M. C., Rossi, P., … Di Fiore, P. P. (2006). Increased ethanol resistance and consumption in Eps8 knockout mice correlates with altered actin dynamics. *Cell*, *127*(1), 213–226. http://dx.doi.org/10.1016/j.cell.2006.09.011.

Olive, M. F., McGeehan, A. J., Kinder, J. R., McMahon, T., Hodge, C. W., Janak, P. H., & Messing, R. O. (2005). The mGluR5 antagonist 6-methyl-2-(phenylethynyl)pyridine decreases ethanol consumption via a protein kinase C epsilon-dependent mechanism. *Molecular Pharmacology*, *67*(2), 349–355. http://dx.doi.org/10.1124/mol.104.003319.

Olive, M. F., Mehmert, K. K., Koenig, H. N., Camarini, R., Kim, J. A., Nannini, M. A., … Hodge, C. W. (2003). A role for corticotropin releasing factor (CRF) in ethanol consumption, sensitivity, and reward as revealed by CRF-deficient mice. *Psychopharmacology*, *165*(2), 181–187. http://dx.doi.org/10.1007/s00213-002-1248-2.

Olive, M. F., Mehmert, K. K., Messing, R. O., & Hodge, C. W. (2000). Reduced operant ethanol self-administration and in vivo mesolimbic dopamine responses to ethanol in PKCepsilon-deficient mice. *The European Journal of Neuroscience*, *12*(11), 4131–4140. Retrieved from, http://www.ncbi.nlm.nih.gov/pubmed/11069609.

Olney, J. J., Sprow, G. M., Navarro, M., & Thiele, T. E. (2014). The protective effects of the melanocortin receptor (MCR) agonist, melanotan-II (MTII), against binge-like ethanol drinking are facilitated by deletion of the MC3 receptor in mice. *Neuropeptides*, *48*(1), 47–51. http://dx.doi.org/10.1016/j.npep.2013.11.001.

Ortega-Alvaro, A., Ternianov, A., Aracil-Fernandez, A., Navarrete, F., Garcia-Gutierrez, M. S., & Manzanares, J. (2015). Role of cannabinoid CB2 receptor in the reinforcing actions of ethanol. *Addiction Biology*, *20*(1), 43–55. http://dx.doi.org/10.1111/adb.12076.

Osterndorff-Kahanek, E., Ponomarev, I., Blednov, Y. A., & Harris, R. A. (2013). Gene expression in brain and liver produced by three different regimens of alcohol consumption in mice: Comparison with immune activation. *PLoS One*, *8*(3), e59870. http://dx.doi.org/10.1371/journal.pone.0059870.

Palmer, A. A., Low, M. J., Grandy, D. K., & Phillips, T. J. (2003). Effects of a Drd2 deletion mutation on ethanol-induced locomotor stimulation and sensitization suggest a role for epistasis. *Behavior Genetics*, *33*(3), 311–324. Retrieved from, http://www.ncbi.nlm.nih.gov/pubmed/12837020.

Panagis, G., Mackey, B., & Vlachou, S. (2014). Cannabinoid regulation of brain reward processing with an emphasis on the role of CB1 receptors: A step back into the future. *Frontiers in Psychiatry*, *5*, 92. http://dx.doi.org/10.3389/fpsyt.2014.00092.

Pascual, M., Balino, P., Alfonso-Loeches, S., Aragon, C. M., & Guerri, C. (2011). Impact of TLR4 on behavioral and cognitive dysfunctions associated with alcohol-induced neuroinflammatory damage. *Brain, Behavior, and Immunity*, *25*(Suppl. 1), S80–S91. http://dx.doi.org/10.1016/j.bbi.2011.02.012.

Pastor, R., Reed, C., Burkhart-Kasch, S., Li, N., Sharpe, A. L., Coste, S. C., … Phillips, T. J. (2011). Ethanol concentration-dependent effects and the role of stress on ethanol drinking in corticotropin-releasing factor type 1 and double type 1 and 2 receptor knockout mice. *Psychopharmacology*, *218*(1), 169–177. http://dx.doi.org/10.1007/s00213-011-2284-6.

Perkins, D. I., Trudell, J. R., Crawford, D. K., Alkana, R. L., & Davies, D. L. (2010). Molecular targets and mechanisms for ethanol action in glycine receptors. *Pharmacology & Therapeutics*, *127*(1), 53–65. http://dx.doi.org/10.1016/j.pharmthera.2010.03.003.

Phillips, T. J., Brown, K. J., Burkhart-Kasch, S., Wenger, C. D., Kelly, M. A., Rubinstein, M., … Low, M. J. (1998). Alcohol preference and sensitivity are markedly reduced in mice lacking dopamine D2 receptors. *Nature Neuroscience*, *1*(7), 610–615. http://dx.doi.org/10.1038/2843.

Poncelet, M., Maruani, J., Calassi, R., & Soubrie, P. (2003). Overeating, alcohol and sucrose consumption decrease in CB1 receptor deleted mice. *Neuroscience Letters*, *343*(3), 216–218. Retrieved from, http://www.ncbi.nlm.nih.gov/pubmed/12770700.

Ponomarev, I., Wang, S., Zhang, L., Harris, R. A., & Mayfield, R. D. (2012). Gene coexpression networks in human brain identify epigenetic modifications in alcohol dependence. *The Journal of Neuroscience*, *32*(5), 1884–1897. http://dx.doi.org/10.1523/JNEUROSCI.3136-11.2012.

Popova, N. K., Vishnivetskaya, G. B., Ivanova, E. A., Skrinskaya, J. A., & Seif, I. (2000). Altered behavior and alcohol tolerance in transgenic mice lacking MAO A: A comparison with effects of MAO A inhibitor clorgyline. *Pharmacology, Biochemistry, and Behavior*, *67*(4), 719–727. Retrieved from, http://www.ncbi.nlm.nih.gov/pubmed/11166062.

Quintanilla, M. E., Israel, Y., Sapag, A., & Tampier, L. (2006). The UChA and UChB rat lines: Metabolic and genetic differences influencing ethanol intake. *Addiction Biology*, *11*(3–4), 310–323. http://dx.doi.org/10.1111/j.1369-1600.2006.00030.x.

Racz, I., Bilkei-Gorzo, A., Toth, Z. E., Michel, K., Palkovits, M., & Zimmer, A. (2003). A critical role for the cannabinoid CB1 receptors in alcohol dependence and stress-stimulated ethanol drinking. *The Journal of Neuroscience*, *23*(6), 2453–2458. Retrieved from, http://www.ncbi.nlm.nih.gov/pubmed/12657705.

Racz, I., Markert, A., Mauer, D., Stoffel-Wagner, B., & Zimmer, A. (2013). Long-term ethanol effects on acute stress responses: Modulation by dynorphin. *Addiction Biology*, *18*(4), 678–688. http://dx.doi.org/10.1111/j.1369-1600.2012.00494.x.

Racz, I., Schurmann, B., Karpushova, A., Reuter, M., Cichon, S., Montag, C., … Zimmer, A. (2008). The opioid peptides enkephalin and beta-endorphin in alcohol dependence. *Biological Psychiatry*, *64*(11), 989–997. http://dx.doi.org/10.1016/j.biopsych.2008.05.008.

Rahman, S., Engleman, E. A., & Bell, R. L. (2014). Nicotinic receptor modulation to treat alcohol and drug dependence. *Frontiers in Neuroscience*, *8*, 426. http://dx.doi.org/10.3389/fnins.2014.00426.

Rao, P. S., Bell, R. L., Engleman, E. A., & Sari, Y. (2015). Targeting glutamate uptake to treat alcohol use disorders. *Frontiers in Neuroscience*, *9*, 144. http://dx.doi.org/10.3389/fnins.2015.00144.

Repunte-Canonigo, V., Berton, F., Cottone, P., Reifel-Miller, A., Roberts, A. J., Morales, M., … Sanna, P. P. (2010). A potential role for adiponectin receptor 2 (AdipoR2) in the regulation of alcohol intake. *Brain Research*, *1339*, 11–17. http://dx.doi.org/10.1016/j.brainres.2010.03.060.

Rewal, M., Jurd, R., Gill, T. M., He, D. Y., Ron, D., & Janak, P. H. (2009). Alpha4-containing GABAA receptors in the nucleus accumbens mediate moderate intake of alcohol. *The Journal of Neuroscience*, *29*(2), 543–549. http://dx.doi.org/10.1523/JNEUROSCI.3199-08.2009.

Risinger, F. O., Doan, A. M., & Vickrey, A. C. (1999). Oral operant ethanol self-administration in 5-HT1b knockout mice. *Behavioural Brain Research*, *102*(1–2), 211–215. Retrieved from, http://www.ncbi.nlm.nih.gov/pubmed/10403028.

Risinger, F. O., Freeman, P. A., Greengard, P., & Fienberg, A. A. (2001). Motivational effects of ethanol in DARPP-32 knock-out mice. *The Journal of Neuroscience*, *21*(1), 340–348. Retrieved from, http://www.ncbi.nlm.nih.gov/pubmed/11150351.

Risinger, F. O., Freeman, P. A., Rubinstein, M., Low, M. J., & Grandy, D. K. (2000). Lack of operant ethanol self-administration in dopamine D2 receptor knockout mice. *Psychopharmacology*, *152*(3), 343–350. Retrieved from, http://www.ncbi.nlm.nih.gov/pubmed/11105945.

Roberts, A. J., Gold, L. H., Polis, I., McDonald, J. S., Filliol, D., Kieffer, B. L., & Koob, G. F. (2001). Increased ethanol self-administration in delta-opioid receptor knockout mice. *Alcoholism, Clinical and Experimental Research*, *25*(9), 1249–1256. Retrieved from, http://www.ncbi.nlm.nih.gov/pubmed/11584142.
Roberts, A. J., McDonald, J. S., Heyser, C. J., Kieffer, B. L., Matthes, H. W., Koob, G. F., & Gold, L. H. (2000). mu-Opioid receptor knockout mice do not self-administer alcohol. *The Journal of Pharmacology and Experimental Therapeutics*, *293*(3), 1002–1008. Retrieved from, http://www.ncbi.nlm.nih.gov/pubmed/10869404.
Robinson, G., Most, D., Ferguson, L. B., Mayfield, J., Harris, R. A., & Blednov, Y. A. (2014). Neuroimmune pathways in alcohol consumption: Evidence from behavioral and genetic studies in rodents and humans. *International Review of Neurobiology*, *118*, 13–39. http://dx.doi.org/10.1016/B978-0-12-801284-0.00002-6.
Rodd, Z. A., Bell, R. L., Sable, H. J., Murphy, J. M., & McBride, W. J. (2004). Recent advances in animal models of alcohol craving and relapse. *Pharmacology, Biochemistry, and Behavior*, *79*(3), 439–450. http://dx.doi.org/10.1016/j.pbb.2004.08.018.
Rosenwasser, A. M., Fixaris, M. C., Crabbe, J. C., Brooks, P. C., & Ascheid, S. (2013). Escalation of intake under intermittent ethanol access in diverse mouse genotypes. *Addiction Biology*, *18*(3), 496–507. http://dx.doi.org/10.1111/j.1369-1600.2012.00481.x.
Ruby, C. L., Vadnie, C. A., Hinton, D. J., Abulseoud, O. A., Walker, D. L., O'Connor, K. M., … Choi, D. S. (2014). Adenosinergic regulation of striatal clock gene expression and ethanol intake during constant light. *Neuropsychopharmacology*, *39*(10), 2432–2440. http://dx.doi.org/10.1038/npp.2014.94.
Ruby, C. L., Walker, D. L., An, J., Kim, J., & Choi, D. S. (2011). Sex-specific regulation of depression, anxiety-like behaviors and alcohol drinking in mice lacking ENT1. *Journal of Addiction Research and Therapy*, *S4*. http://dx.doi.org/10.4172/2155-6105.S4-004.
Saba, L. M., Bennett, B., Hoffman, P. L., Barcomb, K., Ishii, T., Kechris, K., & Tabakoff, B. (2011). A systems genetic analysis of alcohol drinking by mice, rats and men: Influence of brain GABAergic transmission. *Neuropharmacology*, *60*(7–8), 1269–1280. http://dx.doi.org/10.1016/j.neuropharm.2010.12.019.
Saba, L. M., Flink, S. C., Vanderlinden, L. A., Israel, Y., Tampier, L., Colombo, G., … Tabakoff, B. (2015). The sequenced rat brain transcriptome—Its use in identifying networks predisposing alcohol consumption. *The FEBS Journal*, *282*(18), 3556–3578. http://dx.doi.org/10.1111/febs.13358.
Sachs, B. D., Salahi, A. A., & Caron, M. G. (2014). Congenital brain serotonin deficiency leads to reduced ethanol sensitivity and increased ethanol consumption in mice. *Neuropharmacology*, 77, 177–184. http://dx.doi.org/10.1016/j.neuropharm.2013.09.010.
Sakoori, K., & Murphy, N. P. (2008). Endogenous nociceptin (orphanin FQ) suppresses basal hedonic state and acute reward responses to methamphetamine and ethanol, but facilitates chronic responses. *Neuropsychopharmacology*, *33*(4), 877–891. http://dx.doi.org/10.1038/sj.npp.1301459.
Salinas, A. G., Nguyen, C. T., Ahmadi-Tehrani, D., & Morrisett, R. A. (2014). Reduced ethanol consumption and preference in cocaine- and amphetamine-regulated transcript (CART) knockout mice. *Addiction Biology*, *19*(2), 175–184. http://dx.doi.org/10.1111/j.1369-1600.2012.00475.x.
Sanbe, A., Takagi, N., Fujiwara, Y., Yamauchi, J., Endo, T., Mizutani, R., … Tanoue, A. (2008). Alcohol preference in mice lacking the Avpr1a vasopressin receptor. *American Journal of Physiology. Regulatory, Integrative and Comparative Physiology*, *294*(5), R1482–R1490. http://dx.doi.org/10.1152/ajpregu.00708.2007.
Sanchis-Segura, C., Borchardt, T., Vengeliene, V., Zghoul, T., Bachteler, D., Gass, P., … Spanagel, R. (2006). Involvement of the AMPA receptor GluR-C subunit in

alcohol-seeking behavior and relapse. *The Journal of Neuroscience*, *26*(4), 1231–1238. http://dx.doi.org/10.1523/JNEUROSCI.4237-05.2006.

Santos, N., Chatterjee, S., Henry, A., Holgate, J., & Bartlett, S. E. (2013). The alpha5 neuronal nicotinic acetylcholine receptor subunit plays an important role in the sedative effects of ethanol but does not modulate consumption in mice. *Alcoholism, Clinical and Experimental Research*, *37*(4), 655–662. http://dx.doi.org/10.1111/acer.12009.

Savarese, A., Zou, M. E., Kharazia, V., Maiya, R., & Lasek, A. W. (2014). Increased behavioral responses to ethanol in Lmo3 knockout mice. *Genes, Brain, and Behavior*, *13*(8), 777–783. http://dx.doi.org/10.1111/gbb.12176.

Savelieva, K. V., Caudle, W. M., Findlay, G. S., Caron, M. G., & Miller, G. W. (2002). Decreased ethanol preference and consumption in dopamine transporter female knockout mice. *Alcoholism, Clinical and Experimental Research*, *26*(6), 758–764. Retrieved from, http://www.ncbi.nlm.nih.gov/pubmed/12068242.

Schank, J. R., Ryabinin, A. E., Giardino, W. J., Ciccocioppo, R., & Heilig, M. (2012). Stress-related neuropeptides and addictive behaviors: Beyond the usual suspects. *Neuron*, *76*(1), 192–208. http://dx.doi.org/10.1016/j.neuron.2012.09.026.

Sharpe, A. L., Coste, S. C., Burkhart-Kasch, S., Li, N., Stenzel-Poore, M. P., & Phillips, T. J. (2005). Mice deficient in corticotropin-releasing factor receptor type 2 exhibit normal ethanol-associated behaviors. *Alcoholism, Clinical and Experimental Research*, *29*(9), 1601–1609. Retrieved from, http://www.ncbi.nlm.nih.gov/pubmed/16205360.

Siems, W., Maul, B., Krause, W., Gerard, C., Hauser, K. F., Hersh, L. B., … Saria, A. (2000). Neutral endopeptidase and alcohol consumption, experiments in neutral endopeptidase-deficient mice. *European Journal of Pharmacology*, *397*(2-3), 327–334. Retrieved from, http://www.ncbi.nlm.nih.gov/pubmed/10844131.

Sikela, J. M., Maclaren, E. J., Kim, Y., Karimpour-Fard, A., Cai, W. W., Pollack, J., … Miles, M. F. (2006). DNA microarray and proteomic strategies for understanding alcohol action. *Alcoholism, Clinical and Experimental Research*, *30*(4), 700–708. http://dx.doi.org/10.1111/j.1530-0277.2006.00081.x.

Sillaber, I., Rammes, G., Zimmermann, S., Mahal, B., Zieglgansberger, W., Wurst, W., … Spanagel, R. (2002). Enhanced and delayed stress-induced alcohol drinking in mice lacking functional CRH1 receptors. *Science*, *296*(5569), 931–933. http://dx.doi.org/10.1126/science.1069836.

Spanagel, R., Pendyala, G., Abarca, C., Zghoul, T., Sanchis-Segura, C., Magnone, M. C., … Albrecht, U. (2005). The clock gene Per2 influences the glutamatergic system and modulates alcohol consumption. *Nature Medicine*, *11*(1), 35–42. http://dx.doi.org/10.1038/nm1163.

Spanagel, R., Siegmund, S., Cowen, M., Schroff, K. C., Schumann, G., Fiserova, M., … Putzke, J. (2002). The neuronal nitric oxide synthase gene is critically involved in neurobehavioral effects of alcohol. *The Journal of Neuroscience*, *22*(19), 8676–8683. Retrieved from, http://www.ncbi.nlm.nih.gov/pubmed/12351742.

Speca, D. J., Chihara, D., Ashique, A. M., Bowers, M. S., Pierce-Shimomura, J. T., Lee, J., … McIntire, S. L. (2010). Conserved role of unc-79 in ethanol responses in lightweight mutant mice. *PLoS Genetics*, *6*(8). http://dx.doi.org/10.1371/journal.pgen.1001057.

Sperling, R. E., Gomes, S. M., Sypek, E. I., Carey, A. N., & McLaughlin, J. P. (2010). Endogenous kappa-opioid mediation of stress-induced potentiation of ethanol-conditioned place preference and self-administration. *Psychopharmacology*, *210*(2), 199–209. http://dx.doi.org/10.1007/s00213-010-1844-5.

Stacey, D., Bilbao, A., Maroteaux, M., Jia, T., Easton, A. C.,Longueville, S. … IMAGEN Consortium. (2012). RASGRF2 regulates alcohol-induced reinforcement by influencing mesolimbic dopamine neuron activity and dopamine release. *Proceedings of the National Academy of Sciences of the United States of America*, *109*(51), 21128–21133. http://dx.doi.org/10.1073/pnas.1211844110.

Steffensen, S. C., Bradley, K. D., Hansen, D. M., Wilcox, J. D., Wilcox, R. S., Allison, D. W., ... Edwards, J. G. (2011). The role of connexin-36 gap junctions in alcohol intoxication and consumption. *Synapse*, *65*(8), 695–707. http://dx.doi.org/10.1002/syn.20885.

Stephens, D. N., Pistovcakova, J., Worthing, L., Atack, J. R., & Dawson, G. R. (2005). Role of GABAA alpha5-containing receptors in ethanol reward: The effects of targeted gene deletion, and a selective inverse agonist. *European Journal of Pharmacology*, *526*(1–3), 240–250. http://dx.doi.org/10.1016/j.ejphar.2005.09.031.

Stewart, A., Maity, B., Anderegg, S. P., Allamargot, C., Yang, J., & Fisher, R. A. (2015). Regulator of G protein signaling 6 is a critical mediator of both reward-related behavioral and pathological responses to alcohol. *Proceedings of the National Academy of Sciences of the United States of America*, *112*(7), E786–E795. http://dx.doi.org/10.1073/pnas.1418795112.

Stork, O., Kojima, N., Stork, S., Kume, N., & Obata, K. (2002). Resistance to alcohol withdrawal-induced behaviour in Fyn transgenic mice and its reversal by ifenprodil. *Brain Research. Molecular Brain Research*, *105*(1–2), 126–135. Retrieved from, http://www.ncbi.nlm.nih.gov/pubmed/12399115.

Szumlinski, K. K., Ary, A. W., Lominac, K. D., Klugmann, M., & Kippin, T. E. (2008). Accumbens Homer2 overexpression facilitates alcohol-induced neuroplasticity in C57BL/6J mice. *Neuropsychopharmacology*, *33*(6), 1365–1378. http://dx.doi.org/10.1038/sj.npp.1301473.

Szumlinski, K. K., Lominac, K. D., Oleson, E. B., Walker, J. K., Mason, A., Dehoff, M. H., ... Kalivas, P. W. (2005). Homer2 is necessary for EtOH-induced neuroplasticity. *The Journal of Neuroscience*, *25*(30), 7054–7061. http://dx.doi.org/10.1523/JNEUROSCI.1529-05.2005.

Szumlinski, K. K., Toda, S., Middaugh, L. D., Worley, P. F., & Kalivas, P. W. (2003). Evidence for a relationship between Group 1 mGluR hypofunction and increased cocaine and ethanol sensitivity in Homer2 null mutant mice. *Annals of the New York Academy of Sciences*, *1003*, 468–471. Retrieved from, http://www.ncbi.nlm.nih.gov/pubmed/14684491.

Tammimaki, A., Forsberg, M. M., Karayiorgou, M., Gogos, J. A., & Mannisto, P. T. (2008). Increase in free choice oral ethanol self-administration in catechol-o-methyltransferase gene-disrupted male mice. *Basic & Clinical Pharmacology & Toxicology*, *103*(4), 297–304. http://dx.doi.org/10.1111/j.1742-7843.2008.00267.x.

Tanaka, K., Kunishige-Yamamoto, A., Hashimoto, H., Shintani, N., Hayata, A., & Baba, A. (2010). Increased ethanol preference and serotonin 1A receptor-dependent attenuation of ethanol-induced hypothermia in PACAP-deficient mice. *Biochemical and Biophysical Research Communications*, *391*(1), 773–777. http://dx.doi.org/10.1016/j.bbrc.2009.11.136.

Thanos, P. K., Dimitrakakis, E. S., Rice, O., Gifford, A., & Volkow, N. D. (2005). Ethanol self-administration and ethanol conditioned place preference are reduced in mice lacking cannabinoid CB1 receptors. *Behavioural Brain Research*, *164*(2), 206–213. http://dx.doi.org/10.1016/j.bbr.2005.06.021.

Thanos, P. K., Rivera, S. N., Weaver, K., Grandy, D. K., Rubinstein, M., Umegaki, H., ... Volkow, N. D. (2005). Dopamine D2R DNA transfer in dopamine D2 receptor-deficient mice: Effects on ethanol drinking. *Life Sciences*, 77(2), 130–139. http://dx.doi.org/10.1016/j.lfs.2004.10.061.

Thiele, T. E., Koh, M. T., & Pedrazzini, T. (2002). Voluntary alcohol consumption is controlled via the neuropeptide Y Y1 receptor. *The Journal of Neuroscience*, *22*(3), RC208. Retrieved from, http://www.ncbi.nlm.nih.gov/pubmed/11826154.

Thiele, T. E., Marsh, D. J., Ste Marie, L., Bernstein, I. L., & Palmiter, R. D. (1998). Ethanol consumption and resistance are inversely related to neuropeptide Y levels. *Nature*, *396*(6709), 366–369. http://dx.doi.org/10.1038/24614.

Thiele, T. E., Miura, G. I., Marsh, D. J., Bernstein, I. L., & Palmiter, R. D. (2000). Neurobiological responses to ethanol in mutant mice lacking neuropeptide Y or the Y5 receptor. *Pharmacology, Biochemistry, and Behavior*, *67*(4), 683–691. Retrieved from, http://www.ncbi.nlm.nih.gov/pubmed/11166058.

Thiele, T. E., & Navarro, M. (2014). "Drinking in the dark" (DID) procedures: A model of binge-like ethanol drinking in non-dependent mice. *Alcohol*, *48*(3), 235–241. http://dx.doi.org/10.1016/j.alcohol.2013.08.005.

Thiele, T. E., Naveilhan, P., & Ernfors, P. (2004). Assessment of ethanol consumption and water drinking by NPY Y(2) receptor knockout mice. *Peptides*, *25*(6), 975–983. http://dx.doi.org/10.1016/j.peptides.2004.03.009.

Thiele, T. E., Willis, B., Stadler, J., Reynolds, J. G., Bernstein, I. L., & McKnight, G. S. (2000). High ethanol consumption and low sensitivity to ethanol-induced sedation in protein kinase A-mutant mice. *The Journal of Neuroscience*, *20*(10), RC75. Retrieved from, http://www.ncbi.nlm.nih.gov/pubmed/10783399.

Thorsell, A., Schank, J. R., Singley, E., Hunt, S. P., & Heilig, M. (2010). Neurokinin-1 receptors (NK1R:s), alcohol consumption, and alcohol reward in mice. *Psychopharmacology*, *209*(1), 103–111. http://dx.doi.org/10.1007/s00213-010-1775-1.

Trudell, J. R., Messing, R. O., Mayfield, J., & Harris, R. A. (2014). Alcohol dependence: Molecular and behavioral evidence. *Trends in Pharmacological Sciences*, *35*(7), 317–323. http://dx.doi.org/10.1016/j.tips.2014.04.009.

van den Brink, W., Sorensen, P., Torup, L., Mann, K., & Gual, A.Sense Study Group. (2014). Long-term efficacy, tolerability and safety of nalmefene as-needed in patients with alcohol dependence: A 1-year, randomised controlled study. *Journal of Psychopharmacology*, *28*(8), 733–744. http://dx.doi.org/10.1177/0269881114527362.

Vanderlinden, L. A., Saba, L. M., Bennett, B., Hoffman, P. L., & Tabakoff, B. (2015). Influence of sex on genetic regulation of "drinking in the dark" alcohol consumption. *Mammalian Genome*, *26*(1–2), 43–56. http://dx.doi.org/10.1007/s00335-014-9553-8.

Vanhanen, J., Nuutinen, S., Lintunen, M., Maki, T., Ramo, J., Karlstedt, K., & Panula, P. (2013). Histamine is required for H(3) receptor-mediated alcohol reward inhibition, but not for alcohol consumption or stimulation. *British Journal of Pharmacology*, *170*(1), 177–187. http://dx.doi.org/10.1111/bph.12170.

van Huijstee, A. N., & Mansvelder, H. D. (2014). Glutamatergic synaptic plasticity in the mesocorticolimbic system in addiction. *Frontiers in Cellular Neuroscience*, *8*, 466. http://dx.doi.org/10.3389/fncel.2014.00466.

van Rijn, R. M., & Whistler, J. L. (2009). The delta(1) opioid receptor is a heterodimer that opposes the actions of the delta(2) receptor on alcohol intake. *Biological Psychiatry*, *66*(8), 777–784. http://dx.doi.org/10.1016/j.biopsych.2009.05.019.

Vengeliene, V., Bilbao, A., & Spanagel, R. (2014). The alcohol deprivation effect model for studying relapse behavior: A comparison between rats and mice. *Alcohol*, *48*(3), 313–320. http://dx.doi.org/10.1016/j.alcohol.2014.03.002.

Vinod, K. Y., Sanguino, E., Yalamanchili, R., Manzanares, J., & Hungund, B. L. (2008). Manipulation of fatty acid amide hydrolase functional activity alters sensitivity and dependence to ethanol. *Journal of Neurochemistry*, *104*(1), 233–243. http://dx.doi.org/10.1111/j.1471-4159.2007.04956.x.

Vinod, K. Y., Yalamanchili, R., Thanos, P. K., Vadasz, C., Cooper, T. B., Volkow, N. D., & Hungund, B. L. (2008). Genetic and pharmacological manipulations of the CB(1) receptor alter ethanol preference and dependence in ethanol preferring and nonpreferring mice. *Synapse*, *62*(8), 574–581. http://dx.doi.org/10.1002/syn.20533.

Walker, A. W., Smith, C. M., Chua, B. E., Krstew, E. V., Zhang, C., Gundlach, A. L., & Lawrence, A. J. (2015). Relaxin-3 receptor (RXFP3) signalling mediates stress-related alcohol preference in mice. *PLoS One*, *10*(4), e0122504. http://dx.doi.org/10.1371/journal.pone.0122504.

Wallace, M. J., Newton, P. M., Oyasu, M., McMahon, T., Chou, W. H., Connolly, J., & Messing, R. O. (2007). Acute functional tolerance to ethanol mediated by protein kinase Cepsilon. *Neuropsychopharmacology*, *32*(1), 127–136. http://dx.doi.org/10.1038/sj.npp.1301059.

Weinshenker, D., Rust, N. C., Miller, N. S., & Palmiter, R. D. (2000). Ethanol-associated behaviors of mice lacking norepinephrine. *The Journal of Neuroscience*, *20*(9), 3157–3164. Retrieved from, http://www.ncbi.nlm.nih.gov/pubmed/10777779.

Werner, D. F., Blednov, Y. A., Ariwodola, O. J., Silberman, Y., Logan, E., Berry, R. B., … Homanics, G. E. (2006). Knockin mice with ethanol-insensitive alpha1-containing gamma-aminobutyric acid type A receptors display selective alterations in behavioral responses to ethanol. *The Journal of Pharmacology and Experimental Therapeutics*, *319*(1), 219–227. http://dx.doi.org/10.1124/jpet.106.106161.

Werner, C., Raivich, G., Cowen, M., Strekalova, T., Sillaber, I., Buters, J. T., … Hofmann, F. (2004). Importance of NO/cGMP signalling via cGMP-dependent protein kinase II for controlling emotionality and neurobehavioural effects of alcohol. *The European Journal of Neuroscience*, *20*(12), 3498–3506. http://dx.doi.org/10.1111/j.1460-9568.2004.03793.x.

Wyatt, L. R., Finn, D. A., Khoja, S., Yardley, M. M., Asatryan, L., Alkana, R. L., & Davies, D. L. (2014). Contribution of P2X4 receptors to ethanol intake in male C57BL/6 mice. *Neurochemical Research*, *39*(6), 1127–1139. http://dx.doi.org/10.1007/s11064-014-1271-9.

Yaka, R., Tang, K. C., Camarini, R., Janak, P. H., & Ron, D. (2003). Fyn kinase and NR2B-containing NMDA receptors regulate acute ethanol sensitivity but not ethanol intake or conditioned reward. *Alcoholism, Clinical and Experimental Research*, *27*(11), 1736–1742. http://dx.doi.org/10.1097/01.ALC.0000095924.87729.D8.

Zghoul, T., Abarca, C., Sanchis-Segura, C., Albrecht, U., Schumann, G., & Spanagel, R. (2007). Ethanol self-administration and reinstatement of ethanol-seeking behavior in Per1(Brdm1) mutant mice. *Psychopharmacology*, *190*(1), 13–19. http://dx.doi.org/10.1007/s00213-006-0592-z.

Zhou, Z., Karlsson, C., Liang, T., Xiong, W., Kimura, M., Tapocik, J. D., … Goldman, D. (2013). Loss of metabotropic glutamate receptor 2 escalates alcohol consumption. *Proceedings of the National Academy of Sciences of the United States of America*, *110*(42), 16963–16968. http://dx.doi.org/10.1073/pnas.1309839110.

Zindel, L. R., & Kranzler, H. R. (2014). Pharmacotherapy of alcohol use disorders: Seventy-five years of progress. *Journal of Studies on Alcohol and Drugs. Supplement*, *75*(Suppl. 17), 79–88. Retrieved from, http://www.ncbi.nlm.nih.gov/pubmed/24565314.

CHAPTER TEN

Gene Targeting Studies of Hyperexcitability and Affective States of Alcohol Withdrawal in Rodents

G.D. Greenberg[*,†,‡,1], **J.C. Crabbe**[*,†,‡]
[*]Oregon Health & Science University, Portland, OR United States
[†]Portland Alcohol Research Center, Portland, OR, United States
[‡]VA Portland Health Care System, Portland, OR, United States
[1]Corresponding author: e-mail address: greenbeg@ohsu.edu

Contents

Abstract

Genetically engineered rodents can be used to examine the influence of single genes on alcoholism-related phenotypes. We review studies that employed gene targeting with a focus on ethanol withdrawal-associated behaviors. Earlier studies targeted the glutamate and GABA systems as contributors to the underlying hyperexcitable state of convulsions or similar signs of ethanol withdrawal. Over the past decade, many gene-targeting studies have continued to focus on the glutamatergic and GABAergic systems; however, an increasing number of these studies have focused on other withdrawal outcomes such as anxiety-like behavior and escalated ethanol consumption. Although negative affective states may drive escalated ethanol drinking, few reported studies examined the phenotypes together. However, there is significant overlap in the systems

International Review of Neurobiology, Volume 126
ISSN 0074-7742
http://dx.doi.org/10.1016/bs.irn.2016.02.010

that were manipulated in relation to studying the phenotypes individually. These studies reveal common genetic influences on withdrawal-associated anxiety, convulsions, and escalated drinking that may contribute to relapse, setting the stage for the identification of novel medications to jointly target these effects.

1. INTRODUCTION

Alcohol dependence is a highly prevalent psychiatric disorder with a clear genetic component in humans and animals (Cloninger, Bohman, & Sigvardsson, 1981; Crabbe, Mcswigan, & Belknap, 1985). However, like many neuropsychiatric disorders, its occurrence also has environmental influences, and its genetic basis is complex (Goldman & Bergen, 1998). Phenotypes of alcoholism are often the product of multiple genes with small contributions (Mayfield, Harris, & Schuckit, 2008). Additionally, many of the individual genes associated with alcoholism are pleiotropic, meaning those genes can predispose individuals to a variety of traits and/or symptoms across disorders. It is known from human studies that there are shared genetic risk factors for alcoholism and other psychiatric disorders such as bipolar, anxiety, and major depressive disorders (Davis et al., 2008; Kushner, Abrams, & Borchardt, 2000; Winokur et al., 1996). There is also clinical evidence for a high degree of comorbidity of alcohol use and mood disorders (Kessler et al., 2005; Swendsen et al., 2010), which share many symptoms, particularly during alcohol withdrawal. In fact, many symptoms of alcohol withdrawal are similar to mood disorder symptomatology: anxiety, tiredness, sleep disturbances, depressed mood, irritability, and heightened response to stressors (Mossberg, Liljeberg, & Borg, 1985). These symptoms have long been theorized to decrease control over alcohol intake and lead to its escalation (Hershon, 1977). Shared behavioral manifestations of alcoholism and mood disorders can be the product of overlapping neural circuitries, and understanding this overlap may help reveal important genes acting in those pathways.

The neuroadaptive view of alcoholism proposes that a plasticity-induced negative affective state is a driving force behind relapse and alcohol consumption during withdrawal (Koob & Le Moal, 1997, 2001). Similar to neuroadaptive theories of chronic stress, prolonged alcohol use is surmised to induce plasticity within and between whole-brain networks, including stress and reward circuitries (Koob, 2013). These deviations also extend to peripheral systems, creating a new set point for the whole body that requires persistent, escalated consumption of the drug to be maintained; this has been called the allostatic hypothesis of alcohol dependence (Koob & Le Moal, 2001). It is theorized from this framework that a negative emotional state (eg, anxiety) combined with loss of natural reward function (ie, anhedonia) is part of a feed-forward system that produces recurring alcohol consumption during withdrawal. Even in the absence of physical symptoms of withdrawal such as convulsions, this negative affective state can persist into protracted periods of abstinence and increase vulnerability to relapse (Heilig et al., 2010). Further understanding of the neural systems important for the relationship between negative affect and increased alcohol consumption are therefore of great importance, and studies using gene-targeting approaches to address these phenotypes have recently increased.

Genetically engineered models can be used as a tool to examine the influence of single genes on alcoholism-related phenotypes. A 2006 review by Crabbe et al. (2006) included approximately 150 reports on the modifications of nearly 100 genes in mice and their effects on behaviors related to alcohol use disorders, but only a fraction of those studies were applied to studying questions of alcohol withdrawal. The majority (~80%) of alcohol withdrawal studies used the handling-induced convulsion (HIC) or similar physical signs as measurements of withdrawal severity. HICs have been the gold standard for measuring withdrawal severity in mice for over 40 years (Goldstein, 1973); this phenotype is the product of a hyperexcitable state of the central nervous system during the withdrawal period. Therefore, it is sensible that nearly half of the studies that examined alcohol withdrawal-induced HICs targeted either the glutamatergic or GABAergic systems. They are the major excitatory and inhibitory neurotransmitter systems in the brain and are known to contribute to a compensatory hyperexcitable state during alcohol withdrawal. In fact, of the studies reported in 2006 that examined effects of γ-aminobutyric acid (GABA)- or glutamate-related mutations on alcohol withdrawal severity, all of them analyzed only HICs as a measure of withdrawal severity. Just 2 of 14 gene-targeting studies outside of the GABAergic and glutamatergic system measured anxiety-like behavior during withdrawal (Ghozland et al., 2005; Timpl et al., 1998). Two other studies, only one published (Lallemand & De Witte, 2005), examined escalated drinking in postdependent mice. There have been many more studies using genetically engineered rodents to study alcohol withdrawal since 2006. The emphasis on targeting single genes associated with the glutamatergic and GABAergic systems has carried on; however, the scope of withdrawal-induced behavioral outcomes has widened.

In this chapter, we present progress over the last 10 years in understanding the neural circuitry and genetic contributions to alcohol withdrawal from studies using genetically engineered rodents. Alcohol withdrawal phenotypes such as anxiety-like behavior and escalated ethanol consumption have been increasingly tested. The use of mutant mice to identify medications that treat excessive alcohol intake in nondependent animals is covered elsewhere (Mayfield, Arends, Harris, & Blednov, 2016). We will outline recent genetic modifications with a focus on withdrawal-induced changes in escalated drinking, anxiety-like behavior, and HICs. Although it is proposed that postdependent anxiety may influence escalated alcohol drinking (Heilig et al., 2010; Koob & Le Moal, 1997), none of the studies examined both phenotypes in the same mice. However, several studies examined these phenotypes individually, which could set the stage for future experiments that target those systems with medications intended to alleviate the driving forces for drinking and, in turn, reduce drinking itself.

Because neuroadaptation between alcohol reward pathways (eg, opioid, dopamine) and neural systems associated with anxiety (eg, CRF/CRH) is a foundational hypothesis for alcohol dependence, we will begin with mutants for these systems that were reported in 2006 for effects on anxiety-like behavior during alcohol withdrawal. We will then discuss how recent studies have targeted genes within the same systems to examine effects on escalated alcohol consumption. Each section will conclude with other studies that have tested effects on ethanol withdrawal severity based on HICs, as there is an established negative genetic correlation between HIC scores and preference for 10% ethanol in nondependent mice (Metten et al., 1998). Hypotheses for disrupted homeostasis have expanded to include neurotransmitter systems such as glutamate which can signal within the reward and stress pathways, and gene-targeting studies for these systems have also increased with a focus on anxiety and excessive consumption of alcohol during withdrawal. Changes in glutamatergic and GABAergic signaling (ie, neural plasticity) may lead to escalated alcohol consumption through various factors, including increased negative affect (ie, negative reinforcement to relieve the symptoms), loss of inhibition, and reduced threshold for reward (Koob & Le Moal, 1997). We will discuss how connections between the systems and pathways can work together to produce this complex, encompassing syndrome of alcohol withdrawal, and how treatments that target these interconnections can be tested using recently developed behavioral models of excessive ethanol consumption, potentially leading to novel therapies for shared symptoms across mood and alcohol use disorders.

2. CONSIDERATIONS FOR GENE TARGETING STUDIES

The general challenges to gene-targeting studies are well known (Crabbe et al., 2006). We will concentrate here on only three of the multiple aspects of alcohol withdrawal—hyperexcitability, anxiety-like behavior, and escalated drinking. Individual genes affect one or more of these symptoms. Targeted manipulations also are placed on different background strains, which can create conflicting results even following the same targeted manipulations (Phillips, Hen, & Crabbe, 1999); they can also be informative about how gene–gene interactions influence behavior. Background strain influences on ethanol withdrawal hyperexcitability have been documented (Metten & Crabbe, 2005; Metten et al., 2010). For example, DBA/2 (D2) mice have been consistently found to have higher sensitivity to ethanol withdrawal, including increased handling induced convulsions (HICs), relative to other background strains such as the C57BL/6 (B6). Withdrawal severity of background strains also has an inverse relationship with consumption of 10% ethanol in tests of two-bottle preference drinking in nondependent animals (Metten et al., 1998). When possible, we provide information about the background strain.

Studies with similar hypotheses can produce contrasting results based on the method used to induce dependence. Methods range from oral self-administration (choice or forced), to liquid diets where ethanol consumption becomes necessary for nutritional sustenance, to intraperitoneal (i.p.) injection or intragastric (i.g.) infusion, and vapor chamber inhalation (some vapor studies manipulate alcohol metabolism using the alcohol dehydrogenase inhibitor pyrazole, while others do not). These different methods of administration may produce different gene–environment interactions that influence behavioral outcomes, particularly for measurements of anxiety. For example, pyrazole administration can be considered a stressor that elevates plasma corticosterone when administered in conjunction with vapor inhalation chambers (Keith & Crabbe, 1992). Additionally, the different methods themselves could be differentially stressful. Finally, behavioral outcomes could also be influenced by variability in blood ethanol concentrations (BECs) achieved from the method of administration. We will outline the method of administration for each study reviewed and report BECs when available. Studies of targeted genes are summarized in Table 1.

Table 1 Systems Targeted for Effects on Ethanol Consumption or Anxiety

Mutation/Gene	Background and ΔExpression	Reference	EtOH Admin	Ethanol Consumption	Anxiety			HICs and Other Measures
					Baseline	Intoxication	Withdrawal	
CRF/CRH								
$CRHR1^{-/-}$[a,b] *Crhr1*	CD1 eliminated	Timpl et al. (1998)	2–16% free-choice for 14 days, 20% forced drinking for 18 days. BECs ~2.0 mg/mL at end of forced drinking	–	KO < WT L/D latency to enter lit; KO > WT % entries into lit, time in lit; KO = WT entries into dark	–	KO < WT L/D latency to enter lit; KO > WT % entries into lit 12 h after end of forced drinking	KO = WT activity (all wdr mice reduced activity in OFT)
$CRHR1^{-/-}$ *Crhr1*	C57BL/6J eliminated	Chu et al. (2007)	Operant self-admin w/saccharin fading (23 days escalating 5–10%, then 20 days 10%); 2 week liquid diet (2–5%). BECs ~2.0 mg/mL	KO < WT responses for 10% ethanol operant self-admin (escalated presses in WT during wdr vs baseline not observed in KO) for 10 days beginning 1 week into withdrawal	–	–	–	–
$CRHR1^{-/-}$ and $CRHR1^{NestinCre}$ *Crhr1*	CD1 and N2 C57BL/6J eliminated and selective neuronal removal	Molander et al. (2011)	3 months of 2-bottle free choice (8%); 4 cycles CIE (16 h on, 8 h off) with loading dose of 2 g/kg, pyrazole not specified. BECs ~2.2 mg/mL in $CRHR1^{NestinCre}$; BECs ~3.5 mg/mL in $CRHR1^{-/-}$	$CRHR1^{NestinCre}$ < control escalated intake for 4 weeks starting 24 h after CIE and increased PR for 1 week in control mice, not observed in $CRHR1^{NestinCre}$; $CRHR1^{-/-}$ > WT escalated intake in KO 1 and 2 weeks postdependence; only escalation 1 week in WT	–	–	–	–

FKBP5$^{-/-}$ *Fkbp5*	N4 C57BL/6J eliminated	Huang et al. (2014)	Acute: 4 g/kg i.p.; chronic: 3 weeks liquid EtOH diet	–	–	–	–	Acute HIC: EtOH KO > saline KO (AUC 2–12, 24, 25 h), no increase in EtOH WT; chronic HICs: EtOH KO > EtOH WT (AUC)
Opioid								
MOR$^{-/-}$ [b] *Oprm1*	C57BL/6 eliminated	Ghozland et al. (2005)	3 wdr periods following successive 2 weeks of EtOH liquid diet (increasing conc. 0–5%). BECs ~1.6 mg/mL in KO at 2nd wdr; BECs ~2.6 mg/mL in WT. BECs ~2.2 mg/mL in both at 1st wdr	–	KO = WT L/D time in light	KO < WT L/D time in light 20 min after acute 1.25, 1.75 g/kg EtOH	KO < WT L/D time in light 8 h after 2nd wdr period	KO > WT HIC (AUC 2, 4, 6, 8, 11, and 24 h into 1st wdr period); KO = WT after 2nd and 3rd wdr periods
MOR$^{-/-}$ *Oprm1*	C57BL/6J eliminated	Contet et al. (2013)	2 h limited access two-bottle choice (15%, 5 days/week) for 5 weeks, with or without alternating deprivation weeks starting at week 2	Without deprivation: KO < WT (escalated ethanol intake in WT but not KO) With deprivation: KO = WT (escalated ethanol intake in both to same degree)	–	–	–	–

Continued

Table 1 Systems Targeted for Effects on Ethanol Consumption or Anxiety—cont'd

Mutation/Gene	Background and ΔExpression	Reference	EtOH Admin	Ethanol Consumption	Anxiety: Baseline	Anxiety: Intoxication	Anxiety: Withdrawal	HICs and Other Measures
$PDYN^{-/-}$ *Pdyn*	C57BL/6 eliminated	Femenía and Manzanares (2012)	Acute 4 g/kg p.o.	–	–	–	–	KO > WT HICs (1 h into wdr); trend for KO > WT (2–6 h)
$PDYN^{-/-}$ *Pdyn*	C57BL/6J eliminated	Blednov et al. (2006a)	Acute 4 g/kg i.p.	–	–	–	–	KO = WT HIC (AUC)
Glutamate related								
NR1 ($Grin1^{D381N}$) *Grin1*	C57BL/6 modified	Kiefer et al. (2003)	Two-bottle free choice, 16 days (2–16%, increase 2% every 2 days), forced drinking (16%) 24 days, 4 day abstinent period	–	WT = $Grin1^{D381N}$ (OFT center time and EPM time in open arms)	$Grin1^{D381N}$ < WT (OFT/EPM center/open arms time, 3 days into forced drinking; WT had increased time in open arms/center)	$Grin1^{D418N}$ < WT (EPM time in open arms at 12 h vs baseline); $Grin1^{D418N}$ > WT (latency to enter center of open field)	-
$NR1^{+/-}$ *Grin1*	C57BL/6J reduced	Du et al. (2012)	Acute 4 g/kg i.p.	–	$NR^{+/-}$ = WT (time in open arms, EPM)	–	–	$NR^{+/-}$ < WT HIC 7 h post 4 g/kg
$NR2A^{-/-}$ *Grin2a*	N10 C57BL/6J eliminated	Daut et al. (2014)	CIE vapor 4 weeks (16 h on, 8 h off). BECs ~1.75 mg/mL	–	–	–	NR2A = WT, (L/D exploration at 48 h, no anxiety in either genotype)	–

Fyn^{+} [a] *Fyn*	C57BL/ 6 × CBA2 N6 backcross to C57BL/6 increased	Stork et al. (2002)	Two-bottle free choice, 7 days (2% days 1–3, 5% days 4–7), forced drinking 21 days (5% days 8–10, 10% days 11–14, 20% days 15–28). BECs ~0.8 mg/mL	–	fyn^{+} = WT (transitions, activity in light, time in dark of L/D box)	–	fyn^{+} > WT (L/D light activity; WT decreased 16 h into wdr vs naïve WT)	KO > WT forebrain expression of pNR2B
PTPα shRNA *Ptpra*	C57BL/6J and rat selective reduction in dorsal striatum	Hamida et al. (2013)	Two-bottle free choice (20%, intermittent, alternating days, 6 weeks total), then surgery/infusion	PTPα knockdown < controls, intake and PR (3–7 weeks following infusion in weekly 24 h free choice)	–	–	–	PTPα shRNA < control NR2B dorsomedial striatum expression of pNR2B
mGlur2 lentiviral *Grm2*	Rat selective increase in prefrontal cortex	Meinhardt et al. (2013)	CIE vapor (14 h on, 10 h off, 5 days per week, 7 weeks total), no target or pyrazole stated. BECs ~2.5 mg/mL	Lenti-mGluR2 < EGFP control for lever pressing in cue-induced reinstatement 2 weeks after CIE and 1 week after conditioning (10% ethanol)	–	–	–	–
mGluR2$^{-/-}$ *Grm2*	C57BL/6 N3 × CD1 eliminated	Zhou et al. (2013)	3–17% increasing. Two-bottle choice, 80 days, no wdr	KO > WT escalated intake and PR	–	–	–	–
mGlur5$^{-/-}$ *Grm5*	C57BL/6J eliminated	Blednov and Harris (2008)	Acute 4 g/kg i.p.	–	–	–	–	Female KO < WT HIC (AUC 1–15 h)

Continued

Table 1 Systems Targeted for Effects on Ethanol Consumption or Anxiety—cont'd

Mutation/Gene	Background and ΔExpression	Reference	EtOH Admin	Ethanol Consumption	Anxiety			HICs and Other Measures
					Baseline	Intoxication	Withdrawal	
Neuroimmune/other neuropeptides								
Cytokine								
$TLR2^{-/-}$/$TLR4^{-/-}$ *Tlr2/Tlr4*	C57BL/6J eliminated	Pascual et al. (2015)	Forced 10% ethanol 5 months, EtOH replaced with water for 24 h for wdr. BECs ~1.25 mg/mL	–	–	–	KO > WT L/D time in light and EPM time in open arms; 24 h into wdr WT mice had reduced times compared to air	
$TLR4^{-/-}$ *Tlr4*	C57BL/6J eliminated	Pascual et al. (2011)	Forced 10% ethanol 5 months (progressive increase over first weeks 2–10%), EtOH replaced with water for 2 weeks for wdr	–	–	–	KO > WT L/D latency to enter light (2 weeks into wdr, WT mice latency to enter light side shorter than air, no diff. in EtOH-treated KOs)	
$IL\text{-}1ra^{-/-}$ *Il1rn*	C57BL/6J eliminated	Blednov et al. (2015a)	Acute 4 g/kg i.p.	–	KO = WT % open arm entries and time on EPM	KO = WT (1.25 g/kg EtOH increased open arm time and entries in both genotypes)	–	KO > WT HICs

k-ras $^{+/-}$/Nf1$^{+/-}$ *Kras/Nf1*	C57BL/6J reduced and increased k-ras	Repunte-Canonigo et al. (2010a)	Two-bottle choice, 10 days (10% EtOH limited access) then CIE (16 h on, 8 off, 3 days, target BEC 1.5–2.0 mg/mL; pyrazole not stated) ×2 wdr periods	k-ras$^{+/-}$ < WT (escalated postdependent drinking in two-bottle choice after both CIE exposures in WT but not k-ras$^{+/-}$)	–	–	–	–
AdipoR2$^{-/-}$ *Adipor2*	C57BL/6J eliminated	Repunte-Canonigo et al. (2010b)	Two-bottle choice, 10 days (10% EtOH limited access) then CIE (16 h on, 8 off, 3 days, target BEC 1.5–2.0 mg/mL; pyrazole not stated)	KO < WT (escalated postdependent drinking in two-bottle choice after both CIE exposures in WT but not KO)	–	–	–	–
Angiotensin								
TGR(ASrAOGEN) 680 *Agt*	Rat reduced AOGEN-specific antisense RNA placed under GFAP promoter	Sommer et al. (2007)	7 weeks of EtOH vapor with target BECs of 1.5–2.5 mg/mL for 12 h per day. 3-week abstinence period	Transgenic = WT escalated postdependent drinking in two-bottle choice, 19 days (2–4% first 10 days, 6% final 9 days)	–	–	–	–

[a]Previously reviewed in Crabbe and Phillips (2004).
[b]Previously reviewed in Crabbe et al. (2006).
Abbreviations: *AUC*, area under curve; *BEC*, blood ethanol concentration; *CIE*, chronic intermittent ethanol; *EPM*, elevated plus maze; *EtOH*, ethanol; *GFAP*, glial fibrillary acidic protein; *i.p.*, intraperitoneal; *KO*, knockout; *L/D*, light/dark box; *OFT*, open-field test; *p.o.*, per os; *PR*, preference ratio; *wdr*, withdrawal; *WT*, wild type.

3. TRADITIONAL TARGETS OF THE ALLOSTATIC MODEL—THE CRF/CRH AND OPIOID SYSTEMS

3.1 CRF/CRH

One theory deduced from the pleiotropic genetic basis of alcoholism and comorbid mood disorders is that genes underlying susceptibility to stress may predispose individuals to alcoholism (reviewed in Goldowitz et al., 2006; Radke, Pickens, & Holmes, 2014; Pandey, 2014). The corticotropin-releasing factor (CRF) system, in addition to its well-established role in stress responses (Dunn & Berridge, 1990; Heinrichs & Koob, 2004), is a critical component of many phases of alcohol dependence, including withdrawal. Initial pharmacological and neurochemical studies in rats determined that CRF mediates anxiety-like behavior during withdrawal. Ventricular infusion of a CRF receptor antagonist reversed anxiety-like behavior during withdrawal from a 3-week ethanol liquid diet (Baldwin et al., 1991), and CRF immunoreactivity was increased during withdrawal in rats following a similar paradigm of alcohol dependence (Pich et al., 1995). Since then, studies using mice have targeted CRF-related genes in order to examine their function for alcohol withdrawal-like symptoms.

An early study knocked out the gene encoding the CRF1 receptor (*CRHR1* KOs) (Timpl et al., 1998). KOs had reduced withdrawal-induced anxiety in a light/dark (L/D) box compared to wild-type (WT) mice. However, these differences were also present in untreated mice. Effects of *CRHR1* gene manipulations on dependence-induced escalated ethanol intake have also been studied. Escalated ethanol self-administration in WT mice 1 week following liquid ethanol diet was attenuated in *CRHR1* KO mice (Chu et al., 2007). Ethanol-withdrawn *CRHR1* KO mice also had lower ethanol preference than ethanol-withdrawn WT mice near the end of the 10-day operant chamber procedure. Nondependent *CRHR1* KO mice did not differ from nondependent WT mice for lever presses for ethanol or ethanol preference. Results of these two studies suggest a positive relationship between postdependent anxiety and drinking during withdrawal that is mediated by increased CRF1 signaling.

More-recent studies have employed conditional knockout strategies to create mice lacking CRF1 in the central nervous system but leaving expression in the anterior pituitary intact ($CRHR1^{NestinCre}$ KOs) (Molander et al.,

2011). *CRHR1*^{NestinCre} mice, in addition to global KOs (*CRHR1*$^{-/-}$), were made dependent using 4-day chronic intermittent ethanol (CIE) vapor inhalation. Consistent with Chu et al. (2007), *CRHR1*^{NestinCre} mice did not display increased postdependent ethanol intake compared to consumption before vapor exposure. However, *CRHR1*$^{-/-}$ global KO mice from the 2011 study actually displayed *more-pronounced* increases in postdependent drinking than WT mice, and this effect persisted for all 4 weeks of testing. One major difference between Molander et al. (2011) and Chu et al. (2007) is the degree of intoxication achieved for each experiment. *CRHR1*^{NestinCre} mice and control littermates made dependent with ethanol vapor in Molander et al. (2011) had approximate BECs of 2.2 mg/mL at the time they finished the inhalation paradigm, which more closely resemble the BECs (~2.0 mg/mL) obtained from *CRHR1* KO mice near the end of the ethanol liquid diet in Chu et al. (2007). On the other hand, *CRHR1* KO mice and their control littermates made dependent with ethanol vapor in Molander et al. (2011) had BECs that reached nearly 3.5 mg/mL on average by the end of CIE. Other possible explanations for the discrepancy in postdependent drinking between the two studies may be the different drinking paradigms used (operant vs two-bottle choice) and the time point of testing during withdrawal (1 week vs 1 day of abstinence). Although more work is needed, these studies provide support for an importance of dysregulated CRF signaling for ethanol withdrawal-associated anxiety-like behavior and drinking.

In addition to targeting the CRF system directly, other studies have targeted glucocorticoid receptors (GRs) that regulate CRF release for effects on withdrawal hyperexcitability. FK606 binding protein 5 (FKBP5) is an important mediator of negative feedback on the hypothalamic–pituitary–adrenal (HPA) axis (Binder, 2009). Global *FKBP5* KO mice, but not WT mice, had increased HICs following acute 4 g/kg ethanol or chronic ethanol liquid diet for 3 weeks (Huang et al., 2014). When combined with the CRF targeting studies, these results provide evidence that elevated CRF signaling increases ethanol withdrawal severity as measured by three separate components: anxiety-like behavior, escalated drinking, and convulsions.

3.2 Opioid

The recruitment of stress systems during alcohol withdrawal can induce neuroadaptations within reward systems, including dopamine and opioids

(Koob & Le Moal, 1997; Radke et al., 2014). The second study reviewed by Crabbe et al. (2006) that examined withdrawal-induced anxiety involved a mutation of the μ-opioid receptor (MOR). MOR KO ($MOR^{-/-}$) mice had increased anxiety-like behavior in a L/D box compared to WT mice following multiple rounds of liquid ethanol diet (Ghozland et al., 2005). $MOR^{-/-}$ mice also displayed increased withdrawal severity as measured by HICs during the first withdrawal period. Since then, a study has examined $MOR^{-/-}$ mice for ethanol preference drinking following periods of abstinence (Contet et al., 2013). Compared to WT mice, $MOR^{-/-}$ mice that underwent limited access two-bottle choice for 5 consecutive weeks showed less voluntary ethanol intake that did not escalate over time. In separate groups of mice, ethanol intake escalated to a greater extent when drinking weeks were separated by deprivation weeks compared to non-deprived mice from the previous experiment. This effect was present across genotypes, with $MOR^{-/-}$ mice showing escalated ethanol consumption to the same degree as WT mice. It is unclear if the augmented escalation in drinking following deprivation weeks was associated with a withdrawal-like state. Whether $MOR^{-/-}$ mice would display greater increases in drinking following a dependence-inducing paradigm (eg, CIE) would be an interesting follow-up study. Together, these studies support effects of reduced MOR signaling that mirror effects of elevated CRF during withdrawal—increased anxiety-like behavior, convulsions, and ethanol consumption.

In contrast to MORs, κ-opioid (KOR) signaling has been a target for dysphoric states and its hypothesized importance for avoidance of ethanol and anxiogenic effects of ethanol withdrawal (reviewed in Wee & Koob, 2010; Walker et al., 2012). KOR antagonists attenuate ethanol withdrawal-induced escalations of drinking and anxiety-like behavior on an elevated plus maze (EPM) in rats (Walker et al., 2012; Wee & Koob, 2010). Two studies have examined effects of targeting preprodynorphin (PDYN), a precursor for dynorphins that bind KORs, on acute ethanol withdrawal severity and nondependent ethanol drinking. One study found a trend for increased HICs during ethanol withdrawal in PDYN KOs vs WT mice (Femenía & Manzanares, 2012). A second study found no effect on HICs (Blednov et al., 2006a). More work is needed to resolve these conflicting reports and to discern the importance of KOR signaling for ethanol withdrawal severity. Additionally, as this system is linked to dysphoric states, it would be interesting to examine how gene manipulations may influence depression- or anxiety-like behavior during withdrawal.

4. EXPANDED TARGETS OF THE ALLOSTATIC MODEL—GLUTAMATE, PERIPHERAL SYSTEMS, AND GABA

4.1 Glutamate

Both of the systems reviewed earlier in this chapter are traditionally considered to be dysregulated under the allostatic model of drug dependence (Koob & Le Moal, 1997), but other neurotransmitters that work within their neural circuitry are also involved. After prolonged drug use, the stress and reward systems are hijacked, resulting in a two-pronged drive to increase drug consumption, through (1) the need to alleviate negative symptoms (ie, negative reinforcement by self-medication) and (2) to compensate for reduced rewarding properties of the drug. Changes in glutamatergic signaling within these pathways may underlie the neural plasticity that influences the negative affective state, increased drive, and lack of control contributing to alcohol consumption during withdrawal. The α-amino-3-hydroxy-5-methyl-4-isoxazolepropionic acid (AMPA) and *N*-methyl-D-aspartate (NMDA) receptors are the primary ionotropic glutamate receptors in the central nervous system (CNS), and both are involved in synaptic plasticity following administration of ethanol and other drugs of abuse (reviewed in Kalivas, 2009). Of the publications reported in 2006, no studies had targeted glutamatergic signaling in relation to anxiety or drinking during alcohol withdrawal. The majority of studies since the 2006 review have manipulated NMDA signaling in particular and examined ethanol-related effects outside of HICs.

Ethanol can inhibit NMDA receptor signaling by serving as an antagonist at glycine binding sites (occurring through NR1 subunits) or glutamate-binding sites (occurring through NR2 subunits) (Szumlinksi & Woodward, 2014). Rats exposed to 2 weeks of chronic ethanol vapor inhalation had upregulated NR1, NR2A, and NR2B protein expression in central amygdala (Roberto et al., 2005). A transgenic point mutation of *Grin1* (*Grin1*^{D418N}), the gene encoding NR1, disrupts only a single amino acid but reduces glycine affinity for NMDARs by approximately 80% (Kiefer et al., 2003). *Grin1*^{D418N}, but not WTs, had significantly reduced time in the open arms of the EPM during withdrawal from a forced drinking paradigm relative to baseline (no BECs reported), suggesting increased withdrawal-induced anxiety-like behavior in the transgenic mice (Kiefer et al., 2003). In support of this finding, *Grin1*^{D418N} had greater latency to enter the center of an open field compared to WT mice during withdrawal. The induced anxiety-like behavior in *Grin1*^{D418N} is impressive considering these mice were developed on a C57BL/6J background, a strain that appears resistant to CIE-associated anxiety-like behavior in a L/D box (McCool & Chappell, 2015). In contrast, heterozygous NR1 KOs (*NR1*$^{+/-}$) show reduced withdrawal HICs after acute ethanol (Du et al., 2012). This difference was small yet significant even though these mice were on a B6 background, and B6 mice typically show little HIC response (Metten & Crabbe, 2005).

Studies targeting subunits that comprise glutamate-binding sites (ie, NR2A and NR2B) for effects on withdrawal-associated anxiety-like behavior have produced varied results. NR2A KOs did not differ from WTs in a L/D box 2 days following CIE vapor exposure (Daut et al., 2014). However, as mentioned earlier, the background strain used for this study (B6) is particularly resistant to withdrawal severity as measured by convulsions (Metten & Crabbe, 2005), and the authors confirmed the wild type's resistance to withdrawal-induced deficits in the L/D box. In contrast, modulations of NR2 signaling through upstream activation cascades (in transgenic mice on a B6 background that were interbred with CBA2 mice) have produced effects on anxiety-like behavior during withdrawal (Stork et al., 2002). Mice overexpressing fyn (a kinase that phosphorylates NR2A and NR2B subunits) had increased forebrain expression of phosphorylated NR2B and were resistant to withdrawal-associated anxiety-like behavior in a L/D box following 4 weeks of forced drinking (Stork et al., 2002). When mice were treated with a NR2B antagonist, activity in the light and light/dark transitions in *fyn*$^{+}$ mice were reduced to WT mice levels. Thus, increasing NMDA activity in Stork et al. (2002) (indirectly through fyn overexpression) decreased anxiogenic effects of alcohol withdrawal, while decreasing it in Kiefer et al. (2003) (with the *Grin1*^{D418N} point mutation) exaggerated these effects. However, these studies contrast with the null results in NR2A KO mice from Daut et al. (2014), which may possibly be attributed to compensatory effects in the knockout during development. Two additional differences between the transgenic studies and the NR2A KO study are (1) the time of testing for anxiety-like behavior (first day of withdrawal in transgenics vs second day in NR2A KOs) and (2) the method of dependence (liquid diet/forced drinking in transgenics vs CIE vapor exposure in NR2A KOs).

Effects of glutamate signaling on escalated drinking have also been examined with gene targeting. Rats that were trained for excessive ethanol intake over 6 weeks had decreased ethanol consumption and preference 3–7 weeks following selective knockdown of protein tyrosine phosphatase α (PTPα), a protein that activates fyn, in dorsomedial striatum (Hamida et al., 2013). The selective knockdown also decreased ethanol-induced NR2B phosphorylation. However, deprivation did not escalate drinking in animals that received control infusions (ie, nonspecific virus), so whether the same manipulation would attenuate withdrawal-associated escalation of ethanol intake is unknown. If the selective knockdown did reduce escalated ethanol consumption during withdrawal, the results would seem to oppose the fyn overexpression study (Stork et al., 2002) (ie, increased fyn decreases anxiety-like behavior but decreased fyn activation reduces escalated ethanol intake). This contrast could be explained by the regional specificity of the gene manipulations. The Stork et al. study suggested that the anxiolytic effects of fyn overexpression are mediated by increased NMDA activity in inhibitory forebrain neurons; and, therefore, the modification compensates for reduced inhibitory transmission taking place after prolonged alcohol exposure. On the other hand, Hamida et al. (2013) specifically downregulated PTPα in the dorsomedial striatum. Rather than acting on inhibitory neurons in this region, these NMDA receptors in the basal ganglia circuitry are proposed to be on excitatory neurons that mediate goal-related behaviors such as escalated drinking. Therefore, the PTPα-fyn-NMDA pathway exemplifies how the same signaling system may contribute to multiple components of withdrawal severity but can also act in opposing directions depending on the region in which it acts.

Metabotropic glutamate receptors play a role in escalated drinking during withdrawal, but effects are neural-circuit dependent. Increased phasic dopamine release within the mesocorticolimbic dopamine pathway is a characteristic response to several drugs of abuse, and there is evidence this effect is partly mediated by metabotropic glutamatergic plasticity. Glutamatergic projections from several brain regions, including the prefrontal cortex (PFC), synapse onto nucleus accumbens (NAc) neurons to induce dopamine release (reviewed in Spanagel, 2009). It is hypothesized that altered homeostasis through PFC-NAc glutamatergic projections decreases the ability to control drug-seeking behaviors (Spanagel, 2009). This plasticity is partly mediated by decreased mGluR2/3 function on presynaptic PFC dopaminergic neurons projecting to the NAc. Because they are autoreceptors, decreased mGluR2/3 function leads to increased glutamate release from these neurons (ie, removal of inhibition). mGluR2 receptors were selectively increased in rat PFC to examine whether CIE-induced escalation of lever pressing for 10% ethanol could be attenuated

(Meinhardt et al., 2013). After 2 weeks of abstinence and 12 days of reconditioning, rats with mPFC lenti-mGluR2 infusions performed less lever pressing than postdependent rats that did not receive the lenti-mGluR2. Similarly, mice lacking the gene for mGluR2, *Grm2,* displayed greater alcohol preference and escalation of drinking than controls during a two-bottle free choice model over 80 days, although this model did not have a withdrawal component (Zhou et al., 2013). In contrast, mGluR5 signaling can also induce plasticity through receptor upregulation that is associated with withdrawal-induced hyperexcitability (Szumlinksi & Woodward, 2014). Female mGlur5 KOs had reduced HICs compared to WTs following a single 4 g/kg ethanol injection (Blednov & Harris, 2008). Therefore, increasing mGluR2/3 signaling, but decreasing mGlur5 signaling, reduced withdrawal severity.

The studies targeting glutamate-related genes for effects on withdrawal severity over the past 10 years are a major advance in the field. Manipulations of glutamate signaling exemplify how a specific neurotransmitter system can have site-specific effects on two different components hypothesized to mediate escalated drinking during withdrawal through (1) negative affect during withdrawal and (2) increased drive for alcohol drinking. Recent evidence from Daut et al. (2014) suggests the system's importance for anxiety-like behavior may be timing- and receptor subtype-dependent. This hypothesis is supported by pharmacological studies of anxiety-like behavior during alcohol withdrawal. Treating ethanol-withdrawn rats with the specific, NR2B subunit-containing NMDA receptor antagonist ifenprodil resulted in no anxiolytic effects on the EPM or a social interaction test (Knapp et al., 2004). However, experiments using antagonists that competitively block glutamate-binding sites on NMDA receptors, such as 2-amino-7-phosphonoheptanoic acid (AP-7) and D,L-(E)-amino-4-methyl-5-phosphono-3-pentenoic acid (CGP 37849), reduced withdrawal-induced anxiogenic effects on the EPM in rats (Gatch, Wallis, & Lal, 1999). Future studies are needed that screen medications that act on specific receptor subtypes for effects on withdrawal-associated escalated drinking.

4.2 Other Neuropeptides/Neuroimmune/Endocrine

The allostatic hypothesis of drug use disorders suggests that as dependence develops, an increasing number of signaling systems are dysregulated, including endocrine pathways (Koob & Le Moal, 2001). As discussed in Section 3.1, this dysregulation can lead to persistent elevation of neuroendocrine factors. Other signaling systems often with peripheral origins (eg, neuroimmune signaling) offer an intriguing target to treat ethanol withdrawal-associated anxiety-like behavior and escalated drinking.

Toll-like receptors (TLRs) in glial cells are involved in inflammatory responses in the brain (Okun et al., 2009), and TLRs, particularly TLR4, are part of signaling pathways that induce expression of inflammatory cytokines interleukin-1 beta (IL-1β) and tumor necrosis factor alpha (TNFα) in response to pathogens like lipopolysaccharide (LPS) (Bowie & O'neill, 2000). The production of IL-1β and TNFα in microglia treated with ethanol in culture (and the subsequent neuronal degeneration) was blocked in TLR4-deficient mice ($TLR4^{-/-}$) (Fernandez-Lizarbe, Pascual, & Guerri, 2009). $TLR4^{-/-}$ mice have since been shown to not display anxiety-like behavior in the L/D box or on the EPM during withdrawal from forced ethanol (Pascual et al., 2015). However, the total number of crossings was also reduced during both tests in ethanol-withdrawing WT mice, suggesting reduced locomotor activity that was not observed in withdrawing KO mice. In another study, anxiety-like behavior was tested in these mice during protracted abstinence, 2 weeks into withdrawal from 5 months of forced drinking (Pascual et al., 2011). Interestingly, ethanol-withdrawn WT mice had a shorter latency to enter the lighted chamber of a L/D box than untreated controls, suggesting decreased anxiety-like behavior that was not observed in $TLR4^{-/-}$ mice (Pascual et al., 2011). The anxiolytic effects 2 weeks into withdrawal in WT mice are an interesting aberration from the typically observed anxiogenic effects of withdrawal, but this testing was also done much later into protracted abstinence compared to other studies, suggesting a potential biphasic effect of ethanol withdrawal on anxiety-like behavior.

Targeting cytokine receptor genes directly has also produced effects on ethanol withdrawal severity, although seemingly in the opposing direction. KOs for the gene that encodes the receptor for IL-1β and other IL1s *(Il1r1)* had increased withdrawal HICs following acute ethanol injections (Blednov et al., 2015a). Mice also had increased percent time and entries into the open arms of an EPM during acute withdrawal, but to the same degree in KOs and WTs. Ethanol naïve KO and WT animals also did not differ behaviorally in the EPM. These results contrast with the previously discussed gene-targeting studies and evidence for IL-1β being a proconvulsant. Here, reducing cytokine function increased signs of withdrawal as measured by HICs. The other gene manipulations suggested that reducing cytokine function reduces signs of withdrawal as measured by anxiety-like behavior. These differences may be attributed to animals being tested for withdrawal-associated anxiety-like behavior during chronic ethanol withdrawal.

Studies that examined drinking during withdrawal support the hypothesis that reduced cytokine function blunts ethanol withdrawal severity. In Section 4.1, we outlined studies demonstrating the importance of PFC-NAc projections for the loss of control over ethanol consumption following dependence. In addition to glutamate signaling in this pathway, adipocytokines, regulated by ethanol through the g-protein k-ras, also have hypothesized importance for control over ethanol consumption. PFC microdissections from rats repeatedly administered ethanol (3 g/kg, i.p.) have increased *k-ras* gene expression 2 h postinjection but decreased expression 12 h postinjection (Repunte-Canonigo et al., 2010a). Mice with reduced (k-ras$^{+/-}$) and increased (Nf1$^{+/-}$) k-ras function were tested for escalated ethanol intake twice following a first and second 3-day CIE cycle (Repunte-Canonigo et al., 2010a). WT, but not k-ras$^{+/-}$ mice displayed the expected escalations in ethanol drinking 3 days after both CIE exposures. In a separate study, gene expression of the adiponectin receptor, *AdipoR2*, was increased in the PFC of NF1$^{+/-}$ mice compared to k-ras$^{+/-}$ (Repunte-Canonigo et al., 2010b). When AdipoR2 KO mice were administered similar dependence and drinking paradigms as used in the earlier study, KO mice had attenuated withdrawal-induced increases in drinking compared to WTs after both CIE exposures. Combined with the studies on anxiety-like behavior, these experiments suggest that cytokine signaling increases withdrawal severity, as eliminating it attenuated ethanol withdrawal-induced escalations in both ethanol consumption and anxiety-like behavior.

Other signaling systems, often with peripheral origins, such as endocrine systems can have important functions related to alcohol withdrawal. Leptin, insulin, and ghrelin are neuropeptides typically associated with regulation of food intake and appetite but also have importance for ethanol drinking (Addolorato et al., 2009). Similarly, the angiotensin system that is linked to fluid intake has also been examined for ethanol consumption (Fitts, 1993). Angiotensinogen (AOGEN) is a precursor for angiotensin peptides. An AOGEN-specific antisense RNA was used to decrease angiotensin levels in the rat brain, but the genetic modification did not attenuate escalated drinking during withdrawal from a 7-week ethanol vapor exposure (12 h per day) (Sommer et al., 2007).

4.3 γ-Aminobutyric acid

Studies of targeted genes that have not been studied for withdrawal-associated anxiety-like behavior are summarized in Table 2. The majority

Table 2 Systems Targeted for Effects Outside of Anxiety

Mutation/ Gene	Background and ΔExpression	References	EtOH Admin	HICs or Other wdr Measure
GABA related				
$GABA_A$ $\alpha 1^{-/-a}$ *Gabra1*	N6 C57Bl/6J eliminated	Blednov et al. (2003)	11 days liquid ethanol diet (0–6%)	KO = WT HIC (AUC, 0, 2, 4–12, 23, 24 h)
$GABA_A$ α1 KI *Gabra1*	C57Bl/6J reduced ethanol sensitivity	Werner et al. (2009)	CIE vapor 3 days (16 h on, 8 h off) w/ pyrazole 1st day	Male KI > WT HIC (AUC, 1–24 h)
$GABA_A$ β3 KI *Gabrb3*	129/SvJ reduced	Sanchis-Segura et al. (2007)	Continuous (72 h) vapor ethanol vapor (target BEC of 1.5 mg/mL. No pyrazole)	KI > WT HIC (6 h)
$GAD2^{+/-}$ *Gad2*	3 backgrounds: (1) N4 B6, (2) 1 × 129/SvJ (129N1), (3) 2 × 129/SvJ (129N2) eliminated	Blednov et al. (2010)	–	129N2 < WT HIC AUC; B6 and 129N1 = WT HIC AUC
$ENT1^{Slc29a1a}$ *Slc29a1*	C57BL/6J eliminated	Kim et al. (2011)	CIE 3 days (16 h on, 8 h off, target BE 1.5–2.0 mg/mL w/pyrazole 1st day)	KO > WT HIC (8 h and AUC 2–30 h)
$SRD5A1^{-/-}$ *Srd5a1*	C57BL/6J eliminated	Tanchuck-Nipper et al. (2014)	Chronic (72 h) vapor inhalation target BEC 1.5/mg/mL w/pyrazole 1st day	KO = WT, HICs (AUC 0–12, 24, 25 h)
MPDZ-TG and $MPDZ^{+/-}$ *Mpdz*	DBA/2J and C57BL/6J increased and reduced	Milner et al. (2013)	Acute 4 g/kg EtOH i.p.	MPDZ-TG < WT HICs; $MPDZ^{+/-}$ > WT HICs (AUC, 2–12 h)

Continued

Table 2 Systems Targeted for Effects Outside of Anxiety—cont'd

Mutation/ Gene	Background and ΔExpression	References	EtOH Admin	HICs or Other wdr Measure
MPDZ RNAi *Mpdz*	DBA/2J selective reduction in caudolateral SNpr	Kruse, Walter, and Buck (2014)	Acute 4 g/kg EtOH i.p.	MPDZ RNAi > control HICs (AUC, 2–12, 24, 25 h)
Ion channels				
GlyRα1^{M287L} GlyRα1^{Q266I} *Glra1*	C57BL/6J reduced	Blednov et al. (2012)	Acute 4 g/kg i.p.	M287L KI < (Q266I KI = WT) HIC
GlyRα2$^{-/-}$ and GlyRα3$^{-/-}$ *Glra2* and *Glra3*	N2 C57BL/6J eliminated	Blednov et al. (2015b)	Acute 4 g/kg i.p.	KO = WT HIC AUC
GIRK3$^{-/-}$ *Kcnj9*	N6 backcross DBA/D2J eliminated	Kozell et al. (2009)	Acute 4 g/kg i.p.	KO < WT HICs (AUC 2–12 h)
BK β1 and BK β4$^{-/-}$ *Kcnmb1* and *Kcnmb4*	C57BL/6J eliminated	Kreifeldt et al. (2013)	CIE alternating weeks of chamber and two-bottle choice for 8 weeks total	β1 > WT > β4 escalated ethanol intake
Cannabinoids				
CB1$^{-/-}$ [a] *Cnr1*	C57BL/6J eliminated	Racz et al. (2003)	Forced (4–16%) 4 weeks	KO < WT HICs (3 h)
CB1$^{-/-}$ *Cnr1*	(1) N10 C57BL/6 (2) interbred with DBA/D2 (F1) then backcrossed N10 to D2 eliminated	Vinod et al. (2008a)	72 h ethanol vapor w/daily pyrazole	B6 and D2 KO < WT HICs (8 h)
Trpv1$^{-/-}$ *Trpv1*	C57BL/6J eliminated	Blednov and Harris (2009)	Acute 4 g/kg EtOH i.p.	KO = WT AUC HICs

Table 2 Systems Targeted for Effects Outside of Anxiety—cont'd

Mutation/ Gene	Background and ΔExpression	References	EtOH Admin	HICs or Other wdr Measure
FAAH$^{-/-}$ *Faah*	C57BL/6J eliminated	Blednov et al. (2006b)	Acute 4 g/kg EtOH i.p.	KO = WT HIC AUC
FAAH$^{-/-}$ *Faah*	N10 backcross C57BL/6J eliminated	Vinod et al. (2008b)	72 h ethanol vapor w/daily pyrazole	KO < WT HICs (T6, 7, 8, 9 h)
Dopamine				
D2$^{-/-}$ *Drd2*	N21 backcross to C57Bl/6J eliminated	Ting-A-Kee et al. (2009)	4 days liquid diet target BEC 2.0 mg/mL	KO = WT somatic signs (2, 4, 6, and 24 h)

[a]Previously reviewed in Crabbe and Phillips (2004).

of these studies focused on withdrawal hyperexcitability and targeted GABA signaling. $GABA_A$ receptors are a conglomerate of five subunits, including α, β, δ, ε, and γ, which may have differential effects on ethanol withdrawal phenotypes. The effects of manipulating expression of many of these subunits through genetic engineering have been covered in the review by Crabbe et al. (2006), so this chapter serves as an update. The α1 $GABA_A$ subunit has the greatest abundance in the brain, and the subunit's expression is generally decreased across brain regions following chronic ethanol exposure (reviewed in Faingold, N'gouemo, & Riaz, 1998), suggesting neuroadaptations involving α1-containing receptors may have importance for withdrawal. However, α1 KO mice did not differ from WT mice for HIC scores following a chronic liquid ethanol diet (Blednov et al., 2003). More recently, male $GABA_A$ α1 subunit KI mice with a mutation at two amino acid residues (serine 270 to histidine and leucine 277 to alanine, making the receptors ethanol insensitive) had greater HIC scores after a 3-day CIE than withdrawing WT mice (Werner et al., 2009). It may be that the kindling effect of CIE produced a significant elevation of HICs that could not be detected after a continuous liquid ethanol diet in the 2003 (Blednov et al., 2003) study.

Knockout and transgenic models for the $GABA_A$ β3 subunit and presynaptic GABAergic transmission have also been tested for ethanol withdrawal severity. Through a point mutation in a single amino acid residue, $GABA_A$

β3 expression was reduced (Sanchis-Segura et al., 2007). β3(N265M) mice had greater HIC scores than WTs after a 3-day continuous ethanol vapor procedure. These results contrast with the decreased HIC scores in mice with KOs of other subunits comprising extrasynaptic receptors (see Crabbe et al., 2006), but the previous reports used acute ethanol exposure rather than the chronic vapor paradigm. Another possible explanation for these contrasting results may be neuroadaptations during development due to complete gene knockout of the receptor subunits in the earlier studies which may be absent in the more-recent transgenic knockin study. The mice from the knockin study were also developed purely on a 129 background, while the KO studies had mice generated from mixed 129 and B6 backgrounds. As B6 mice are less sensitive to withdrawal severity, the background could have contributed to this contrast. The importance of background strain is also apparent in mice with mutations for the enzyme responsible for converting glutamate to GABA, glutamic acid decarboxylase (GAD). *Gad2* KO mice were tested from three different strain combinations: a B6 background, an additional cross with 129S1/SvJ (129N1), or from two additional crosses (129N2) (Blednov et al., 2010), which revealed 129N2 $Gad2^{-/-}$ mice had decreased HIC scores compared to WT mice after acute ethanol.

Many other genes functionally linked to GABAergic signaling have been examined for their contribution to hyperexcitability during withdrawal. Adenosine can induce GABA release and inhibit glutamate release, thus having an overall inhibitory effect. Studies reported in 2006 examined effects of adenosine manipulations on HICs and ethanol preference in ethanol-naïve mice. KOs, for the transporter of adenosine, equilibrative nucleoside transporter 1 (ENT1), more recently displayed increased withdrawal HICs than WT mice following 3-day CIE (Kim et al., 2011). Other genes are linked to the GABAergic system through steroid mechanisms. In addition to contributing to withdrawal hyperexcitability, GABAergic steroids, such as allopregnanolone, have importance for withdrawal-induced anxiety in humans and rodent models (Finn et al., 2010). Global KO of 5α-reductase type 1 was used to eliminate expression of the enzyme that produces allopregnanolone (Tanchuck-Nipper et al., 2014). Following a continuous 72 h vapor inhalation paradigm, KOs did not differ from WTs for HIC severity. An acute 2 g/kg injection of ethanol produced anxiolytic effects on an EPM in WT mice but not KOs. These mice were not tested for anxiety-like behavior during withdrawal; however, it could be hypothesized that KOs would show decreased withdrawal-induced anxiogenesis if compensatory responses of $GABA_A$ receptors are indeed playing a role in the phenotype.

A final indirect connection to GABAergic signaling comes from the multiple PDZ domain protein (MPDZ), a scaffolding protein that interacts with GABA signaling among other systems. Transgenic mice overexpressing MPDZ (MPDZ-TG) on a predominant D2 background and MPDZ KO heterozygotes on a B6 background were given acute ethanol injections (Milner et al., 2013). MPDZ-TG mice had reduced HIC scores but $MPDZ^{+/-}$ mice had greater HIC scores compared to their WT mice. $MPDZ^{+/-}$ mice also consumed less ethanol from 6% to 20% compared to WT mice during two-bottle choice, supporting a negative relationship between the two behavioral phenotypes. A more-recent study employed an RNAi approach to selectively knockdown *Mpdz* expression in the caudolateral substantia nigra pars reticulata of DBA/2 J (D2) mice (Kruse et al., 2014). Consistent with the heterozygous KO mice, MPDZ RNAi mice had augmented HICs compared to controls given scrambled shRNA after acute ethanol (4 g/kg).

5. OTHER SYSTEMS

Other classes of neurotransmitters and receptors have been studied for effects on ethanol withdrawal HICs and ethanol drinking but with no anxiety component (Blednov & Harris, 2009; Blednov et al., 2006b, 2012, 2015b; Kozell et al., 2009; Kreifeldt et al., 2013; Racz et al., 2003; Ting-A-Kee et al., 2009; Vinod et al., 2008a, 2008b). We have compiled these references into Table 2. In the interests of space, we do not discuss these studies separately.

6. CONCLUSIONS

Recent studies represent a shift in the field away from gene-targeting studies that focused only on ethanol withdrawal hyperexcitability. No gene-targeting studies published over 10 years ago examined escalated drinking during withdrawal, and very few examined anxiety-like behavior. The development of novel behavioral paradigms (eg, CIE), genetic techniques (eg, site-specific and conditional approaches), and hypotheses for dysregulated signaling that underlie escalated drinking have contributed to the evolution of these experimental designs.

The CIE paradigm and its potential for medications screening is discussed elsewhere (Becker & Lopez, 2016). Importantly, CIE produces a post-dependent state that models many withdrawal symptoms of alcohol use disorders, and therefore has immense potential for the development of pharmacological treatments (reviewed in Meinhardt & Sommer, 2015). CIE was used by numerous studies in this chapter in concert with gene-targeting approaches to identify genes important for escalated ethanol consumption during withdrawal—of the nine studies that examined escalated drinking and were published since 2004, six employed the CIE vapor paradigm. Three other studies used CIE with gene targeting for effects on HICs, and one study examined anxiety-like behavior during withdrawal. Thus, experiments combining CIE with gene targeting have spanned all three areas of interest in this chapter.

Although CIE has the potential to draw in emotionality and other aspects of motivational plasticity into medications screening for escalated drinking, exploring multiple single gene manipulation responses in a single study has been somewhat constrained by a background strain issue. Earlier studies found CIE consistently produced increased withdrawal HIC severity, but all of this work was conducted in C3H/HeNCrl mice. In contrast, its effects on escalated drinking of unadulterated ethanol have used B6. This strain shows nearly the lowest signs of withdrawal hyperexcitability after CIE observed in mice (Metten et al., 2010); and, as discussed earlier in this chapter, also appears resistant to withdrawal-associated anxiety-like behavior. Following CIE, D2 mice, rather than B6 mice, show increased anxiety-like behavior (McCool & Chappell, 2015) and c-fos expression in limbic brain areas (Chen et al., 2009). The vast majority of genetically engineered rodents were tested on a predominant B6 background, which may explain why only one study (Daut et al., 2014) tested for anxiety-like behavior following CIE. Yet, through gene targeting, researchers have managed to elicit withdrawal responses such as HICs (Du et al., 2012) and increased anxiety-like behavior (Kiefer et al., 2003) after disrupting the same gene (ie, NR1 subunit) in mice from predominantly B6 backgrounds, overcoming the possible limitation that particular withdrawal phenotypes might be specific to certain strains.

While no studies we found examined all three phenotypes together (HICs, anxiety-like behavior, and escalated drinking) following CIE, it may be possible to design future studies that combine testing for these phenotypes. Some genetic manipulations, such as those for NR1, produced withdrawal-associated differences in the same direction but with different dependence models and behavioral tests. Another example comes from targeting CRHR1, where selective neuronal removal attenuated escalated ethanol intake in mice after CIE (Molander et al., 2011); and, in a separate study, knocking it out decreased withdrawal-associated anxiety-like behavior in a L/D box procedure during withdrawal from forced drinking (Timpl et al., 1998). Another gene manipulation (FKBP5 KO) that impairs feedback onto the HPA axis increased severity of HICs following acute injection of ethanol and chronic liquid ethanol diet (Huang et al., 2014). Therefore, mice could potentially be tested for all three phenotypes using a constant dependence-induction method (eg, CIE) in combination with a medication that can restore HPA axis function (eg, SNRI and SSRI antidepressants, reviewed in Maric & Adzic, 2013). Some drugs of this type have already been tested for effects on escalated ethanol consumption. Similar to the CRHR1 KO, systemic administration of the CRHR1 antagonist antalarmin attenuated postdependent ethanol self-administration (Chu et al., 2007). Could antalarmin also reduce withdrawal-associated anxiety and HICs at similar timepoints in mice from the same genetic background? FKBP5 has also recently become a drug target for use in animal models of stress-induced anxiety-like behavior, and a small molecule drug that targets it has produced anxiolytic effects when administered systematically (Hartmann et al., 2012). Other currently prescribed antidepressants shown to restore FKBP function after stress, such as the SNRI duloxetine in rats (Guidotti et al., 2013), could also be screened for effects on ethanol withdrawal phenotypes after CIE.

Another rodent model that has shown some relevant results used the drinking in the dark (DID) procedure. High DID (HDID) mice selectively bred for high BEC after an acute 4 h binge showed elevated withdrawal HICs (Crabbe et al., 2014) after drinking. After 4 cycles of CIE, they also show gradually escalating drinking (Crabbe et al., 2012). B6 mice undergoing a modified DID procedure (2-h 20% ethanol sessions, 5 days per week for 6 weeks) displayed increased anxiety-like behavior compared to water-drinking controls on the EPM and in the L/D box during a battery of tests conducted 1 day (acute) or 21 days (protracted) into abstinence, with effects that were more pronounced during the earlier time point (Lee et al., 2015). Ethanol-drinking mice also had greater immediate early gene (EGR1) cell counts in limbic areas, including the CeA and dorsal BNST, and these differences were speculated to be due to neuroadaptations in glutamate signaling which occur following prolonged ethanol exposure. These behavioral and neurochemical findings suggest the DID procedure could potentially be used to study both the positive motivational aspects and negative affective states contributing to ethanol consumption, and that the HDID model could be advantageous for screening medications or genetic manipulations acting on glutamatergic signaling.

Of the systems covered in this review, studies of glutamate-related signaling by far outnumbered other systems for testing escalated ethanol consumption or anxiety-like behavior during withdrawal (seven studies total, with the neuroimmune system being the next closest with four studies). The effects of these gene modifications, when combined with pharmacological and neurochemical results from HDID studies, further suggest the potential use of this model for screening medications that work on glutamate signaling for putative effects on both positive motivational aspects and negative affective states contributing to ethanol consumption. The particular receptors and receptor subunits that were modified overlap with reported effects of glutamate signaling and DID. We reviewed experiments where mGlur5 KOs had reduced acute ethanol withdrawal HICs and where selective knockdown of PTPα in rat dorsal striatum (which reduced NR2B phosphorylation) attenuated escalated ethanol intake (Blednov & Harris, 2008; Hamida et al., 2013). In the same vein, B6 mice undergoing 30 consecutive days of DID had greater protein levels of mGlur5 and NR2B in the NAc shell 1 day into abstinence compared to water-drinking mice, and an mGluR5 antagonist infused into the NAc reduced ethanol intake during DID (Cozzoli et al., 2012). The same group further linked the increases in mGluR5 to the central amygdala (Cozzoli et al., 2014), an extralimbic region with proposed importance for negative affective states during ethanol withdrawal (Koob & Le Moal, 2001). Thus, the brain regions and receptor subtypes modified during DID parallel the gene-targeting effects on withdrawal severity and escalated drinking.

Specificity is one issue with screening medications that act on the glutamate system. The neurotransmitter is widespread throughout the brain, but there are many different receptor subtypes with different subunit components. Ethanol acts on most of these receptors in many different brain regions (reviewed in Gass & Olive, 2008). We have reviewed conditional studies that have contributed to our knowledge of region-specific effects of some of these targets (Hamida et al., 2013; Molander et al., 2011), yet applying cannulation surgeries to deliver drugs in humans is not feasible. Various antiglutamatergic drugs have already been screened in clinical studies for effects on depressed mood and escalated alcohol consumption during alcohol withdrawal. Two different drugs that work on ionotropic glutamate receptors, memantine and topiramate (NMDA receptor and AMPA receptor antagonists, respectively) reduced withdrawal-associated depression more than placebo in alcohol-dependent men during abstinence (Krupitsky et al., 2007). These effects parallel results from animal models. Topiramate reduced ethanol preference in B6 mice (Gabriel & Cunningham, 2005). Acamprosate, an NMDA receptor modulator, dose dependently reduced ethanol intake in B6 mice undergoing DID (Gupta et al., 2008). Acamprosate also attenuates escalated ethanol drinking in rats following CIE and is currently FDA-approved to treat humans with alcohol use disorders (Rimondini et al., 2002). Drugs that work on mGluR5 and mGlurR2/3 are also an exciting avenue for medication screenings. *N*-acetylcysteine (NAC) that activates mGluR2 is currently available over-the-counter and is being tested for effects on ethanol withdrawal severity in animal models. In rats undergoing withdrawal from 30 days ethanol by oral gavage, 4-day NAC administration blocked anxiety-like behavior in an open field and reduced elevated corticosterone levels (Schneider et al., 2015). MPEP, an mGluR5 antagonist, reduced ethanol intake in B6 mice during DID (Gupta et al., 2008). Medications that act on mGluR5 have only been recently screened for antidepressant effects and are currently undergoing clinical testing (Lindemann et al., 2015). Whether these medications could reduce ethanol drinking in models such as CIE or DID would be interesting future studies.

With the increased emphasis on personal genomics, understanding how compounds can be used to treat different effects of alcohol withdrawal depending on an individual's genetic make-up is of great importance. Combining gene-targeting models with newer methods of alcohol administration could help to tease apart pharmacogenetic targets for clinical practice. Thus, manipulations of specific genes, especially if the stain's background is held constant, will continue to offer powerful approaches to improve our understanding of alcohol dependence and developing novel therapeutic medications for its treatment.

ACKNOWLEDGMENTS

The authors declare no conflicts of interest. This work was supported in part by grants 101BX000313 from the US Department of Veterans Affairs, and grants AA020246, AA013519, AA10760, and AA007468 from the NIH. The views expressed herein are solely those of the authors and do not necessarily represent those of the funding agencies.

REFERENCES

Addolorato, G., et al. (2009). Hormones and drinking behaviour: New findings on ghrelin, insulin, leptin and volume-regulating hormones. An ESBRA Symposium report. *Drug and Alcohol Review*, *28*(2), 160–165.

Baldwin, H., et al. (1991). CRF antagonist reverses the "anxiogenic" response to ethanol withdrawal in the rat. *Psychopharmacology*, *103*(2), 227–232.

Becker, H. C., & Lopez, M. F. (2016). An animal model of alcohol dependence to screen medications for treating alcoholism. In R. L. Bell & S. Rahman (Eds.), *Animal models for medications screening to treat addiction*. Amsterdam: Academic Press.
Binder, E. B. (2009). The role of FKBP5, a co-chaperone of the glucocorticoid receptor in the pathogenesis and therapy of affective and anxiety disorders. *Psychoneuroendocrinology*, *34*, S186–S195.
Blednov, Y. A., & Harris, R. (2008). Metabotropic glutamate receptor 5 (mGluR5) regulation of ethanol sedation, dependence and consumption: Relationship to acamprosate actions. *The International Journal of Neuropsychopharmacology*, *11*(06), 775–793.
Blednov, Y., & Harris, R. (2009). Deletion of vanilloid receptor (TRPV1) in mice alters behavioral effects of ethanol. *Neuropharmacology*, *56*(4), 814–820.
Blednov, Y. A., et al. (2003). $GABA_A$ receptor $\alpha 1$ and $\beta 2$ subunit null mutant mice: Behavioral responses to ethanol. *Journal of Pharmacology and Experimental Therapeutics*, *305*(3), 854–863.
Blednov, Y. A., et al. (2006a). Reduced alcohol consumption in mice lacking preprodynorphin. *Alcohol*, *40*(2), 73–86.
Blednov, Y. A., et al. (2006b). Role of endocannabinoids in alcohol consumption and intoxication: Studies of mice lacking fatty acid amide hydrolase. *Neuropsychopharmacology*, *32*(7), 1570–1582.
Blednov, Y. A., et al. (2010). Preclinical study: Mice lacking Gad2 show altered behavioral effects of ethanol, flurazepam and gabaxadol. *Addiction Biology*, *15*(1), 45–61.
Blednov, Y. A., et al. (2012). Behavioral characterization of knockin mice with mutations M287L and Q266I in the glycine receptor $\alpha 1$ subunit. *Journal of Pharmacology and Experimental Therapeutics*, *340*(2), 317–329.
Blednov, Y. A., et al. (2015a). Role of interleukin-1 receptor signaling in the behavioral effects of ethanol and benzodiazepines. *Neuropharmacology*, *95*, 309–320.
Blednov, Y. A., et al. (2015b). Glycine receptors containing $\alpha 2$ or $\alpha 3$ subunits regulate specific ethanol-mediated behaviors. *Journal of Pharmacology and Experimental Therapeutics*, *353*(1), 181–191.
Bowie, A., & O'neill, L. (2000). The interleukin-1 receptor/Toll-like receptor superfamily: Signal generators for pro-inflammatory interleukins and microbial products. *Journal of Leukocyte Biology*, *67*(4), 508–514.
Chen, G., et al. (2009). Differential activation of limbic circuitry associated with chronic ethanol withdrawal in DBA/2 J and C57BL/6J mice. *Alcohol*, *43*(6), 411–420.
Chu, K., et al. (2007). Dependence-induced increases in ethanol self-administration in mice are blocked by the CRF_1 receptor antagonist antalarmin and by CRF_1 receptor knockout. *Pharmacology Biochemistry and Behavior*, *86*(4), 813–821.
Cloninger, C. R., Bohman, M., & Sigvardsson, S. (1981). Inheritance of alcohol abuse: Cross-fostering analysis of adopted men. *Archives of General Psychiatry*, *38*(8), 861–868.
Contet, C., et al. (2013). μ-Opioid receptors mediate the effects of chronic ethanol binge drinking on the hippocampal neurogenic niche. *Addiction Biology*, *19*(5), 770–780.
Cozzoli, D. K., et al. (2012). Nucleus accumbens mGluR5-associated signaling regulates binge alcohol drinking under drinking-in-the-dark procedures. *Alcoholism: Clinical and Experimental Research*, *36*(9), 1623–1633.
Cozzoli, D. K., et al. (2014). Binge alcohol drinking by mice requires intact Group 1 metabotropic glutamate receptor signaling within the central nucleus of the amygdale. *Neuropsychopharmacology*, *39*(2), 435–444.
Crabbe, J. C., Mcswigan, J. D., & Belknap, J. (1985). The role of genetics in substance abuse. In M. Galizio & S. A. Maisto (Eds.), *Determinants of substance abuse* (pp. 13–64). New York: Springer.
Crabbe, J. C., & Phillips, T. J. (2004). Review: Pharmacogenetic studies of alcohol self-administration and withdrawal. *Psychopharmacology*, *174*(4), 539–560.

Crabbe, J. C., et al. (2006). Review: Alcohol-related genes: Contributions from studies with genetically engineered mice. *Addiction Biology, 11*(3–4), 195–269.

Crabbe, J. C., et al. (2012). Ethanol withdrawal-associated drinking and drinking in the dark: Common and discrete genetic contributions. *Addiction Genetics, 1*, 3–11.

Crabbe, J. C., et al. (2014). Progress in a replicated selection for elevated blood ethanol concentrations in HDID mice. *Genes, Brain and Behavior, 13*(2), 236–246.

Daut, R. A., et al. (2014). Tolerance to ethanol intoxication after chronic ethanol: Role of GluN2A and PSD-95. *Addiction Biology, 20*(2), 259–262.

Davis, L., et al. (2008). Major depression and comorbid substance use disorders. *Current Opinion in Psychiatry, 21*(1), 14–18.

Du, X., et al. (2012). Heterozygous deletion of NR1 subunit of the NMDA receptor alters ethanol-related behaviors and regional expression of NR2 subunits in the brain. *Neurotoxicology and Teratology, 34*(1), 177–186.

Dunn, A. J., & Berridge, C. W. (1990). Physiological and behavioral responses to corticotropin-releasing factor administration: Is CRF a mediator of anxiety or stress responses? *Brain Research Reviews, 15*(2), 71–100.

Faingold, C., N'gouemo, P., & Riaz, A. (1998). Ethanol and neurotransmitter interactions—From molecular to integrative effects. *Progress in Neurobiology, 55*(5), 509–535.

Femenía, T., & Manzanares, J. (2012). Increased ethanol intake in prodynorphin knockout mice is associated to changes in opioid receptor function and dopamine transmission. *Addiction Biology, 17*(2), 322–337.

Fernandez-Lizarbe, S., Pascual, M., & Guerri, C. (2009). Critical role of TLR4 response in the activation of microglia induced by ethanol. *The Journal of Immunology, 183*(7), 4733–4744.

Finn, D. A., et al. (2010). Manipulation of GABAergic steroids: Sex differences in the effects on alcohol drinking-and withdrawal-related behaviors. *Hormones and Behavior, 57*(1), 12–22.

Fitts, D. A. (1993). Angiotensin and captopril increase alcohol intake. *Pharmacology Biochemistry and Behavior, 45*(1), 35–43.

Gabriel, K. I., & Cunningham, C. L. (2005). Effects of topiramate on ethanol and saccharin consumption and preferences in C57BL/6J mice. *Alcoholism: Clinical and Experimental Research, 29*(1), 75–80.

Gass, J. T., & Olive, M. F. (2008). Glutamatergic substrates of drug addiction and alcoholism. *Biochemical Pharmacology, 75*(1), 218–265.

Gatch, M. B., Wallis, C. J., & Lal, H. (1999). Effects of NMDA antagonists on ethanol-withdrawal induced "anxiety" in the elevated plus maze. *Alcohol, 19*(3), 207–211.

Ghozland, S., et al. (2005). Lack of stimulant and anxiolytic-like effects of ethanol and accelerated development of ethanol dependence in mu-opioid receptor knockout mice. *Neuropharmacology, 49*(4), 493–501.

Goldman, D., & Bergen, A. (1998). General and specific inheritance of substance abuse and alcoholism. *Archives of General Psychiatry, 55*(11), 964–965.

Goldowitz, D., et al. (2006). Progress in using mouse inbred strains, consomics, and mutants to identify genes related to stress, anxiety, and alcohol phenotypes. *Alcoholism: Clinical and Experimental Research, 30*(6), 1066–1078.

Goldstein, D. B. (1973). Inherited differences in intensity of alcohol withdrawal reactions in mice. *Nature, 245*(5421), 154–156.

Guidotti, G., et al. (2013). Glucocorticoid receptor and FKBP5 expression is altered following exposure to chronic stress: Modulation by antidepressant treatment. *Neuropsychopharmacology, 38*(4), 616–627.

Gupta, T., et al. (2008). Acute effects of acamprosate and MPEP on ethanol drinking-in-the-dark in male C57BL/6J mice. *Alcoholism: Clinical and Experimental Research, 32*(11), 1992–1998.

Hamida, S. B., et al. (2013). Protein tyrosine phosphatase α in the dorsomedial striatum promotes excessive ethanol-drinking behaviors. *The Journal of Neuroscience*, *33*(36), 14369–14378.

Hartmann, J., et al. (2012). The involvement of FK506-binding protein 51 (FKBP5) in the behavioral and neuroendocrine effects of chronic social defeat stress. *Neuropharmacology*, *62*(1), 332–339.

Heilig, M., et al. (2010). Review: Acute withdrawal, protracted abstinence and negative affect in alcoholism: Are they linked? *Addiction Biology*, *15*(2), 169–184.

Heinrichs, S. C., & Koob, G. F. (2004). Corticotropin-releasing factor in brain: A role in activation, arousal, and affect regulation. *Journal of Pharmacology and Experimental Therapeutics*, *311*(2), 427–440.

Hershon, H. I. (1977). Alcohol withdrawal symptoms and drinking behavior. *Journal of Studies on Alcohol and Drugs*, *38*(05), 953.

Huang, M.-C., et al. (2014). FKBP5 moderates alcohol withdrawal severity: Human genetic association and functional validation in knockout mice. *Neuropsychopharmacology*, *39*(8), 2029–2038.

Kalivas, P. W. (2009). The glutamate homeostasis hypothesis of addiction. *Nature Reviews Neuroscience*, *10*(8), 561–572.

Keith, L. D., & Crabbe, J. C. (1992). Specific and nonspecific effects of ethanol vapor on plasma corticosterone in mice. *Alcohol*, *9*(6), 529–533.

Kessler, R. C., et al. (2005). Prevalence, severity, and comorbidity of 12-month DSM-IV disorders in the National Comorbidity Survey Replication. *Archives of General Psychiatry*, *62*(6), 617–627.

Kiefer, F., et al. (2003). Involvement of NMDA receptors in alcohol-mediated behavior: Mice with reduced affinity of the NMDA R1 glycine binding site display an attenuated sensitivity to ethanol. *Biological Psychiatry*, *53*(4), 345–351.

Kim, J.-H., et al. (2011). Functional role of the polymorphic 647 T/C variant of ENT1 (*SLC29A1*) and its association with alcohol withdrawal seizures. *PLoS One*, *6*(1), e16331.

Knapp, D. J., et al. (2004). SB242084, flumazenil, and CRA1000 block ethanol withdrawal–induced anxiety in rats. *Alcohol*, *32*(2), 101–111.

Koob, G. F. (2013). Theoretical frameworks and mechanistic aspects of alcohol addiction: Alcohol addiction as a reward deficit disorder. In W. H. Sommer & R. Spanagel (Eds.), *Behavioral Neurobiology of Alcohol Addiction* (pp. 3–30). Berlin: Springer.

Koob, G. F., & Le Moal, M. (1997). Drug abuse: Hedonic homeostatic dysregulation. *Science*, *278*(5335), 52–58.

Koob, G. F., & Le Moal, M. (2001). Drug addiction, dysregulation of reward, and allostasis. *Neuropsychopharmacology*, *24*(2), 97–129.

Kozell, L. B., et al. (2009). Mapping a barbiturate withdrawal locus to a 0.44 Mb interval and analysis of a novel null mutant identify a role for Kcnj9 (GIRK3) in withdrawal from pentobarbital, zolpidem, and ethanol. *The Journal of Neuroscience*, *29*(37), 11662–11673.

Kreifeldt, M., et al. (2013). BK channel β1 and β4 auxiliary subunits exert opposite influences on escalated ethanol drinking in dependent mice. *Frontiers in Integrative Neuroscience*, 7, e105.

Krupitsky, E. M., et al. (2007). Antiglutamatergic strategies for ethanol detoxification: Comparison with placebo and diazepam. *Alcoholism: Clinical and Experimental Research*, *31*(4), 604–611.

Kruse, L., Walter, N., & Buck, K. (2014). Mpdz expression in the caudolateral substantia nigra pars reticulata is crucially involved in alcohol withdrawal. *Genes, Brain and Behavior*, *13*(8), 769–776.

Kushner, M. G., Abrams, K., & Borchardt, C. (2000). The relationship between anxiety disorders and alcohol use disorders: A review of major perspectives and findings. *Clinical Psychology Review*, *20*(2), 149–171.

Lallemand, F., & De Witte, P. (2005). Ethanol induces higher BEC in CB1 cannabinoid receptor knockout mice while decreasing ethanol preference. *Alcohol and Alcoholism*, *40*(1), 54–62.

Lee, K. M., et al. (2015). Binge alcohol drinking elicits persistent negative affect in mice. *Behavioural Brain Research, 291*, 385–398.
Lindemann, L., et al. (2015). Pharmacology of basimglurant (RO4917523, RG7090), a unique metabotropic glutamate receptor 5 negative allosteric modulator in clinical development for depression. *Journal of Pharmacology and Experimental Therapeutics, 353*(1), 213–233.
Maric, N. P., & Adzic, M. (2013). Pharmacological modulation of HPA axis in depression—New avenues for potential therapeutic benefits. *Psychiatria Danubina, 25*(3), 299–305.
Mayfield, J., Arends, M. A., Harris, R. A., & Blednov, Y. A. (2016). Genes and alcohol consumption: Studies with mutant mice. In R. L. Bell & S. Rahman (Eds.), *Animal models for medications screening to treat addiction*. Amsterdam: Academic Press.
Mayfield, R., Harris, R., & Schuckit, M. (2008). Genetic factors influencing alcohol dependence. *British Journal of Pharmacology, 154*(2), 275–287.
McCool, B. A., & Chappell, A. M. (2015). Chronic intermittent ethanol inhalation increases ethanol self-administration in both C57BL/6J and DBA/2 J mice. *Alcohol, 49*(2), 111–120.
Meinhardt, M. W., & Sommer, W. H. (2015). Postdependent state in rats as a model for medication development in alcoholism. *Addiction Biology, 20*(1), 1–21.
Meinhardt, M. W., et al. (2013). Rescue of infralimbic mGluR2 deficit restores control over drug-seeking behavior in alcohol dependence. *The Journal of Neuroscience, 33*(7), 2794–2806.
Metten, P., & Crabbe, J. C. (2005). Alcohol withdrawal severity in inbred mouse (*Mus musculus*) strains. *Behavioral Neuroscience, 119*(4), 911.
Metten, P., et al. (1998). High genetic susceptibility to ethanol withdrawal predicts low ethanol consumption. *Mammalian Genome, 9*(12), 983–990.
Metten, P., et al. (2010). Withdrawal severity after chronic intermittent ethanol in inbred mouse strains. *Alcoholism: Clinical and Experimental Research, 34*(9), 1552–1564.
Milner, L. C., et al. (2013). Novel MPDZ/MUPP1 transgenic and knockdown models confirm Mpdz's role in ethanol withdrawal and support its role in voluntary ethanol consumption. *Addiction Biology, 20*(1), 143–147.
Molander, A., et al. (2011). Brain-specific inactivation of the Crhr1 gene inhibits post-dependent and stress-induced alcohol intake, but does not affect relapse-like drinking. *Neuropsychopharmacology, 37*(4), 1047–1056.
Mossberg, D., Liljeberg, P., & Borg, S. (1985). Clinical conditions in alcoholics during long-term abstinence: A descriptive, longitudinal treatment study. *Alcohol, 2*(3), 551–553.
Okun, E., et al. (2009). Toll-like receptors in neurodegeneration. *Brain Research Reviews, 59*(2), 278–292.
Pandey, S. C. (2014). Anxiety and alcohol use disorders. In A. Noronha, et al. (Eds.), *Neurobiology of alcohol dependence*. Amsterdam: Elsevier.
Pascual, M., et al. (2011). Impact of TLR4 on behavioral and cognitive dysfunctions associated with alcohol-induced neuroinflammatory damage. *Brain, Behavior, and Immunity, 25*, S80–S91.
Pascual, M., et al. (2015). Cytokines and chemokines as biomarkers of ethanol-induced neuroinflammation and anxiety-related behavior: Role of TLR4 and TLR2. *Neuropharmacology, 89*, 352–359.
Phillips, T. J., Hen, R., & Crabbe, J. C. (1999). Complications associated with genetic background effects in research using knockout mice. *Psychopharmacology, 147*(1), 5–7.
Pich, E. M., et al. (1995). Increase of extracellular corticotropin-releasing factor-like immunoreactivity levels in the amygdala of awake rats during restraint stress and ethanol withdrawal as measured by microdialysis. *The Journal of Neuroscience, 15*(8), 5439–5447.
Racz, I., et al. (2003). A critical role for the cannabinoid CB1 receptors in alcohol dependence and stress-stimulated ethanol drinking. *The Journal of Neuroscience, 23*(6), 2453–2458.
Radke, A., Pickens, C., & Holmes, A. (2014). The effects of stress on measures of alcohol drinking in rodents. In A. Noronha, et al. (Eds.), *Neurobiology of alcohol dependence*. Amsterdam: Elsevier.

Repunte-Canonigo, V., et al. (2010a). Genome-wide gene expression analysis identifies K-ras as a regulator of alcohol intake. *Brain Research, 1339*, 1–10.

Repunte-Canonigo, V., et al. (2010b). A potential role for adiponectin receptor 2 (AdipoR2) in the regulation of alcohol intake. *Brain Research, 1339*, 11–17.

Rimondini, R., et al. (2002). Long-lasting increase in voluntary ethanol consumption and transcriptional regulation in the rat brain after intermittent exposure to alcohol. *The FASEB Journal, 16*(1), 27–35.

Roberto, M., et al. (2005). Chronic ethanol exposure and protracted abstinence alter NMDA receptors in central amygdala. *Neuropsychopharmacology, 31*(5), 988–996.

Sanchis-Segura, C., et al. (2007). Etomidate and propofol-hyposensitive $GABA_A$ receptor β3 (N265M) mice show little changes in acute alcohol sensitivity but enhanced tolerance and withdrawal. *Neuroscience Letters, 416*(3), 275–278.

Schneider, R., et al. (2015). N-acetylcysteine prevents behavioral and biochemical changes induced by alcohol cessation in rats. *Alcohol, 49*(3), 259–263.

Sommer, W., et al. (2007). Plasticity and impact of the central renin–angiotensin system during development of ethanol dependence. *Journal of Molecular Medicine, 85*(10), 1089–1097.

Spanagel, R. (2009). Alcoholism: A systems approach from molecular physiology to addictive behavior. *Physiological Reviews, 89*(2), 649–705.

Stork, O., et al. (2002). Resistance to alcohol withdrawal-induced behaviour in Fyn transgenic mice and its reversal by ifenprodil. *Molecular Brain Research, 105*(1), 126–135.

Swendsen, J., et al. (2010). Mental disorders as risk factors for substance use, abuse and dependence: Results from the 10-year follow-up of the National Comorbidity Survey. *Addiction, 105*(6), 1117–1128.

Szumlinksi, K., & Woodward, J. (2014). Glutamate signaling in alcohol abuse and dependence. In A. Noronha, et al. (Eds.), *Neurobiology of alcohol dependence*. Amsterdam: Elsevier.

Tanchuck-Nipper, M. A., et al. (2014). Sex differences in ethanol's anxiolytic effect and chronic ethanol withdrawal severity in mice with a null mutation of the 5α-reductase type 1 gene. *Behavior Genetics, 45*(3), 354–367.

Timpl, P., et al. (1998). Impaired stress response and reduced anxiety in mice lacking a functional corticotropin-releasing hormone receptor 1. *Nature Genetics, 19*(2), 162–166.

Ting-A-Kee, R., et al. (2009). $GABA_A$ receptors mediate the opposing roles of dopamine and the tegmental pedunculopontine nucleus in the motivational effects of ethanol. *European Journal of Neuroscience, 29*(6), 1235–1244.

Vinod, K. Y., et al. (2008a). Genetic and pharmacological manipulations of the CB1 receptor alter ethanol preference and dependence in ethanol preferring and nonpreferring mice. *Synapse, 62*(8), 574–581.

Vinod, K. Y., et al. (2008b). Manipulation of fatty acid amide hydrolase functional activity alters sensitivity and dependence to ethanol. *Journal of Neurochemistry, 104*(1), 233–243.

Walker, B. M., et al. (2012). Targeting dynorphin/kappa opioid receptor systems to treat alcohol abuse and dependence. *Alcohol, 46*(4), 359–370.

Wee, S., & Koob, G. F. (2010). The role of the dynorphin-κ opioid system in the reinforcing effects of drugs of abuse. *Psychopharmacology, 210*(2), 121–135.

Werner, D. F., et al. (2009). Alcohol-induced tolerance and physical dependence in mice with ethanol insensitive α1 $GABA_A$ receptors. *Alcoholism: Clinical and Experimental Research, 33*(2), 289–299.

Winokur, G., et al. (1996). Familial alcoholism in manic-depressive (bipolar) disease. *American Journal of Medical Genetics, 67*(2), 197–201.

Zhou, Z., et al. (2013). Loss of metabotropic glutamate receptor 2 escalates alcohol consumption. *Proceedings of the National Academy of Sciences, 110*(42), 16963–16968.

CHAPTER ELEVEN

Abstinence-Conflict Model: Toward an Optimal Animal Model for Screening Medications Promoting Drug Abstinence

J.A. Peck[1]

State University of New York College at Cortland, Cortland, NY, United States

[1]Corresponding author: e-mail address: joshua.peck@cortland.edu

Contents

Abstract

Drug addiction is a significant health and societal problem for which there is no highly effective long-term behavioral or pharmacological treatment. A rising concern are the use of illegal opiate drugs such as heroin and the misuse of legally available pain relievers that have led to serious deleterious health effects or even death. Therefore, treatment strategies that prolong opiate abstinence should be the primary focus of opiate treatment. Further, because the factors that support abstinence in humans and laboratory animals are similar, several animal models of abstinence and relapse have been developed. Here, we review a few animal models of abstinence and relapse and evaluate their validity and utility in addressing human behavior that leads to long-term drug abstinence. Then, a novel abstinence "conflict" model that more closely mimics human drug-seeking episodes by incorporating negative consequences for drug seeking (as are typical in humans, eg, incarceration and job loss) and while the drug remains readily available is discussed. Additionally, recent research investigating both cocaine and heroin seeking in rats using the animal conflict model is presented and the implications for heroin treatments are examined. Finally, it is argued that the use of animal abstinence/relapse models that more closely approximate human drug addiction, such as

International Review of Neurobiology, Volume 126
ISSN 0074-7742
http://dx.doi.org/10.1016/bs.irn.2016.02.015

the abstinence-conflict model, could lead to a better understanding of the neurobiological and environmental factors that support long-term drug abstinence. In turn, this will lead to the development of more effective environmental and pharmacotherapeutic interventions to treat opiate addiction and addiction to other drugs of abuse.

Drug addiction is a serious and growing epidemic in the United States and costs Americans hundreds of billions of dollars each year, when considering the combined medical, economic, criminal, and social impact (www.nida.nih.gov, retrieved 9/15/2015). Of particular concern is the rising use of illegal opiate drugs such as heroin and the misuse of legally available pain relievers such as oxycodone and hydrocodone that have led to serious negative health effects. Perhaps, even more alarming is the reporting of many people abusing legally prescribed analgesics before they start to use heroin (www.samhsa.gov, retrieved 10/08/15). According to SAMHSA's recent "Behavioral Health Trends in the United States: Results from the 2014 National Survey on Drug Use and Health" report, 4.8 million people have used heroin at some point in their lives and approximately 435,000 people were regular (past-month) users of heroin. Further, SAMSHA reported that although the percentage of some drug use has decreased since 2002 (eg, nicotine and cocaine), nonmedical prescription pain medication and heroin were two of the categories where use has increased.

Long-term heroin (and other opiates) abuse can alter the brain's structure and function, resulting in neurological changes that persist while using the drug and long after drug use has ceased. Current pharmacotherapies for heroin addiction (eg, buprenorphine and methadone) have shown promise compared to no treatment in reducing heroin use. However, they continue to fall short of being highly effective when the goal is long-term abstinence and relapse prevention (Koob, Lloyd, & Mason, 2009; Kreek, LaForge, & Butelman, 2002; Peck & Ranaldi, 2014). Therefore, a serious problem for treatment of heroin addiction is relapse to heroin use after prolonged periods of abstinence (O'Brien, 2005; Peck, Galaj, Eshak, Newman, & Ranaldi, 2015). Thus, both clinical and animal research should focus on increasing our understanding about which neurobiological and environmental factors determine successful long-term heroin abstinence.

In heroin addicts, craving and relapse during abstinence (as well as that for other drugs of abuse) are often triggered by acute reexposure to the drug, drug-associated cues, or stress (de Wit & Stewart, 1983; O'Brien, 2005; Peck & Ranaldi, 2014; Sinha, Shaham, & Heilig, 2011). Further, because some of the factors that induce heroin relapse in humans and laboratory animals are similar, several animal models have been developed to study the repeating cycle of heroin abstinence and relapse. Here, we review a few popular animal models of abstinence and relapse and evaluate their validity in addressing human behavior that leads to long-term heroin abstinence. Then, I will discuss a novel abstinence "conflict" model that more closely mimics human drug addiction. Additionally, recent research investigating both cocaine and heroin seeking in rats using the animal conflict model will be presented and the implications for heroin treatments discussed. It is my view that using animal models that more closely mimic human drug abstinence will lead to a better understanding of the neurobiological, environmental, and behavioral factors that support long-term abstinence.

1. COMMON ANIMAL MODELS OF DRUG ABSTINENCE AND RELAPSE

When evaluating the validity of different animal models of abstinence and relapse I will compare the similarities between the animal models discussed and the natural phenomenon of human drug abstinence and relapse. For example, an animal model that uses instrumental extinction training (withholding of the drug) alone instead of inducing abstinence in the presence of the drug (as is often the case in humans) would have less validity. The latter of which may better capture the neural and environmental mechanisms of abstinence in humans. The animal models of abstinence and relapse that will be discussed include: reinstatement, abstinence model, punishment-based model, and abstinence-conflict model.

1.1 Reinstatement Model

Researchers have primarily studied the repeating cycle of drug abstinence and relapse using the reinstatement model (Carroll & Comer, 1996; Shaham & Miczek, 2003). In this model, laboratory animals are trained to self-administer drug which is accompanied by a discrete stimulus (eg, tone, light), usually by pressing a lever. Then, after extinction of the drug-taking response, by withholding the drug reinforcer and in the absence of the discrete stimulus, non-reinforced reinstatement of responding is induced by either acute exposure to the discrete cue, drug priming, contextual cues, or stress (Crombag, Bossert, Koya, & Shaham, 2008; de Wit & Stewart, 1983; Feltenstein & See, 2008; Meil & See, 1996; Shaham & Stewart, 1995). For example, studies with rats have shown that after extinction, cocaine or heroin seeking are reliably reinstated by acute injections of the drug or cues (discrete, discriminative, or contextual) that are associated with the drug (Crombag et al., 2008; Feltenstein & See, 2008). The validity of the reinstatement model is strong when considering the clinical scenario of human relapse that is elicited by acute reexposure to the drug, drug-associated cues, or stress during abstinence (de Wit & Stewart, 1983; O'Brien, 2005; Sinha et al., 2011).

However, like most animal models, the reinstatement model has limitations (Epstein, Preston, Stewart, & Shaham, 2006; Katz & Higgins, 2003). For example, in the reinstatement model there are no aversive consequences of drug seeking and abstinence is forced through extinction, whereas in humans abstinence often results from the aversive consequences that coincide with drug seeking rather than extinction (Cooper, Barnea-Ygael, Levy, Shaham, & Zangen, 2007; Epstein & Preston, 2003; Epstein et al., 2006; Peck & Ranaldi, 2014; Peck, Wercberger, Kariyeva, & Ranaldi, 2013). That is, some of the aversive consequences that may occur while drug seeking are hiding from law enforcement, family and friends, loss of employment, and securing the funds for obtaining the drug. Therefore, human drug-seeking episodes during abstinence often involve a "conflict" situation, which usually entails a choice between experiencing the positive effects of the drug and the aversive consequences of drug seeking (Cooper et al., 2007). Thus, an abstinence/relapse model that captures these features of addiction—the negative consequences occurring during drug seeking—may be useful in further characterizing human drug addiction.

1.2 Abstinence Model

Another popular animal model used to investigate the drug addiction cycle of abstinence and relapse is the abstinence model. In this animal model, subjects are first trained to perform an instrumental response (eg, lever pressing) with drug reinforcement in the presence of a drug cue. After training, subjects remain in their home cages or an alternate context, which contains no instrumental manipulandum. The experimenter-imposed drug-free period (forced abstinence) can range from days to months (Grimm, 2002; Pickens et al., 2011). On the test day, subjects are reexposed to the drug-paired context in the presence of the previously paired drug cue, which results in robust drug seeking.

In the abstinence model, the fact that subjects do not undergo explicit extinction training before testing provides some validity to this model. However, the procedural use of forced removal from the drug context to induce abstinence diminishes the model's validity when considering human drug abstinence. That is, in humans, abstinence is often self-imposed, not forced, and instead results from the aversive consequences that coincide with drug seeking (Cooper et al., 2007; Epstein & Preston, 2003; Epstein et al., 2006; Peck et al., 2013).

2. HOW ARE DRUG ABSTINENCE AND EXTINCTION DIFFERENT

Extinction is an important process of learning that is typically defined as the removal of reinforcers and/or biologically relevant stimuli from a previously established stimulus relationship that results in a reduction in conditioned responding (Pavlov, 1927). Further, extinction of a drug instrumental response involves an organism responding for the drug when there is no drug delivery upon the drug-taking response. Abstinence is different from responding under extinction, because under extinction conditions, the organism cannot take the drug as it is not available. However, abstinence is said to occur when an organism *refrains* from taking a drug when the *drug remains available* (unless abstinence is forced). Further, unlike extinction conditions, abstinence requires a choice between continued drug use and refraining from it (Epstein et al., 2006; Katz & Higgins, 2003).

Therefore, in the animal models of abstinence and relapse previously discussed, they attempt to either measure abstinence using environmental extinction or by removing the organism from the drug-paired context altogether. Consequently, neither of these procedures allow for the organism to make a choice (take drug vs not take drug). A true abstinence paradigm would allow the organism to take the drug after some behavioral response (as is the case in humans). Further, abstinence in humans occurs because the drug's rewarding effects are outweighed by the aversive consequences of drug seeking or drug taking (Barnea-Ygael, Yadid, Yaka, Ben-Shahar, & Zangen, 2012; Cooper et al., 2007; Panlilio, Thorndike, & Schindler, 2003, 2005). However, in most animal models of abstinence and relapse, abstinence is achieved either through extinction or forced removal from the drug context, not by the aversive consequences of drug seeking or taking. Thus, the use of extinction or forced abstinence decreases the validities between the current animal models of abstinence/relapse and human drug abstinence. Therefore, in the study of drug abstinence what is needed is an animal model that more closely resembles the human drug abstinence condition (Cooper et al., 2007; Epstein et al., 2006; Panlilio et al., 2003, 2005; Peck et al., 2013).

2.1 Punishment Model

There are some earlier studies that have presented aversive consequences during drug *taking* with animals. For example, punishment-based relapse models (Panlilio et al., 2003, 2005) have been developed in which drug-reinforced lever pressing is suppressed by the delivery of an electric shock immediately after drug infusions. The punishment model corresponds to the typical human abstinence drug situation, where the negative consequences of drug *taking* outweigh its hedonic (ie, rewarding) effects (Panlilio et al., 2005). In this model, abstinence is not a result of extinction or forced removal from the drug context. Thus, the punishment model has improved validity when compared to other animal models discussed (eg, reinstatement model) that use either extinction or abstinence by forced removal from the drug context (eg, abstinence model).

However, in humans, some of the aversive consequences related to drug use are those that occur during drug seeking, such as hiding from law enforcement, family and friends, or securing the funds to obtain the drug. As previously mentioned, human drug-seeking episodes during abstinence often involve a conflict situation, which usually involves a choice between pursuing the path that leads to experiencing the positive effects of drug(s) accompanied with aversive consequences and the path that avoids the aversive consequences of drug seeking (Cooper et al., 2007; Epstein & Preston, 2003; Peck et al., 2013). In this case the aversive consequences are not contingent on the drug-taking response and therefore do not occur after drug intake, but instead are present during drug seeking and can occur before drug taking. Therefore, an abstinence-conflict model where the aversive consequences occur during drug seeking (conflict situation) would be useful in further characterizing the different aspects involved in human drug abstinence.

2.2 Abstinence-Conflict Model

Cooper et al. (2007) developed a conflict-based abstinence/relapse model where aversive consequences occur during cocaine *seeking*. In this model, rats are trained to lever press for intravenous cocaine infusions paired with a discrete light stimulus. An electric barrier is then introduced by electrifying the floor area near the levers (Fig. 1) while the drug is available; thus, the animals could continue to self-administer drug but doing so necessitated enduring electric shock (a conflict situation). The researchers increased the electric shock intensities daily until the rats stopped emitting the cocaine-taking (ie, lever pressing) response for three consecutive sessions

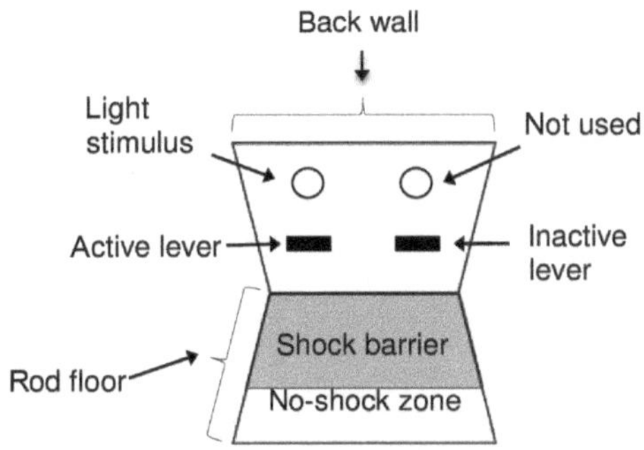

Fig. 1 Schematic diagram of the animal conflict model apparatus that includes the back wall and floor of operant conditioning chambers.

(abstinence). In a relapse test with the electric barrier remaining activated, the effect of cocaine cue presentations led to the resumption of lever pressing (relapse). Thus, this model focuses on features similar to those seen in many human addicts who abstain because of the negative consequences of continued drug use and who face the conflict situation of choosing to seek the drug or to abstain from this.

Recently, we (Peck et al., 2013) extended the use of the abstinence/relapse conflict model by examining both cocaine and heroin seeking in rats. We found that after suppression of heroin self-administration by the electric barrier, noncontingent presentation of a drug cue (light stimulus, every 5 min during a 30-min relapse session) while the electric barrier remained activated resulted in resumption of active lever pressing in all 10 heroin-seeking rats tested. When we repeated this exact same procedure with cocaine, we observed that only 40% (3 out of 8) of the rats resumed active lever pressing during the relapse test. The cocaine results are similar to those reported by both Cooper et al. (2007) and Barnea-Ygael et al. (2012) using a similar animal conflict model that tested cue-induced relapse in cocaine-trained rats. These authors reported that in animals that had achieved abstinence, noncontingent cue presentations led to active lever responding for just under 60% and 40% of cocaine-seeking rats (respectively) in the relapse tests. Thus, identical procedures for cocaine and heroin resulted in significantly greater relapse rates in heroin-experienced rats than in cocaine-experienced rats (Fig. 2).

The reasons for the substantially higher relapse in heroin vs cocaine in the abstinence-conflict model are currently unknown. One possibility is perhaps heroin use produces certain neuroadaptive or increased motivational effects (ie, craving) that make relapse more likely. For example, research has shown

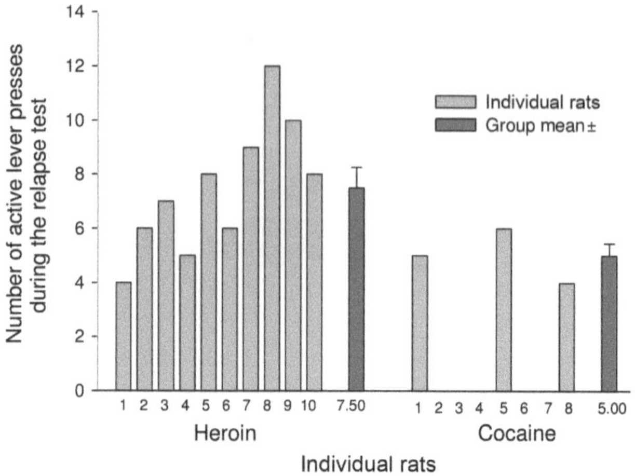

Fig. 2 Individual and mean (±SEM) active lever presses for heroin ($n=10$) and cocaine-seeking rats ($n=8$) presented in the same order as their final shock intensities during the 30-min relapse test.

that there are some critical differences between heroin and cocaine with regard to the neurobiological and behavioral mechanisms in reward and drug addiction. These include, but are not limited to, the following: dopamine plays a significant role in psychostimulant reward (Wise & Rompre, 1989) but not in heroin reward (Ettenberg, Pettit, Bloom, & Koob, 1982); cocaine produces both approach and avoidance motivational states whereas heroin produces only an approach-like motivational state (Ettenberg & Geist, 1993); cocaine context-induced reinstatement is attenuated by reversible inactivation of the dorsal medial prefrontal cortex (mPFC) but not the ventral mPFC, whereas the opposite is the case for heroin (Bossert et al., 2011); reinstatement of heroin seeking seems to involve a greater number of brain sites than that of cocaine seeking (McFarland & Kalivas, 2001; Rogers, Ghee, & See, 2008). Further, the inactivation of additional brain regions that attenuate reinstatement to heroin seeking as compared to reinstatement of cocaine seeking, implies a more diffuse network underlying relapse to heroin (Badiani, Belin, Epstein, Calu, & Shaham, 2011). Therefore, it might be that higher relapse rates observed in heroin-seeking rats is in part because heroin-seeking engages a more diffuse neural network, which therefore might be more readily accessible by relapse-inducing stimuli. Our findings using the abstinence-conflict model further suggest that different relapse vulnerabilities between heroin- and cocaine-experienced rats are a result of differences in the behavioral and

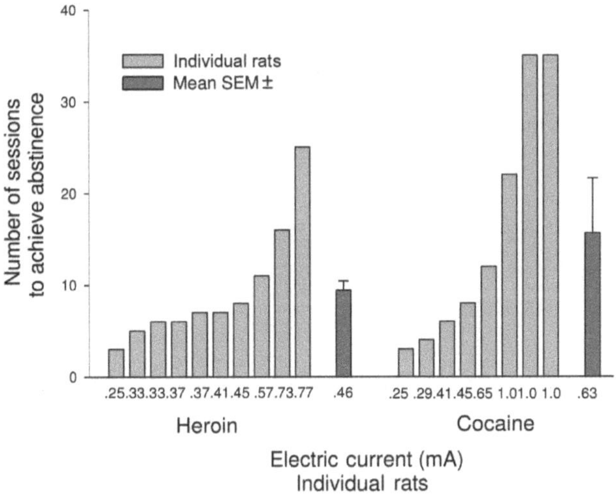

Fig. 3 Individual and mean (±SEM) final shock intensities for heroin ($n = 10$) and cocaine-seeking rats ($n = 10$) that led to 3 days of zero presses on either lever.

neurochemical effects between opiates and psychostimulants (Peck et al., 2013).

Interestingly, the use of the abstinence-conflict model may provide the ability to study compulsive drug use in animals during abstinence (ie, electric barrier phase). That is, drug-seeking performance during abstinence may constitute a measure of compulsivity; the extent to which individual rats will endure escalating aversive consequences over sessions (eg, increasing electric shock intensity) in order to continue drug consumption (Fig. 3). Moreover, the ability to study compulsive use in animals provides the abstinence-conflict model with strong construct validity because it approximates some of the defining features of human drug addiction. In humans, addiction is not just the taking of drugs but the compulsive drug use maintained despite adverse consequences for the user (DSM-5, 2013). Therefore, the abstinence-conflict model may provide insight into how to identify subjects at risk for subsequent development of severe heroin or other drugs of abuse addiction (compulsive use).

3. MOVING CLOSER TO HUMAN DRUG ADDICTION

The abstinence-conflict model has important abstinence features that are not present in the other commonly used animal models for drug addiction described here. As already mentioned, some of the aversive consequences related to drug use are those that occur during drug seeking. However, in current animal models (except for the punishment model), there are no aversive consequences for drug seeking. Further, in the vast majority of drug addiction cases, abstinence is achieved by choice; addicts choose to abstain from drug taking while the opportunity exists, because the aversive consequences of drug seeking and taking outweigh the drug's rewarding effects (Cooper et al., 2007; Epstein & Preston, 2003; Peck et al., 2013; Vanderschuren & Everitt, 2004). However, in other animal models of abstinence and relapse (eg, reinstatement model), abstinence is not achieved by choice; instead, abstinence is a result of either an extinction procedure or the forced removal from the drug context. In contrast, the abstinence-conflict model incorporates these important characteristics of human drug addiction, where abstinence occurs while the drug is readily available and the animal must endure aversive consequences for its pursuit and consumption of the drug. Further, the model demonstrates how the aversive consequences of drug use play an integral part in the initiation and maintenance of drug abstinence and perhaps, in the prevention of relapse.

Importantly, the abstinence-conflict model may be useful for studying the environmental and neural mechanisms underlying long-term abstinence in drug addiction (Peck & Ranaldi, 2014; Saunders, Yager, & Robinson, 2013). For example, the abstinence-conflict model would seem suitable for testing and further developing pharmacotherapeutic strategies for opiate addiction (eg, buprenorphine and naltrexone) that support long-term abstinence in humans. Recently, we (Peck et al., 2015) found that the implementation of environmental enrichment (EE) supported heroin abstinence in the abstinence-conflict model. Further, the current use of the abstinence-conflict model to investigate EE as a strategy to facilitate heroin abstinence may help in the development of effective treatments for human heroin addicts by bringing together the positive consequences of abstinent behavior in an enriched environment with the aversive consequences of heroin seeking (ie, electric barrier). Therefore, this model might serve an important complimentary role in drug abuse research by emphasizing features of the human addiction cycle that are not emphasized by other animal models.

4. CONCLUSIONS

Drug addiction is a significant health and societal problem for which there are no effective and well-accepted long-term behavioral or pharmacological treatments. Of particular concern is the rising use of illegal opiate drugs such as heroin and the misuse of legally available pain relievers that have led to serious deleterious health effects or even death. Therefore, treatment strategies that prolong heroin abstinence should be the primary focus of heroin treatment research. Further, because the factors that support long-term drug abstinence in humans are similar to those in animals, the field might benefit from greater use of animal models of drug abstinence that incorporate negative consequences for drug seeking and drug taking while the drug remains readily available. Thus, animal abstinence models that mimic human drug abstinence, such as the abstinence-conflict model, could lead to a better understanding of the neurobiological and environmental factors that support long-term abstinence. In turn, this will lead to the development of more effective environmental and pharmacotherapeutic interventions to treat opiate addiction and addiction to other drugs of abuse.

REFERENCES

American Psychiatric Association. (2013). *Diagnostic and Statistical Manual of Mental Disorders (DSM-5®)*. Virginia, USA: American Psychiatric Publishing.

Badiani, A., Belin, D., Epstein, D., Calu, D., & Shaham, Y. (2011). Opiate versus psychostimulant addiction: The differences do matter. *Nature Reviews. Neuroscience*, *12*(11), 685–700.

Barnea-Ygael, N., Yadid, G., Yaka, R., Ben-Shahar, O., & Zangen, A. (2012). Cue-induced reinstatement of cocaine seeking in the rat "conflict": Effect of prolonged home-cage confinement. *Psychopharmacology*, *219*, 875–883.

Bossert, J. M., Stern, A. L., Theberge, F. R., Cifani, C., Koya, E., Hope, B. T., et al. (2011). Ventral medial prefrontal cortex neuronal ensembles mediate context-induced relapse to heroin. *Nature Neuroscience*, *14*(4), 420–422.

Carroll, M. E., & Comer, S. D. (1996). Animal models of relapse. *Experimental and Clinical Psychopharmacology*, *4*, 11–18.

Cooper, A., Barnea-Ygael, N., Levy, D., Shaham, Y., & Zangen, A. (2007). A conflict rat model of cue-induced relapse to cocaine seeking. *Psychopharmacology*, *194*, 117–125.

Crombag, H. S., Bossert, J. M., Koya, E., & Shaham, Y. (2008). Context-induced relapse to drug seeking: A review. *Philosophical Transactions of the Royal Society of London. Series B, Biological Sciences*, *363*, 3233–3243.

de Wit, H., & Stewart, J. (1983). Drug reinstatement of heroin reinforced responding in the rat. *Psychopharmacology*, *79*, 29–31.

Epstein, D. H., & Preston, K. L. (2003). The reinstatement model and relapse prevention: A clinical perspective. *Psychopharmacology*, *168*, 31–41.

Epstein, D. H., Preston, K. L., Stewart, J., & Shaham, Y. (2006). Toward a model of drug relapse: An assessment of the validity of the reinstatement procedure. *Psychopharmacology*, *189*, 1–16.

Ettenberg, A., & Geist, T. D. (1993). Qualitative and quantitative differences in the operant runway behavior of rats working for cocaine and heroin reinforcement. *Pharmacology, Biochemistry, and Behavior*, *44*(1), 191–198.

Ettenberg, A., Pettit, H. O., Bloom, F. E., & Koob, G. F. (1982). Heroin and cocaine intravenous self-administration in rats: Mediation by separate neural systems. *Psychopharmacology*, *78*(3), 204–209.

Feltenstein, M. W., & See, R. E. (2008). The neurocircuitry of addiction: An overview. *British Journal of Pharmacology*, *154*(2), 261–274.
Grimm, J. W. (2002). Effect of cocaine and sucrose withdrawal period on extinction behavior, cue-induced reinstatement, and protein levels of the dopamine transporter and tyrosine hydroxylase in limbic and cortical areas in rats. *Behavioral Pharmacology*, *13*, 379–388.
Katz, J. L., & Higgins, S. T. (2003). The validity of the reinstatement model of craving and relapse to drug use. *Psychopharmacology*, *168*(1–2), 21–30.
Koob, G. F., Lloyd, G. K., & Mason, B. J. (2009). Development of pharmacotherapies for drug addiction: A Rosetta Stone approach. *Nature Reviews. Drug Discovery*, *8*, 500–515.
Kreek, M. J., LaForge, K. S., & Butelman, E. (2002). Pharmacotherapy of addictions. *Nature Reviews. Drug Discovery*, *1*(9), 710–726.
McFarland, K., & Kalivas, P. W. (2001). The circuitry mediating cocaine-induced reinstatement of drug-seeking behavior. *The Journal of Neuroscience*, *21*(21), 8655–8663.
Meil, W. M., & See, R. E. (1996). Conditioned cue recovery of responding following prolonged withdrawal from self-administered cocaine in rats: An animal model of relapse. *Behavioral Pharmacology*, 7, 754–763.
O'Brien, C. P. (2005). Anticraving medications for relapse prevention: A possible new class of psychoactive medications. *The American Journal of Psychiatry*, *162*, 1423–1431.
Panlilio, L. V., Thorndike, E. B., & Schindler, C. W. (2003). Reinstatement of punishment-suppressed opioid self-administration in rats: An alternative model of relapse to drug abuse. *Psychopharmacology*, *168*, 229–235.
Panlilio, L. V., Thorndike, E. B., & Schindler, C. W. (2005). Lorazepam reinstates punishment-suppressed remifentanil self-administration in rats. *Psychopharmacology*, *179*, 374–382.
Pavlov, I. P. (1927). *Conditioned reflexes*. London: Oxford University Press.
Peck, J. A., Galaj, E., Eshak, S., Newman, K. L., & Ranaldi, R. (2015). Environmental enrichment induces early heroin abstinence in an animal conflict model. *Pharmacology, Biochemistry, and Behavior*, *138*, 20–25.
Peck, J. A., & Ranaldi, R. (2014). Drug abstinence: Exploring animal models and human strategies. *Psychopharmacology*, *231*, 2045–2058.
Peck, J., Wercberger, R., Kariyeva, E., & Ranaldi, R. (2013). Cue-induced resumption of heroin and cocaine seeking in rats using a conflict model of abstinence and relapse. *Psychopharmacology (Berl)*, *228*, 651–658.
Pickens, C. L., Airavaara, M., Theberge, F., Fanous, S., Hope, B. T., & Shaham, Y. (2011). Neurobiology of incubation of drug craving. *Trends in Neurosciences*, *34*, 411–420.
Rogers, J. L., Ghee, S., & See, R. E. (2008). The neural circuitry underlying reinstatement of heroin-seeking behavior in an animal model of relapse. *Neuroscience*, *151*(2), 579–588.
Saunders, B. T., Yager, L. M., & Robinson, T. E. (2013). Cue-evoked cocaine "craving": Role of dopamine in the accumbens core. *The Journal of Neuroscience*, *33*(35), 13989–14000.
Shaham, Y., & Miczek, K. A. (2003). Reinstatement-toward a model of relapse. *Psychopharmacology*, *168*, 1–2.
Shaham, Y., & Stewart, J. (1995). Stress reinstates heroin-seeking in drug-free animals: An effect mimicking heroin, not withdrawal. *Psychopharmacology*, *119*, 334–341.
Sinha, R., Shaham, Y., & Heilig, M. (2011). Translational and reverse translational research on the role of stress in drug craving and relapse. *Psychopharmacology*, *218*, 69–82.
Vanderschuren, L. J., & Everitt, B. J. (2004). Drug seeking becomes compulsive after prolonged cocaine self-administration. *Science*, *305*(5686), 1017–1019.
Wise, R. A., & Rompre, P. P. (1989). Brain dopamine and reward. *Annual Review of Psychology*, *40*, 191–225.

CHAPTER TWELVE

Prairie Voles as a Model to Screen Medications for the Treatment of Alcoholism and Addictions

A.E. Ryabinin[1], C.M. Hostetler
Oregon Health & Science University, Portland, OR, United States
[1]Corresponding author: e-mail address: ryabinin@ohsu.edu

Contents

Abstract

Most preclinical studies of medications to treat addictions are performed in mice and rats. These two rodent species belong to one phylogenetic subfamily, which narrows the likelihood of identifying potential mechanisms regulating addictions in other species, ie, humans. Expanding the genetic diversity of organisms modeling alcohol and drug abuse enhances our ability to screen for medications to treat addiction. Recently, research laboratories adapted the prairie vole model to study mechanisms of alcohol and drugs of abuse. This development not only expanded the diversity of genotypes used to screen medications, but also enhanced capabilities of such screens. Prairie voles belong to 3–5% of mammalian species exhibiting social monogamy. This unusual trait is reflected in their ability to form lasting long-term affiliations between adult individuals. The prairie vole animal model has high predictive validity for mechanisms regulating human social behaviors. In addition, these animals exhibit high alcohol intake and preference. In laboratory settings, prairie voles are used to model social influences on drug reward and alcohol consumption as well as effects of addictive substances on social bonding. As a result, this species can be adapted to screen medications whose effectiveness could be (a) resistant to social influences promoting excessive drug taking, (b) dependent on the presence of social support, and (c) medications affecting harmful social consequences of alcohol and drug abuse. This report reviews the literature on

International Review of Neurobiology, Volume 126
ISSN 0074-7742
http://dx.doi.org/10.1016/bs.irn.2016.02.019

studies of alcohol and psychostimulants in prairie voles and discusses capabilities of this animal model as a screen for novel medications to treat alcoholism and addictions.

1. INTRODUCTION

In addition to being the third leading factor of preventable mortality (CDC, 2014), alcohol abuse and dependence have a tremendous impact on social behaviors. On one hand, alcohol is often consumed as a "social lubricant" (Carrigan & Randall, 2003; Sayette et al., 2012; Steele & Josephs, 1990). On the other hand, alcohol abuse contributes to incidence of aggression, intimate partner violence, and marital disintegration (Homish, Leonard, Kozlowski, & Cornelius, 2009; Justus, Finn, & Steinmetz, 2000; Leonard, 2005; McLeod, 1993). In turn, the social environment can also influence alcohol drinking and abuse. For example, while peer pressure can facilitate drinking, social support contributes to recovery from alcohol dependence (Bond, Kaskutas, & Weisner, 2003; Charney, Zikos, & Gill, 2010; Fisher, Miles, Austin, Camargo, & Colditz, 2007; Gordon & Zrull, 1991; Kelly, Stout, Magill, & Tonigan, 2011; Park, Sher, & Krull, 2008; Wu & Witkiewitz, 2008). One may argue that some of these bidirectional effects are related to social practices, expectations, and economic factors that could be uniquely relevant for humans. However, research performed over the last decade has increased our appreciation for the fact that biological factors also regulate human social behaviors (Insel, 2010; Meyer-Lindenberg, Domes, Kirsch, & Heinrichs, 2011; Walum et al., 2008). As a result, evolutionary predecessors of social aspects of alcoholism can be observed in animals and used to model these problems encountered in humans. Such modeling would provide important insights informing policies, interventions, and treatments of social problems associated with alcohol use disorder.

Not surprisingly, therefore, researchers using mice, rats, and nonhuman primates developed a number of models of alcohol-induced aggression, dominance, isolation, and other social aspects of alcohol drinking (reviewed in Anacker & Ryabinin, 2010). As research identifies increasingly complex social phenomena, for example, empathy and helping behaviors (Bartal, Decety, & Mason, 2011; Knapska, Mikosz, Werka, & Maren, 2010; Langford et al., 2006; Panksepp & Lahvis, 2011), these models could become more and more sophisticated and elegant. However, some human social behaviors have been extremely difficult to model in standard laboratory animals. Specifically, rats and mice are promiscuous species and do not form long-term attachments between adult individuals. This quality does not prevent mice and rats from preferring social environment over isolation or being attentive to distress of their cagemates (Bartal et al., 2011; Panksepp & Lahvis, 2011), but it does contrast these species to humans who tend to form social attachments to specific individuals.

Fortunately, advancement of social neurobiology has helped identify species in which such behaviors are easier to model than in mice, rats, or nonhuman primates. In particular, prairie voles emerged as a superb species to study a number of social behaviors (Carter, 1998). Prairie voles (*Microtus ochrogaster)* belong to the *Cricetidae* subfamily of rodents. Animals of this species are unusual in that they tend to be socially monogamous, as male and female voles form a long-term emotional attachment, a "pair bond," following mating (Carter & Getz, 1993). These pair bonds do not prevent animals from occasional mating outside of the bond, a quality indicated by the word "social" preceding the word "monogamous" in this terminology. As a frequent characteristic of monogamous species, these animals are biparental, meaning that both males and females take care of the young (Carter & Getz, 1993). Moreover, in the wild, litters of these animals often do not disperse and older siblings remaining in the parental nest engage in alloparental behaviors and form affiliations with siblings (Carter & Getz, 1993). These qualities of prairie voles mimic typical human behaviors much better than those of standard laboratory species.

Importantly, the prairie vole model not only has high face validity, but also high construct and predictive validity for mechanisms regulating social attachment. Several mechanisms regulating pair bonding identified first in prairie voles, involving specific vasopressin, oxytocin, dopamine and their respective receptors, were later shown to play a role in human social attachment and bonding (Hammock & Young, 2005; Knafo et al., 2008; Meyer-Lindenberg et al., 2011; Walum et al., 2008; Young, Gobrogge, Liu, & Wang, 2011). Emotional attachments between romantic partners, social affiliations between adult individuals and parent–children relationships influence and are known to be influenced by alcohol and drug addiction. Prairie voles, therefore, appear to be an ideal species to study many of the social aspects of alcohol and drug dependence.

2. PRAIRIE VOLES AS A MODEL OF DRUG AND ALCOHOL TAKING

The laboratory of Dr. Zuoxin Wang demonstrated that administration of amphetamine can disrupt pair bonding (Liu et al., 2010; Young, Liu, et al., 2011), and on the other hand, that pair bonding modulates rewarding properties of this drug (Liu, Young, Curtis, Aragona, & Wang, 2011). These studies provided first crucial insights into molecular mechanisms underlying bidirectional effects between social affiliations and addictive drugs. However, one disadvantage of these studies is that amphetamine was administered by an experimenter and not voluntarily self-administered by prairie voles. This is an important caveat, as voluntary and nonvoluntary modes of administration are known to engage different neural mechanisms.

Our laboratory has demonstrated that animals of this species, if given a chance, voluntarily consume ethanol solutions and in fact, demonstrate preference for ethanol-containing solutions over water (Anacker, Loftis, Kaur, & Ryabinin, 2011). Moreover, they consume alcohol in quantities sufficient to produce central effects (Anacker, Loftis, Kaur, et al., 2011). Voluntary alcohol self-administration in voles is not trivial since the majority of rat and mouse strains avoid alcohol. Thus, only strains related to inbred C57BL/6J mice as well as selectively bred lines of mice and rats prefer alcohol solutions over water. Inbreeding and selective breeding decrease genetic diversity of animals leading to a relatively low number of genotypes that can be identified as important for regulation of alcohol drinking. In contrast, prairie voles, being recently derived from wild populations, are genetically diverse. As a result, alcohol drinking in individual prairie voles can range from relatively low (~5 g/kg/day) to extremely high (over 30 g/kg/day) (Anacker, Loftis, Kaur, et al., 2011). This makes prairie voles an excellent model of genetic variability in predisposition to high alcohol consumption in humans.

It is not yet clear why prairie voles exhibit high alcohol consumption compared to the majority of mouse and rat strains. One intriguing hypothesis is that neurocircuitries regulating social monogamy, drug addiction, and alcohol consumption could be strongly overlapping, such that the tendency for strong social bonding predisposes these animals to self-administration of addictive substances (Burkett & Young, 2012, but see Hostetler & Ryabinin, 2012). This hypothesis is in agreement with higher amphetamine-induced dopamine release in prairie voles vs meadow voles, a promiscuous species (Aragona, Detwiler, & Wang, 2007). Also in agreement with this idea is the finding that meadow voles exhibited lower intakes of ethanol than prairie voles (Anacker, Ahern, Young, & Ryabinin, 2012). However, alcohol intake in meadow voles is still higher than in most mouse and rat strains. This interpretation is also complicated by the fact that meadow voles have slightly slower ethanol metabolism than prairie voles indicating that even if the difference in intake between these species is statistically significant, it could be driven by noncentral mechanisms (Anacker et al., 2012). A simpler interpretation of these data is that both monogamous prairie voles and promiscuous meadow voles can voluntarily consume large quantities of alcohol. Therefore, it seems to be more important to understand why both monogamous and promiscuous species of voles demonstrate high alcohol consumption.

Another hypothetical explanation for high alcohol consumption in voles is that wild-derived species could be predisposed to high consumption of any highly caloric substance. This hypothesis is in agreement with occurrence of high alcohol-consuming mice inhabiting wineries and distilleries. However, alcohol consumption in these animals could be a trait selected by a combination of hunger encountered in wild populations and availability of alcohol as the only source of calories. We favor another hypothesis that higher alcohol consumption in voles vs mice and rats is due to slightly higher proportion of plants in their diet than in the diets of mice and rats (Tamarin, 1985), leading to a higher possibility of consumption of fermented plant substances. This interpretation would also be in agreement with more frequent encounter of high alcohol consumption in mouse strains over rat strains.

Whatever the reason for high alcohol consumption in voles, the combination of high voluntary consumption and wide genotypic diversity makes them an attractive model for alcohol researchers trying to understand mechanisms regulating alcohol intake. Moreover, perhaps an even greater attraction lies in the possibility to model various social aspects of alcohol use disorder.

3. SOCIAL FACILITATION OF DRINKING

In humans, alcohol consumption is a frequent attribute of socializing. Increased drinking in social settings is frequently described in both scientific and popular literature. Therefore, it is not surprising, that enhanced consumption in social vs isolate conditions is described in several rodent species, including voles (see references in Anacker & Ryabinin, 2010; Tomie, Samuel, Sprung, Malul, & Yu, 2015). Our laboratory demonstrated that if pair-housed same-sex (male or female) voles are introduced to a two-bottle choice between a solution of ethanol and water, they will on average consume more ethanol than if they were introduced to alcohol in isolation (Anacker, Loftis, Kaur, et al., 2011; Hostetler, Anacker, Loftis, & Ryabinin, 2012).

The experiments comparing drinking in isolate vs social conditions can be classified into two types. In the first type of conditions, social housing entails simply cohousing two or more animals in the same cage. The consumption of ethanol per individual animal is then calculated by dividing total amount of ethanol consumed per cage by the number of animals in the cage. In the second type of conditions, the animals are housed together, but are separated by a mesh partition permitting a certain degree of social interaction and, at the same time, allowing exact measurements of fluid consumption per animal. The first type of condition has the disadvantage of assuming that ethanol intake will be equal between all cagemates. This assumption can be problematic, however, since subordinate voles can drink more alcohol than dominant voles, when in isolation (Anacker, Smith, & Ryabinin, 2014). It has been argued also that in this condition, drinking can be lower than in isolate housing because social interaction and aggressive behaviors can distract cagemates from drinking alcohol (Tomie et al., 2015). This concern is not applicable to voles, as they do not consume less alcohol in social housing. On the other hand, it can be argued that the second type of social housing (separation by a mesh) could prevent some of the social interactions between cagemates, and perhaps thereby create an environmental confound (for example, anxiety) that, in turn, would affect drinking. Since prairie voles are extremely sensitive to isolation stress (Grippo, Wu, Hassan, & Carter, 2008), this idea contradicts the observation that isolated voles drink less than voles separated by partition. Moreover, we have observed higher alcohol consumption in voles that are either housed together without a partition (Anacker & Ryabinin, unpublished data) or housed separated by a partition (Anacker, Loftis, Kaur, et al., 2011; Hostetler et al., 2012) than isolated voles. Therefore, both types of social housing in voles appear to model social facilitation of alcohol drinking.

The social facilitation of drinking in voles is reminiscent of effects of peer influence observed in human social groups. This facilitation is not dependent on the sex of voles as it occurs in both male–male and female–female pairs of animals (Anacker, Loftis, Kaur, et al., 2011; Hostetler et al., 2012). However, it is not observed in male–female pairs (Hostetler et al., 2012), an observation suggesting that it is not the strength of pair bonding (which is considered to be stronger in male–female than in same-sex pairs), but a specific quality of social interactions between individuals that contributes to this facilitation. The latter idea is in agreement with the observation that same-sex pairs of nonmonogamous meadow voles also exhibited facilitation of drinking in social conditions (Anacker et al., 2012). Although meadow voles are not monogamous, they can form preference for individuals of the same sex (Beery & Zucker, 2010). It needs to be noted that experiments on effects of social context on alcohol drinking in meadow voles have not been performed using the same experimental conditions as experiments in prairie voles, and therefore, warrant future investigations. Nevertheless, the currently available data suggest that studies of social facilitation of alcohol drinking do not need to be performed only in monogamous species of voles.

4. SOCIAL INHIBITION OF DRINKING

While peer pressure and social settings can promote alcohol abuse, social support can aid in recovery from alcoholism (Charney et al., 2010; McCrady et al., 2006). Prairie voles, capable of individual-specific social bonds, can also be used to model social inhibition of alcohol consumption and dependence. The prairie vole models of social facilitation of drinking differ from models of inhibition of drinking in the social context under which the first introduction to alcohol occurs. Thus, if animals are introduced to alcohol while they are pair housed, they show higher alcohol intake and preference than isolated voles, ie, social facilitation of drinking (Anacker, Loftis, Kaur, et al., 2011). In contrast, if animals are introduced to alcohol when they are in isolation, their subsequent pair housing can lead to social inhibition of alcohol consumption (Anacker, Loftis, & Ryabinin, 2011).

In a typical social inhibition of drinking experiment, isolate-housed prairie voles are introduced to a two-bottle choice between a solution of 10% ethanol and water for 4 days. During this time their alcohol intake stabilizes and they can be categorized into high and low drinkers. During the next 4 days the voles are rehoused to pair-housed conditions (being separated by a mesh partition), such that each high drinker is paired with a low drinker. Under these conditions, a significant proportion of high-drinking voles will decrease intake to match intake of low-drinking voles (Anacker, Loftis, & Ryabinin, 2011). Importantly, the decrease of drinking does not occur if high-drinking voles are pair housed with high-drinking voles (Anacker, Loftis, & Ryabinin, 2011), indicating that this effects is not due to a regression to mean, but is indeed due to the social influence. Moreover, this effect is specific to ethanol, since no change in drinking is observed when voles are consuming sweetened solutions, instead of alcohol (Anacker, Loftis, & Ryabinin, 2011). As with the experiments demonstrating facilitation of drinking, the social inhibition is observed in same-sex, but not male–female pairs of voles (Hostetler et al., 2012).

A detailed analysis of these inhibitory effects using lickometer approaches suggested that the decrease of alcohol drinking does not occur via mimicking behaviors of low-drinking voles by the high-drinking voles (Anacker & Ryabinin, 2013). Instead, it showed that the subpopulation of high-drinking voles that decreases its drinking during social housing seems to be predisposed to this change by exhibiting lower alcohol preference and less drinking bouts (but not less total intake of alcohol) prior to pair housing. This observation is reminiscent of the phenomenon of "readiness to change" described in human literature (Heather, Rollnick, & Bell, 1993; Merrill, Wardell, & Read, 2015; Rollnick, Heather, Gold, & Hall, 1992). Importantly, the inhibition of alcohol drinking in high-drinking voles during the social housing was indicative of a prolonged change in drinking, as it was maintained even if these voles were again placed in isolation following pair housing (Anacker & Ryabinin, 2013).

The experiments described above relied on modeling alcohol drinking in subjects with relatively short drinking histories. This could be considered a drawback since problem drinking in humans typically develops over time, and patients who require treatment typically have long history of alcohol abuse. This caveat is overcome in a study of inhibitory effects of social housing on alcohol consumption following prolonged history of alcohol drinking in voles (Hostetler & Ryabinin, 2014). In this experiment, single-housed male and female prairie voles were exposed to the two-bottle choice procedure for 4 weeks. Following this exposure, they were not given access to ethanol for 3 days and then again were introduced to the two-bottle choice procedure. This reintroduction to alcohol resulted in increased alcohol consumption compared to baseline (Hostetler & Ryabinin, 2014). This phenomenon of increased drinking following lack of access to alcohol is known as the alcohol deprivation effect (ADE). This effect is typically observed after longer intervals of alcohol exposure than 4 weeks in other rodent species and is considered to be one of the hallmarks of alcohol dependence (McBride & Li, 1998; Sinclair & Senter, 1968; Vengeliene, Bilbao, & Spanagel, 2014). Importantly, in this study ADE was only observed if voles were re-introduced to alcohol in isolation, but not when they were housed socially (separated by partition) with either alcohol–naïve or alcohol-consuming same-sex voles (Hostetler & Ryabinin, 2014). Thus, social partners are able to inhibit not only short-term drinking, but also dependence-like long-term alcohol drinking, a finding that underscores validity of the prairie vole model of social inhibition of alcohol drinking.

5. MODELING EFFECTS OF MEDICATIONS ON ALCOHOL DRINKING

The first study testing medications to decrease alcohol drinking had the goal of testing the construct validity of the prairie vole model (Anacker & Ryabinin, 2010). Such validation is based on the idea that if similar compounds decrease alcohol consumption in humans and in voles, then similar mechanisms must be involved in regulation of alcohol drinking in both species. Naltrexone is an FDA-approved medication to curb craving for alcohol. Naltrexone an antagonist of pan-opioid receptors and is thought to decrease alcohol intake by blocking actions of endogenous opioids (Froehlich & Li, 1993; Volpicelli, Alterman, Hayashida, & O'Brien, 1992). Although instances of low efficacy of naltrexone have been described, it is the most commonly prescribed medication to decrease alcohol abuse (Donoghue et al., 2015; Garbutt et al., 2014). Therefore, the choice of naltrexone for a validation study was obvious. Under same-sex, mesh-divided social housing conditions, male and female prairie voles demonstrated sensitivity to intraperitoneal (i.p.) administration of 8 mg/kg of naltrexone by significantly decreasing alcohol preference in a 2-h limited access session (Anacker & Ryabinin, 2010). Importantly, this finding not only provided evidence for the validity of the prairie vole model of excessive alcohol use, but also suggested that novel compounds, not yet in use in the clinical settings, could be screened in prairie voles.

Voles can be used to screen pharmacological agents not yet approved for human use. This was first demonstrated in a recent study in which an antagonist of growth hormone secretagogue receptor 1a (GHS-R1a), was administered to prairie voles to test its effectiveness in decreasing alcohol consumption (Stevenson et al., 2015). GHS-R1a is the endogenous receptor for the orexigenic peptide ghrelin. The ghrelin system was originally considered a promising target for development of medications to treat obesity (Dhillo & Bloom, 2001; Muccioli et al., 2002). More recently, it came to the attention of alcohol researchers following numerous reports of relationships between ghrelin levels, alcohol craving, and alcohol consumption in humans (Addolorato et al., 2006; Calissendorff, Danielsson, Brismar, & Rojdmark, 2005; Leggio et al., 2012). The effectiveness of ghrelin antagonists to decrease alcohol consumption was demonstrated in rats and mice (Jerlhag et al., 2009; Kaur & Ryabinin, 2010). It has been debated whether actions of these antagonists are via GHS-R1a receptors in the ventral tegmental and laterodorsal tegmental areas (Jerlhag et al., 2007; Jerlhag, Janson, Waters, & Engel, 2012), the centrally projecting Edinger–Westphal nucleus (Kaur & Ryabinin, 2010) or the central nucleus of amygdala (Cruz, Herman, Cote, Ryabinin, & Roberto, 2013). Nevertheless, the recent demonstration that ghrelin antagonists are effective in mouse and rat models of dependence (Gomez et al., 2015; Suchankova, Steensland, Fredriksson, Engel, & Jerlhag, 2013), make this system an attractive candidate for development of medications to treat alcoholism.

Stevenson and colleagues (2015) administered GHS-R1a antagonist JMV2959 (12 mg/kg, i.p.) to isolate-housed female prairie voles prior to a 2 h two-bottle choice drinking session. The antagonist significantly reduced ethanol intake and preference at both doses. Interestingly, it did not affect consumption of sucrose, suggesting selectivity of ghrelin antagonists against alcohol consumption over natural rewards. These findings further confirmed the promise of the ghrelin system as a potential target for development of pharmacotherapies to treat alcoholism and the prairie voles as a potential screening tool for identification of medications for this disorder. It needs to be noted that published studies in voles so far only tested one of the existing ghrelin antagonists and only a single injection of this drug. Different ghrelin antagonists can be differentially effective in decreasing alcohol intake and some of them lose effectiveness with repeated administration in other rodent species (Gomez et al., 2015; Gomez & Ryabinin, 2014; Jerlhag et al., 2009). Moreover, our preliminary studies with [D-Lys(3)]-GHRP-6, one of the less specific GHS-R1a antagonists, indicated toxicity in voles at doses effective in mice and rats (Hostetler & Ryabinin, unpublished data). These results suggest the need to expand the number of species used for screening of potential pharmacotherapies, and that prairie voles would be a useful addition to this number.

6. TESTING SOCIAL MODULATION OF MEDICATION'S EFFICACY

Medications may be effective in preclinical settings or in experiments in clinical laboratories, but can fail when the patients return from treatment centers into the community. This failure could be due to effects of a patients' social network overcoming the efficacy of the medication. On the other hand, a medication may be more effective in the presence of social support than without it. Historically, preclinical experiments of alcohol pharmacotherapies primarily involved studies in isolated animals and did not evaluate whether the medications are more or less effective in social settings. Moreover, as described in the introduction of this paper, although social environment plays a tremendous role in rats and mice, animals of these laboratory species do not form long-term individual-specific social attachments. Therefore, it is difficult to use these animals to model effects of social environment on alcohol and effects of social environment on effectiveness of medications.

In contrast, prairie voles can be used to model social influence of specific partners to both facilitate and inhibit drinking. Therefore, the prairie vole model allows comparison of efficacy of medications in isolation vs under the influence of a drinking conspecific. As a consequence, two types of screens of medications can be envisioned. First, the effectiveness of these medications can be assessed in animals introduced to the two-bottle choice alcohol drinking procedure as a pair. Under these conditions, the drinking in the pair is higher than in isolation (Anacker, Loftis, Kaur, et al., 2011; Hostetler et al., 2012). An ideal medication would be effective in both settings: in isolated and socially housed voles. On the other hand, if the medication is effective only in isolated animals, but not in pair-housed voles, this outcome would suggest that the medication is effective in typical preclinical housing conditions, but could lose its efficacy if used under social context in humans. Our laboratory is currently involved in testing potential medications in this paradigm.

Second, the efficacy of potential medications can be tested in animals that have been introduced to alcohol solution when they were isolated and later placed into pair housing in the presence of alcohol. In this case, voles initially characterized as high drinkers decrease alcohol drinking matching the intake of the low drinkers (Anacker, Loftis, & Ryabinin, 2011; Anacker & Ryabinin, 2013). Administration of the potential medication to high drinkers during pair housing could enhance the effectiveness of the presence of low drinkers to decrease alcohol intake in the high drinkers. Medications effective in these settings would be promising for administering to patients who receive social support in their attempt to decrease alcohol consumption.

7. MODELS OF SOCIAL CONSEQUENCES OF ALCOHOL ABUSE AND THEIR POTENTIAL TREATMENTS

It is typically thought that the main purpose of medications in the treatment of alcoholism is to decrease alcohol consumption or to treat alcohol-induced organ damage. However, alcoholism is a multifaceted disorder, accompanied not only by severe alcohol-induced health disturbances, but also by increased social problems. Thus, increased violence, for example, against an intimate partner, and marital disintegration are well-known consequences of alcohol abuse (Leonard & Eiden, 2007; Leonard & Rothbard, 1999; McLeod, 1993). It is worth considering whether medications can be developed to treat alcohol-induced social problems. Obviously pair bonding in prairie voles is an incomplete model of human marriage, but it does serve well to model social emotional attachments involved in romantic relationships and their biological substrates. Therefore, alcohol's effects on pair bonding in this socially monogamous species deserve further investigation.

Typically, cohabitation of a male and sexually receptive female prairie vole is sufficient to establish a pair bond, which is manifested in that each member of this newly established pair will prefer to spend time with its "partner" rather than with an unfamiliar "stranger" vole of the opposite sex. Such partner preference is not observed in nonmonogamous rodent species and serves as a laboratory proxy for pair bond formation (Ahern & Young, 2009; Carter, DeVries, & Getz, 1995; Lim et al., 2004; Modi & Young, 2012; Young, Liu, & Wang, 2008). In a recent series of studies, our laboratory tested effects of alcohol on formation of partner preference in male and female prairie voles (Anacker, Ahern, et al., 2014).

In brief, male–female pairs of prairie voles were placed for 24 h in cohabitation conditions during which they had access to 10% ethanol and water. Following the cohabitation, the animals were tested for their preference of their partner over a stranger vole. While male voles that were drinking water during cohabitation exhibited robust partner preference, this partner preference was absent in males that were drinking alcohol. Thus, alcohol blocked formation of partner preference, and hence pair bonding, in males. An opposite effect was observed in females. Sexually unreceptive female voles exhibited significant partner preference only if they had access to alcohol, but not if they consumed only water. Thus, alcohol facilitated pair bonding in female voles (Anacker, Ahern, et al., 2014). These effects of alcohol on partner preference were not mediated by potential effects of this drug on mating, aggression, locomotor activity or memory of the partner, indicating specificity of effects of alcohol on social bonding. Remarkably, the lack of effect on aggression differs from the way repeated injections of amphetamine disrupt pair bonding in voles (Gobrogge, Liu, Young, & Wang, 2009; Liu et al., 2011; Young, Liu et al., 2011), as administration of this drug does promote nonselective aggression in this species (Anacker, Ahern, et al., 2014).

In humans, there are situations where inhibition of bonding as well as facilitation of bonding can be harmful to the social wellbeing of an individual. For example, inhibition of bonding could disrupt a previously healthy relationship, while facilitation of bonding could lead to a continued relationship with an abusive person. Therefore, it is intriguing that the prairie vole animal model could for the first time provide a screen for potential approaches and medications to improve relationships affected by alcohol abuse. Such a screen is at this point impossible to develop in other animal models. The appropriateness of such an intervention would need to be evaluated on an individual basis and this direction developed with considerations of ethical implications. However, given the pervasiveness of alcohol's effects on social attachments, the possibility of modeling such intervention is important.

8. CONCLUSIONS AND FUTURE DIRECTIONS

To summarize, the prairie vole model of alcohol drinking provides the following advantages over other animal models for screening medications to treat alcoholism. First, it allows testing these medications in another rodent species besides standard laboratory rodents. Prairie voles belong to another subfamily of rodents from (Old World) mice and rats, and their genetic makeup is closer to lemmings, hamsters, and deer mice. Identifying a potential medication that works (or is lacking dangerous side effects) across mice, rats, and voles increases the likelihood that it will be applicable to humans. Second, it allows testing medications in a genetically diverse population of subjects. This advantage is based on the fact that prairie voles are not inbred or selectively bred rodents, in contrast to strains of mice and rats in which such medications are typically tested. Third, it allows testing medications which effectiveness could be resistant to social influences promoting excessive drinking. Such medications could be identified in models of social facilitation of drinking. Fourth, it allows testing medications which effectiveness could be dependent on the presence of social support. Such medications could be identified in the models of social inhibition of drinking. Finally, it potentially allows testing medications affecting harmful social consequences of excessive drinking. Such medications could be identified using the partner preference test.

Taken together, the prairie vole model is an excellent addition to the arsenal of screening tools for novel medications to treat alcoholism. One caveat worth keeping in mind is the underdevelopment of genetic tools in voles vs mice. While the prairie vole genome has been sequenced and annotated by the National Center for Biotechnology Information, it is clearly less studied than that of mice or humans. This is a serious drawback in light of the need to develop individualized medicines. Researchers using prairie voles are adapting transgenic and viral technologies keeping up with mouse genetics, which should in the future allow bridging this technical gap. With this view in mind, it is worth noting that because of the genetic diversity, polymorphisms that could affect effectiveness of medications are more likely to be encountered in voles than in the currently used mouse and rat strains.

Finally, alcohol is only one of many addictive drugs consumed by humans. As noted earlier, researchers began to adapt prairie voles for the studies of other addictive substances, such as psychostimulants and to a lesser degree, opiates. Repeated administration of amphetamine can lead to conditioned place preference (Aragona et al., 2007), a model of reinforcing properties of addictive drugs, in prairie voles. Intriguingly, amphetamine is more effective in establishing conditioned place preference in nonpair bonded than in pair-bonded male prairie voles, an effect involving adaptations in the dopamine system (Liu et al., 2011). Thus, it is possible to test whether any potential medications affect reinforcing properties of psychostimulants using the conditioned place preference procedure, and whether their effectiveness is sensitive to the social context. Moreover, our laboratory has also demonstrated that prairie voles voluntarily consume methamphetamine (Hostetler, Phillips, & Ryabinin, 2014), opening an avenue for studies on pharmacotherapy of psychostimulant abuse. As researchers using prairie voles adapt self-administration procedures to their studies, this animal model could be used to screen effective medications not only for the treatment of alcoholism, but also for other addictive substances.

REFERENCES

Addolorato, G., Capristo, E., Leggio, L., Ferrulli, A., Abenavoli, L., Malandrino, N., et al. (2006). Relationship between ghrelin levels, alcohol craving, and nutritional status in current alcoholic patients. *Alcoholism, Clinical and Experimental Research*, *30*(11), 1933–1937.

Ahern, T. H., & Young, L. J. (2009). The impact of early life family structure on adult social attachment, alloparental behavior, and the neuropeptide systems regulating affiliative behaviors in the monogamous prairie vole (microtus ochrogaster). *Frontiers in Behavioral Neuroscience*, *3*, 17.

Anacker, A. M., Ahern, T. H., Hostetler, C. M., Dufour, B. D., Smith, M. L., Cocking, D. L., et al. (2014). Drinking alcohol has sex-dependent effects on pair bond

formation in prairie voles. *Proceedings of the National Academy of Sciences of the United States of America, 111*(16), 6052–6057.
Anacker, A. M., Ahern, T. H., Young, L. J., & Ryabinin, A. E. (2012). The role of early life experience and species differences in alcohol intake in microtine rodents. *PLoS One, 7*(6), e39753.
Anacker, A. M., Loftis, J. M., Kaur, S., & Ryabinin, A. E. (2011a). Prairie voles as a novel model of socially facilitated excessive drinking. *Addiction Biology, 16*(1), 92–107.
Anacker, A. M., Loftis, J. M., & Ryabinin, A. E. (2011b). Alcohol intake in prairie voles is influenced by the drinking level of a peer. *Alcoholism, Clinical and Experimental Research, 35*(10), 1884–1890.
Anacker, A. M., & Ryabinin, A. E. (2010). Biological contribution to social influences on alcohol drinking: Evidence from animal models. *International Journal of Environmental Research and Public Health, 7*(2), 473–493.
Anacker, A. M., & Ryabinin, A. E. (2013). Identification of subpopulations of prairie voles differentially susceptible to peer influence to decrease high alcohol intake. *Frontiers in Pharmacology, 4*, 84.
Anacker, A. M., Smith, M. L., & Ryabinin, A. E. (2014). Establishment of stable dominance interactions in prairie vole peers: Relationships with alcohol drinking and activation of the paraventricular nucleus of the hypothalamus. *Social Neuroscience, 9*(5), 484–494.
Aragona, B. J., Detwiler, J. M., & Wang, Z. (2007). Amphetamine reward in the monogamous prairie vole. *Neuroscience Letters, 418*(2), 190–194.
Bartal, I. B.-A., Decety, J., & Mason, P. (2011). Empathy and pro-social behavior in rats. *Science, 334*(6061), 1427–1430.
Beery, A. K., & Zucker, I. (2010). Oxytocin and same-sex social behavior in female meadow voles. *Neuroscience, 169*(2), 665–673.
Bond, J., Kaskutas, L. A., & Weisner, C. (2003). The persistent influence of social networks and alcoholics anonymous on abstinence. *Journal of Studies on Alcohol, 64*(4), 579–588.
Burkett, J. P., & Young, L. J. (2012). The behavioral, anatomical and pharmacological parallels between social attachment, love and addiction. *Psychopharmacology, 224*(1), 1–26.
Calissendorff, J., Danielsson, O., Brismar, K., & Rojdmark, S. (2005). Inhibitory effect of alcohol on ghrelin secretion in normal man. *European Journal of Endocrinology, 152*(5), 743–747.
Carrigan, M. H., & Randall, C. L. (2003). Self-medication in social phobia: A review of the alcohol literature. *Addictive Behaviors, 28*(2), 269–284.
Carter, C. S. (1998). Neuroendocrine perspectives on social attachment and love. *Psychoneuroendocrinology, 23*(8), 779–818.
Carter, C. S., DeVries, A. C., & Getz, L. L. (1995). Physiological substrates of mammalian monogamy: The prairie vole model. *Neuroscience and Biobehavioral Reviews, 19*(2), 303–314.
Carter, C. S., & Getz, L. L. (1993). Monogamy and the prairie vole. *Scientific American, 268*(6), 100–106.
CDC. (2014). *Fact sheets: Alcohol use and your health.* Retrieved from, http://www.cdc.gov/alcohol/fact-sheets/alcohol-use.htm.
Charney, D. A., Zikos, E., & Gill, K. J. (2010). Early recovery from alcohol dependence: Factors that promote or impede abstinence. *Journal of Substance Abuse Treatment, 38*(1), 42–50.
Cruz, M. T., Herman, M. A., Cote, D. M., Ryabinin, A. E., & Roberto, M. (2013). Ghrelin increases GABAergic transmission and interacts with ethanol actions in the rat central nucleus of the amygdala. *Neuropsychopharmacology, 38*(2), 364–375.
Dhillo, W. S., & Bloom, S. R. (2001). Hypothalamic peptides as drug targets for obesity. *Current Opinion in Pharmacology, 1*(6), 651–655.

Donoghue, K., Elzerbi, C., Saunders, R., Whittington, C., Pilling, S., & Drummond, C. (2015). The efficacy of acamprosate and naltrexone in the treatment of alcohol dependence, Europe versus the rest of the world: A meta-analysis. *Addiction, 110*(6), 920–930.

Fisher, L. B., Miles, I. W., Austin, S. B., Camargo, C. A., Jr., & Colditz, G. A. (2007). Predictors of initiation of alcohol use among US adolescents: Findings from a prospective cohort study. *Archives of Pediatrics & Adolescent Medicine, 161*(10), 959–966.

Froehlich, J. C., & Li, T. K. (1993). Recent developments in alcoholism: Opioid peptides. *Recent Developments in Alcoholism, 11*, 187–205.

Garbutt, J. C., Greenblatt, A. M., West, S. L., Morgan, L. C., Kampov-Polevoy, A., Jordan, H. S., et al. (2014). Clinical and biological moderators of response to naltrexone in alcohol dependence: A systematic review of the evidence. *Addiction, 109*(8), 1274–1284.

Gobrogge, K. L., Liu, Y., Young, L. J., & Wang, Z. (2009). Anterior hypothalamic vasopressin regulates pair bonding and drug-induced aggression in a monogamous rodent. *Proceedings of the National Academy of Sciences of the United States of America, 106*(45), 19144–19149.

Gomez, J. L., Cunningham, C. L., Finn, D. A., Young, E. A., Helpenstell, L. K., Schuette, L. M., et al. (2015). Differential effects of ghrelin antagonists on alcohol drinking and reinforcement in mouse and rat models of alcohol dependence. *Neuropharmacology, 97*, 182–193.

Gomez, J. L., & Ryabinin, A. E. (2014). The effects of ghrelin antagonists [D-Lys(3)]-GHRP-6 or JMV2959 on ethanol, water, and food intake in C57BL/6 J mice. *Alcoholism, Clinical and Experimental Research, 38*(9), 2436–2444.

Gordon, A. J., & Zrull, M. (1991). Social networks and recovery: One year after inpatient treatment. *Journal of Substance Abuse Treatment, 8*(3), 143–152.

Grippo, A. J., Wu, K. D., Hassan, I., & Carter, C. S. (2008). Social isolation in prairie voles induces behaviors relevant to negative affect: Toward the development of a rodent model focused on co-occurring depression and anxiety. *Depression and Anxiety, 25*(6), E17–E26.

Hammock, E. A., & Young, L. J. (2005). Microsatellite instability generates diversity in brain and sociobehavioral traits. *Science, 308*, 1630–1634.

Heather, N., Rollnick, S., & Bell, A. (1993). Predictive validity of the readiness to change questionnaire. *Addiction, 88*(12), 1667–1677.

Homish, G. G., Leonard, K. E., Kozlowski, L. T., & Cornelius, J. R. (2009). The longitudinal association between multiple substance use discrepancies and marital satisfaction. *Addiction, 104*(7), 1201–1209.

Hostetler, C. M., Anacker, A. M., Loftis, J. M., & Ryabinin, A. E. (2012). Social housing and alcohol drinking in male-female pairs of prairie voles (Microtus ochrogaster). *Psychopharmacology, 224*(1), 121–132.

Hostetler, C. M., Phillips, T. J., & Ryabinin, A. E. (2014). Opposing effects of social housing on alcohol and methamphetamine intake in the prairie vole. *Alcoholism, Clinical and Experimental Research, 38*(S1), 169A.

Hostetler, C. M., & Ryabinin, A. E. (2012). Love and addiction: The devil is in the differences: A commentary on "the behavioral, anatomical and pharmacological parallels between social attachment, love and addiction". *Psychopharmacology, 224*(1), 27–29.

Hostetler, C. M., & Ryabinin, A. E. (2014). Social partners prevent alcohol relapse behavior in prairie voles. *Psychoneuroendocrinology, 39*, 152–157.

Insel, T. R. (2010). The challenge of translation in social neuroscience: A review of oxytocin, vasopressin, and affiliative behavior. *Neuron, 65*(6), 768–779.

Jerlhag, E., Egecioglu, E., Dickson, S. L., Douhan, A., Svensson, L., & Engel, J. A. (2007). Ghrelin administration into tegmental areas stimulates locomotor activity and increases extracellular concentration of dopamine in the nucleus accumbens. *Addiction Biology, 12*(1), 6–16.

Jerlhag, E., Egecioglu, E., Landgren, S., Salome, N., Heilig, M., Moechars, D., et al. (2009). Requirement of central ghrelin signaling for alcohol reward. *Proceedings of the National Academy of Sciences of the United States of America*, *106*(27), 11318–11323.

Jerlhag, E., Janson, A. C., Waters, S., & Engel, J. A. (2012). Concomitant release of ventral tegmental acetylcholine and accumbal dopamine by ghrelin in rats. *PLoS One*, 7(11), e49557.

Justus, A. N., Finn, P. R., & Steinmetz, J. E. (2000). The influence of traits of disinhibition on the association between alcohol use and risky sexual behavior. *Alcoholism, Clinical and Experimental Research*, *24*(7), 1028–1035.

Kaur, S., & Ryabinin, A. E. (2010). Ghrelin receptor antagonism decreases alcohol consumption and activation of perioculomotor urocortin-containing neurons. *Alcoholism, Clinical and Experimental Research*, *34*(9), 1525–1534.

Kelly, J. F., Stout, R. L., Magill, M., & Tonigan, J. S. (2011). The role of alcoholics anonymous in mobilizing adaptive social network changes: A prospective lagged mediational analysis. *Drug and Alcohol Dependence*, *114*(2–3), 119–126.

Knafo, A., Israel, S., Darvasi, A., Bachner-Melman, R., Uzefovsky, F., Cohen, L., et al. (2008). Individual differences in allocation of funds in the dictator game associated with length of the arginine vasopressin 1a receptor RS3 promoter region and correlation between RS3 length and hippocampal mRNA. *Genes, Brain, and Behavior*, 7(3), 266–275.

Knapska, E., Mikosz, M., Werka, T., & Maren, S. (2010). Social modulation of learning in rats. *Learning & Memory*, *17*(1), 35–42.

Langford, D. J., Crager, S. E., Shehzad, Z., Smith, S. B., Sotocinal, S. G., Levenstadt, J. S., et al. (2006). Social modulation of pain as evidence for empathy in mice. *Science*, *312*(5782), 1967–1970.

Leggio, L., Ferrulli, A., Cardone, S., Nesci, A., Miceli, A., Malandrino, N., et al. (2012). Ghrelin system in alcohol-dependent subjects: Role of plasma ghrelin levels in alcohol drinking and craving. *Addiction Biology*, *17*(2), 452–464.

Leonard, K. E. (2005). Alcohol and intimate partner violence: When can we say that heavy drinking is a contributing cause of violence? *Addiction*, *100*(4), 422–425.

Leonard, K. E., & Eiden, R. D. (2007). Marital and family processes in the context of alcohol use and alcohol disorders. *Annual Review of Clinical Psychology*, *3*, 285–310.

Leonard, K. E., & Rothbard, J. C. (1999). Alcohol and the marriage effect. *Journal of Studies on Alcohol. Supplement*, *13*, 139–146.

Lim, M. M., Wang, Z., Olazabal, D. E., Ren, X., Terwilliger, E. F., & Young, L. J. (2004). Enhanced partner preference in a promiscuous species by manipulating the expression of a single gene. *Nature*, *429*(6993), 754–757.

Liu, Y., Aragona, B. J., Young, K. A., Dietz, D. M., Kabbaj, M., Mazei-Robison, M., et al. (2010). Nucleus accumbens dopamine mediates amphetamine-induced impairment of social bonding in a monogamous rodent species. *Proceedings of the National Academy of Sciences of the United States of America*, *107*(3), 1217–1222.

Liu, Y., Young, K. A., Curtis, J. T., Aragona, B. J., & Wang, Z. (2011). Social bonding decreases the rewarding properties of amphetamine through a dopamine D1 receptor-mediated mechanism. *The Journal of Neuroscience*, *31*(22), 7960–7966.

McBride, W. J., & Li, T. K. (1998). Animal models of alcoholism: Neurobiology of high alcohol-drinking behavior in rodents. *Critical Reviews in Neurobiology*, *12*(4), 339–369.

McCrady, B. S., Zucker, R. A., Molina, B. S., Ammon, L., Ames, G. M., & Longabaugh, R. (2006). Social environmental influences on the development and resolution of alcohol problems. *Alcoholism, Clinical and Experimental Research*, *30*(4), 688–699.

McLeod, J. D. (1993). Spouse concordance for alcohol dependence and heavy drinking: Evidence from a community sample. *Alcoholism, Clinical and Experimental Research*, *17*(6), 1146–1155.

Merrill, J. E., Wardell, J. D., & Read, J. P. (2015). Is readiness to change drinking related to reductions in alcohol use and consequences? A week-to-week analysis. *Journal of Studies on Alcohol and Drugs*, *76*(5), 790–798.

Meyer-Lindenberg, A., Domes, G., Kirsch, P., & Heinrichs, M. (2011). Oxytocin and vasopressin in the human brain: Social neuropeptides for translational medicine. *Nature Reviews. Neuroscience*, *12*(9), 524–538.

Modi, M. E., & Young, L. J. (2012). The oxytocin system in drug discovery for autism: Animal models and novel therapeutic strategies. *Hormones and Behavior*, *61*(3), 340–350.

Muccioli, G., Tschop, M., Papotti, M., Deghenghi, R., Heiman, M., & Ghigo, E. (2002). Neuroendocrine and peripheral activities of ghrelin: Implications in metabolism and obesity. *European Journal of Pharmacology*, *440*(2-3), 235–254.

Panksepp, J. B., & Lahvis, G. P. (2011). Rodent empathy and affective neuroscience. *Neuroscience and Biobehavioral Reviews*, *35*(9), 1864–1875.

Park, A., Sher, K. J., & Krull, J. L. (2008). Risky drinking in college changes as fraternity/sorority affiliation changes: A person-environment perspective. *Psychology of Addictive Behaviors*, *22*(2), 219–229.

Rollnick, S., Heather, N., Gold, R., & Hall, W. (1992). Development of a short 'readiness to change' questionnaire for use in brief, opportunistic interventions among excessive drinkers. *British Journal of Addiction*, *87*(5), 743–754.

Sayette, M. A., Creswell, K. G., Dimoff, J. D., Fairbairn, C. E., Cohn, J. F., Heckman, B. W., et al. (2012). Alcohol and group formation: A multimodal investigation of the effects of alcohol on emotion and social bonding. *Psychological Science*, *23*(8), 869–878.

Sinclair, J. D., & Senter, R. J. (1968). Development of an alcohol-deprivation effect in rats. *Quarterly Journal of Studies on Alcohol*, *29*(4), 863–867.

Steele, C. M., & Josephs, R. A. (1990). Alcohol myopia. Its prized and dangerous effects. *The American Psychologist*, *45*(8), 921–933.

Stevenson, J. R., Buirkle, J. M., Buckley, L. E., Young, K. A., Albertini, K. M., & Bohidar, A. E. (2015). GHS-R1A antagonism reduces alcohol but not sucrose preference in prairie voles. *Physiology & Behavior*, *147*, 23–29.

Suchankova, P., Steensland, P., Fredriksson, I., Engel, J. A., & Jerlhag, E. (2013). Ghrelin receptor (GHS-R1A) antagonism suppresses both alcohol consumption and the alcohol deprivation effect in rats following long-term voluntary alcohol consumption. *PLoS One*, *8*(8), e71284.

Tamarin, R. (1985). *Biology of New World Microtus (Special Publication No. 8 ed.).* Shippensburg, PA: American Society of Mammalogists.

Tomie, A., Samuel, A. G., Sprung, D. M., Malul, Y., & Yu, L. (2015). Effects of number of cagemates on home cage ethanol drinking during proximal cagemate drinking (PCD) procedures in male and female CD-1 mice. *Progress in Neuro-Psychopharmacology & Biological Psychiatry*, *57*, 1–10.

Vengeliene, V., Bilbao, A., & Spanagel, R. (2014). The alcohol deprivation effect model for studying relapse behavior: A comparison between rats and mice. *Alcohol*, *48*(3), 313–320.

Volpicelli, J. R., Alterman, A. I., Hayashida, M., & O'Brien, C. P. (1992). Naltrexone in the treatment of alcohol dependence. *Archives of General Psychiatry*, *49*(11), 876–880.

Walum, H., et al. (2008). Genetic variation in the vasopressin receptor 1a gene (AVPR1A) associates with pair bonding behavior in humans. *Proceedings of the National Academy of Sciences of the United States of America*, *105*(37), 14153–14156.

Wu, J., & Witkiewitz, K. (2008). Network support for drinking: An application of multiple groups growth mixture modeling to examine client-treatment matching. *Journal of Studies on Alcohol and Drugs*, *69*(1), 21–29.

Young, K. A., Gobrogge, K. L., Liu, Y., & Wang, Z. (2011). The neurobiology of pair bonding: Insights from a socially monogamous rodent. *Frontiers in Neuroendocrinology*, *32*(1), 53–69.

Young, K. A., Liu, Y., Gobrogge, K. L., Dietz, D. M., Wang, H., & Kabbaj, M. (2011). Amphetamine alters behavior and mesocorticolimbic dopamine receptor expression in the monogamous female prairie vole. *Brain Research*, *1367*, 213–222.

Young, K. A., Liu, Y., & Wang, Z. (2008). The neurobiology of social attachment: A comparative approach to behavioral, neuroanatomical, and neurochemical studies. *Comparative Biochemistry and Physiology Part C: Toxicology & Pharmacology*, *148*(4), 401–410.

CHAPTER THIRTEEN

Animal Models for Medication Development and Application to Treat Fetal Alcohol Effects

S. Barron*^,1^, A. Hawkey*, L. Fields*, J.M. Littleton*^,†^

*University of Kentucky, Lexington, KY, United States
†Naprogenix, Inc., Lexington, KY, United States
[1]Corresponding author: e-mail address: sbarron@uky.edu

Contents

Abstract

Ethanol consumption during pregnancy can have lifelong consequences for the offspring, their family and society. Fetal alcohol spectrum disorders (FASD) include a range of physical and behavioral effects with the most significant impact occurring as a result of the effects of ethanol on the developing central nervous system (CNS). To date, there are no FDA approved drugs that have been tested that prevent/reduce or specifically treat the symptoms of FASD. There are several promising lines of research from rodent models aimed at reducing the neurotoxic effects of ethanol on the developing CNS or in treating the resulting behavioral impairments but these have not yet moved to clinical testing. The current review discusses some of the most promising targets for intervention and provides a review of the past and ongoing efforts to develop and screen pharmacological treatments for reducing the effects of prenatal ethanol exposure.

International Review of Neurobiology, Volume 126
ISSN 0074-7742
http://dx.doi.org/10.1016/bs.irn.2016.02.002

1. INTRODUCTION: FAS AND FASD, INCIDENCE AND COSTS, ETC.

Throughout history anecdotal reports have suggested that ethanol consumption during pregnancy could have adverse effects for the developing offspring (see Rosett, 1984); however, it was not until 1973 in the United States (and 1968 in France) that a syndrome with specific physical characteristics was described in the literature and has since been termed fetal alcohol syndrome (FAS) (Jones, Smith, Ulleland, & Streissguth, 1973; Lemoine, Harousseau, Borteyru, & Menuet, 1968). FAS is characterized by prenatal and/or postnatal growth retardation, evidence of central nervous system (CNS) abnormalities (including microcephaly, mental retardation, attentional deficit hyperactivity disorder (ADHD), or other behavioral problems), and characteristic craniofacial anomalies, see http://www.cdc.gov/ncbddd/fasd/diagnosis.html. The incidence of FAS in the western world has been reported to be from 1 to 3 per 1000 live births (Stratton, Howe, & Battaglia, 1996) although there are subpopulations with a much higher incidence. FAS is only the "tip of the iceberg" with a far greater number of children affected by prenatal ethanol exposure but not showing all of the required criteria for a diagnosis of FAS. These children exhibit a subset of behavioral and/or physical effects referred to as fetal alcohol spectrum disorders (FASD). FASD is not a diagnosis but rather an umbrella term that incorporates a range of effects caused by prenatal ethanol exposure and includes alcohol-related neurodevelopmental disorders and/or alcohol-related birth defects (see Doyle & Mattson, 2015). The incidence of FASD is much higher than FAS, with reports of 3–5 per 100 live births (May et al., 2009), however, even these estimates may be low. In a recent study assessing a large clinical sample, 86.5% of adolescent/teens with an FASD were either never diagnosed or misdiagnosed (Chasnoff, Wells, & King, 2015).

Estimated costs of FASDs in any country are extremely difficult to calculate. Included are various direct health care costs such as screening, diagnosis, and specialized treatments that address motor, social, and/or psychiatric issues. However, children with FAS and FASD are also at greater risk for a variety of secondary disabilities including issues with the law, financial problems, substance abuse, and difficulty maintaining employment (Moore & Riley, 2015; Streissguth et al., 2004). In one of the more detailed efforts, Popova and colleagues through the Centre for Addiction and Mental Health have included many of the cost variables associated with FASD in Canada. In a country with a population of approximately 34.5 million people, their cost estimates for FASD were approximately 1.3 billion Canadian dollars in 2013 (Popova, Lange, Burd, & Rehm, 2016). With the 2013 US population estimate of more than 315 million and with an assumption that the incidence rates of individuals with an FASD and costs being somewhat similar between the United States and Canada, an extrapolation estimate would be between 8 and 9 billion US dollars annually (using current exchange rates). These numbers provide strong support for the pressing need to improve pharmacotherapies to help individuals with an FASD and, clearly, interventions that could improve outcome for individuals with an FASD would have individual, familial, and societal cost benefits.

2. APPROACHES FOR INTERVENTION

There are at least two types of intervention strategies when considering medication development to treat the effects of prenatal ethanol exposure. First, interventions could take place during the perinatal period—during pregnancy, parturition, and/or shortly after birth. This approach would target the immediate damaging effects of ethanol on the CNS and the novel medication could reduce or block ethanol's effects—thus producing neuroprotection. Alternatively, the focus could be on developing novel therapeutics for treating the symptoms displayed by an individual with FASD. The primary focus in this review will be on the first approach with a briefer discussion of the second. This is due, in part, because more work has been done on the former than the latter but both approaches are still rather limited.

Currently, there are no effective clinically tested and approved pharmacotherapies specifically for the treatment of damage caused by ethanol exposure during early development or as a later intervention as the child matures. To date, the most successful clinical approaches for early intervention include providing necessary developmental interventions (physical or other types of therapy), a well-structured, supportive and stable postnatal environment, and an understanding by family, teachers, etc. of the special needs of the child with FAS/FASD (Pruett, Waterman, & Caughey, 2013).

3. RODENT MODELS USED IN MEDICATION DEVELOPMENT FOR REDUCING FASD AND FASD SYMPTOMS

Approaches to develop novel therapeutics for FAS/FASD often start with a variety of in vitro screens followed by in vivo animal models typically using either mice or rats. Rodent models of prenatal ethanol exposure emerged very quickly after the first clinical publication describing FAS (Abel, 1974), and the rodent studies of fetal alcohol effects (FAE) have shown remarkable face validity in terms of the CNS regions that are particularly sensitive to prenatal ethanol exposure and in modeling many of the behaviors associated with FASDs. With behavioral similarities including hyperactivity, attentional problems, deficits in learning and memory, balance, and problems with tasks that assess executive function (Driscoll, Streissguth, & Riley, 1990; Goodlett, Horn, & Zhou, 2005; Patten, Fontaine, & Christie, 2014), the overlap in symptomology observed in the rodent models, and the clinical population may help significantly in medication development approaches.

Various exposure paradigms have been used with rodent models of FAE and these vary in terms of a number of factors including developmental age when exposure occurs and the route and dose of administration. These variables are important factors to consider when selecting the appropriate model. The exposure window in which ethanol is administered depends, in part, on the underlying mechanism or mechanisms of the novel agent being tested. While the sequence of CNS development across species shares numerous similarities, the time of parturition, relative to CNS development differs. In humans, the third trimester is associated with a period of rapid neuronal growth and proliferation, often referred to as the "brain growth spurt" (Bayer, Altman, Russo, & Zhang, 1993; Dobbing & Sands, 1979). In rodents, this "brain growth spurt" occurs during the first postnatal weeks after birth. Therefore, ethanol administration to pregnant rats or mice serves as a model for "first and second" trimester exposure while ethanol treatment during early neonatal life represents an exposure model that overlaps the human third trimester of pregnancy. There are also a limited number of rodent studies in which ethanol is administered both pre- and neonatally as a model of ethanol exposure during all three trimesters (eg, Tran, Cronise, Marino, Jenkins, & Kelly, 2000). Not surprisingly, the timing of ethanol exposure can result in different types of damage. For example, early ethanol exposure during gastrulation in mice can produce global impairments in brain growth (eg, Sulik, 2005), while later exposure can have more selective damage on particular CNS structures (O'Leary-Moore, Parnell, Lipinski, & Sulik, 2011). Neonatal ethanol exposure appears to have more selective effects on hippocampal, cerebellar, and prefrontal cortex (PFC) function—all regions that continue to undergo growth and development during the neonatal period (eg, Gil-Mohapel, Boehme, Kainer, & Christie, 2010; Luo, 2012).

In addition to the timing of exposure, the route of administration requires careful consideration. The blood ethanol curve and peak blood ethanol concentration (BEC) are very different if ethanol is consumed voluntarily (by the pregnant dam) or if ethanol is administered in a single bolus by oral intubation. Data from rodent models have shown that peak BEC appears more damaging than total daily ethanol exposure (Maier & West, 2001) and so binge-type exposure as produced by a bolus oral intubation of ethanol produces higher peak BECs and produces more robust developmental deficits than a more continuous exposure with a higher daily dose but lower peak BECs (eg, Bonthius, Goodlett, & West, 1988). Voluntary self-administration paradigms have value for strong face validity, although few rodent strains other than selectively bred or C57Bl/6J mice will voluntarily consume ethanol to reach similar levels of BECs as that achieved by human alcoholics.

4. TARGETS FOR PRENATAL/PERINATAL PHARMACOLOGICAL INTERVENTIONS FOR FAE

Ethanol withdrawal (WD) has been shown to play an important role in ethanol's effects on the developing brain. Both in vitro and in vivo studies have shown the consequences of ethanol WD (Barron, Mulholland, Littleton, & Prendergast, 2008; Hoffman & Tabakoff, 1994; Idrus, McGough, Riley, & Thomas, 2014; Thomas & Riley, 1998) and so one opportunity for pharmacological intervention is to reduce the consequences during ethanol WD. The use of such an agent is particularly appealing because it could be potentially a one-time treatment that would protect the fetus or newborn during ethanol WD (at the time of birth). This type of treatment would be far more likely to be used successfully than a drug that would have to be given throughout pregnancy.

4.1 Reducing Glutamate Activity

Ethanol-induced changes in glutamatergic (GLU) activity and function play a significant role in the effects of ethanol and ethanol WD on the developing rodent brain (see Hoffman & Tabakoff, 1994; Olney, Wozniak, Jevtovic-Todorovic, & Ikonomidou, 2001). Some early studies implicating glutamate-induced excitotoxicity in the effects of ethanol WD on the developing CNS stems from the work by Thomas and colleagues who showed that the *N*-methyl-D-aspartate (NMDA) GLU receptor (R) antagonist MK-801 (dizocilpine) administered at specific times during ethanol WD reduced some of the damaging effects of ethanol in both in vitro and in vivo models (Thomas, Fleming, & Riley, 2001, 2002; Thomas, Weinert, Sharif, & Riley, 1997). While these studies were extremely useful in establishing the role of NMDAr activity as a mechanism contributing to ethanol's effects on the developing CNS, MK-801 has only limited usefulness as a pharmacotherapy due to its abuse potential (Layer, Kaddis, & Wallace, 1993), and potential neurotoxicity by increasing apoptosis (Bender, de Olmos, Bueno, de Olmos, & Lorenzo, 2010; Ikonomidou et al., 1999).

There are alternative ways to reduce GLU-induced neurotoxicity during ethanol WD that may have more promise as an intervention. For a number of years we have been interested in modulating polyamines as a way to reduce ethanol's effects on the developing brain. Polyamines, in particular, spermine and spermidine, are small cationic compounds derived from the amino acid arginine. Polyamines have affinity for many neurotransmitter receptor subtypes and ion channels (Williams, 1997) including the modulation of NMDAr function which is the primary reason for our interest. Polyamines are essential for a variety of functions in the developing organism. Polyamine levels are high in the developing brain and play an important role in cell proliferation, growth, and differentiation (Slotkin & Bartolome, 1986; Slotkin, Ferguson, Cada, McCook, & Seidler, 2000). Certain NMDAr subtypes such as the NR2B are particularly sensitive to polyamine potentiation and are also in high density in the developing brain (Williams, Zappia, Pritchett, Shen, & Molinoff, 1994).

While polyamines are important for CNS development, they also have a negative role in the neurotoxicity associated with ethanol WD. Polyamine levels are elevated during ethanol WD in various CNS regions including the hippocampus, cerebellum, and cortex (Davidson & Wilce, 1998; Gibson et al., 2003), and this increased polyamine activity has been positively correlated with ethanol WD severity symptoms (Davidson & Wilce, 1998; Davidson, Wilce, & Shanley, 1993). Drugs that reduce polyamine activity by blocking the polyamine site on the NMDAr, or block polyamine synthesis reduce ethanol WD symptoms in vivo and ethanol-induced neurotoxicity in vitro using organotypic hippocampal slice cultures (OHSCs). These studies using OHSCs use hippocampal slices derived from neonatal pups; thus, these studies have direct relevance for screening agents with the potential of reducing ethanol's effects on the developing brain.

We and others have shown that reducing the impact of polyamines during ethanol WD improves outcome following neonatal ethanol exposure in both in vitro and in vivo rodent models. Antagonists that are relatively selective for the NMDA NR2Br (the polyamine sensitive NMDAr) including agmatine or CP101,606 (traxoprodil) can reduce ethanol-related behavioral deficits and OHSC ethanol WD-induced neurotoxicity (Lewis, Wellmann, & Barron, 2007; Lewis et al., 2012; Wellmann, Lewis, & Barron, 2010) in rodents following neonatal ethanol exposure. A single treatment with difluromethylornithine (DFMO), which blocks the rate limiting step in the synthesis of polyamines (blocks ornithine decarboxylase), produced long-term improvement in some of the behavioral deficits typically associated with "third trimester" ethanol exposure and also reduced ethanol WD-induced neurotoxicity in OHSC (Gibson et al., 2003; Rubin et al., 2009). However, DFMO is an unlikely candidate as a pharmacotherapy because blocking the synthesis of polyamines has its own potential for neurotoxicity (Schweitzer, Robbins, & Slotkin, 1989; Slotkin & Bartolome, 1986). The NMDAr antagonists are more interesting from a clinical perspective because agmatine and traxoprodil have been tested clinically for a variety of other conditions including neuropathic pain, medication resistant depression, and traumatic brain injury. Although the clinical findings are mixed, their safety in human trials has already been established (Hashimoto, Malchow, Falkai, & Schmitt, 2013; Keynan, Mirovsky, Dekel, Gilad, & Gilad, 2010; Moretti et al., 2014; Preskorn et al., 2008). There is a potential issue with these NMDAr antagonists, however. In addition to their primary action on the NMDAr, they also have effects on other systems in the CNS which could be problematic for the developing brain.

An alternative approach to blocking specific polyamine sensitive NMDAr or polyamine synthesis would be to develop novel drugs that act as inhibitory modulators of the polyamine site on NMDArs. These agents allow for normal GLU activity but do not allow for overexcitation of NMDArs mediated by excess polyamine release that occurs during ethanol WD (Gibson et al., 2003; Littleton et al., 2001). As part of an NIAAA-funded medication development project, agmatine was used as the lead compound, and thousands of novel aryliminoguanidines were synthesized and screened in a molecular screen for their ability to reduce spermidine potentiation of [3H]MK-801 binding (without affecting binding in the absence of polyamines). One of the most potent compounds, JR 220, has been shown to be effective in reducing ethanol WD-induced neurotoxicity in OHSC from neonatal rats and improving behavioral outcomes in our third trimester ethanol exposure model (Barron et al., 2012). Further work is planned or underway to assess the value of JR 220 in improving cognitive function and more complex function in our rodent models of FAE or hypoxia.

4.2 Reducing the Effects of Reactive Oxygen Species

Data from a number of studies have suggested that prenatal ethanol exposure causes oxidative stress in the developing brain (see Brocardo, Gil-Mohapel, & Christie, 2011; Heaton, Paiva, Madorsky, & Shaw, 2003). Depending on the agent, the timing of administration and other parameters there are data suggesting that antioxidants can reduce some of the damaging effects of prenatal ethanol exposure (Joya, Garcia-Algar, Salat-Batlle, Pujades, & Vall, 2015). Supplements like ascorbic acid (vitamin C) or vitamin C in combination with vitamin E can reduce some of the effects of prenatal ethanol exposure in a number of in vivo and in vitro models (Nash, Ibram, Dringenberg, Reynolds, & Brien, 2007; Peng et al., 2005), although there are discrepancies in the literature (see Joya et al., 2015 for review). Another natural antioxidant found in blackberries, cyanidin-3-glucoside (C3G), reduced ethanol-induced activation of caspase-3, suggesting a reduction in ethanol-induced apoptosis (Ke et al., 2011), although it is not yet known whether C3G can ameliorate behavioral effects of fetal ethanol exposure.

4.3 Treatment with Choline or Cholinergic Agonists

There has also been considerable interest in manipulating cholinergic function to reduce the effects of prenatal ethanol exposure. Choline is the precursor for the synthesis of acetylcholine (ACh) and acts as an essential amino acid that plays a variety of important roles in CNS development. Some of the earliest attempts to reduce the behavioral effects of prenatal ethanol exposure in rodents focused on manipulating cholinergic activity, in part, because many of the behavioral deficits observed in the early rodent models could be explained as response inhibition deficits and appeared similar to those observed in rats treated with drugs that block ACh function. Interest in the role of ACh was increased after a series of studies showed improvement in ethanol exposed rats on measures of learning and memory when choline was added as a neonatal supplement following either prenatal or neonatal ethanol exposure (Thomas, Garrison, & O'Neill, 2004; Thomas, La Fiette, Quinn, & Riley, 2000). Since choline has a greater range of effects beyond ACh synthesis and does not easily cross the blood–brain barrier, other cholinergic agonists such as nicotine or drugs working on specific nicotinic receptor subtypes have also been studied to a limited extent. α7 nAChR agonists or partial agonists are particularly interesting because they also are being tested for their potential for neuroprotection and/or improvement from a variety of other conditions including Alzheimer's disease (Godyń, Jończyk, Panek, & Malawska, 2016) and cognitive aspects of schizophrenia (Lieberman et al., 2013). Using the in vitro OHSC model as a preliminary screen for the developing brain, chronic nicotine treatment reduces ethanol WD-induced neurotoxicity in OHSC (Prendergast, Harris, Mayer, & Littleton, 2000) and compounds that act more specifically on the α7 nicotinic receptor (nAChR) such as DMXB are neuroprotective during ethanol WD in fetal hippocampal slices, in primary newborn rat hippocampal neuronal cultures and in our in vivo "third trimester" rodent model (de Fiebre & de Fiebre, 2003; Fields et al., 2014–2015; Li et al., 2002). Further work examining potential α7 nAChR agents is underway. Naprogenix Inc., here at the University of Kentucky, has identified novel α7 nAChR products through a differential pharmacological high throughput screen applied to a plant extract library of approximately 1000 native KY species (Littleton, Rogers, & Falcone, 2005). *Solidago nemoralis* (gray goldenrod) was one plant identified with α7 selectivity and an additional screen of a pure flavonoid library identified a number of flavonoids including the dietary flavonoid rhamnetin which has selectivity for the α7 nAChR. Rhamnetin is neuroprotective during ethanol WD in OHSC (Lutz, Carter, Fields, Barron, & Littleton, 2015), and research is planned and/or ongoing with both of these natural products to assess their potential in reducing FAE behavioral effects.

Choline has also been administered at various times after ethanol exposure to try to reduce some of the long-term consequences of FAE with some promising results. Chronic choline administration during the first weeks after either prenatal or neonatal ethanol exposure resulted in significant improvement in behavioral outcome of rodents (Monk, Leslie, & Thomas, 2012; Thomas et al., 2004). Due to these preclinical data, a recent double-blind placebo-controlled pilot study was conducted to assess both safety and efficacy of daily choline supplements in children with FASD. These children were given choline daily for 9 months with minimal side effects reported. There was also some suggestion of age-dependent improvement in cognitive function in the choline-treated FASD offspring so further clinical studies are warranted (Wozniak et al., 2015).

5. TREATING LATER SYMPTOMS

The pharmacotherapeutic approach for treating FAS/FASD symptomology relies on prescribing drugs to manage these symptoms (see Koren, 2015). For the most part, these drugs have not been tested specifically for their effectiveness in FAS/FASD children/adolescents/adults nor do they necessarily target the CNS regions implicated following fetal ethanol exposure. So, for example, drugs prescribed for treating attentional/hyperactivity issues with FASD children are the same as those used for ADHD. While an ADHD diagnosis is often comorbid with FASD, it is unlikely that the underlying etiology is the same. Another set of symptoms associated with FASD as individuals mature includes externalizing and disruptive behavior that is often diagnosed as oppositional defiant/conduct disorder (ODD/CD) although again, the underlying etiology is unlikely to be the same for someone with FASD and someone with ODD/CD from unspecified causes. Other behavioral issues and drugs prescribed for FAS/FASD include anxiety (mood-stabilizers) and depression (eg, SSRI's) all with mixed results (Coe, Sidders, Riley, Waltermire, & Hagerman, 2001; Doig, McLennan, & Gibbard, 2008; O'Connor et al., 2002; Pruett et al., 2013). Since the underlying causes for these symptoms are due to fetal ethanol exposure, drugs specifically for treating FASD might have better results than the standard drugs currently used but very little research has been done in this area.

There is one area that has received limited study assessing pharmacotherapy following fetal ethanol exposure and this focuses on treating symptoms of ADHD/attentional issues. Since the comorbidity of FAS/FASD with ADHD-like symptoms is so high (Burd, Klug, Martsolf, & Kerbeshian, 2003), a limited number of clinical and preclinical studies have been conducted to assess the effectiveness of standard treatments for ADHD in offspring with a history of prenatal ethanol exposure. Preclinical studies suggest that FAE exposed rats are hyperreactive to methylphenidate (Means, Medlin, Hughes, & Gray, 1984; Ulug & Riley, 1983), and a number of clinical studies have also suggested that methylphenidate may also be contraindicated for children with FAS/FASD. One very small clinical study ($n=4$ children with FAS) suggests methylphenidate was effective for treating hyperactivity but not attentional issues (Oesterheld et al., 1998), and other studies suggest a poorer or more varied response to methylphenidate in FAS/FASD offspring (O'Malley & Hagerman, 1998; also

O'Malley & Nanson, 2002; Snyder, Nanson, Snyder, & Block, 1997). An additional argument against the use of psychostimulants for FASD populations stems from data showing that adolescents with FASD are at greater risk for later drug and alcohol-related problems (Streissguth et al., 2004; Yates, Cadoret, Troughton, Stewart, & Giunta, 1998).

We have recently looked at lobeline as a potential novel therapeutic for improving outcome and performance following neonatal ethanol exposure in our rodent model. Lobeline is a major alkaloid of the *Lobelia inflata* (Indian tobacco) plant. Lobeline has a variety of pharmacological actions including acting on presynaptic proteins in dopaminergic terminals including the dopamine transporter (DAT) and the vesicular monoamine transporter-2 (VMAT2), which regulate intracellular and extracellular DA concentrations (Santha et al., 2000) as well as noradrenergic function (Dwoskin & Crooks, 2002). Lobeline also has mixed actions on $\alpha 4\beta 2$ and $\alpha 7$ nicotinic ACh receptors (Briggs & McKenna, 1998; Kaniaková, Lindovský, Krůšek, Adámek, & Vyskočil, 2011). In rodent studies, lobeline reduces voluntary ethanol consumption (Bell, Eiler, Cook, & Rahman, 2009; Farook, Lewis, Gaddis, Littleton, & Barron, 2009), psychostimulant-induced hyperactivity (Miller et al., 2003; Polston, Cunningham, Rodvelt, & Miller, 2006), depressive-like behavior (Roni & Rahman, 2013), and protects against MPTP-induced neurotoxicity (Li, Zhao, Shi, & Zhang, 2014). Importantly, lobeline does not produce hyperactivity, nor is it self-administered by rats, suggesting a low abuse potential relative to the psychostimulants used to treat ADHD (Dwoskin & Crooks, 2002; Harrod, Dwoskin, Green, Gehrke, & Bardo, 2003). A "proof of concept" pilot clinical study assessing lobeline in an adult ADHD population has been published and their findings suggested minor side effects (nausea) with modest improvements in working memory but little improvement in attention. It should be noted that this adult ADHD population was heterogenous with no attempt to identify individuals that might actually have FASD (Martin et al., 2013).

We have reported that lobeline administered 30 min prior to testing reduced hyperactivity following neonatal ethanol exposure and normalized DAT function in the PFC using our third trimester ethanol exposure rodent model, with no effect on controls (Smith et al., 2012). Further study with rodent models to assess its potential for attention and cognitive function is clearly warranted; and, since this alkaloid has been tested for a variety of human conditions, the possibility of moving forward in clinical trials as a potential treatment for FAS/FASD has promise.

6. SUMMARY AND FUTURE DIRECTIONS

In this review, we have discussed research focusing on a number of potential mechanisms for the effects of prenatal ethanol exposure and pharmacological attempts to reduce these damaging effects using either established drugs or describing screens for novel therapeutics with some promising preliminary results. This field of research is still at a very early stage and more work is needed to assess standard and/or novel treatments for their efficacy in FAS/FASD populations. A recent meta-analysis reported 428 comorbid conditions observed in FASD populations (Popova, Lange, Shield, et al., 2016) yet we currently do not know the best approach for treating any of these for individuals with FASD.

There are a number of issues that need to be addressed in order to help this field move forward. First, it would be very beneficial to have criterion for FASD and the Center for Disease Control and Prevention and FAE researchers are currently working on guidelines for this (Doyle & Mattson, 2015; Hoyme et al., 2005). FASD is not a diagnosis but rather an umbrella term used to characterize the consequences of fetal ethanol exposure. Still, a better way to identify these individuals is really important. Related to this issue, we need to improve our ability to differentiate ADHD of unknown or other etiology from FASD, and there has been some progress in this area as well (Crocker, Vaurio, Riley, & Mattson, 2009; Ware et al., 2012). Being able to distinguish between these individuals is critical if we want to better understand and test potential pharmacotherapies for ethanol exposed offspring. With the potential number of children, adolescents and adults affected by prenatal ethanol exposure, medication development, and its potential translation to clinical trials must be a high priority.

ACKNOWLEDGMENTS

This work was funded, in part, by SBIR AA021038 and a NIAAA contract to JML and a Research Grant from the Vice President for Research at University of Kentucky to SB.

Conflict of interest: John Littleton is CEO of Naprogenix and owns stock in the company. He was not involved in collection, analysis, or interpretation of any of the data presented in this article. No other conflicts of interest exist.

REFERENCES

Abel, E. L. (1974). Alcohol ingestion in lactating rats: Effects on mothers and offspring. I. *Archives Internationales de Pharmacodynamie et de Thérapie*, *210*, 121–127.

Barron, S., Lewis, B., Wellmann, K., Carter, M., Farook, J., Ring, J., et al. (2012). Polyamine modulation of NMDARs as a mechanism to reduce effects of alcohol dependence. *Recent Patents on CNS Drug Discovery*, 7, 129–144.

Barron, S., Mulholland, P. J., Littleton, J. M., & Prendergast, M. A. (2008). Age and gender differences in response to neonatal ethanol withdrawal and polyamine challenge in organotypic hippocampal cultures. *Alcoholism, Clinical and Experimental Research, 32*, 929–936.

Bayer, A., Altman, J., Russo, R. J., & Zhang, X. (1993). Timetables of neurogenesis in the human brain based on experimentally determined patterns in the rat. *Neurotoxicology, 14*, 83–144.

Bell, R. L., Eiler, W. J. A., II, Cook, J. B., & Rahman, S. (2009). Nicotinic receptor ligands reduce ethanol intake by high alcohol-drinking HAD-2 rats. *Alcohol, 43*, 581–592.

Bender, C., de Olmos, S., Bueno, A., de Olmos, J., & Lorenzo, A. (2010). Comparative analyses of the neurodegeneration induced by the non-competitive NMDA-receptor-antagonist drug MK-801 in mice and rats. *Neurotoxicology and Teratology, 32*, 542–550.

Bonthius, D. J., Goodlett, C. R., & West, J. R. (1988). Blood alcohol concentration and severity of microencephaly in neonatal rats depend on the pattern of alcohol administration. *Alcohol, 5*, 209–214.

Briggs, C. A., & McKenna, D. G. (1998). Activation and inhibition of the human alpha7 nicotinic acetylcholine receptor by agonists. *Neuropharmacology, 37*, 1095–1102.

Brocardo, P. S., Gil-Mohapel, J., & Christie, B. R. (2011). The role of oxidative stress in fetal alcohol spectrum disorders. *Brain Research Reviews, 67*, 209–225.

Burd, L., Klug, M. G., Martsolf, J. T., & Kerbeshian, J. (2003). Fetal alcohol syndrome: Neuropsychiatric phenomics. *Neurotoxicology and Teratology, 25*, 697–705.

Chasnoff, I. J., Wells, A. M., & King, L. (2015). Misdiagnosis and missed diagnoses in foster and adopted children with prenatal alcohol exposure. *Pediatrics, 135*, 264–270.

Coe, J., Sidders, J., Riley, K., Waltermire, J., & Hagerman, R. (2001). A survey of medication responses in children and adolescents with fetal alcohol syndrome. *Mental Health Aspects of Developmental Disabilities, 4*, 148–155.

Crocker, N., Vaurio, L., Riley, E. P., & Mattson, S. N. (2009). Comparison of adaptive behavior in children with heavy prenatal alcohol exposure or attention-deficit/hyperactivity disorder. *Alcoholism, Clinical and Experimental Research, 33*, 2015–2023.

Davidson, M., & Wilce, P. (1998). Chronic ethanol treatment leads to increased ornithine decarboxylase activity: Implications for a role of polyamines in ethanol dependence and withdrawal. *Alcoholism, Clinical and Experimental Research, 22*, 1205–1211.

Davidson, M. D., Wilce, P., & Shanley, B. C. (1993). Increased sensitivity of the hippocampus in ethanol-dependent rats to toxic effect of N-methyl-D-aspartic acid in vivo. *Brain Research, 606*, 5–9.

de Fiebre, N. C., & de Fiebre, C. M. (2003). α7 Nicotinic acetylcholine receptor-mediated protection against ethanol-induced neurotoxicity. *Alcohol, 31*, 149–153.

Dobbing, J., & Sands, J. (1979). Comparative aspects of the brain growth spurt. *Early Human Development, 3*, 79–83.

Doig, J., McLennan, J. D., & Gibbard, W. B. (2008). Medication effects on symptoms of attention-deficit/hyperactivity disorder in children with fetal alcohol spectrum disorder. *Journal of Child and Adolescent Psychopharmacology, 18*, 365–371.

Doyle, L. R., & Mattson, S. N. (2015). Neurobehavioral disorder associated with prenatal alcohol exposure (ND-PAE): Review of evidence and guidelines for assessment. *Current Developmental Disorders Reports, 2*, 175–186.

Driscoll, C. D., Streissguth, A. P., & Riley, E. P. (1990). Prenatal alcohol exposure: Comparability of effects in humans and animal models. *Neurotoxicology and Teratology, 12*, 231–237.

Dwoskin, L. P., & Crooks, P. A. (2002). A novel mechanism of action and potential use for lobeline as a treatment for psychostimulant abuse. *Biochemical Pharmacology, 63*, 89–98.

Farook, J. M., Lewis, B., Gaddis, J. G., Littleton, J. M., & Barron, S. (2009). Lobeline, a nicotinic partial agonist attenuates alcohol consumption and preference in male C57BL/6J mice. *Physiology and Behavior, 97*, 503–506.

Fields, L., Carter, M., Hawkey, A., Lutz, J.A., Littleton, J., & Barron, S. DMXB-A reduces damage following "third trimester" ethanol exposure in both an in vivo and in vitro rodent model. Data presented at RSA, 2014, 2015 meetings. Abstracts *in Alcoholism: Clinical and Experimental Research*, 2014–2015.

Gibson, D. A., Harris, B. R., Prendergast, M. A., Hart, S. R., Blanchard, J. A., 2nd, Holley, R. C., et al. (2003). Polyamines contribute to ethanol withdrawal-induced neurotoxicity in rat hippocampal slice cultures through interactions with the NMDA receptor. *Alcoholism, Clinical and Experimental Research*, *27*, 1099–1106.

Gil-Mohapel, J., Boehme, F., Kainer, L., & Christie, B. R. (2010). Hippocampal cell loss and neurogenesis after fetal alcohol exposure: Insights from different rodent models. *Brain Research Reviews*, *64*, 283–303.

Godyń, J., Jończyk, J., Panek, D., & Malawska, B. (2016). Therapeutic strategies for Alzheimer's disease in clinical trials. *Pharmacological Reports*, *8*, 127–138.

Goodlett, C. R., Horn, K. H., & Zhou, F. C. (2005). Alcohol teratogenesis: Mechanisms of damage and strategies for intervention. *Experimental Biology and Medicine*, *230*, 394–406.

Harrod, S. B., Dwoskin, L. P., Green, T. A., Gehrke, B. J., & Bardo, M. T. (2003). Lobeline does not serve as a reinforcer in rats. *Psychopharmacology*, *165*, 397–404.

Hashimoto, K., Malchow, B., Falkai, P., & Schmitt, A. (2013). Glutamate modulators as potential therapeutic drugs in schizophrenia and affective disorders. *European Archives of Psychiatry and Clinical Neuroscience*, *263*, 367–377.

Heaton, M. B., Paiva, M., Madorsky, I., & Shaw, G. (2003). Ethanol effects on neonatal rat cortex: Comparative analyses of neurotrophic factors, apoptosis-related proteins, and oxidative processes during vulnerable and resistant periods. *Developmental Brain Research*, *145*, 249–262.

Hoffman, P. L., & Tabakoff, B. (1994). The role of the NMDA receptor in ethanol withdrawal. *EXS*, *71*, 61–70.

Hoyme, H. E., May, P. A., Kalberg, W. O., Kodituwakku, P., Gossage, J. P., Trujillo, P. M., et al. (2005). A practical clinical approach to diagnosis of fetal alcohol spectrum disorders: Clarification of the 1996 institute of medicine criteria. *Pediatrics*, *115*, 39–47.

Idrus, N. M., McGough, N. N., Riley, E. P., & Thomas, J. D. (2014). Administration of memantine during withdrawal mitigates overactivity and spatial learning impairments associated with neonatal alcohol exposure in rats. *Alcoholism, Clinical and Experimental Research*, *38*, 529–537.

Ikonomidou, C., Bosch, F., Miksa, M., Bittigau, P., Vöckler, J., Dikranian, K., et al. (1999). Blockade of NMDA receptors and apoptotic neurodegeneration in the developing brain. *Science*, *283*, 70–74.

Jones, K. L., Smith, D. W., Ulleland, C. N., & Streissguth, P. (1973). Pattern of malformation in offspring of chronic alcoholic mothers. *Lancet*, *1*, 1267–1271.

Joya, X., Garcia-Algar, O., Salat-Batlle, J., Pujades, C., & Vall, O. (2015). Advances in the development of novel antioxidant therapies as an approach for fetal alcohol syndrome prevention. *Birth Defects Research A: Clinical and Molecular Teratology*, *103*, 163–177.

Kaniaková, M., Lindovský, J., Krůšek, J., Adámek, S., & Vyskočil, F. (2011). Dual effect of lobeline on α4β2 rat neuronal nicotinic receptors. *European Journal of Pharmacology*, *658*, 108–113.

Ke, Z., Liu, Y., Wang, X., Fan, Z., Chen, G., Xu, M., et al. (2011). Cyanidin-3-glucoside ameliorates ethanol neurotoxicity in the developing brain. *Journal of Neuroscience Research*, *89*, 1676–1684.

Keynan, O., Mirovsky, Y., Dekel, S., Gilad, V. H., & Gilad, G. M. (2010). Safety and efficacy of dietary agmatine sulfate in lumbar disc-associated radiculopathy. An open-label, dose-escalating study followed by a randomized, double-blind, placebo-controlled trial. *Pain Medicine*, *11*, 356–368.

Koren, G. (2015). Pharmacological treatment of disruptive behavior in children with fetal alcohol spectrum disorder. *Pediatric Drugs*, *17*, 179–184.

Layer, R. T., Kaddis, F. G., & Wallace, L. J. (1993). The NMDA receptor antagonist MK-801 elicits conditioned place preference in rats. *Pharmacology, Biochemistry, and Behavior*, *44*, 245–247.

Lemoine, P., Harousseau, H., Borteyru, J. P., & Menuet, J. C. (1968). Les Enfants de Parents Alcooliques. Anomalies Observees a Propos de 127 Cas [Children of alcoholic parents: Abnormalities observed in 127 cases]. *Ouest Medical*, *21*, 476–482.

Lewis, B., Wellmann, K. A., & Barron, S. (2007). Agmatine reduces balance deficits in a rat model of third trimester binge-like ethanol exposure. *Pharmacology, Biochemistry, and Behavior*, *88*, 114–121.

Lewis, B., Wellmann, K. A., Kehrberg, A. M., Carter, M. L., Baldwin, T., Cohen, M., et al. (2012). Behavioral deficits and cellular damage following developmental ethanol exposure in rats are attenuated by CP-101,606, an NMDAR antagonist with unique NR2B specificity. *Pharmacology, Biochemistry, and Behavior*, *100*, 545–553.

Li, Y., Meyer, E. M., Walker, D. W., Millard, W. J., He, Y., & King, M. A. (2002). Alpha 7 nicotinic receptor activation inhibits ethanol-induced mitochondrial dysfunction, cytochrome C release and neurotoxicity in primary rat hippocampal neuronal cultures. *Journal of Neurochemistry*, *81*, 853–858.

Li, C. Y., Zhao, L. M., Shi, X. W., & Zhang, J. D. (2014). Lobeline shows protective effects against MPTP-induced dopaminergic neuron death and attenuates behavior deficits in animals. *Experimental and Therapeutic Medicine*, 7, 375–378.

Lieberman, J. A., Dunbar, G., Segreti, A. C., Girgis, R. R., Seoane, F., Beaver, J. S., et al. (2013). A randomized exploratory trial of an α-7 nicotinic receptor agonist (TC-5619) for cognitive enhancement in schizophrenia. *Neuropsychopharmacology*, *38*, 968–975.

Littleton, J. M., Lovinger, D., Liljequist, S., Ticku, R., Matsumoto, I., & Barron, S. (2001). Role of polyamines and NMDA receptors in ethanol dependence and withdrawal. *Alcoholism, Clinical and Experimental Research*, *25*(5 Suppl. ISBRA), 132S–136S.

Littleton, J., Rogers, T., & Falcone, D. (2005). Novel approaches to plant drug discovery based on high throughput pharmacological screening and genetic manipulation. *Life Sciences*, *78*, 467–475.

Luo, J. (2012). Mechanisms of ethanol-induced death of cerebellar granule cells. *The Cerebellum*, *11*, 145–154.

Lutz, J. A., Carter, M., Fields, L., Barron, S., & Littleton, J. M. (2015). The dietary flavonoid rhamnetin inhibits both inflammation and excitotoxicity during ethanol withdrawal in rat organotypic hippocampal slice cultures. *Alcoholism, Clinical and Experimental Research*, *39*, 2345–2353.

Maier, S. E., & West, J. R. (2001). Patterns and alcohol-related birth defects. *Alcohol Research and Health*, *25*, 168–174.

Martin, C. A., Nuzzo, P. A., Ranseen, J. D., Kleven, M. S., Guenthner, G., Williams, Y., et al. (August 21, 2013). Lobeline effects on cognitive performance in adult ADHD. *Journal of Attentional Disorders*. http://dx.doi.org/10.1177/1087054713497791, PMID:23966351.

May, P. A., Gossage, J. P., Kalberg, W. O., Robinson, L. K., Buckley, D., Manning, M., et al. (2009). Prevalence and epidemiologic characteristics of FASD from various research methods with an emphasis on recent in-school studies. *Developmental Disabilities Research Reviews*, *15*, 176–192.

Means, L. W., Medlin, C. W., Hughes, V. D., & Gray, S. L. (1984). Hyperresponsiveness to methylphenidate in rats following prenatal ethanol exposure. *Neurobehavioral Toxicology and Teratology*, *6*, 187–192.

Miller, D. K., Harrod, S. B., Green, T. A., Wong, M. Y., Bardo, M. T., & Dwoskin, L. P. (2003). Lobeline attenuates locomotor stimulation induced by repeated nicotine administration in rats. *Pharmacology, Biochemistry, and Behavior*, *74*, 279–286.

Monk, B. R., Leslie, F. M., & Thomas, J. D. (2012). The effects of perinatal choline supplementation on hippocampal cholinergic development in rats exposed to alcohol during the brain growth spurt. *Hippocampus*, *22*, 1750–1757.

Moore, E. M., & Riley, E. P. (2015). What happens when children with fetal alcohol spectrum disorders become adults? *Current Developmental Disorders Reports*, *2*, 219–227.

Moretti, M., Matheus, F. C., de Oliveira, P. A., Neis, V. B., Ben, J., Walz, R., et al. (2014). Role of agmatine in neurodegenerative diseases and epilepsy. *Frontiers in Bioscience*, *6*, 341–359.

Nash, C. M., Ibram, F., Dringenberg, H. C., Reynolds, J. N., & Brien, J. F. (2007). Effects of maternal administration of vitamins C and E on ethanol neurobehavioral teratogenicity in the guinea pig. *Alcohol*, *41*, 577–586.

Oesterheld, J. R., Kofoed, L., Tervo, R., Fogas, B., Wilson, A., & Fiechtner, H. (1998). Effectiveness of methylphenidate in native American children with fetal alcohol syndrome and attention deficit/hyperactivity disorder: A controlled pilot study. *Journal of Child and Adolescent Psychopharmacology*, *8*, 39–48.

O'Malley, K. D., & Hagerman, R. J. (1998). Developing clinical practice guidelines for pharmacological interventions with alcohol-affected children. In Center for Disease Control and Prevention (Ed.), *Intervening with children affected by prenatal alcohol exposure: Proceedings of a special focus session of the interagency coordinating committee on fetal alcohol syndrome* (pp. 145–177). Chevy Chase, MD: NIAAA.

O'Malley, K. D., & Nanson, J. (2002). Clinical implications of a link between fetal alcohol spectrum disorder and attention-deficit hyperactivity disorder. *Canadian Journal of Psychiatry*, *47*, 349–354.

O'Connor, M. J., Shah, B., Whaley, S., Cronin, P., Gunderson, B., & Graham, J. (2002). Psychiatric illness in a clinical sample of children with prenatal alcohol exposure. *American Journal of Drug and Alcohol Abuse*, *28*, 743–754.

O'Leary-Moore, S. K., Parnell, S. E., Lipinski, R. J., & Sulik, K. K. (2011). Magnetic resonance-based imaging in animal models of fetal alcohol spectrum disorder. *Neuropsychology Review*, *21*, 167–185.

Olney, J. W., Wozniak, D. F., Jevtovic-Todorovic, V., & Ikonomidou, C. (2001). Glutamate signaling and the fetal alcohol syndrome. *Mental Retardation and Developmental Disabilities Research Reviews*, *7*, 267–275.

Patten, A. R., Fontaine, C. J., & Christie, B. R. (2014). A comparison of the different animal models of fetal alcohol spectrum disorders and their use in studying complex behaviors. *Frontiers in Pediatrics*, *2*, 1–19 (article 93).

Peng, Y., Kwok, K. H., Yang, P. H., Ng, S. S., Liu, J., Wong, O. G., et al. (2005). Ascorbic acid inhibits ROS production, NF-kappa B activation and prevents ethanol-induced growth retardation and microencephaly. *Neuropharmacology*, *48*, 426–434.

Polston, J. E., Cunningham, C. S., Rodvelt, K. R., & Miller, D. K. (2006). Lobeline augments and inhibits cocaine-induced hyperactivity in rats. *Life Sciences*, *79*, 981–990.

Popova, S., Lange, S., Burd, L., & Rehm, J. (2016). The economic burden of fetal alcohol spectrum disorder in Canada in 2013. *Alcohol and Alcoholism* (Advance Access published October 2015), PMID:26493100.

Popova, S., Lange, S., Shield, K., Mihic, A., Chudley, A. E., Mukherjee, R. A. S., et al. (2016). Comorbidity of fetal alcohol spectrum disorder: A systematic review and meta-analysis. *The Lancet* (Published online January 2016), PMID:26777270 (in press).

Prendergast, M. A., Harris, B. R., Mayer, S., & Littleton, J. M. (2000). Chronic, but not acute, nicotine exposure attenuates ethanol withdrawal-induced hippocampal damage in vitro. *Alcoholism, Clinical and Experimental Research*, *24*, 1583–1592.

Preskorn, S. H., Baker, B., Kolluri, S., Menniti, F. S., Krams, M., & Landen, J. W. (2008). An innovative design to establish proof of concept of the antidepressant effects of the NR2B subunit selective N-methyl-D-aspartate antagonist, CP-101,606, in patients with

treatment-refractory major depressive disorder. *Journal of Clinical Psychopharmacology, 28*, 631–637.
Pruett, D., Waterman, H., & Caughey, A. (2013). Fetal alcohol exposure: Consequences, diagnosis, and treatment. *Obstetrical and Gynecological Surgery, 68*, 62–69.
Roni, M. A., & Rahman, S. (2013). Antidepressant-like effects of lobeline in mice: Behavioral, neurochemical, and neuroendocrine evidence. *Progress in Neuro-Psychopharmacology & Biological Psychiatry, 41*, 44–51.
Rosett, H. L. (1984). *Alcohol and the fetus: A clinical perspective* (1st ed). New York, Oxford: Oxford University Press.
Rubin, M. A., Wellmann, K. A., Lewis, B., Overgaauw, B. J., Littleton, J. M., & Barron, S. (2009). Difluoromethylornithine (DFMO) reduces deficits in isolation-induced ultrasonic vocalizations and balance following neonatal ethanol exposure in rats. *Pharmacology, Biochemistry, and Behavior, 92*, 44–50.
Santha, E., Sperlagh, B., Zelles, T., Zsilla, G., Toth, P. T., Lendvai, B., et al. (2000). Multiple cellular mechanisms mediate the effect of lobeline on the release of norepinephrine. *Journal of Pharmacology and Experimental Therapeutics, 294*, 302–307.
Schweitzer, L., Robbins, A. J., & Slotkin, T. A. (1989). Dendritic development of Purkinje and granule cells in the cerebellar cortex of rats treated postnatally with alpha-difluoromethylornithine. *Journal of Neuropathology & Experimental Neurology, 48*, 11–22.
Slotkin, T., & Bartolome, J. (1986). Role of ornithine decarboxylase and the polyamines in nervous system development: A review. *Brain Research Bulletin, 17*, 307–320.
Slotkin, T. A., Ferguson, S. A., Cada, A. M., McCook, E. C., & Seidler, F. J. (2000). Neonatal polyamine depletion by alpha-difluoromethylornithine: Effects on adenylyl cyclase cell signaling are separable from effects on brain region growth. *Brain Research, 887*, 16–22.
Smith, A. M., Wellmann, K. A., Lundblad, T. M., Carter, M. L., Barron, S., & Dwoskin, L. P. (2012). Lobeline attenuates neonatal ethanol-mediated changes in hyperactivity and dopamine transporter function in the prefrontal cortex in rats. *Neuroscience, 206*, 245–254.
Snyder, J., Nanson, J., Snyder, R., & Block, G. (1997). *Stimulant efficacy in children with FAS.* Seattle, WA: University of Washington Press.
Stratton, K., Howe, C., & Battaglia, F. (Eds.), (1996). *Fetal alcohol syndrome: Diagnosis, epidemiology, prevention, and treatment.* Washington, DC: National Academy Press.
Streissguth, A. P., Bookstein, F. L., Barr, H. M., Sampson, P. D., O'Malley, K., & Young, J. K. (2004). Risk factors for adverse life outcomes in fetal alcohol syndrome and fetal alcohol effects. *Journal of Developmental & Behavioral Pediatrics, 25*, 228–238.
Sulik, K. K. (2005). Genesis of alcohol-induced craniofacial dysmorphism. *Experimental Biology and Medicine, 230*, 266–275.
Thomas, J. D., Fleming, S. L., & Riley, E. P. (2001). MK-801 can exacerbate or attenuate behavioral alterations associated with neonatal alcohol exposure in the rat, depending on the timing of administration. *Alcoholism, Clinical and Experimental Research, 25*, 764–773.
Thomas, J. D., Fleming, S. L., & Riley, E. P. (2002). Administration of low doses of MK-801 during ethanol withdrawal in the developing rat pup attenuates alcohol's teratogenic effects. *Alcoholism, Clinical and Experimental Research, 26*, 1307–1313.
Thomas, J. D., Garrison, M., & O'Neill, T. M. (2004). Perinatal choline supplementation attenuates behavioral alterations associated with neonatal alcohol exposure in rats. *Neurotoxicology and Teratology, 26*, 35–45.
Thomas, J. D., La Fiette, M. H., Quinn, V. R., & Riley, E. P. (2000). Neonatal choline supplementation ameliorates the effects of prenatal alcohol exposure on a discrimination learning task in rats. *Neurotoxicology and Teratology, 22*, 703–711.
Thomas, J. D., & Riley, E. P. (1998). Fetal alcohol syndrome: Does alcohol withdrawal play a role? *Alcohol Health and Research World, 22*, 47–53.

Thomas, J. D., Weinert, S. P., Sharif, S., & Riley, E. P. (1997). MK-801 administration during ethanol withdrawal in neonatal rat pups attenuates ethanol-induced behavioral deficits. *Alcoholism, Clinical and Experimental Research, 21*, 1218–1225.

Tran, T. D., Cronise, K., Marino, M. D., Jenkins, W. J., & Kelly, S. J. (2000). Critical periods for the effects of alcohol exposure on brain weight, body weight, activity and investigation. *Behavioural Brain Research, 116*, 99–110.

Ulug, S., & Riley, E. P. (1983). The effect of methylphenidate on overactivity in rats prenatally exposed to alcohol. *Neurobehavioral Toxicology and Teratology, 5*, 35–39.

Ware, A. L., Crocker, N., O'Brien, J. W., Deweese, B. N., Roesch, S. C., Coles, C. D., et al. (2012). Executive function predicts adaptive behavior in children with histories of heavy prenatal alcohol exposure and attention-deficit/hyperactivity disorder. *Alcoholism, Clinical and Experimental Research, 36*, 1431–1441.

Wellmann, K., Lewis, B., & Barron, S. (2010). Agmatine reduces ultrasonic vocalization deficits in female rat pups exposed neonatally to ethanol. *Neurotoxicology and Teratology, 32*, 158–163.

Williams, K. (1997). Modulation and block of ion channels: A new biology of polyamines. *Cellular Signalling, 9*, 1–13.

Williams, K., Zappia, A., Pritchett, D., Shen, Y., & Molinoff, P. (1994). Sensitivity of the N-methyl-D-aspartate receptor to polyamines is controlled by NR2 subunits. *Molecular Pharmacology, 45*, 803–809.

Wozniak, J. R., Fuglestad, A. J., Eckerle, J. K., Fink, B. A., Hoecker, H. L., Boys, C. J., et al. (2015). Choline supplementation in children with fetal alcohol spectrum disorders: A randomized, double-blind, placebo-controlled trial. *American Journal of Clinical Nutrition, 102*, 1113–1125.

Yates, W. R., Cadoret, R. J., Troughton, E. P., Stewart, M., & Giunta, T. S. (1998). Effect of fetal alcohol exposure on adult symptoms of nicotine, alcohol, and drug dependence. *Alcoholism, Clinical and Experimental Research*, 914–920.

CHAPTER FOURTEEN

Using In Vitro Electrophysiology to Screen Medications: Accumbal Plasticity as an Engram of Alcohol Dependence

R. Renteria, Z.M. Jeanes, R.A. Mangieri, E.Y. Maier, D.M. Kircher, T.R. Buske, R.A. Morrisett[1]

University of Texas at Austin, Austin, TX, United States
[1]Corresponding author: e-mail address: ramorris@mail.utexas.edu

Contents

Abstract

The nucleus accumbens (NAc) is a central component of the mesocorticolimbic reward system. Increasing evidence strongly implicates long-term synaptic neuroadaptations in glutamatergic excitatory activity of the NAc shell and/or core medium spiny neurons in response to chronic drug and alcohol exposure. Such neuroadaptations likely play a

International Review of Neurobiology, Volume 126
ISSN 0074-7742
http://dx.doi.org/10.1016/bs.irn.2016.02.018

critical role in the development and expression of drug-seeking behaviors. We have observed unique cell-type-specific bidirectional changes in NAc synaptic plasticity (metaplasticity) following acute and chronic intermittent ethanol exposure. Other investigators have also previously observed similar metaplasticity in the NAc following exposure to psychostimulants, opiates, and amazingly, even following an anhedonia-inducing experience. Considering that the proteome of the postsynaptic density likely contains hundreds of biochemicals, proteins and other components and regulators, we believe that there is a large number of potential molecular sites through which accumbal metaplasticity may be involved in chronic alcohol abuse. Many of our companion laboratories are now engaged in identifying and screening medications targeting candidate genes and its products previously linked to maladaptive alcohol phenotypes. We hypothesize that if manipulation of such target genes and their products change NAc plasticity, then that observation constitutes an important validation step for the development of novel therapeutics to treat alcohol dependence.

1. INTRODUCTION: THE NUCLEUS ACCUMBENS IN DRUG REWARD PROCESSING

The mesocorticolimbic system is composed of dopaminergic neurons of the ventral tegmental area (VTA) and its projection sites including the nucleus accumbens (NAcs), prefrontal cortex, hippocampus, and amygdala (Sesack & Grace, 2010). The mesocorticolimbic system is often referred to as the reward system of the brain and is critical for reward and reinforcement processing, motivation, and goal-directed behaviors (Wise, 2004). It is well accepted that most drugs of abuse, including ethanol, activate the mesocorticolimbic dopamine system and lead to an increase in extracellular dopamine concentrations in the NAc (Di Chiara & Imperato, 1988; Doyon et al., 2003; Imperato & Di Chiara, 1986; Weiss, Lorang, Bloom, & Koob, 1993).

Depending on the pharmacological class they belong to, all drugs of abuse elicit, to different extents, an incentive arousal state due to their ability to increase extracellular dopamine in the NAc shell. This incentive arousal state facilitates the rate of current instrumental behavior, the acquisition and expression of secondary reinforcement, and the reinstatement of previously extinguished instrumental responding. In addition, it facilitates the consolidation of mnemonic traces of salient stimuli that are associated with affective states (Di Chiara et al., 2004). In short, dopamine in the NAc alerts the animal to the significance of a certain stimulus, so that appropriate associations can be made between the surroundings and behaviors that preceded the result of that stimulus. Repeated noncontingent exposure to drugs of abuse is thought to cause maladaptive activation of dopamine transmission. This could result in aberrant motivational behaviors typical of addiction–compulsive perseverance on obtaining drugs and drug-related stimuli at the expense of more conventional nondrug rewards. Thus, although it had been believed that dopamine directly mediated the rewarding effects of natural and drug reinforcers, more recent hypotheses focus on the role of mesolimbic dopamine as (1) a motivational learning signal (Spanagel & Weiss, 1999); (2) a signal of pathological associative learning in addiction (Di Chiara, 2002); (3) a neural substrate of incentive salience (Robinson & Berridge, 2003); or (4) a signal that informs the predictability of reward-related cues associated with previous drug availability (Fiorillo, Tobler, & Schultz, 2003).

Medium spiny neurons of the NAc therefore function to integrate this aforementioned dopamine signal in the context of cognitive, sensory, and emotional states. As such, cortical neurons are the likely promoters of goal-directed behaviors, with (1) the ventral subiculum of the hippocampus providing spatial and contextual information; (2) the prefrontal cortex supplying executive control, including task switching and response inhibition; and (3) the basolateral and central amygdala communicating information regarding conditioned associations as well as affective drive (Ambroggi, Ishikawa, Fields, & Nicola, 2008; Gruber, Hussain, & O'Donnell, 2009; Ito, Robbins, Pennartz, & Everitt, 2008; Kalivas, Volkow, & Seamans, 2005; Wolf, 2002).

The NAc shell subregion, in particular, has been associated with aspects of drug reward (Carlezon, Devine, & Wise, 1995; Ikemoto, 2007; Rodd-Henricks, McKinzie, Li, Murphy, & McBride, 2002; Sellings & Clarke, 2003). Evidence suggests that the transition to an addicted state follows adaptations first in the NAc shell, then NAc core, and finally the dorsal striatum. In nonhuman primates, the NAc shell and core subregions are organized in a series of parallel circuits linked in an ascending spiral to the dorsal striatum in a manner that could account for the transition from goal-directed to habitual behaviors during the development of addiction (Haber, Fudge, & McFarland, 2000; Sesack & Grace, 2010). Neuroadaptations of the mesocorticolimbic system are thought to underlie both the development and expression of addiction to ethanol and a variety of other reinforcers (Luscher & Malenka, 2011). Thus, understanding the exact sequence and cell-type specificity of these synaptic plasticity changes, within the mesocorticolimbic system, is critical for formulating model systems to investigate new therapeutic targets.

2. FORMS OF SYNAPTIC PLASTICITY IN THE SHELL AND CORE OF THE NUCLEUS ACCUMBENS

The bidirectional ability of neurons to increase and decrease synaptic strength is referred to as synaptic metaplasticity. As such, plasticity of neural circuits allows for the structural and functional reorganization of synapses in response to different stimuli. Long-term depression (LTD) and long-term potentiation (LTP) are the best-characterized mechanisms for modulating synaptic strength in an experience-dependent manner—a long-lasting decrease or increase in synaptic strength, respectively. Both processes are thought to be involved in information storage, important in learning and memory, and other physiological processes. In the NAc, several distinct forms of plasticity have been described that can result in either a decrease or an increase in synaptic strength.

2.1 NMDA Receptor-Dependent Plasticity

NMDA receptor (NMDAR)-dependent LTP and LTD are two forms of synaptic plasticity that require coincident activity of pre- and postsynaptic neurons. At resting membrane potentials, NMDARs are blocked by Mg^{2+}. However with sufficient depolarization, the Mg^{2+} block is relieved and glutamate activation of NMDARs allows Ca^{2+} influx and subsequently activates intracellular signaling cascades that are responsible for altering synaptic strength.

The best-characterized form of postsynaptic plasticity in the NAc is NMDAR-dependent LTD (Thomas, Malenka, & Bonci, 2000). This form of LTD is induced by prolonged low-frequency stimulation (LFS, 1–3 Hz) of presynaptic terminals paired with postsynaptic membrane depolarization, resulting in weak activation of postsynaptic NMDARs. This leads to a moderate amount of calcium influx through NMDARs and this intermediate elevation in postsynaptic calcium results in the activation of calcineurin and protein phosphatase 1 (Mulkey, Herron, & Malenka, 1993). The decrease in synaptic strength is due to the removal of AMPA receptors (AMPARs) from the postsynaptic membrane. Clathrin adaptor proteins bind to the C-terminus of the GluA2 subunit which leads to clathrin-mediated endocytosis of GluA2-containing AMPARs (Beattie et al., 2000; Malenka, 2003). The interaction of adaptor proteins with the GluA2 subunit is critical for the expression of NMDAR-dependent LTD (Brebner et al., 2005; Jeanes, Buske, & Morrisett, 2014; Scholz et al., 2010).

NMDAR-dependent LTP is induced by high-frequency stimulation (100 Hz) and results in stronger activation of NMDARs and a larger Ca^{2+} influx compared to the induction of LTD. The larger increase of postsynaptic Ca^{2+} leads to a distinct intracellular signaling cascade that includes the activation of Ca^{2+}-calmodulin kinase type II (CAMKII) ultimately resulting in the insertion of postsynaptic AMPARs (Malenka & Nicoll, 1999). NMDAR-dependent LTP has been described in the NAc of both rats and mice (Kombian & Malenka, 1994; Li & Kauer, 2004; Pascoli, Turiault, & Lüscher, 2012; Pennartz, Ameerun, Groenewegen, & Lopes da Silva, 1993; Schotanus & Chergui, 2008).

2.2 Endocannabinoid-Mediated Plasticity

Endocannabinoid (eCB)-mediated plasticity is well documented in the dorsal striatum (Lovinger & Mathur, 2012) and has been described in the NAc as well (Robbe, Kopf, Remaury, Bockaert, & Manzoni, 2002). The most common expression mechanism of eCB-mediated plasticity in the NAc involves activation of presynaptic CB1 receptors. CB1 receptors are G-protein-coupled receptors (GPCRs) and are one of the most abundant GPCRs expressed in the central nervous system. Activation of group 1 metabotropic glutamate receptors ($mGluR_{1/5}$), or a rise in postsynaptic Ca^{2+}, leads to the production of eCBs, which then act as a retrograde signal and activate presynaptic CB1 receptors. Activation of CB1 receptors results in a $G_{\alpha i/o}$-dependent reduction in adenylyl cyclase and protein kinase A activity that suppresses neurotransmitter release. Induction protocols for eCB LTD vary widely between brain regions (Heifets & Castillo, 2009). For example, a prolonged moderate stimulation (13 Hz) results in eCB-mediated LTD in the NAc (Hoffman, Oz, Caulder, & Lupica, 2003; Robbe et al., 2002).

Transient receptor potential vanilloid 1 receptor (TRPV1) is a non-selective cation channel that is widely expressed in the peripheral nervous system and has been found to be activated by the eCB anandamide (Ross, 2003). In the central nervous system, expression of TRPV1 and its functional significance have been observed in various brain regions (Brown, Chirila, Schrank, & Kauer, 2013; Gibson, Edwards, Page, Van Hook, & Kauer, 2008; Kauer & Gibson, 2009; Marinelli, Pascucci, Bernardi, Puglisi-Allegra, & Mercuri, 2005; Musella et al., 2009;

Zschenderlein, Gebhardt, von Bohlen Und Halbach, Kulisch, & Albrecht, 2011). Similar to NMDAR-dependent LTD, eCB-mediated TRPV1-dependent LTD results in the removal of postsynaptic AMPARs and has been observed in both the NAc core (Grueter, Brasnjo, & Malenka, 2010) and the NAc shell (Renteria, Jeanes, & Morrisett, 2014).

3. EVIDENCE IMPLICATING NMDAR-DEPENDENT POSTSYNAPTIC AMPAR ENDOCYTOSIS (LTD) IN RESPONSE TO DRUGS OF ABUSE

3.1 Cocaine and Other Psychostimulants

Drug-induced disruption of NMDAR-dependent LTD in the NAc was first described by Thomas and colleagues in 2001 (Thomas, Beurrier, Bonci, & Malenka, 2001). Daily intraperitoneal injections of cocaine resulted in the sensitization of the locomotor response to cocaine. In brain slices taken from mice that show behavioral sensitization, there was a decrease in the AMPA/NMDA ratio and a decrease in AMPAR-mediated miniature excitatory postsynaptic current (mEPSC) amplitude, suggesting a decrease in the number of postsynaptic AMPARs present at the synapse. The cocaine-induced internalization of AMPARs resulted in an occlusion of the expression of NMDAR-dependent LTD. It was later shown that the effects of cocaine on NAc plasticity are dependent on the history of cocaine exposure. Mice exhibiting an occlusion of LTD expression were treated with daily injections of cocaine but did not receive a challenge dose to test for behavioral sensitization (Kourrich, Rothwell, Klug, & Thomas, 2007). Instead, mice in extended cocaine withdrawal showed an increase in AMPA/NMDA ratio and an increase in mEPSC amplitude. These findings suggest an increase in AMPAR signaling following only extended withdrawal from cocaine exposure.

Other groups have shown similar alterations in AMPAR-mediated signaling in response to amphetamine exposure in which the drug-induced internalization of GluA2-containing AMPARs may be necessary for the expression of amphetamine locomotor sensitization (Brebner et al., 2005; Choi, Ahn, Wang, & Phillips, 2014). Sensitization to amphetamine treatment was blocked by the use of the HIV-1 TAT protein to deliver a peptide that mimics the C-terminus of the GluA2 subunit, which occludes GluA2 subunit-containing AMPAR internalization. Application of this peptide directly into the NAc abolished stereotypies that had developed with psychomotor sensitization. However, these stereotypies remained unchanged when the peptide was injected into the VTA. These findings are significant in that it is one of the first examples of altering a drug-induced phenotype by disrupting the drug-induced changes in excitatory signaling in the NAc.

The importance of the expression of a drug-induced LTD-like state is further supported by a study of cocaine self-administration in rats (Kasanetz et al., 2010). In rats that self-administered cocaine, only a subset of animals showed the persistent drug-seeking characteristics similar to those seen in human cocaine dependents (Deroche-Gamonet, Belin, & Piazza, 2004). Rats were separated into groups termed "addicted" or "nonaddicted" based on their score from three addiction-like behaviors paralleling those defined by the DSM-IV. Two weeks after the last self-administration session, LTD was impaired in all animals but gradually recovered in "nonaddicts." In "addicts," LTD was persistently impaired which suggests that the long-lasting impairment of LTD may be important for the transition from drug seeking to dependence.

3.2 Ethanol

Ethanol is a potent modulator of plasticity. Both, in vitro and in vivo exposure to ethanol can disrupt the expression of synaptic plasticity (McCool, 2011; Morrisett & Swartzwelder, 1993). Work from our laboratory has shown that in vitro ethanol can inhibit NMDARs in the NAc (Maldve et al., 2002; Zhang, Hendricson, & Morrisett, 2005) and block the expression of NMDAR-dependent LTD, while chronic intermittent in vivo exposure results in a reversal in the expression of plasticity (Jeanes, Buske, & Morrisett, 2011). Chronic intermittent ethanol exposure (CIE) is a widely accepted model used to induce ethanol dependence in C57Bl/6 mice (Becker & Lopez, 2004). Mice are exposed to ethanol vapor during four consecutive daily 16-h sessions to elicit intoxication (target blood ethanol concentrations (BECs) of approximately 40–50 mM with coadministration of pyrazole). This protocol is normally repeated weekly two to four times, and elevates two-bottle choice (2BC) ethanol preference and intake by approximately 50%. Twenty-four hours after CIE, the same induction protocol that results in LTD in ethanol–naïve mice will induce the expression of LTP in brain slices from ethanol-exposed C57Bl/6J mice. The disruption of NMDAR-dependent LTD persists for up to 72 h into withdrawal from ethanol vapor and results in the absence of both, LTD and LTP. Interestingly, in the hippocampal formation, we previously reported a lack of ethanol inhibition of low-frequency NMDAR-dependent LTD suggesting significant differences in the mechanisms mediating this form of plasticity between hippocampal and accumbal brain regions (Hendricson, Miao, Lippmann, & Morrisett, 2002).

Behavioral sensitization to ethanol also alters NMDAR-dependent NAc LTD in mice. In Swiss Webster mice treated with daily ethanol injections, only a portion showed locomotor sensitization, and only this portion displayed a disruption of NMDAR-dependent accumbal LTD (Abrahao et al., 2013). This effect was the result of dampened NMDAR function as Western blot analysis showed a measured decrease of the NR1 subunit. In another study, the expression of NMDAR-dependent LTD was found to be hampered in the NAc of ethanol-dependent rats. In rats consuming an ethanol-containing diet for 20 days, alterations in plasticity and MSN morphology were observed 12 h into ethanol withdrawal. The ethanol-containing diet led to the loss of long thin dendritic spines as well as a decrease in NMDAR function (Spiga et al., 2014).

4. LENTIVIRAL OCCLUSION OF GluA2 SUBUNIT INTERNALIZATION MODULATES CIE-ENHANCED ETHANOL INTAKE

Given the importance of neuroadaptations of AMPARs in the NAc for the expression of drug-induced behaviors, the use of the synthetic peptide "$GluA2_{3Y}$" derived from the rat GluA2 carboxyl tail (^{869}YKEGYNVYG^{877}) could prove to be effective for the attenuation of ethanol-induced behaviors as well. As previously described, Wang and colleagues' (Brebner et al., 2005) $GluA2_{3Y}$ peptide saturates the cellular machinery that binds to GluA2 and thus prevents the clathrin-mediated internalization of AMPARs and the subsequent formation of LTD (see Fig. 1). To study the effect of LTD occlusion in chronically ethanol-consuming animals, we designed a virus that causes long-term overexpression of the $GluA2_{3Y}$ peptide. As predicted, NAc MSNs of animals injected with the virus overexpressing the $GluA2_{3Y}$ peptide showed an occlusion of LTD in slice (Maier et al., 2015). However, this change in synaptic plasticity did not alter ethanol drinking in animal models of moderate ethanol consumption (ie, operant self-administration and 2BC). Interestingly, CIE exposure, a model that leads to prolonged high BECs resulting in increased ethanol consumption, did not affect drinking in mice injected with the $GluA2_{3Y}$ peptide. These findings support the notion that AMPAR trafficking plays an important role in alcohol reinforcement during intense levels of ethanol exposure.

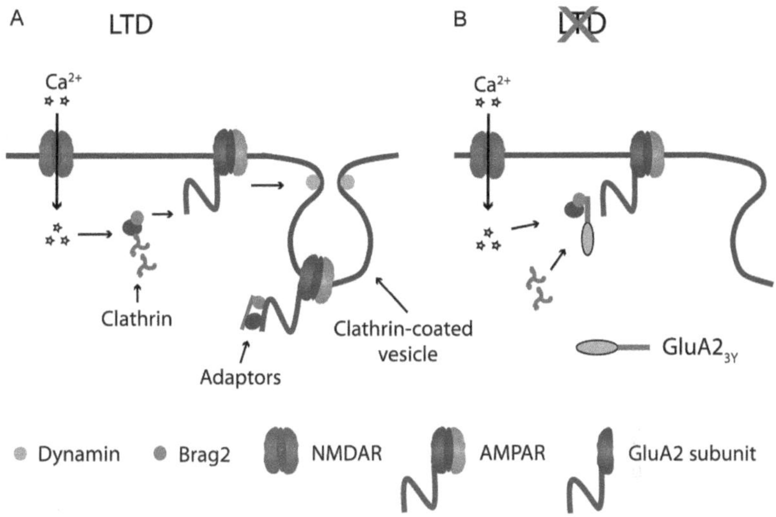

Fig. 1 (A) Low-frequency conditioning stimulation paired with postsynaptic depolarization produces LTD of AMPAR-mediated EPSCs. This form of LTD relies on activation of NMDARs and influx of Ca^{2+}. Ca^{2+} influx recruits clathrin and adaptor proteins to bind to the intracellular tail of the GluA2 subunit C-terminal, resulting in internalization of GluA2-containing AMPARs. (B) The artificial peptide $GluA2_{3Y}$ mimics this tail of the GluA2 subunit, saturates adaptor proteins, prevents AMPAR internalization, and therefore occludes formation of LTD. *This figure was derived from one generously provided by Dr. Yu Tian Wang.*

5. DIFFERENTIAL INVOLVEMENT OF D1 DOPAMINE RECEPTOR-EXPRESSING VS D2 DOPAMINE RECEPTOR-EXPRESSING MEDIUM SPINY NEURONS OF THE SHELL OF THE NUCLEUS ACCUMBENS IN NMDAR-LTD

In both the core and shell subregions of the NAc, there are two major subtypes of MSNs depending on the dopamine receptor expression. Medium spiny neurons that express the D1 dopamine receptor coexpress the signaling peptides dynorphin and substance P (D1 MSN), while MSNs that express the D2 dopamine receptor also express enkephalin (D2 MSN) (Lu, Ghasemzadeh, & Kalivas, 1998; Zhou, Furuta, & Kaneko, 2003). D1 and D2 MSNs of the NAc have similar projection patterns to the direct and indirect pathways of the dorsal striatum although the segregation of pathways is not as clearly distinct, and therefore that terminology is no longer applied to the accumbal projections (Humphries & Prescott, 2010;

Kupchik et al., 2015; Sesack & Grace, 2010; Smith, Lobo, Spencer, & Kalivas, 2013). In the NAc core and shell, D1 MSNs project to the substantia nigra pars reticulata and the VTA, respectively, whereas shell D1 MSNs also project to the ventral pallidum. D2 MSNs from the core and shell project primarily to the ventral pallidum.

The development of transgenic mice that express a fluorophore (such as enhanced green fluorescent protein (eGFP) or tdTomato) under the control of either the D1 or D2 dopamine receptor promoter (Matamales et al., 2009; Valjent, Bertran-Gonzalez, Herve, Fisone, & Girault, 2009) has allowed for greater insight as to how these two populations of MSNs differ. In the dorsal striatum, the electrophysiological properties of D1 MSNs differ from those of D2 MSNs (Cepeda et al., 2008; Day, Wokosin, Plotkin, Tian, & Surmeier, 2008; Gertler, Chan, & Surmeier, 2008). Similarly in the NAc core, D1 and D2 MSNs have different electrophysiological and synaptic properties (Grueter et al., 2010). Given these differences in basal electrophysiological properties, it might also be expected that D1 and D2 MSNs would display differences in the expression of plasticity. Work from our laboratory using *Drd1a*-eGFP transgenic mice confirms the differential expression of plasticity such that conventional LTD conditioning stimuli elicit NMDAR-dependent LTD only in the shell D1 MSNs and not in D1-lacking MSNs (Jeanes et al., 2014).

6. CHRONIC INTERMITTENT ETHANOL EXPOSURE DIFFERENTIALLY MODULATES D1 VS D2 MSN PLASTICITY

CIE exposure is a well-established model for inducing ethanol dependence and increasing volitional ethanol intake in mice (Becker & Lopez, 2004; Griffin, Lopez, & Becker, 2009; Griffin, Lopez, Yanke, Middaugh, & Becker, 2009; Griffin, 2014; Lopez & Becker, 2005). As described above, we have previously reported that CIE vapor exposure disrupts NMDAR-dependent LTD in the NAc shell of C57Bl/6J mice (Jeanes et al., 2011). For this initial characterization of ethanol-induced metaplasticity in wild-type C57Bl/6J mice, we did not have the ability to selectively record from the D1 or D2 MSN subtypes. To determine if ethanol exposure differentially modulates plasticity in D1 or D2 MSNs we used *Drd1a*-eGFP transgenic mice on a Swiss Webster background (Jeanes et al., 2014). First, we found that NMDAR-dependent LTD was expressed only in eGFP-positive (presumed D1) MSNs in the NAc shell and that the LTD conditioning protocol had no long-term effect on EPSC amplitudes in eGFP-negative (presumed D2) MSNs. Second, 24 h after CIE, there was a reversal in the expression of plasticity; the LTD protocol resulted in LTD in D2 MSNs and had no effect on D1 MSNs. Over a 2-week period, the expression of plasticity is gradually restored in D1 MSNs and abolished in D2 MSNs.

Given the discrepancy in the polarity of plasticity after ethanol vapor exposure between C57Bl/6J and *Drd1a*-eGFP on the Swiss Webster background, the ethanol-induced alterations of plasticity seem to be specific to the strain of mouse. The CIE-induced increase in volitional ethanol intake has been characterized primarily in C57Bl/6J mice. Interpretation of our data collected from Swiss Webster mice may be limited by the fact that the latter strain does not voluntarily drink significant amounts of ethanol. To investigate how CIE modulates plasticity in D1 and D2 MSNs of a mouse strain that has a well-documented drinking phenotype, we used *Drd1a*-tdTomato transgenic mice on a C57Bl/6J background (Ade, Wan, Chen, Gloss, & Calakos, 2011). Similar to what we observed in transgenic mice on the Swiss Webster background, LTD was expressed only in D1 MSNs of ethanol–naïve mice (Renteria, Maier, Buske, & Morrisett, 2016). Twenty-four hours after CIE treatment, the pairing protocol resulted in LTP in D1 MSNs and LTD in D2 MSNs. This alteration in the expression of plasticity was accompanied by a change in NMDAR function, measured using NMDA/AMPA ratios as well as an input output curve of isolated NMDAR currents. We found an increase in NMDAR function in D1MSNs and a decrease in D2 MSNs. These findings are illustrated in Fig. 2.

7. STRATEGY FOR TARGET VALIDATION USING ACCUMBAL PLASTICITY

A variety of screening measures can frequently implicate particular genes or proteins in drug-related behaviors or experience. Determining whether such novel targets might indeed develop into fruitful avenues of therapeutic development is a very daunting task. Nevertheless, we and others have observed very consistent plasticity changes in the NAc across different drugs of abuse as well as between different mouse strains and even cell-specific changes across these different mouse lines. We feel that such consistent observations strongly implicate accumbal plasticity as an

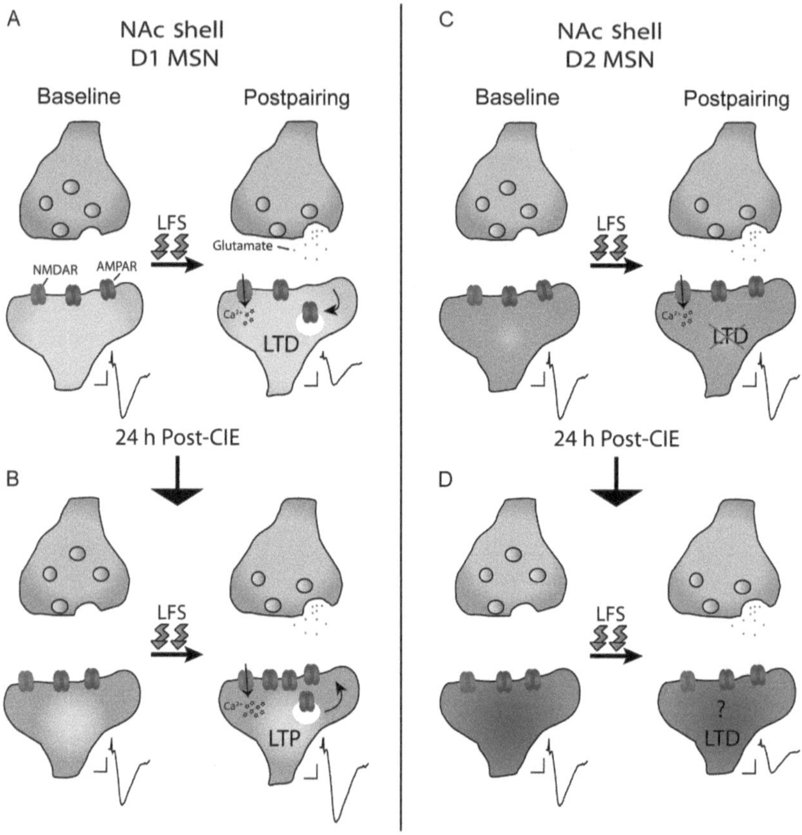

Fig. 2 CIE-induced metaplasticity of glutamatergic transmission differs between NAc shell D1 MSNs (*left panel*) and D2 MSNs (*right panel*). (A) In slices from ethanol–naïve mice, NMDAR-dependent LTD is observed in D1 MSNs following low-frequency stimulation (LFS) paired with postsynaptic depolarization to −50 mV. LTD is expressed via removal of AMPARs from the postsynaptic membrane. (B) Following in vivo CIE exposure the same LTD induction protocol that was applied to ethanol–naïve slices induces LTP of glutamatergic transmission onto D1 MSNs, presumably via insertion of AMPARs into the postsynaptic membrane. (C) LFS paired with postsynaptic depolarization does not elicit LTD in D2 MSNs. (D) Following in vivo CIE exposure, however, this protocol does elicit LTD in D2 MSNs. It is not yet known the underlying mechanism mediating this form of LTD. For all panels, traces next to postsynaptic neurons represent an average amplitude-evoked EPSC during baseline or postpairing periods; corresponding scale bars indicate 5 ms (horizontal) and 50 pA (vertical). "?" signifies the unknown mechanism of LTD in D2 MSNs as stated in (D).

important neuroadaptive process underlying drug-induced behaviors, particularly in the early developmental stages of drug seeking. Therefore, we envision that when modulation of a particular target alters accumbal plasticity in NAc D1 MSNs, such a target would constitute an especially strong candidate for further investigation. Herein, we summarize our work in this area in Table 1, and describe in greater detail in the following sections two examples of such an approach focusing on targets identified by a number of laboratories from animal models of excessive ethanol consumption.

7.1 Studies on the HDID and HS/Npt Mouse Lines

Drinking in the dark (DID) is a rodent model of binge ethanol consumption. A single ethanol tube (20% (v/v)) is inserted in the home cage 3 h into the dark cycle for 2 h for 3 consecutive days, followed by a 4-h exposure on the fourth day. This model has been shown to produce intoxicating levels of consumption in mice (Rhodes, Best, Belknap, Finn, & Crabbe, 2005). Crabbe et al. (2009) developed an outbred strain termed HS/Npt from systematic crossings of multiple inbred strains. HS/Npt mice underwent DID procedures and those mice with the highest BECs following DID were selectively bred together, producing mice termed high drinking in the dark (HDID) mice. By the eleventh generation the HDID BECs and ethanol consumption, in the DID paradigm, were significantly higher than that of the progenitor HS/Npt mice (Crabbe et al., 2009).

Subsequent testing of the HDID mice in comparison to the progenitor HS/Npt mice indicated that HDID mice exhibit increased sensitivity to ethanol. Later generations of HDID mice retained greater BECs compared to HS/Npts following DID, as well as consumed modestly more ethanol in a limited access 2BC paradigm. HDID mice, however, do not consume greater quantities of ethanol than HS/Npt mice during a 24-h 2BC drinking paradigm. These findings indicate that selective breeding for high BECs selects for alleles distinct from those that confer high consumption in continuous access preference paradigms (Crabbe, Spence, Brown, & Metten, 2011). Regarding differences in drinking, HDID mice show greater ethanol-stimulated locomotor activity and greater sensitivity to some but not all ethanol-sensitive behavioral tasks as compared to HS/Npt mice (Crabbe et al., 2012). These findings indicate that some ethanol responses share common genetic control with the ability to reach high BEC after DID procedures. The HDID mice also exhibit a decreased sensitivity to ethanol-induced hypothermia compared to the HS/Npts (Crabbe et al., 2012).

As previously discussed in this chapter, we maintain that alterations in synaptic plasticity within the NAc following ethanol exposure are critical for the expression of ethanol-related behaviors. Since there is a strong

Table 1 Accumbal Plasticity Screening

Background Strain	Manipulation		Ethanol	Postpairing EPSC Amplitude
Line	**Genetic**	**In Vitro**	**Exposure**	**(% Baseline)**
C57BL/6J				
			Naïve	65[a]
			CIE	125[a]
	CD14 KO		Naïve	45[a]
	BKβ1 KO		Naïve	55[a]
Drd1a-tdTomato			Naïve	50
			CIE	125
Drd1a-tdTomato		0 nM TAE684	Naïve	55
		5 nM TAE684	Naïve	75
		25 nM TAE684	Naïve	95
Drd1a-tdTomato x *Alk*	ALK WT		Naïve	45
	ALK KO		Naïve	75
Swiss Webster				
Drd1a-eGFP			Naïve	60
			CIE	95
HS/Npt[b]				
HS/Npt control stock	No selection		Naïve	50
			CIE	50
HDID	Selective		Naïve	50
	Breeding of line		CIE	95

[a]Experiments were performed on unidentified NAc shell MSNs. Ethanol–naïve, wild-type C57Bl/6J mice show NMDAR-LTD that corresponds to a postpairing EPSC amplitude of ~65% of baseline (first row). As described in Jeanes et al. (2014), this population of neurons can be divided into those that show "large LTD" or "small LTD" (average postpairing EPSC amplitudes of ~55% or 90%), presumed to be D1 or D2 MSNs, respectively. For BKβ1 and CD14 KO experiments, unidentified MSNs were classified as having large or small LTD, and values reported in table represent the averages of large LTD MSNs.
[b]Experiments were performed on D1 MSNs identified by a fluorescent retrograde tracer (cholera toxin subunit B conjugated to Alexa Fluor 555 injected into the VTA, see text).
EPSC amplitudes are the average of the 20- to 30-min postpairing period during which LTD is expressed and is reported as a percentage of the baseline average and rounded to the nearest 5%. Neurons were identified as D1 MSNs by epifluorescent illumination of a genetically encoded fluorescent reporter (ie, tdTomato or eGFP), with the following exceptions.

difference in ethanol consumption phenotypes, we hypothesized that a difference in synaptic plasticity would be observed in NAc shell D1 MSNs between the HDID and HS/Npt mice. Thus, whole-cell patch clamp recordings were conducted on putative D1 MSNs identified by injection of a fluorescent tracer (cholera toxin subunit B) into the VTA to retrogradely label VTA-projecting D1 MSNs from the NAc shell. For ethanol–naïve HDID and HS/Npt mice, no differences were observed in LTD induction between the two lines (see Table 1). We then examined if these mice exhibit differential LTD expression 24 h after a 4-day bout of CIE exposure. The HDID mice exhibited a loss of LTD similar to what had been observed in C57Bl/6J mice (Jeanes et al., 2011, 2014; Renteria et al., under revision). Interestingly, the HS/Npt mice did not exhibit a reduction in LTD magnitude following CIE. We are currently conducting more experiments to better understand differences in accumbal synaptic plasticity relative to ethanol exposure between these two lines of mice. Nevertheless, these findings suggest that selective breeding for high BECs does not impact accumbal plasticity or CIE-induced metaplasticity, but the converse is not true such that following CIE HS/Npt mice display an electrophysiological phenotype unique from mice that drink greater amounts of ethanol (Jeanes et al., 2011, 2014; Renteria et al., under revision).

7.2 Studies Concerning the Receptor Tyrosine Kinase ALK

Another molecular target that we have screened for its role in accumbal synaptic plasticity is anaplastic lymphoma kinase (ALK). ALK is a receptor tyrosine kinase that was discovered and initially characterized for its role in non-Hodgkin's, anaplastic large cell lymphoma (Iwahara et al., 1997; Morris et al., 1994, 1997). These early studies of ALK found its mRNA and protein to be located specifically in neural tissues of mice and humans, with expression highest during embryogenesis then declining postnatally until reaching and maintaining a low level of expression into adulthood (Iwahara et al., 1997; Morris et al., 1997; Pulford et al., 1997). This spatiotemporal pattern of expression pointed toward a role for ALK in normal nervous system development, which was supported by subsequent investigations of ALK in a number of species (Degoutin, Brunet-de Carvalho, Cifuentes-Diaz, & Vigny, 2009; Hurley, Clary, Copie, & Lefcort, 2006; Liao, Hung, Abrams, & Zhen, 2004; Rohrbough & Broadie, 2010; Yao et al., 2013). *Alk* transcription is regulated by LIM domain-only (LMO) proteins, which interact with DNA-binding elements to repress *Alk* transcription (Lasek, Gesch, Giorgetti, Kharazia, & Heberlein,

2011; Lasek, Lim, et al., 2011; Lasek, Giorgetti, Berger, Tayor, & Heberlein, 2011), and it is this relationship that brought ALK to the attention of alcohol researchers.

The laboratory of Ulrike Heberlein has long championed *Drosophila melanogaster* (the fruit fly) as a model organism to screen for gene mutations that alter responses to drugs and alcohol (Devineni et al., 2011; Heberlein, Tsai, Kapfhamer, & Lasek, 2009). Using such genetic screening methods, mutations in the fly gene for LMO that affect behavioral sensitivity to cocaine were identified (Tsai, Bainton, Blau, & Heberlein, 2004). In a subsequent series of elegant papers, Lasek and colleagues reported that LMO proteins also affect responses to cocaine in mice, and to ethanol in both flies and mice (Lasek, Giorgetti, et al., 2011; Lasek et al., 2010; Savarese, Zou, Kharazia, Maiya, & Lasek, 2014). ALK was then identified as a transcriptional target of LMO proteins which might mediate the observed effects of cocaine and ethanol (Lasek, Gesch, et al., 2011; Lasek, Lim, et al., 2011); *Alk* mRNA expression in the striatum is inversely correlated with ethanol consumption in the BXD line of inbred mice, and mice with a genetic deletion in a portion of the *Alk* coding sequence (ALK KO) drink more ethanol relative to ALK WT mice when tested in the DID paradigm. More recently, the Lasek group has identified how ethanol affects ALK activity and ALK-regulated intracellular signaling cascades. In both cell culture and mouse brain, ethanol treatment activates ALK and ERK signaling, as evidenced by increased phosphorylation of these proteins (He, Chen, Muramatsu, & Lasek, 2015). Moreover, pretreatment with an inhibitor of ALK, TAE684 (Galkin et al., 2007), prevents the ethanol-induced activation of ERK. As ERK has been shown by several groups to modulate ethanol-related behaviors (Agoglia et al., 2015; Faccidomo, Besheer, Stanford, & Hodge, 2009; Faccidomo, Salling, Galunas, & Hodge, 2015), these latest findings point toward a possible biochemical mechanism for ALK's influence on ethanol consumption.

To summarize the findings from these cell and animal models, preexisting variations in *Alk* expression influence ethanol consumption and associated behavioral sensitivity, while ethanol acutely activates ALK and ALK-dependent signaling pathways that influence ethanol-related behaviors. Thus, given that ALK–ethanol interactions play a role in regulating ethanol consumption alongside the previously discussed relationship between NMDAR-dependent LTD and ethanol experience, our lab has collaborated with Dr. Lasek to investigate the potential involvement of ALK in the induction of LTD by NAc shell D1 MSNs (Mangieri & Morrisett, 2015).

In order to specifically examine plasticity induction in D1 MSNs, we crossed the ALK KO line of mice with *Drd1a*-tdTomato mice (Ade et al., 2011) to generate ALK WT and ALK KO mice that were also hemizygous for *Drd1a*-tdTomato. When we compared the magnitude of depression induced in D1 MSNs of each genotype, we found LTD to be markedly reduced, but not absent, in the ALK KO mice. To determine if acute, pharmacological inhibition of ALK activity would similarly affect LTD induction in *Drd1a*-tdTomato mice, we pretreated brain slices with TAE684, and observed a concentration-dependent attenuation of LTD. The findings of the plasticity experiments are summarized in Table 1. We also recorded spontaneous excitatory postsynaptic currents in D1 MSNs, and found that the event amplitude, but not frequency, was significantly elevated for ALK KO mice and ALK WT slices pretreated with TAE684. All together, these observations suggest that one function of ALK in the adult mouse brain may be to regulate postsynaptic AMPAR trafficking.

In conclusion, the combined genetic, biochemical, electrophysiological, and behavioral approaches to studying ALK in model systems have highlighted it as a target needing further investigation for the development of novel drugs to treat alcohol use disorders. Indeed, there are at least two reports of findings in human subjects that support this idea. One, from the Lasek group, examined the *ALK* gene sequence in humans and discovered two single-nucleotide polymorphisms (SNPs) in the coding sequence that were significantly associated with responses to ethanol (subjective high and lateral body sway) by subjects in a laboratory setting (Lasek, Lim, et al., 2011). The second report, by a different group, performed a meta-analysis of two genome-wide association studies and found strong association between an *ALK* SNP and alcohol dependence in humans (Wang et al., 2011). Although it is unknown as to whether any of these SNPs confer differences in ALK function, these findings nevertheless are consistent with the idea that ALK is a promising target for modulating responses to ethanol in humans. Collaborative experiments are ongoing between our lab and that of Dr. Lasek's to further validate and explore ALK as a target for modulating effects of ethanol in the brain.

8. CONCLUSIONS

In this review, nine experiments were chosen based on either an animal model of enhanced ethanol intake or sensitivity, or on some type of screening mechanism that identified potential targets for medication development. As documented in Table 1, we have investigated the expression of baseline plasticity or its alteration following some type of manipulation. In multiple lines of mice, we have consistently observed that NMDAR-dependent LTD of glutamatergic synaptic transmission is differentially expressed between D1 and D2 MSNs of the NAc shell. This form of synaptic plasticity is highly sensitive to intoxicating concentrations of ethanol applied acutely to these neurons and is markedly altered (ie, metaplastic) following CIE exposure. We have tested three different targets (ALK, CD14, and BKβ1 subunit), implicated by numerous other laboratories using some ethanol-relevant screening protocol, for alterations in accumbal plasticity, but we only observed very prominent and consistent effects following genetic and pharmacological manipulation of ALK. Moreover, work on ALK continues to be a major focus of our laboratory, and we are preparing to investigate the effects of FDA-approved ALK inhibitors on this metaplasticity and ethanol drinking as well. Therefore, taken together, our findings suggest that there is significant predictive value in accumbal plasticity studies to screen and validate the repurposing of medications to treat alcohol abuse and dependence.

REFERENCES

Abrahao, K. P., Ariwodola, O. J., Butler, T. R., Rau, A. R., Skelly, M. J., Carter, E., et al. (2013). Locomotor sensitization to ethanol impairs NMDA receptor-dependent synaptic plasticity in the nucleus accumbens and increases ethanol self-administration. *The Journal of Neuroscience*, *33*(11), 4834–4842. http://dx.doi.org/10.1523/JNEUROSCI.5839-11.2013.

Ade, K. K., Wan, Y., Chen, M., Gloss, B., & Calakos, N. (2011). An improved BAC transgenic fluorescent reporter line for sensitive and specific identification of striatonigral medium spiny neurons. *Frontiers in Systems Neuroscience*, *5*, 32. http://dx.doi.org/10.3389/fnsys.2011.00032.

Agoglia, A. E., Sharko, A. C., Psilos, K. E., Holstein, S. E., Reid, G. T., & Hodge, C. W. (2015). Alcohol alters the activation of ERK1/2, a functional regulator of binge alcohol drinking in adult C57BL/6J mice. *Alcoholism, Clinical and Experimental Research*, *39*(3), 463–475. http://dx.doi.org/10.1111/acer.12645.

Ambroggi, F., Ishikawa, A., Fields, H. L., & Nicola, S. M. (2008). Basolateral amygdala neurons facilitate reward-seeking behavior by exciting nucleus accumbens neurons. *Neuron*, *59*(4), 648–661. http://dx.doi.org/10.1016/j.neuron.2008.07.004.

Beattie, E. C., Carroll, R. C., Yu, X., Morishita, W., Yasuda, H., von Zastrow, M., et al. (2000). Regulation of AMPA receptor endocytosis by a signaling mechanism shared with LTD. *Nature Neuroscience*, *3*(12), 1291–1300. http://dx.doi.org/10.1038/81823.

Becker, H. C., & Lopez, M. F. (2004). Increased ethanol drinking after repeated chronic ethanol exposure and withdrawal experience in C57BL/6 mice. *Alcoholism, Clinical and Experimental Research*, *28*(12), 1829–1838.

Brebner, K., Wong, T. P., Liu, L., Liu, Y., Campsall, P., Gray, S., et al. (2005). Nucleus accumbens long-term depression and the expression of behavioral sensitization. *Science*, *310*(5752), 1340–1343. http://dx.doi.org/10.1126/science.1116894.

Brown, T. E., Chirila, A. M., Schrank, B. R., & Kauer, J. A. (2013). Loss of interneuron LTD and attenuated pyramidal cell LTP in Trpv1 and Trpv3 KO mice. *Hippocampus, 23*(8), 662–671. http://dx.doi.org/10.1002/hipo.22125.

Carlezon, W. A., Jr., Devine, D. P., & Wise, R. A. (1995). Habit-forming actions of nomifensine in nucleus accumbens. *Psychopharmacology, 122*(2), 194–197. http://dx.doi.org/10.1007/BF02246095.

Cepeda, C., Andre, V. M., Yamazaki, I., Wu, N., Kleiman-Weiner, M., & Levine, M. S. (2008). Differential electrophysiological properties of dopamine D1 and D2 receptor-containing striatal medium-sized spiny neurons. *The European Journal of Neuroscience, 27*(3), 671–682. http://dx.doi.org/10.1111/j.1460-9568.2008.06038.x.

Choi, F. Y., Ahn, S., Wang, Y. T., & Phillips, A. G. (2014). Interference with AMPA receptor endocytosis: Effects on behavioural and neurochemical correlates of amphetamine sensitization in male rats. *Journal of Psychiatry & Neuroscience, 39*(3), 189–199.

Crabbe, J. C., Kruse, L. C., Colville, A. M., Cameron, A. J., Spence, S. E., Schlumbohm, J. P., et al. (2012). Ethanol sensitivity in high drinking in the dark selectively bred mice. *Alcoholism: Clinical and Experimental Research, 36*(7), 1162–1170. http://dx.doi.org/10.1111/j.1530-0277.2012.01735.x.

Crabbe, J. C., Metten, P., Rhodes, J. S., Yu, C. H., Brown, L. L., Phillips, T. J., et al. (2009). A line of mice selected for high blood ethanol concentrations shows drinking in the dark to intoxication. *Biological Psychiatry, 65*(8), 662–670. http://dx.doi.org/10.1016/j.biopsych.2008.11.002.

Crabbe, J. C., Spence, S. E., Brown, L. L., & Metten, P. (2011). Alcohol preference drinking in a mouse line selectively bred for high drinking in the dark. *Alcohol, 45*(5), 427–440. http://dx.doi.org/10.1016/j.alcohol.2010.12.001.

Day, M., Wokosin, D., Plotkin, J. L., Tian, X., & Surmeier, D. J. (2008). Differential excitability and modulation of striatal medium spiny neuron dendrites. *The Journal of Neuroscience, 28*(45), 11603–11614. http://dx.doi.org/10.1523/JNEUROSCI.1840-08.2008.

Degoutin, J., Brunet-de Carvalho, N., Cifuentes-Diaz, C., & Vigny, M. (2009). ALK (Anaplastic Lymphoma Kinase) expression in DRG neurons and its involvement in neuron-Schwann cells interaction. *The European Journal of Neuroscience, 29*(2), 275–286. http://dx.doi.org/10.1111/j.1460-9568.2008.06593.x.

Deroche-Gamonet, V., Belin, D., & Piazza, P. V. (2004). Evidence for addiction-like behavior in the rat. *Science, 305*(5686), 1014–1017. http://dx.doi.org/10.1126/science.1099020.

Devineni, A. V., McClure, K. D., Guarnieri, D. J., Corl, A. B., Wolf, F. W., Eddison, M., et al. (2011). The genetic relationships between ethanol preference, acute ethanol sensitivity, and ethanol tolerance in Drosophila melanogaster. *Fly, 5*(3), 191–199.

Di Chiara, G. (2002). Nucleus accumbens shell and core dopamine: Differential role in behavior and addiction. *Behavioural Brain Research, 137*(1–2), 75–114. http://dx.doi.org/10.1016/S0166-4328(02)00286-3.

Di Chiara, G., Bassareo, V., Fenu, S., De Luca, M. A., Spina, L., Cadoni, C., et al. (2004). Dopamine and drug addiction: The nucleus accumbens shell connection. *Neuropharmacology, 47*(Suppl. 1), 227–241. http://dx.doi.org/10.1016/j.neuropharm.2004.06.032.

Di Chiara, G., & Imperato, A. (1988). Drugs abused by humans preferentially increase synaptic dopamine concentrations in the mesolimbic system of freely moving rats. *Proceedings of the National Academy of Sciences of the United States of America, 85*(14), 5274–5278.

Doyon, W. M., York, J. L., Diaz, L. M., Samson, H. H., Czachowski, C. L., & Gonzales, R. A. (2003). Dopamine activity in the nucleus accumbens during consummatory phases of oral ethanol self-administration. *Alcoholism, Clinical and Experimental Research, 27*(10), 1573–1582. http://dx.doi.org/10.1097/01.ALC.0000089959.66222.B8.

Faccidomo, S., Besheer, J., Stanford, P. C., & Hodge, C. W. (2009). Increased operant responding for ethanol in male C57BL/6J mice: Specific regulation by the ERK1/2, but not JNK, MAP kinase pathway. *Psychopharmacology, 204*(1), 135–147. http://dx.doi.org/10.1007/s00213-008-1444-9.

Faccidomo, S., Salling, M. C., Galunas, C., & Hodge, C. W. (2015). Operant ethanol self-administration increases extracellular-signal regulated protein kinase (ERK) phosphorylation in reward-related brain regions: Selective regulation of positive reinforcement in the prefrontal cortex of C57BL/6J mice. *Psychopharmacology, 232*(18), 3417–3430. http://dx.doi.org/10.1007/s00213-015-3993-z.

Fiorillo, C. D., Tobler, P. N., & Schultz, W. (2003). Discrete coding of reward probability and uncertainty by dopamine neurons. *Science, 299*, 1898–1902. http://dx.doi.org/10.1126/science.1077349.

Galkin, A. V., Melnick, J. S., Kim, S., Hood, T. L., Li, N., Li, L., et al. (2007). Identification of NVP-TAE684, a potent, selective, and efficacious inhibitor of NPM-ALK. *Proceedings of the National Academy of Sciences of the United States of America, 104*(1), 270–275. http://dx.doi.org/10.1073/pnas.0609412103.

Gertler, T. S., Chan, C. S., & Surmeier, D. J. (2008). Dichotomous anatomical properties of adult striatal medium spiny neurons. *The Journal of Neuroscience, 28*(43), 10814–10824. http://dx.doi.org/10.1523/JNEUROSCI.2660-08.2008.

Gibson, H. E., Edwards, J. G., Page, R. S., Van Hook, M. J., & Kauer, J. A. (2008). TRPV1 channels mediate long-term depression at synapses on hippocampal interneurons. *Neuron, 57*(5), 746–759. http://dx.doi.org/10.1016/j.neuron.2007.12.027.

Griffin, W. C. (2014). Alcohol dependence and free-choice drinking in mice. *Alcohol (Fayetteville, NY), 48*(3), 287–293. http://dx.doi.org/10.1016/j.alcohol.2013.11.006.

Griffin, W. C., 3rd., Lopez, M. F., & Becker, H. C. (2009a). Intensity and duration of chronic ethanol exposure is critical for subsequent escalation of voluntary ethanol drinking in mice. *Alcoholism, Clinical and Experimental Research, 33*(11), 1893–1900. http://dx.doi.org/10.1111/j.1530-0277.2009.01027.x.

Griffin, W. C., 3rd., Lopez, M. F., Yanke, A. B., Middaugh, L. D., & Becker, H. C. (2009b). Repeated cycles of chronic intermittent ethanol exposure in mice increases voluntary ethanol drinking and ethanol concentrations in the nucleus accumbens. *Psychopharmacology, 201*(4), 569–580. http://dx.doi.org/10.1007/s00213-008-1324-3.

Gruber, A. J., Hussain, R. J., & O'Donnell, P. (2009). The nucleus accumbens: A switchboard for goal-directed behaviors. *PloS One, 4*(4), e5062. http://dx.doi.org/10.1371/journal.pone.0005062.

Grueter, B. A., Brasnjo, G., & Malenka, R. C. (2010). Postsynaptic TRPV1 triggers cell type-specific long-term depression in the nucleus accumbens. *Nature Neuroscience, 13*(12), 1519–1525. http://dx.doi.org/10.1038/nn.2685.

Haber, S. N., Fudge, J. L., & McFarland, N. R. (2000). Striatonigrostriatal pathways in primates form an ascending spiral from the shell to the dorsolateral striatum. *The Journal of Neuroscience, 20*(6), 2369–2382.

He, D., Chen, H., Muramatsu, H., & Lasek, A. W. (2015). Ethanol activates midkine and anaplastic lymphoma kinase signaling in neuroblastoma cells and in the brain. *Journal of Neurochemistry, 135*(3), 508–521. http://dx.doi.org/10.1111/jnc.13252.

Heberlein, U., Tsai, L. T.-Y., Kapfhamer, D., & Lasek, A. W. (2009). Drosophila, a genetic model system to study cocaine-related behaviors: A review with focus on LIM-only proteins. *Neuropharmacology, 56*(Suppl. 1), 97–106. http://dx.doi.org/10.1016/j.neuropharm.2008.07.023.

Heifets, B. D., & Castillo, P. E. (2009). Endocannabinoid signaling and long-term synaptic plasticity. *Annual Review of Physiology, 71*, 283–306. http://dx.doi.org/10.1146/annurev.physiol.010908.163149.

Hendricson, A. W., Miao, C. L. A., Lippmann, M. J., & Morrisett, R. A. (2002). Ifenprodil and ethanol enhance NMDA receptor-dependent long-term depression. *The Journal of Pharmacology and Experimental Therapeutics, 301*(3), 938–944.

Hoffman, A. F., Oz, M., Caulder, T., & Lupica, C. R. (2003). Functional tolerance and blockade of long-term depression at synapses in the nucleus accumbens after chronic cannabinoid exposure. *The Journal of Neuroscience*, *23*(12), 4815–4820. doi: 0270-6474/03/234815-06 [pii].

Humphries, M. D., & Prescott, T. J. (2010). The ventral basal ganglia, a selection mechanism at the crossroads of space, strategy, and reward. *Progress in Neurobiology*, *90*(4), 385–417. http://dx.doi.org/10.1016/j.pneurobio.2009.11.003.

Hurley, S. P., Clary, D. O., Copie, V., & Lefcort, F. (2006). Anaplastic lymphoma kinase is dynamically expressed on subsets of motor neurons and in the peripheral nervous system. *The Journal of Comparative Neurology*, *495*(2), 202–212. http://dx.doi.org/10.1002/cne.20887.

Ikemoto, S. (2007). Dopamine reward circuitry: Two projection systems from the ventral midbrain to the nucleus accumbens-olfactory tubercle complex. *Brain Research Reviews*, *56*(1), 27–78. http://dx.doi.org/10.1016/j.brainresrev.2007.05.004.

Imperato, A., & Di Chiara, G. (1986). Preferential stimulation of dopamine release in the nucleus accumbens of freely moving rats by ethanol. *The Journal of Pharmacology and Experimental Therapeutics*, *239*(1), 219–228.

Ito, R., Robbins, T. W., Pennartz, C. M., & Everitt, B. J. (2008). Functional interaction between the hippocampus and nucleus accumbens shell is necessary for the acquisition of appetitive spatial context conditioning. *The Journal of Neuroscience*, *28*(27), 6950–6959. http://dx.doi.org/10.1523/JNEUROSCI.1615-08.2008.

Iwahara, T., Fujimoto, J., Wen, D., Cupples, R., Bucay, N., Arakawa, T., et al. (1997). Molecular characterization of ALK, a receptor tyrosine kinase expressed specifically in the nervous system. *Oncogene*, *14*(4), 439–449. http://dx.doi.org/10.1038/sj.onc.1200849.

Jeanes, Z. M., Buske, T. R., & Morrisett, R. A. (2011). In vivo chronic intermittent ethanol exposure reverses the polarity of synaptic plasticity in the nucleus accumbens shell. *The Journal of Pharmacology and Experimental Therapeutics*, *336*(1), 155–164. http://dx.doi.org/10.1124/jpet.110.171009.

Jeanes, Z. M., Buske, T. R., & Morrisett, R. A. (2014). Cell type-specific synaptic encoding of ethanol exposure in the nucleus accumbens shell. *Neuroscience*, *277*, 184–195. http://dx.doi.org/10.1016/j.neuroscience.2014.06.063.

Kalivas, P. W., Volkow, N., & Seamans, J. (2005). Unmanageable motivation in addiction: A pathology in prefrontal-accumbens glutamate transmission. *Neuron*, *45*(5), 647–650. http://dx.doi.org/10.1016/j.neuron.2005.02.005.

Kasanetz, F., Deroche-Gamonet, V., Berson, N., Balado, E., Lafourcade, M., Manzoni, O., et al. (2010). Transition to addiction is associated with a persistent impairment in synaptic plasticity. *Science*, *328*(5986), 1709–1712. http://dx.doi.org/10.1126/science.1187801.

Kauer, J. A., & Gibson, H. E. (2009). Hot flash: TRPV channels in the brain. *Trends in Neurosciences*, *32*(4), 215–224. http://dx.doi.org/10.1016/j.tins.2008.12.006.

Kombian, S. B., & Malenka, R. C. (1994). Simultaneous LTP of non-NMDA- and LTD of NMDA-receptor-mediated responses in the nucleus accumbens. *Nature*, *368*(6468), 242–246. http://dx.doi.org/10.1038/368242a0.

Kourrich, S., Rothwell, P. E., Klug, J. R., & Thomas, M. J. (2007). Cocaine experience controls bidirectional synaptic plasticity in the nucleus accumbens. *The Journal of Neuroscience*, *27*(30), 7921–7928. http://dx.doi.org/10.1523/JNEUROSCI.1859-07.2007.

Kupchik, Y. M., Brown, R. M., Heinsbroek, J. A., Lobo, M. K., Schwartz, D. J., & Kalivas, P. W. (2015). Coding the direct/indirect pathways by D1 and D2 receptors is not valid for accumbens projections. *Nature Neuroscience*, *18*(9), 1230–1232. http://dx.doi.org/10.1038/nn.4068.

Lasek, A. W., Gesch, J., Giorgetti, F., Kharazia, V., & Heberlein, U. (2011). Alk is a transcriptional target of LMO4 and ERα that promotes cocaine sensitization and reward.

The Journal of Neuroscience, *31*(40), 14134–14141. http://dx.doi.org/10.1523/JNEUROSCI.3415-11.2011.

Lasek, A. W., Giorgetti, F., Berger, K. H., Tayor, S., & Heberlein, U. (2011). Lmo genes regulate behavioral responses to ethanol in Drosophila melanogaster and the mouse. *Alcoholism, Clinical and Experimental Research*, *35*(9), 1600–1606. http://dx.doi.org/10.1111/j.1530-0277.2011.01506.x.

Lasek, A. W., Kapfhamer, D., Kharazia, V., Gesch, J., Giorgetti, F., & Heberlein, U. (2010). Lmo4 in the nucleus accumbens regulates cocaine sensitivity. *Genes, Brain, and Behavior*, *9*(7), 817–824. http://dx.doi.org/10.1111/j.1601-183X.2010.00620.x.

Lasek, A. W., Lim, J., Kliethermes, C. L., Berger, K. H., Joslyn, G., Brush, G., et al. (2011). An evolutionary conserved role for anaplastic lymphoma kinase in behavioral responses to ethanol. *PloS One*, *6*(7), e22636. http://dx.doi.org/10.1371/journal.pone.0022636.

Li, Y., & Kauer, J. A. (2004). Repeated exposure to amphetamine disrupts dopaminergic modulation of excitatory synaptic plasticity and neurotransmission in nucleus accumbens. *Synapse (New York)*, *51*(1), 1–10. http://dx.doi.org/10.1002/syn.10270.

Liao, E. H., Hung, W., Abrams, B., & Zhen, M. (2004). An SCF-like ubiquitin ligase complex that controls presynaptic differentiation. *Nature*, *430*(6997), 345–350. http://dx.doi.org/10.1038/nature02647.

Lopez, M. F., & Becker, H. C. (2005). Effect of pattern and number of chronic ethanol exposures on subsequent voluntary ethanol intake in C57BL/6J mice. *Psychopharmacology (Berl)*, *181*(4), 688–696. http://dx.doi.org/10.1007/s00213-005-0026-3.

Lovinger, D. M., & Mathur, B. N. (2012). Endocannabinoids in striatal plasticity. *Parkinsonism & Related Disorders*, *18*(Suppl. 1), S132–S134. http://dx.doi.org/10.1016/S1353-8020(11)70041-4.

Lu, X. Y., Ghasemzadeh, M. B., & Kalivas, P. W. (1998). Expression of D1 receptor, D2 receptor, substance P and enkephalin messenger RNAs in the neurons projecting from the nucleus accumbens. *Neuroscience*, *82*(3), 767–780.

Luscher, C., & Malenka, R. C. (2011). Drug-evoked synaptic plasticity in addiction: From molecular changes to circuit remodeling. *Neuron*, *69*(4), 650–663. http://dx.doi.org/10.1016/j.neuron.2011.01.017.

Maier, E. Y., Kircher, D. M., Renteria, R., Lasek, A. W., Wang, Y. T., & Morrisett, R. A. (2015). Lentivirus-induced accumbal overexpression of a GluA2 C-terminal peptide disrupts synaptic plasticity and alters operant and CIE-induced ETOH drinking. In *38th Annual scientific meeting of the research society on alcoholism, Abstract #198, 39(Issue S1), 61A.*

Maldve, R. E., Zhang, T. A., Ferrani-Kile, K., Schreiber, S. S., Lippmann, M. J., Snyder, G. L., et al. (2002). DARPP-32 and regulation of the ethanol sensitivity of NMDA receptors in the nucleus accumbens. *Nature Neuroscience*, *5*(7), 641–648. http://dx.doi.org/10.1038/nn877.

Malenka, R. C. (2003). Synaptic plasticity and AMPA receptor trafficking. *Annals of the New York Academy of Sciences*, *1003*, 1–11.

Malenka, R. C., & Nicoll, R. A. (1999). Long-term potentiation—A decade of progress? *Science (New York)*, *285*(5435), 1870–1874. http://dx.doi.org/10.1126/science.285.5435.1870.

Mangieri, R. A., & Morrisett, R. A. (2015). Disruption of anaplastic lymphoma kinase activity attenuates LTD induction in medium spiny neurons (MSNs) in the nucleus accumbens shell. In *38th Annual scientific meeting of the research society on alcoholism*, Abstract #216, 39(Issue S1), 65A. http://dx.doi.org/10.111/acer.12741.

Marinelli, S., Pascucci, T., Bernardi, G., Puglisi-Allegra, S., & Mercuri, N. B. (2005). Activation of TRPV1 in the VTA excites dopaminergic neurons and increases chemical- and noxious-induced dopamine release in the nucleus accumbens. *Neuropsychopharmacology*, *30*(5), 864–870. http://dx.doi.org/10.1038/sj.npp.1300615.

Matamales, M., Bertran-Gonzalez, J., Salomon, L., Degos, B., Deniau, J. M., Valjent, E., et al. (2009). Striatal medium-sized spiny neurons: Identification by nuclear staining and study of neuronal subpopulations in BAC transgenic mice. *PloS One*, *4*(3), e4770. http://dx.doi.org/10.1371/journal.pone.0004770.

McCool, B. A. (2011). Ethanol modulation of synaptic plasticity. *Neuropharmacology*, *61*(7), 1097–1108. http://dx.doi.org/10.1016/j.neuropharm.2010.12.028.

Morris, S. W., Kirstein, M. N., Valentine, M. B., Dittmer, K. G., Shapiro, D. N., Look, A. T., et al. (1994). Fusion of a kinase gene, ALK, to a nucleolar protein gene, NPM, in non-Hodgkin's lymphoma. *Science (New York)*, *267*(5151), 1281–1284.

Morris, S. W., Naeve, C., Mathew, P., James, P. L., Kirstein, M. N., Cui, X., et al. (1997). ALK, the chromosome 2 gene locus altered by the t(2;5) in non-Hodgkin's lymphoma, encodes a novel neural receptor tyrosine kinase that is highly related to leukocyte tyrosine kinase (LTK). *Oncogene*, *14*, 2175–2188.

Morrisett, R. A., & Swartzwelder, H. S. (1993). Attenuation of hippocampal long-term potentiation by ethanol: A patch-clamp analysis of glutamatergic and GABAergic mechanisms. *The Journal of Neuroscience*, *13*(5), 2264–2272.

Mulkey, R. M., Herron, C. E., & Malenka, R. C. (1993). An essential role for protein phosphatases in hippocampal long-term depression. *Science*, *261*(5124), 1051–1055.

Musella, A., De Chiara, V., Rossi, S., Prosperetti, C., Bernardi, G., Maccarrone, M., et al. (2009). TRPV1 channels facilitate glutamate transmission in the striatum. *Molecular and Cellular Neurosciences*, *40*(1), 89–97. http://dx.doi.org/10.1016/j.mcn.2008.09.001.

Pascoli, V., Turiault, M., & Lüscher, C. (2012). Reversal of cocaine-evoked synaptic potentiation resets drug-induced adaptive behaviour. *Nature*, *481*(7379), 71–75. http://dx.doi.org/10.1038/nature10709.

Pennartz, C. M. A., Ameerun, R. F., Groenewegen, H. J., & Lopes da Silva, F. H. (1993). Synaptic plasticity in an in vitro slice preparation of the rat nucleus accumbens. *The European Journal of Neuroscience*, *5*, 107–117.

Pulford, K., Lamant, L., Morris, S. W., Butler, L. H., Wood, K. M., Stroud, D., et al. (1997). Detection of anaplastic lymphoma kinase (ALK) and nucleolar protein nucleophosmin (NPM)-ALK proteins in normal and neoplastic cells with the monoclonal antibody ALK1. *Blood*, *89*(4), 1394–1404.

Renteria, R., Jeanes, Z. M., & Morrisett, R. A. (2014). Ethanol attenuation of long-term depression in the nucleus accumbens can be overcome by activation of TRPV1 receptors. *Alcoholism, Clinical and Experimental Research*, *38*(11), 2763–2769. http://dx.doi.org/10.1111/acer.12542.

Renteria, R., Maier, E. Y., Buske, T. R., & Morrisett, R. A. (2016). Selective alterations of NMDAR function and plasticity in D1 and D2 medium spiny neurons in the nucleus accumbens shell following chronic intermittent ethanol exposure. *Neuropharmacology*. http://dx.doi.org/10.1016/j.neuropharm.2016.03.004.

Rhodes, J. S., Best, K., Belknap, J. K., Finn, D. A., & Crabbe, J. C. (2005). Evaluation of a simple model of ethanol drinking to intoxication in C57BL/6J mice. *Physiology and Behavior*, *84*(1), 53–63. http://dx.doi.org/10.1016/j.physbeh.2004.10.007.

Robbe, D., Kopf, M., Remaury, A., Bockaert, J., & Manzoni, O. J. (2002). Endogenous cannabinoids mediate long-term synaptic depression in the nucleus accumbens. *Proceedings of the National Academy of Sciences of the United States of America*, *99*(12), 8384–8388. http://dx.doi.org/10.1073/pnas.122149199.

Robinson, T. E., & Berridge, K. C. (2003). Addiction. *Annual Review of Psychology*, *54*, 25–53. http://dx.doi.org/10.1146/annurev.psych.54.101601.145237.

Rodd-Henricks, Z. A., McKinzie, D. L., Li, T.-K., Murphy, J. M., & McBride, W. J. (2002). Cocaine is self-administered into the shell but not the core of the nucleus accumbens of Wistar rats. *The Journal of Pharmacology and Experimental Therapeutics*, *303*(3), 1216–1226. http://dx.doi.org/10.1124/jpet.102.038950.

Rohrbough, J., & Broadie, K. (2010). Anterograde Jelly belly ligand to Alk receptor signaling at developing synapses is regulated by Mind the gap. *Development (Cambridge, England), 137*(20), 3523–3533. http://dx.doi.org/10.1242/dev.047878.

Ross, R. A. (2003). Anandamide and vanilloid TRPV1 receptors. *British Journal of Pharmacology, 140*(5), 790–801. http://dx.doi.org/10.1038/sj.bjp.0705467.

Savarese, A., Zou, M. E., Kharazia, V., Maiya, R., & Lasek, A. W. (2014). Increased behavioral responses to ethanol in Lmo3 knockout mice. *Genes, Brain, and Behavior, 13*(8), 777–783. http://dx.doi.org/10.1111/gbb.12176.

Scholz, R., Berberich, S., Rathgeber, L., Kolleker, A., Köhr, G., & Kornau, H. C. (2010). AMPA receptor signaling through BRAG2 and Arf6 critical for long-term synaptic depression. *Neuron, 66*(5), 768–780. http://dx.doi.org/10.1016/j.neuron.2010.05.003.

Schotanus, S. M., & Chergui, K. (2008). Long-term potentiation in the nucleus accumbens requires both NR2A- and NR2B-containing N-methyl-d-aspartate receptors. *European Journal of Neuroscience, 27*(8), 1957–1964. http://dx.doi.org/10.1111/j.1460-9568.2008.06173.x.

Sellings, L. H., & Clarke, P. B. (2003). Segregation of amphetamine reward and locomotor stimulation between nucleus accumbens medial shell and core. *The Journal of Neuroscience, 23*(15), 6295–6303. doi: 0270-6474/03/236295-09 [pii].

Sesack, S. R., & Grace, A. A. (2010). Cortico-basal ganglia reward network: Microcircuitry. *Neuropsychopharmacology, 35*(1), 27–47. http://dx.doi.org/10.1038/npp.2009.93.

Smith, R. J., Lobo, M. K., Spencer, S., & Kalivas, P. W. (2013). Cocaine-induced adaptations in D1 and D2 accumbens projection neurons (a dichotomy not necessarily synonymous with direct and indirect pathways). *Current Opinion in Neurobiology, 23*(4), 546–552. http://dx.doi.org/10.1016/j.conb.2013.01.026.

Spanagel, R., & Weiss, F. (1999). The dopamine hypothesis of reward: Past and current status. *Trends in Neurosciences, 22*(11), 521–527. http://dx.doi.org/10.1016/S0166-2236(99)01447-2.

Spiga, S., Talani, G., Mulas, G., Licheri, V., Fois, G. R., Muggironi, G., et al. (2014). Hampered long-term depression and thin spine loss in the nucleus accumbens of ethanol-dependent rats. *Proceedings of the National Academy of Sciences of the United States of America, 111*(35), E3745–E3754. http://dx.doi.org/10.1073/pnas.1406768111.

Thomas, M. J., Beurrier, C., Bonci, A., & Malenka, R. C. (2001). Long-term depression in the nucleus accumbens: A neural correlate of behavioral sensitization to cocaine. *Nature Neuroscience, 4*(12), 1217–1223. http://dx.doi.org/10.1038/nn757.

Thomas, M. J., Malenka, R. C., & Bonci, A. (2000). Modulation of long-term depression by dopamine in the mesolimbic system. *The Journal of Neuroscience, 20*(15), 5581–5586.

Tsai, L. T.-Y., Bainton, R. J., Blau, J., & Heberlein, U. (2004). Lmo mutants reveal a novel role for circadian pacemaker neurons in cocaine-induced behaviors. *PLoS Biology, 2*(12), e408. http://dx.doi.org/10.1371/journal.pbio.0020408.

Valjent, E., Bertran-Gonzalez, J., Herve, D., Fisone, G., & Girault, J. A. (2009). Looking BAC at striatal signaling: Cell-specific analysis in new transgenic mice. *Trends in Neurosciences, 32*(10), 538–547. http://dx.doi.org/10.1016/j.tins.2009.06.005.

Wang, K.-S., Liu, X., Zhang, Q., Pan, Y., Aragam, N., & Zeng, M. (2011). A meta-analysis of two genome-wide association studies identifies 3 new loci for alcohol dependence. *Journal of Psychiatric Research, 45*(11), 1419–1425.

Weiss, F., Lorang, M. T., Bloom, F. E., & Koob, G. F. (1993). Oral alcohol self-administration stimulates dopamine release in the rat nucleus accumbens: Genetic and motivational determinants. *The Journal of Pharmacology and Experimental Therapeutics, 267*(1), 250–258.

Wise, R. A. (2004). Dopamine, learning and motivation. *Nature Reviews Neuroscience, 5*(6), 483–494. http://dx.doi.org/10.1038/nrn1406.

Wolf, M. E. (2002). Addiction: Making the connection between behavioral changes and neuronal plasticity in specific pathways. *Molecular Interventions*, *2*(3), 146–157. http://dx.doi.org/10.1124/mi.2.3.146.

Yao, S., Cheng, M., Zhang, Q., Wasik, M., Kelsh, R., & Winkler, C. (2013). Anaplastic lymphoma kinase is required for neurogenesis in the developing central nervous system of zebrafish. *PloS One*, *8*(5), e63757. http://dx.doi.org/10.1371/journal.pone.0063757.

Zhang, T. A., Hendricson, A. W., & Morrisett, R. A. (2005). Dual synaptic sites of D(1)-dopaminergic regulation of ethanol sensitivity of NMDA receptors in nucleus accumbens. *Synapse (New York)*, *58*(1), 30–44. http://dx.doi.org/10.1002/syn.20181.

Zhou, L., Furuta, T., & Kaneko, T. (2003). Chemical organization of projection neurons in the rat accumbens nucleus and olfactory tubercle. *Neuroscience*, *120*(3), 783–798. http://dx.doi.org/10.1016/S0306-4522(03)00326-9.

Zschenderlein, C., Gebhardt, C., von Bohlen Und Halbach, O., Kulisch, C., & Albrecht, D. (2011). Capsaicin-induced changes in LTP in the lateral amygdala are mediated by TRPV1. *PloS One*, *6*(1), e16116. http://dx.doi.org/10.1371/journal.pone.0016116.

CHAPTER FIFTEEN

The Zebrafish, a Novel Model Organism for Screening Compounds Affecting Acute and Chronic Ethanol-Induced Effects

S. Tran[1], A. Facciol, R. Gerlai[1]
University of Toronto Mississauga, Mississauga, ON, Canada
[1]Corresponding authors: e-mail address: stevenhuy.tran@mail.utoronto.ca; robert_gerlai@yahoo.com

Contents

Abstract

Alcohol addiction is a major unmet medical and economic issue for which very few efficacious pharmacological treatment options are currently available. The development and identification of new compounds and drugs to treat alcohol addiction is hampered by the high costs and low amenability of traditional laboratory rodents to high-throughput behavioral screens. The zebrafish represents an excellent compromise between systems complexity and practical simplicity by overcoming many limitations inherent in these rodent models. In this chapter, we review current advances in the behavioral and neurochemical characterization of ethanol-induced changes in zebrafish. We also discuss the basic principles and methods of and the most recent advances in using paradigms with which one can screen for compounds altering acute and chronic ethanol-induced effects in zebrafish.

International Review of Neurobiology, Volume 126
ISSN 0074-7742
http://dx.doi.org/10.1016/bs.irn.2016.02.016

1. THE COST OF ALCOHOL ADDICTION IN HUMANS

Alcohol (ethanol or ethyl alcohol) abuse and binge drinking has increased in the United States, and is associated with numerous significant adverse health effects (Naimi et al., 2003). In 2012, alcohol-impaired driving led to more than 10,000 motor vehicle fatalities (Jewett, Shults, Banerjee, & Bergen, 2015). The lifetime prevalence of alcohol dependence is estimated to be as high as 17% (Haberstick et al., 2014), and the annual cost associated with excessive alcohol consumption approached $250 billion in 2010 in the United States alone (Sacks, Gonzales, Bouchery, Tomedi, & Brewer, 2015). Despite alcohol addiction being a major unmet medical need and a substantial economic issue, there are only a few efficacious pharmacological treatment options available (Franck & Jayaram-Lindstrom, 2013). The development of new medications to treat alcohol addiction is hampered by the fact that the mechanisms underlying alcohol's effects in the brain and thus the factors behind the development of alcohol addiction (ie, the transition from casual to compulsive drinking) are poorly understood.

2. CURRENTLY AVAILABLE PHARMACOTHERAPIES FOR ALCOHOL ADDICTION

Alcohol is known to act on multiple biological targets and engage numerous biochemical pathways making the study of its mechanisms difficult. Although the exact mechanisms through which alcohol acts on the brain and the subsequent development of addiction are still unclear, several molecular targets have been implicated in the initiation and maintenance of compulsive alcohol consumption through rodent research (Vengeliene, Bilbao, Molander, & Spanagel, 2008). Although mechanistically complex, alcohol addiction has been recently reconceptualized as a treatable chronic relapsing brain disease (National Institute on Drug Abuse, 2008) with some pharmacological treatment options already available.

The Food and Drug Administration (FDA) approved drugs employed for the treatment of alcohol addiction generally fall under one of three categories. (1) Compounds that reduce the rewarding effects of alcohol, (2) compounds that reduce the negative effects associated with alcohol withdrawal, and (3) compounds that induce an aversive response following alcohol consumption.

Naltrexone is an FDA-approved opioid receptor antagonist which has been shown to reduce alcohol consumption and decrease relapse rates in rodents (Parkes & Sinclair, 2000) as well as in humans during treatment via diminishing the rewarding effects of this substance (Srisurapanont & Jarusuraisin, 2005). Naltrexone is thought to exert its effect by reducing the release of dopamine in the nucleus accumbens. It antagonizes opioid receptors in the ventral tegmental area (VTA), a brain region known to mediate the rewarding effects of alcohol (Vengeliene et al., 2008). However, once naltrexone is discontinued, no long-term beneficial effects remain detectable (Anton et al., 2006; Mann et al., 2013). Acamprosate is an FDA-approved *N*-methyl-D-aspartate receptor (NMDA-R) modulator which has been shown to increase the rates of abstinence from alcohol by reducing the negative effects associated with alcohol withdrawal. Although its mechanism of action is not completely understood, it is thought to exert its effect by antagonizing NMDA-Rs, glutamate-gated postsynaptic neurotransmitter receptors believed to be hyperactivated during alcohol withdrawal (Hammarberg, Jayaram-Lindstrom, Beck, Franck, & Reid, 2009). However, the effect size of this drug on reducing drinking is also small if found at all (Mann et al., 2013). Lastly, Disulfiram is an FDA approved drug which interferes with the breakdown of alcohol's metabolite and thus it increases acetaldehyde concentration, which is associated with aversive symptoms including facial flushing, headaches, nausea, and vomiting. When individuals take Disulfiram and subsequently consume alcohol, the aversive reaction is expected to serve as a deterrent for future episodes of drinking.

One of the main issues identified with using the above drugs as treatment has been lack of compliance (Fuller et al., 1986; Volpicelli, Rhines, & Rhines, 1997). A large meta-analysis determined that Disulfiram was significantly better at decreasing the total number of drinks during treatment compared to naltrexone, acamprosate, and placebo, with compliance being a crucial factor leading to the reduction of drinking (Skinner, Lahmek, Pham, & Aubin, 2014). These studies suggest short-term benefits only when medication compliance is high, ie, under well-controlled trial conditions, but with few to no long-term benefits under real-life conditions. Combining multiple medications also does not appear to increase treatment efficacy (Anton et al., 2006; Mann et al., 2013; Petrakis et al., 2005).

In summary, due to the lack of available efficacious pharmacological treatments, screening new compounds to treat alcohol addiction is an urgent necessity. However, given the complex mode of action of alcohol, such screens are often prohibitively expensive with traditional laboratory model organisms, the rat and the house mouse. Most recently, the zebrafish has gained traction as a novel animal model to overcome this limitation.

3. ZEBRAFISH AS AN ANIMAL MODEL FOR STUDYING ALCOHOL ADDICTION

The zebrafish has been a favorite of geneticists for the past several decades, especially for developmental biology (Grunwald & Streisinger, 1992; Streisinger, Singer, Walker, Knauber, & Dower, 1986). However, more recently, this small teleost has started to be utilized in the field of behavioral neuroscience with increasing popularity. Zebrafish are thought to strike an excellent compromise between systems complexity and practical simplicity (Gerlai, 2012, 2014). Briefly, it is an evolutionarily ancient and simple vertebrate that is easy to keep and breed in large numbers in the laboratory, yet it possesses several fundamental features, evolutionarily conserved characteristics shared across all major vertebrate taxa. The zebrafish genome has been sequenced and approximately 70% of genes in the human genome have at least one zebrafish orthologue (Howe et al., 2013). Furthermore, several zebrafish genes have been found to have high (70–90%) nucleotide sequence homology to mammalian (and human) genes. Although genetic approaches to studying alcohol addiction in zebrafish are becoming increasingly feasible due to the continued development of efficient forward and reverse genetic methods (Bedell et al., 2012; Hwang et al., 2013), we argue that one of the main advantages of this small vertebrate lies in its amenability for high-throughput drug screens. Zebrafish are highly prolific, capable of laying 100–300 eggs per clutch every other day upon reaching sexual development at approximately 3 months of age (Gerlai, 2014; Goldsmith, 2004). The social nature of zebrafish allows high density housing in system racks, with facilities capable of housing approximately 20,000 zebrafish in a standard 40 m^2 vivarium room. These qualities allow efficient breeding and housing of subjects while minimizing associated costs, important factors when deciding on an animal model for high-throughput chemical screens.

One of the main advantages of using zebrafish for alcohol addiction research is the drug administration procedure. Water-soluble compounds such as alcohol can be directly mixed with the tank water, and the immersed zebrafish can take up the compound through its skin and gills in a noninvasive manner (Dlugos & Rabin, 2003; Gerlai, Lahav, Guo, & Rosenthal, 2000; Mathur, Berberoglu, & Guo, 2011). Unlike rodents in which alcohol levels rise and subsequently decline following a stressful injection, blood and brain alcohol levels in zebrafish remain at a steady concentration due to the constant external alcohol bath. Furthermore, detailed time-dependent changes in brain (Dlugos & Rabin, 2003; Mathur, Berberoglu, et al., 2011; Tran, Chatterjee, & Gerlai, 2015) and blood alcohol concentrations (Echevarria, Toms, & Jouandot, 2011) have been reported following alcohol exposure in zebrafish. Water-insoluble compounds can also be mixed with the tank water using a vehicle such as dimethyl sulfoxide allowing zebrafish to take up these drugs in a noninvasive manner (Nowicki, Tran, Muraleetharan, Markovic, & Gerlai, 2014; Sackerman et al., 2010; Tran, Nowicki, Muraleetharan, & Gerlai, 2015).

Despite the clear advantages of zebrafish for screening drugs to treat alcohol addiction, there is a current bottleneck in this research. The zebrafish is a relative novice in behavioral pharmacology. For example, the phenotypical characterization of alcohol-induced functional changes in the brain and the behavioral paradigms required to screen for drugs that may alter alcohol effects or the addictive properties of this substance are still being developed. However, several aspects of alcohol addiction have already been examined in zebrafish. For example, the behavioral effects of acute and chronic ethanol exposure have already been quite well characterized in zebrafish.

4. ACUTE ETHANOL EXPOSURE IN ZEBRAFISH

Genetic variation among humans is known to alter how individuals respond to pharmacological compounds including alcohol (Pinto & Annsseau, 2009; Wang, Kapoor, & Goate, 2012). In humans, individual variation in how people respond to alcohol's stimulant effect has been identified as a risk factor for the development of addiction (Holdstock, King, & de Wit, 2000). The concentration-dependent locomotor activating effects of alcohol observed in animals parallel the time and concentration-dependent euphoric and stimulant effects of ethanol in humans. These similarities suggest alcohol-induced locomotor activity in animals can be used to study the rewarding effects of alcohol, which may underlie the development of alcohol addiction in humans (Phillips & Shen, 1996). Numerous animal models have been established to study alcohol-induced locomotor activity, yet the mechanisms underlying this behavioral response remain unknown (Phillips & Shen, 1996; Tran & Gerlai, 2014; Vengeliene et al., 2008).

The first use of zebrafish to study alcohol addiction and the effects of alcohol on locomotor activity was reported in 2000. Gerlai et al. (2000)

demonstrated that zebrafish of an outbred strain responded to acute ethanol exposure (ie, a single short-term exposure) in a biphasic, ie, inverted U-shape-dependent, manner. Low-to-moderate concentrations of ethanol (0.25 and 0.5% v/v) were found to increase locomotor activity while higher concentrations (1% v/v) were found to inhibit locomotor activity, similarly to the known biphasic stimulant and sedative effects of alcohol described in humans. The biphasic effect of acute ethanol exposure on locomotor activity of outbred zebrafish has also been replicated by others, and was correlated with increases in brain alcohol concentrations (Rosemberg et al., 2012).

Notably, the above inverted U-shaped dose response was not always observed in zebrafish. Zebrafish of the AB strain were found to show the highest level of locomotor response when exposed to the highest alcohol concentration (1% v/v) (Tran, Chatterjee, et al., 2015; Tran & Gerlai, 2013), a finding that demonstrates a strain-dependent idiosyncratic aspect of the effect of acute alcohol exposure. Differences in the effect of acute alcohol exposure between inbred and outbred strains of zebrafish have since been demonstrated in multiple studies (de Esch et al., 2012; Dlugos & Rabin, 2003; Gerlai, Chatterjee, Pereira, Sawashima, & Krishnannair, 2009; Lockwood, Bjerke, Kobayashi, & Guo, 2004; Pannia, Tran, Rampersad, & Gerlai, 2014), confirming the role of genes in mediating acute alcohol effects. Furthermore, individual variation in how zebrafish respond to alcohol's locomotor stimulant effect has also been reported in zebrafish (Tran, Nowicki, Muraleetharan, Chatterjee, & Gerlai, 2015b). These studies reinforce the validity of using zebrafish to study the locomotor stimulant effects of alcohol. Characterizing the locomotor stimulant effects of alcohol in zebrafish is the first step toward screening for compounds that could potentially reduce the rewarding effects of alcohol.

Although studying alcohol's locomotor stimulant effects (the positively reinforcing aspect of alcohol) in animal models may represent a reasonable way to investigate factors underlying alcohol addiction or to identify compounds that may alter addiction (Phillips & Shen, 1996; Robinson & Berridge, 2008) opponents of this approach emphasize negative reinforcement as the critical factor in the development of addiction (Koob, 2009). Alcohol has anxiolytic properties at low-to-moderate concentrations both in animals and in humans. Not surprisingly, there is a high rate of comorbidity for anxiety disorders and alcohol addiction (Robinson, Sareen, Cox, & Bolton, 2009). Similar to alcohol-induced locomotor activity, the neurochemical mechanisms mediating alcohol-induced anxiolysis is also poorly understood. Alcohol's anxiolytic effects have been demonstrated in numerous animal models, including mice (Radwanska & Kaczmarek, 2012), rats (Varlinskaya & Spear, 2015), and zebrafish (Echevarria et al., 2011). In zebrafish, acute exposure to low and moderate doses of alcohol has been shown to reduce anxiety-like behaviors, including reduction of erratic movement (Egan et al., 2009), freezing (Blaser & Penalosa, 2011), predator-induced avoidance (Pannia et al., 2014), and bottom dwelling (Wong et al., 2010). Characterizing the behavioral profile of alcohol-induced anxiolysis will allow for screens to identify compounds that could potentially reduce stress and anxiety associated with alcohol addiction.

5. CHRONIC ETHANOL EXPOSURE IN ZEBRAFISH

According to the negative reinforcement model of addiction (Koob, 2009), following repeated and chronic consumption of alcohol, individuals begin to develop tolerance to the effects of alcohol requiring larger quantities to achieve the same rewarding effects (Novier, Diaz-Granados, & Matthews, 2015). However, once alcohol consumption ceases, individuals experience the aversive effects of withdrawal, which may motivate them to reinstate the consumption of the substance (Fukushiro et al., 2012). Both tolerance and withdrawal may thus contribute to the development and maintenance of alcohol addiction in humans. Fortunately, paradigms to study alcohol tolerance (Damodaran, Dlugos, Wood, & Rabin, 2006; Dlugos & Rabin, 2003; Gerlai et al., 2009; Tran, Chatterjee, et al., 2015) and also alcohol withdrawal have been developed for zebrafish using chronic ethanol exposure paradigms (Cachat et al., 2010; Holcombe, Schalomon, & Hamilton, 2014; Mathur & Guo, 2011; Tran, Chatterjee, et al., 2015), which can be used to screen for drugs that may affect alcohol addiction.

Chronic ethanol exposure in zebrafish has been performed using two different protocols. First, zebrafish can be acutely exposed to alcohol repeatedly and intermittently over a period of time, usually over the span of a few weeks with alcohol doses reaching as high as 3% v/v concentration (Chacon & Luchiar, 2014; Mathur & Guo, 2011; Pittman & Ichikawa, 2013). The advantage of this protocol is that it mimics the repeated and intermittent alcohol consumption observed among humans thus providing translational relevance. The second alcohol exposure protocol involves continuous alcohol exposure by administering alcohol directly to the housing tanks of zebrafish for an extended period of time with alcohol doses reaching 0.5% v/v concentration applied continuously for as long as 10 weeks (Damodaran et al., 2006; Dlugos & Rabin, 2003; Egan et al., 2009;

Gerlai et al., 2009; Tran, Chatterjee, et al., 2015). For reasons of brevity, this second ethanol exposure protocol will henceforth be referred to as continuous alcohol exposure. Although continuous alcohol exposure in zebrafish does not parallel the temporal pattern of alcohol consumption in humans, the continuity and stability of the alcohol concentration applied throughout is viewed as a simplifying reductionist approach, one which allows the removal of the potentially complicating effects of repeated withdrawal from the drug inherent in the repeated intermittent protocol. Both protocols have been shown to induce tolerance to ethanol's locomotor stimulant, anxiolytic, and anxiogenic effects (Damodaran et al., 2006; Dlugos & Rabin, 2003; Luchiari, Salajan, & Gerlai, 2015; Tran, Chatterjee, et al., 2015; Tran, Nowicki, Chatterjee, & Gerlai, 2015). Notably, continuous alcohol exposure appears to lead to a more robust level of tolerance compared to repeated intermittent alcohol exposure (Tran, Nowicki, Chatterjee, et al., 2015).

Finally, following both repeated intermittent and continuous alcohol exposure in zebrafish, testing subjects in regular system water without alcohol allows for the investigation of effects of withdrawal from alcohol. The validity of this withdrawal paradigm is reinforced by the observed increase of anxiety-like behavioral responses following withdrawal both after repeated intermittent (Holcombe et al., 2014; Mathur & Guo, 2011; Pittman & Hylton, 2015) and continuous alcohol exposure in zebrafish (Cachat et al., 2010; Luchiari et al., 2015; Tran, Chatterjee, et al., 2015; Tran & Gerlai, 2013).

6. ALTERNATIVE ALCOHOL EXPOSURE PARADIGMS TO STUDY ALCOHOL ADDICTION

The alcohol exposure paradigms developed to study alcohol addiction in zebrafish discussed thus far represent a large majority of the published studies in the literature (Tran & Gerlai, 2014). However, there are other paradigms that are currently being used to study alcohol addiction in zebrafish, many of which have been adapted from the rodent literature (Collier, Khan, Caramillo, Mohn, & Echevarria, 2014; Guo, Wagle, & Mathur, 2012; Stewart et al., 2011).

Although intermittent repeated exposure to ethanol has been reported to induce mild tolerance in zebrafish (Tran, Nowicki, Chatterjee, et al., 2015), it is important to note that zebrafish can also become sensitized or develop reverse tolerance to the effects of ethanol (Blaser, Koid, & Poliner, 2010;

Tran & Gerlai, 2014). According to the incentive sensitization theory, repeated exposure to drugs of abuse can hypersensitize brain circuits responsible for attributing saliency and desirability to drugs and drug-associated stimuli. These long-term alterations increase the motivational value or "wanting" of the drug, leading to compulsive drug-seeking behavior (Robinson & Berridge, 2008). Sensitization of alcohol's locomotor stimulant effect has been well established in the rodent literature (Phillips, Roberts, & Lessov, 1997) and has already been demonstrated in zebrafish as well (Blaser et al., 2010).

The effect of intermittent repeated alcohol exposure has also been studied using conditioned place preference (CPP), a paradigm that is aimed at quantifying rewarding properties of drugs and changes in such properties after different exposure regimens (Collier & Echevarria, 2013; Collier et al., 2014). In this paradigm, alcohol can be repeatedly paired with a neutral environment during multiple training sessions. Over time, subjects will exhibit a preference for the alcohol-paired environment compared to a nonalcohol-paired environment, which is used as a measure of the rewarding or motivational effects of alcohol. Zebrafish may be exceptionally well suited for CPP, since a significant increase in preference for the alcohol-paired environment may be achieved following a single 20 min exposure (Mathur, Berberoglu, et al., 2011; Mathur, Lau, & Guo, 2011), a necessity for efficient high-throughput behavioral screens. Notably, however, the common problem with CPP paradigms in alcohol research is that these tasks have a significant memory component: the subject needs to acquire and consolidate the association between the spatial stimuli representing the rewarded location and the reward (alcohol). Alcohol is known to interact with numerous molecular targets and neurobiological mechanisms that subserve neuronal plasticity, learning, and memory. Thus, exposure to alcohol may exert effects on performance in CPP tasks in two ways, one via altering the motivational valence of the reward and two by disrupting memory, factors that are not possible to dissociate using CPP. For example, if the rewarding value of alcohol increases due to prior chronic exposure to the drug, the subject is expected to show increased place preference, but as a result of exposure to alcohol during the acquisition and/or consolidation phase of the task, memory may be impaired, effects that may cancel, or interact with, each other. Thus, if one wants to study the rewarding properties of alcohol, it is better to measure such effects directly, ie, by using a choice task. One of these tasks is the alcohol self-administration paradigm employed with rodents.

Alcohol self-administration has been difficult to achieve in zebrafish. Presently, most zebrafish studies use involuntary alcohol exposure (alcohol is mixed directly with the tank water). Thus, paradigms allowing the quantification of alcohol choice in zebrafish have been limited. However, a novel model of voluntary alcohol consumption in zebrafish has been developed in which zebrafish can choose between consuming gelatin containing 20% alcohol compared to gelatin containing 0% alcohol (Sterling, Karatayev, Chang, Algava, & Leibowitz, 2015). Although still in early stages of development, these alcohol-related behavioral tasks will help us investigate the neurobiological mechanisms underlying the actions of this complex drug.

7. TARGETS FOR MEDICATION TO TREAT ALCOHOL ADDICTION

Analysis of the effects of alcohol in zebrafish has already generated a substantial amount of data, which may serve as a starting point for screening compounds to treat alcohol addiction. For example, analysis of the effects of acute ethanol exposure on the zebrafish brain found numerous alterations including neurochemicals such as monoamines (Gerlai et al., 2009; Guo et al., 2015; Puttonen, Sundvik, Rozov, Chen, & Panula, 2013), amino acids (Chatterjee, Shams, & Gerlai, 2014; Tran, Chatterjee, et al., 2015), hormones (Oliveira et al., 2013; Tran, Chatterjee, et al., 2015), peptides (Sterling et al., 2015), enzyme activity (Chatterjee, Shams, & Gerlai, 2014; Rico, Rosemberg, Dias, Bogo, & Bonan, 2007), gene expression (Puttonen et al., 2013; Rico et al., 2007), and changes in signal transduction pathways (Peng et al., 2009). Similarly, following repeated and/or continuous alcohol exposure, numerous changes were detected at the level of mRNA expression (Kily et al., 2008; Pan, Kaiguo, Razak, Westwood, & Gerlai, 2011; Rico et al., 2011) and protein expression in the zebrafish brain (Damodaran et al., 2006) as well as in enzymatic activity in the liver (Tran, Nowicki, Chatterjee, et al., 2015).

Some studies have focused on specific molecular mechanisms associated with alcohol-induced functional changes in the brain. For example, locomotor activity is often investigated in the context of the dopaminergic system in rodents (Phillips & Shen, 1996) and recently in zebrafish (Tran & Gerlai, 2014). In zebrafish, acute alcohol exposure alters locomotor and dopaminergic activity in a correlated dose-dependent manner (Guo et al., 2015; Puttonen et al., 2013; Tran, Chatterjee, et al., 2015). Alterations to the dopaminergic system include changes in whole-brain levels of dopamine, in the levels of dopamine's metabolite 3,4-dihydroxyphenylacetic acid (DOPAC) (Guo et al., 2015; Puttonen et al., 2013; Tran, Chatterjee, et al., 2015), in tyrosine hydroxylase enzymatic activity (Chatterjee et al., 2014), as well as in tyrosine hydroxylase gene expression (Puttonen et al., 2013). In line with the effects of acute alcohol exposure, following continuous alcohol exposure, zebrafish develop tolerance to both the locomotor and dopaminergic altering effects of alcohol (Chatterjee et al., 2014; Tran, Chatterjee, et al., 2015). The link between alcohol-induced locomotor and dopaminergic activity is well supported by the incentive salience theory of addiction. The proposed circuit mediating incentive salience in mammals is the mesolimbic dopaminergic pathway originating from the VTA to the nucleus accumbens (Robinson & Berridge, 2008), and alcohol is thought to hijack this pathway (Vengeliene et al., 2008). Although the zebrafish brain is structurally different from the rodent brain, data discussed earlier suggest a similar involvement of the dopaminergic system in alcohol-related functional effects in the brain of zebrafish.

Alcohol's rewarding effects are strongly associated with changes in dopaminergic activity; however, the anxiety-altering effects of this substance may be mediated by another neurotransmitter system, the serotonergic system. Similar to mammals, the role of the serotonergic neurotransmitter system is often examined in the context of anxiety-like behavioral responses in zebrafish (Herculano & Maximino, 2014). In zebrafish, acute ethanol exposure has been shown to alter anxiety-like behavioral responses (Blaser & Penalosa, 2011; Egan et al., 2009; Wong et al., 2010), as well as serotonergic activity including whole-brain levels of serotonin and of serotonin's metabolite 5-hydroxyindoleacetic acid (5-HIAA), an effect that is alcohol dose--dependent (Chatterjee et al., 2014; Guo et al., 2015; Tran, Chatterjee, et al., 2015). Likewise, changes in serotonergic activity are also observed during withdrawal from alcohol (Chatterjee et al., 2014; Tran, Chatterjee, et al., 2015), which results in increased anxiety-like behavioral responses in zebrafish (Cachat et al., 2010; Pittman & Hylton, 2015; Tran, Chatterjee, et al., 2015). The link between alcohol-induced withdrawal and serotonergic activity is also well supported in the rodent literature (Uzbay, 2008) as well as in humans (Clapp, Bhave, & Hoffman, 2008). The data currently available on alcohol-induced dopaminergic and serotonergic changes represent a strong starting point for the development and screening of compounds that may treat alcohol addiction. These results may be used to develop zebrafish screening paradigms to identify compounds that reduce the rewarding effects of alcohol and/or diminish the negative effects of withdrawal from alcohol.

8. CURRENT ADVANCES IN HIGH-THROUGHPUT SCREENS OF COMPOUNDS TO TREAT ALCOHOL ADDICTION

The establishment of alcohol-related paradigms in zebrafish has allowed investigators to screen for compounds to treat alcohol addiction at the preclinical stage. Although a large-scale screen of numerous compounds to treat alcohol addiction has yet to be conducted with zebrafish, investigators have begun using a targeted approach to screen specific compounds likely to block alcohol-induced locomotor activity and withdrawal.

As mentioned previously, the dopaminergic system is a likely candidate for alcohol's reinforcing effects in humans (Charlet, Beck, & Heinz, 2013) and locomotor stimulant effects in zebrafish (Nowicki, Tran, Chatterjee, & Gerlai, 2015; Puttonen et al., 2013). Recently, several screens for dopaminergic compounds were conducted in zebrafish to block alcohol-induced locomotor activity. A selective phosphorylated tyrosine hydroxylase inhibitor (tetrahydropapaveroline) was found to attenuate alcohol-induced locomotor activity and abolish alcohol-induced increase of whole-brain dopamine levels (Nowicki et al., 2015). The dopamine D_1 receptor antagonist (SCH-23390) was found to inhibit both locomotor and dopaminergic activity independent of alcohol exposure, suggesting alcohol-induced locomotor activity is independent of dopamine D_1-like receptor activation (Tran, Nowicki, Muraleetharan, Chatterjee, & Gerlai, 2015a). A subsequent follow-up study found that the D_2-like receptor antagonist (Haloperidol) attenuated alcohol-induced locomotor activity (unpublished data). Together, these studies suggest that alcohol-induced locomotor activity in zebrafish may be driven by the phosphorylation of tyrosine hydroxylase to increase dopamine synthesis, which leads to activation of dopamine D_2-like receptors. It is possible that the rewarding effects of alcohol in humans may occur in a similar manner. Alternatively, compounds have also been screened to block alcohol-induced sedation, quantified by a decrease in locomotor activity. Acute alcohol exposure has been shown to alter whole-brain levels of the amino sulfonic acid taurine in zebrafish (Chatterjee et al., 2014). Rosemberg et al. (2012) found that exogenous administration of taurine abolished ethanol-induced sedation and attenuated the increase in brain ethanol levels in zebrafish, identifying a non-dopaminergic pathway mediating ethanol's locomotor altering effects.

Screens for compounds in zebrafish to treat anxiety-like behaviors associated with alcohol-induced withdrawal have also been conducted. In zebrafish, withdrawal from chronic alcohol exposure has been shown to increase anxiety-like behavioral responses (Cachat et al., 2010; Holcombe et al., 2014; Pittman & Hylton, 2015; Tran, Chatterjee, et al., 2015) as well as both serotonergic and glutamatergic activity in the brain (Chatterjee et al., 2014; Tran, Chatterjee, et al., 2015). The serotonin reuptake inhibitor (fluoxetine) and the glutamatergic NMDA receptor antagonist (ketamine) have been shown to block the anxiogenic effects of withdrawal from alcohol and increases in serotonin synthesis (Pittman & Hylton, 2015). Although studies screening for medications to block alcohol-induced CPP in zebrafish have not been conducted yet, several investigators have identified compounds attenuating nicotine- (Ponzoni et al., 2014) and cocaine-induced conditioned placed preference in zebrafish (Darland et al., 2012).

An increasing number of studies screening for compounds in zebrafish with the goal of identifying mechanisms underlying and treating alcohol addiction have been published over the past several years, reinforcing the utility of zebrafish in the field of behavioral pharmacology. These studies have established a foundation for the use of zebrafish at the preclinical stage in pharmacological screening of compounds to treat not just alcohol addiction, but diseases and disorders in general. The advantages offered by this small vertebrate including its small size, prolific nature, and low cost of maintenance overcome many limitations inherent in traditional rodent models. We forecast that zebrafish will be perhaps the best vertebrate system for high-throughput screens with which one can identify compounds for the treatment of alcohol addiction.

ACKNOWLEDGMENTS

Supported by NSERC (Discovery Grant 311637) and NIH/NIAAA (R01 AA14357-01A2).

REFERENCES

Anton, R. F., O'Malley, S. S., Ciraulo, D. A., Ciraulo, D. A., Cisler, R. A., Couper, D., et al. (2006). Combined pharmacotherapies and behavioural interventions for alcohol dependence. *JAMA*, *295*, 2003–2017.

Bedell, V. M., Wang, Y., Campbell, J. M., Poshusta, T. L., Starker, C. G., Krug, R. G., 2nd, et al. (2012). In vivo genome editing using a high-efficiency TALEN system. *Nature*, *491*, 114–118.

Blaser, R. E., Koid, A., & Poliner, R. M. (2010). Context-dependent sensitization to ethanol in zebrafish (Danio rerio). *Pharmacology, Biochemistry, and Behavior*, *95*, 278–284.

Blaser, R. E., & Penalosa, Y. M. (2011). Stimuli affecting zebrafish (Danio rerio) behavior in the light/dark preference test. *Physiology & Behavior, 104*, 831–837.

Cachat, J., Canavello, P., Elegante, M., Bartels, B., Hart, P., Bergner, C., et al. (2010). Modeling withdrawal syndrome in zebrafish. *Behavioural Brain Research, 208*, 371–376.

Chacon, D. M., & Luchiar, A. C. (2014). A dose for the wiser is enough: The alcohol benefits for associative learning in zebrafish. *Progress in Neuro-Psychopharmacology & Biological Psychiatry, 53*, 109–115.

Charlet, K., Beck, A., & Heinz, A. (2013). The dopamine system in mediating alcohol effects in humans. *Current Topics in Behavioral Neurosciences, 13*, 461–488.

Chatterjee, D., Shams, S., & Gerlai, R. (2014). Chronic and acute alcohol administration induced neurochemical changes in the brain: Comparison of distinct zebrafish populations. *Amino Acids, 46*, 921–930.

Clapp, P., Bhave, S. V., & Hoffman, P. L. (2008). How adaptation of the brain to alcohol leads to dependence: A pharmacological perspective. *Alcohol Research & Health, 31*, 310–339.

Collier, A. D., & Echevarria, D. J. (2013). The utility of the zebrafish model in conditioned place preference to assess the rewarding effects of drugs. *Behavioural Pharmacology, 24*, 375–383.

Collier, A. D., Khan, K. M., Caramillo, E. M., Mohn, R. S., & Echevarria, D. J. (2014). Zebrafish and conditioned place preference: A translational model of drug reward. *Progress in Neuro-Psychopharmacology & Biological Psychiatry, 55*, 16–25.

Damodaran, S., Dlugos, C. A., Wood, T. D., & Rabin, R. A. (2006). Effects of chronic ethanol administration on brain protein levels: A proteomic investigation using 2-D DIGE system. *European Journal of Pharmacology, 547*, 75–82.

Darland, T., Mauch, J. T., Meier, E. M., Hagan, S. J., Dowling, J. E., & Darland, D. C. (2012). Sulpiride, but not SCH23390, modifies cocaine-induced conditioned place preference and expression of tyrosine hydroxylase and elongation factor 1α in zebrafish. *Pharmacology, Biochemistry, and Behavior, 103*, 157–167.

de Esch, C., van der Linde, H., Slieker, R., Willemsen, R., Wolterbeek, A., Woutersen, R., et al. (2012). Locomotor activity assay in larvae: Influence of age, strain and ethanol. *Neurotoxicology and Teratology, 34*, 425–433.

Dlugos, C. A., & Rabin, R. A. (2003). Ethanol effects on three strains of zebrafish: Model system for genetic investigations. *Pharmacology, Biochemistry, and Behavior, 74*, 471–480.

Echevarria, D. J., Toms, C. N., & Jouandot, D. J. (2011). Alcohol-induced behavioral change in zebrafish models. *Reviews in Neuroscience, 22*, 85–93.

Egan, R. J., Bergner, C. L., Hart, P. C., Cachat, J. M., Canavello, P. R., Elegante, M. F., et al. (2009). Understanding behavioral and physiological phenotypes of stress and anxiety in zebrafish. *Behavioural Brain Research, 205*, 38–44.

Franck, J., & Jayaram-Lindstrom, N. (2013). Pharmacotherapy for alcohol dependence: Status of current treatments. *Current Opinions in Neurobiology, 23*, 692–699.

Fukushiro, D. F., Saito, L. P., Mari-Kawamoto, E., Aramini, T. C. F., Costa, J. M., Josino, F. S., et al. (2012). Withdrawal from repeated treatment with ethanol induces a protracted decrease in novelty-seeking behavior and enhancement of environmental habituation in mice. *Pharmacology, Biochemistry, and Behavior, 101*, 132–137.

Fuller, R. K., Branchey, L., Brightwell, D. R., Derman, R. M., Emrick, C. D., Iber, F. L., et al. (1986). Disulfiram treatment of alcoholism: A veterans administration cooperative study. *JAMA, 256*, 1449–1455.

Gerlai, R. (2012). Using zebrafish to unravel the genetics of complex brain disorders. *Current Topics in Behavioral Neuroscience, 12*, 3–24.

Gerlai, R. (2014). Social behavior of zebrafish: From synthetic images to biological mechanisms of shoaling. *Journal of Neuroscience Methods, 234*, 59–65.

Gerlai, R., Chatterjee, D., Pereira, T., Sawashima, T., & Krishnannair, R. (2009). Acute and chronic alcohol dose: Population differences in behavior and neurochemistry of zebrafish. *Genes, Brain, and Behavior, 8*, 586–599.

Gerlai, R., Lahav, M., Guo, S., & Rosenthal, A. (2000). Drinks like a fish: Zebra fish (Danio rerio) as a behavior genetic model to study alcohol effects. *Pharmacology, Biochemistry, and Behavior, 67*, 773–782.

Goldsmith, P. (2004). Zebrafish as a pharmacological tool: The how, why and when. *Current Opinion in Pharmacology, 4*, 504–512.

Grunwald, D. J., & Streisinger, G. (1992). Induction of mutations in the zebrafish with ultraviolet light. *Genetics Research, 59*, 93–101.

Guo, N., Lin, J., Peng, X., Chen, H., Zhang, Y., Liu, X., et al. (2015). Influences of acute ethanol exposure on locomotor activities of larvae under different illumination. *Alcohol, 49*, 727–737.

Guo, S., Wagle, M., & Mathur, P. (2012). Toward molecular and genetic dissection of neural circuits for emotional and motivational behaviors. *Developmental Neurobiology, 72*, 358–365.

Haberstick, B. C., Young, S. E., Zeiger, J. S., Lessem, J. M., Kewitt, K. L., & Hopfer, C. J. (2014). Prevalence and correlates of alcohol and cannabis use disorders in the United States: Results from the national longitudinal study of adolescent health. *Drug and Alcohol Dependence, 136*, 158–161.

Hammarberg, A., Jayaram-Lindstrom, N., Beck, O., Franck, J., & Reid, M. S. (2009). The effects of acamprosate on alcohol-cue reactivity and alcohol priming in dependent patients: A randomized controlled trial. *Psychopharmacology, 205*, 53–62.

Herculano, A. M., & Maximino, C. (2014). Serotonergic modulation of zebrafish behavior: Towards a paradox. *Progress in Neuro-Psychopharmacology & Biological Psychiatry, 55*, 50–66.

Holcombe, A., Schalomon, M., & Hamilton, T. J. (2014). A novel method of drug administration to multiple zebrafish (Danio rerio) and the quantification of withdrawal. *Journal of Visualized Experiments, 93*, e51851.

Holdstock, L., King, A. C., & de Wit, H. (2000). Subjective and objective responses to ethanol in moderate/heavy and light social drinkers. *Alcoholism: Clinical and Experimental Research, 24*, 789–794.

Howe, K., Clark, M. D., Torroja, C. F., Torrance, J., Berthelot, C., Muffato, M., et al. (2013). The zebrafish reference genome sequence and its relationship to the human genome. *Nature, 496*, 498–503.

Hwang, W. Y., Fu, Y., Reyon, D., Maeder, M. L., Tsai, S. Q., Sander, J. D., et al. (2013). Efficient genome editing in zebrafish using a CRISPR-Cas system. *Nature Biotechnology, 31*, 227–229.

Jewett, A., Shults, R. A., Banerjee, T., & Bergen, G. (2015). Alcohol-impaired driving among adults—United States, 2012. *Morbidity and Mortality Weekly Report, 64*, 814–817.

Kily, L. J., Cowe, Y. C., Hussain, O., Patel, S., McElwaine, S., Cotter, F. E., et al. (2008). Gene expression changes in a zebrafish model of drug dependency suggest conservation of neuro-adaptation pathways. *Journal of Experimental Biology, 211*, 1623–1634.

Koob, G. F. (2009). Neurobiological substrates for the dark side of compulsivity in addiction. *Neuropharmacology, 56*, 18–31.

Lockwood, B., Bjerke, S., Kobayashi, K., & Guo, S. (2004). Acute effects of alcohol on larval zebrafish: A genetic system for large-scale screening. *Pharmacology, Biochemistry, and Behavior, 77*, 647–654.

Luchiari, A. C., Salajan, D. C., & Gerlai, R. (2015). Acute and chronic alcohol administration: Effects on performance of zebrafish in a latent learning task. *Behavioural Brain Research, 282*, 76–83.

Mann, K., Lemenager, T., Hoffmann, S., Reinhard, I., Hermann, D., Batra, A., et al. (2013). Results of a double-blind, placebo-controlled pharmacotherapy trial in alcoholism

conducted in Germany and comparison with the US COMBINE study. *Addiction Biology, 18*, 937–946.
Mathur, P., Berberoglu, M. A., & Guo, S. (2011). Preference for ethanol in zebrafish following a single exposure. *Behavioural Brain Research, 217*, 128–133.
Mathur, P., & Guo, S. (2011). Differences of acute versus chronic ethanol exposure on anxiety-like behavioral responses in zebrafish. *Behavioural Brain Research, 219*, 234–239.
Mathur, P., Lau, B., & Guo, S. (2011). Conditioned place preference behavior in zebrafish. *Nature Protocols, 6*, 338–345.
Naimi, T. S., Brewer, R. D., Mokdad, A., Denny, C., Serdula, M. K., & Marks, J. S. (2003). Bing drinking among US adults. *JAMA, 289*, 70–75.
National Institute on Drug Abuse. (2008). *Drugs, brains, and behaviour: The science of addiction. (revised).* Washington DC: National Institute on Drug Abuse.
Novier, A., Diaz-Granados, J. L., & Matthews, D. B. (2015). Alcohol use across the lifespan: An analysis of adolescent and aged rodents and humans. *Pharmacology, Biochemistry, and Behavior, 33*, 65–82.
Nowicki, M., Tran, S., Chatterjee, D., & Gerlai, R. (2015). Inhibition of phosphorylated tyrosine hydroxylase attenuates ethanol-induced hyperactivity in adult zebrafish (Danio rerio). *Pharmacology, Biochemistry, and Behavior, 138*, 32–39.
Nowicki, M., Tran, S., Muraleetharan, A., Markovic, S., & Gerlai, R. (2014). Serotonin antagonists induce anxiolytic and anxiogenic-like behavior in zebrafish in a receptor subtype dependent manner. *Pharmacology, Biochemistry, and Behavior, 126*, 170–180.
Oliveira, T. A., Koalski, G., Kreutz, L. C., Ferreira, D., da Rosa, J. G., de Abreu, M. S., et al. (2013). Alcohol impairs predation risk response and communication in zebrafish. *PloS One, 8*, e75780.
Pan, Y., Kaiguo, M., Razak, Z., Westwood, J. T., & Gerlai, R. (2011). Chronic alcohol exposure induced gene expression changes in the zebrafish brain. *Behavioural Brain Research, 216*, 66–76.
Pannia, E., Tran, S., Rampersad, M., & Gerlai, R. (2014). Acute ethanol exposure induces behavioral differences in two zebrafish (Danio rerio) strains: A time course analysis. *Behavioural Brain Research, 259*, 174–185.
Parkes, J. H., & Sinclair, J. D. (2000). Reduction of alcohol drinking and upregulation of opioid receptors by oral naltrexone in AA rats. *Alcohol, 21*, 215–221.
Peng, J., Wagle, M., Mueller, T., Mathur, P., Lockwood, B. L., & Guo, S. (2009). Ethanol-modulated camouflage response screen in zebrafish uncovers a novel role for cAMP and extracellular signal-related kinase signaling in behavioral sensitivity to ethanol. *The Journal of Neuroscience, 29*, 8408–8418.
Petrakis, I. L., Poling, J., Levinson, C., Nich, C., Carroll, K., & Rounsaville, B. (2005). Naltrexone and disulfiram in patients with alcohol dependence and comorbid psychiatric disorders. *Biological Psychiatry, 57*, 1128–1137.
Phillips, T. J., Roberts, A. J., & Lessov, C. N. (1997). Behavioral sensitization to ethanol: Genetics and the effects of stress. *Pharmacology, Biochemistry, and Behavior, 57*, 487–493.
Phillips, T. J., & Shen, E. H. (1996). Neurochemical bases of locomotion and ethanol stimulant effects. *International Review of Neurobiology, 243*, 243–282.
Pinto, E., & Annsseau, M. (2009). Genetic factors of alcohol dependence. *Encephale, 35*, 461–469.
Pittman, J., & Hylton, A. (2015). Behavioral, endocrine, and neuronal alterations in zebrafish (Danio rerio) following sub-chronic coadministration of fluoxetine and ketamine. *Pharmacology, Biochemistry and Behavior, 139*, 158–162.
Pittman, J. T., & Ichikawa, K. M. (2013). iPhone applications as versatile video tracking tools to analyze behavior in zebrafish (Danio rerio). *Pharmacology, Biochemistry and Behavior, 106*, 137–142.
Ponzoni, L., Braida, D., Pucci, L., Andrea, D., Fasioli, F., Manfredi, I., et al. (2014). The cytosine derivatives, CC4 and CC26, reduce nicotine-induced conditioned place

preference in zebrafish by acting on heteromeric neuronal nicotinic acetylcholine receptors. *Psychopharmacology*, *231*, 4681–4693.

Puttonen, H. A., Sundvik, M., Rozov, S., Chen, Y. C., & Panula, P. (2013). Acute ethanol treatment upregulates Th1, Th2, and Hdc in larval zebrafish in stable networks. *Frontiers in Neural Circuits*, 7, 1–10.

Radwanska, K., & Kaczmarek, L. (2012). Characterization of an alcohol addiction-prone phenotype in mice. *Addiction Biology*, *17*, 601–612.

Rico, E. P., Rosemberg, D. B., Dias, R. D., Bogo, M. R., & Bonan, C. D. (2007). Ethanol alters acetylcholinesterase activity and gene expression in zebrafish brain. *Toxicology Letters*, *174*, 25–30.

Rico, E. P., Rosemberg, D. B., Langoni Ada, S., Souto, A. A., Dias, R. D., Bogo, M. R., et al. (2011). Chronic ethanol treatment alters purine nucleotide hydrolysis and nucleotidase gene expression pattern in zebrafish brain. *Neurotoxicology*, *32*, 871–878.

Robinson, T. E., & Berridge, K. C. (2008). The incentive sensitization theory of addiction: Some current issues. *Philosophical Transactions of the Royal Society of London B: Biological Sciences*, *363*, 3137–3146.

Robinson, J., Sareen, J., Cox, B. J., & Bolton, J. (2009). Self-medication of anxiety disorders with alcohol and drugs: Results from a nationally representative sample. *Journal of Anxiety Disorders*, *23*, 38–45.

Rosemberg, D. B., Braga, M. M., Rico, E. P., Loss, C. M., Cordova, S. D., Mussulini, B. H., et al. (2012). Behavioral effects of taurine pretreatment in zebrafish acute exposed to ethanol. *Neuropharmacology*, *63*, 613–623.

Sackerman, J., Donegan, J. J., Cunningham, C. S., Nguyen, N. N., Lawless, K., Long, A., et al. (2010). Zebrafish behavior in novel environments: Effects of acute exposure to anxiolytic compounds and choice of Danio rerio line. *International Journal of Comparative Psychology*, *23*, 43–61.

Sacks, J. J., Gonzales, K. R., Bouchery, E. E., Tomedi, L. E., & Brewer, R. D. (2015). 2010 national and state costs of excessive alcohol consumption. *American Journal of Preventive Medicine*, *49*, e73–e79.

Skinner, M. D., Lahmek, P., Pham, H., & Aubin, H. J. (2014). Disulfiram efficacy in the treatment of alcohol dependence: A meta-analysis. *PloS One*, *9*, e87366.

Srisurapanont, M., & Jarusuraisin, N. (2005). Naltrexone for the treatment of alcoholism: A meta-analysis of randomized controlled trials. *International Journal of Neuropharmacology*, *8*, 267–280.

Sterling, M. E., Karatayev, O., Chang, G. Q., Algava, D. B., & Leibowitz, S. F. (2015). Model of voluntary ethanol intake in zebrafish: Effect on behavior and hypothalamic orexigenic peptides. *Behavioural Brain Research*, *278*, 29–39.

Stewart, A., Wong, K., Cachat, J., Gaikwad, S., Kyzar, E., Wu, N., et al. (2011). Zebrafish models to study drug abuse-related phenotypes. *Reviews in the Neurosciences*, *22*, 95–105.

Streisinger, G., Singer, F., Walker, C., Knauber, D., & Dower, N. (1986). Segregation analyses and gene-centromere distances in zebrafish. *Genetics*, *112*, 311–319.

Tran, S., Chatterjee, D., & Gerlai, R. (2015). An integrative analysis of ethanol tolerance and withdrawal in zebrafish (Danio rerio). *Behavioural Brain Research*, *276*, 161–170.

Tran, S., & Gerlai, R. (2013). Time-course of behavioral changes induced by ethanol in zebrafish (Danio rerio). *Behavioural Brain Research*, *252*, 204–213.

Tran, S., & Gerlai, R. (2014). Recent advances with a novel model organism: Alcohol tolerance and sensitization in zebrafish (Danio rerio). *Progress in Neuro-Psychopharmacology & Biological Psychiatry*, *55*, 87–93.

Tran, S., Nowicki, M., Chatterjee, D., & Gerlai, R. (2015). Acute and chronic ethanol exposure differentially alters alcohol dehydrogenase and aldehyde dehydrogenase activity in the zebrafish liver. *Progress in Neuro-Psychopharmacology & Biological Psychiatry*, *56*, 221–226.

Tran, S., Nowicki, M., Muraleetharan, A., Chatterjee, D., & Gerlai, R. (2015b). Neurochemical factors underlying individual differences in locomotor activity and anxiety-like behavioral responses in zebrafish. *Progress in Neuro-Psychopharmacology & Biological Psychiatry*, *65*, 25–33.

Tran, S., Nowicki, M., Muraleetharan, A., Chatterjee, D., & Gerlai, R. (2015a). Differential effects of acute administration of SCH-23390, a D1 receptor antagonist, and of ethanol on swimming activity, anxiety-related responses, and neurochemistry of zebrafish. *Psychopharmacology*, *232*, 3709–3718.

Tran, S., Nowicki, M., Muraleetharan, A., & Gerlai, R. (2015). Differential effects of dopamine D1 and D2/3 receptor antagonism on motor responses. *Psychopharmacology*, *232*, 795–806.

Uzbay, I. T. (2008). Serotonergic anti-depressants and ethanol withdrawal syndrome: A review. *Alcohol and Alcoholism*, *43*, 15–24.

Varlinskaya, E. I., & Spear, L. P. (2015). Social consequences of ethanol: Impact of age, stress, and prior history of ethanol exposure. *Physiology & Behavior*, *148*, 145–150.

Vengeliene, V., Bilbao, A., Molander, A., & Spanagel, R. (2008). Neuropharmacology of alcohol addiction. *British Journal of Pharmacology*, *154*, 299–315.

Volpicelli, J. R., Rhines, K. C., & Rhines, J. S. (1997). Naltrexone and alcohol dependence: Role of subject compliance. *Archives of General Psychiatry*, *54*, 737–742.

Wang, J. C., Kapoor, M., & Goate, A. M. (2012). The genetics of substance dependence. *Annual Review of Genomics and Human Genetics*, *13*, 241–261.

Wong, K., Elegante, M., Bartels, B., Elkhayat, S., Tien, D., Roy, S., et al. (2010). Analyzing habituation responses to novelty in zebrafish (Danio rerio). *Behavioural Brain Research*, *208*, 450–457.

INDEX

Note: Page numbers followed by "*f*" indicate figures, and "*t*" indicate tables.

H

I

CONTENTS OF RECENT VOLUMES

Volume 39

Volume 40

Volume 41

Volume 42

Volume 43

Volume 44

Volume 45

Volume 46

Volume 47

Volume 48

Volume 49

Volume 50

Volume 51

Volume 52

Volume 53

Volume 54

Volume 55

Volume 56

Volume 57

Volume 58

Volume 59

Volume 60

Volume 61

Volume 62

Volume 63

Volume 64

Volume 65

Volume 66

Volume 67

Volume 68

Volume 69

Volume 70

Volume 71

Volume 72

Volume 73

Volume 74

Volume 75

Volume 76

Volume 77

Volume 78

Volume 79

Volume 80

Volume 81

Volume 82

Volume 83

Volume 84

Volume 85

Volume 86

Volume 87

Volume 88

Volume 89

Volume 90

Volume 91

Volume 92

Volume 93

Volume 94

Volume 95

Volume 96

Volume 97

Volume 98

Volume 99

Volume 100

Volume 101

Volume 102

Volume 103

Volume 104

Volume 105

Volume 106

Volume 107

Volume 108

Volume 109

Volume 110

Volume 111

Volume 112

Volume 113

Volume 114

Volume 115

Volume 116

Volume 117

Volume 118

Volume 119

Volume 120

Volume 121

Volume 122

Volume 123

Volume 124

Volume 125

FSC
www.fsc.org
MIX
Papier | Fördert gute Waldnutzung
FSC® C083411

Kolibri 360 GmbH
Ferdinand-Jühlke-Straße 7
99095 Erfurt, Deutschland
produktsicherheit@kolibri360.de